Precalculus and Trigonometry
Explorations

Paul A. Foerster

Project Editor: Christopher David
Editorial Assistant: Lori Dixon
Production Director: Diana Jean Ray
Production Editor: Stephanie Tanaka
Copyeditor: Elliot Simon
Production Coordinator: Mike Hurtik
Text Designer and Compositor: GTS Graphics
Technical Artist: Lineworks, Inc.
Art and Design Coordinator: Kavitha Becker
Cover Designer: Todd Bushman
Printer: Data Reproductions

Executive Editor: Casey FitzSimons
Publisher: Steven Rasmussen

Key Curriculum Press
1150 65th Street
Emeryville, CA 94608
510-595-7000
editorial@keypress.com
http://www.keypress.com

Printed in the United States of America

10 9 8 7 6 5 4 3 2   15 14 13 12 11 10

ISBN 1-55953-653-5

Contents

To the Instructor   vii

To the Student   ix

Explorations

Functions and Mathematical Models

1	Finding Functions: Paper Cup Analysis	1
2	Names of Functions	2
3	Restricted Domains and Boolean Variables	3
4	Translations and Dilations, Numerically	4
5	Translations and Dilations, Algebraically	5
6	Transformations from Graphs	6
7	Transformation Review	7
8	Composition of Functions	8
9	Inverses of Functions	9
10	Translation, Dilation, and Reflection	10

Periodic Functions and Right Triangle Problems

11	Transformed Periodic Functions	12
12	Reference Angles	13
13	Definitions of Sine and Cosine	14
14	uv-Graphs and θy-Graphs of Sinusoids	15
15	Parent Sinusoids	16
16	Values of the Six Trigonometric Functions	17
17	Direct Measurement of Function Values	18
18	Measurement of Right Triangles	19
19	Accurate Right Triangle Practice	20
20	Right Triangles and the Empire State Building	21

Applications of Trigonometric and Circular Functions

21	Periodic Daily Temperatures	22
22	Sine and Cosine Graphs, Manually	23
23	Transformed Sinusoid Graphs	24
24	Sinusoidal Equations from Graphs	25
25	Tangent and Secant Graphs	26
26	Transformed Tangent and Secant Graphs	27
27	Radian Measure of Angles 1	28
28	Radian Measure of Angles 2	29
29	Circular Function Parent Graphs	30
30	Sinusoids, Given y, Find x Numerically	31
31	Sinusoids, Given y, Find x Algebraically	32
32	Using Sinusoids to Model Chemotherapy	33
33	Modeling Geological Formations with Sinusoids	34

Trigonometric Function Properties, Identities, and Parametric Functions

34	Properties of Trigonometric Functions	35
35	Transforming an Expression	37
36	Trigonometric Transformations	38

37	Trigonometric Identities	39
38	Arccosine, Arcsine, and Arctangent	40
39	Trigonometric Equations	41
40	Parametric Function Pendulum Problem	42
41	Parametric Equations for Ellipses	43
42	Graphs of Inverse Trigonometric Relations	44
43	Principal Branches of Inverse Trigonometric Relations	45
44	Inverse Trigonometric Relation Values	47

Properties of Combined Sinusoids

45	Linear Combination of Cosine and Sine	48
46	Cosine of a Difference	49
47	Composite Argument Property Proof	50
48	Sum or Product of Sinusoids with Unequal Periods	51
49	Harmonic Analysis	52
50	Harmonic Analysis Practice	53
51	More Harmonic Analysis Practice	54
52	Equivalence of Sinusoid Sums and Products	55
53	Double Argument Properties	56
54	Algebraic Transformations	57

Triangle Trigonometry

55	Introduction to Oblique Triangles	59
56	Derivation of the Law of Cosines	60
57	Angles by the Law of Cosines	61
58	Area of a Triangle	62
59	Hero's Formula	63
60	The Law of Sines	65
61	The Law of Sines for Angles	66
62	The Ambiguous Case, SSA	67
63	Golf and the Ambiguous Case	68
64	Introduction to Vectors	69
65	Navigation Vectors	70
66	The Ship's Path Problem	71
67	Area of a Regular Polygon	72

Properties of Elementary Functions

68	Graphical Patterns in Functions	73
69	Numerical Patterns in Function Values	75
70	Patterns for Quadratic Functions	76
71	Equations from Given Values Practice	77
72	Introduction to Logarithmic Functions	78
73	The Logistic Function for Population Growth	79

Fitting Functions to Data

74	Introduction to Linear Regression	80
75	Sums of Squares of Residuals	81
76	The Correlation Coefficient	82
77	Formula for the Correlation Coefficient	83
78	Violent Crimes and Regression Inaccuracies	85
79	Coffee Data Residual Plot	86

80 Airplane Fuel and Non-Linear Regression … 87
81 Carbon Dioxide and Regression on Residuals … 88

Probability, and Functions of a Random Variable

82 Counting Principles for "And" or "Or" … 89
83 Probability of Various Permutations … 90
84 Properties of Probability … 91
85 The Binomial Distribution … 92
86 Mathematical Expectation … 93

Three-Dimensional Vectors

87 Introduction to Three-Dimensional Vectors … 94
88 Introduction to the Scalar Product of Two Vectors … 95
89 Projections of Vectors … 96
90 Summary of Vector Properties … 98
91 Equation of a Plane Normal to a Vector … 100
92 Introduction to the Cross Product … 101
93 Dot and Cross Product, and Their Uses … 103
94 Direction Angles and Direction Cosines … 104
95 Vector Equation of a Line in Space … 105
96 Intersection of a Line and a Plane in Space … 106
97 Other Equations of a Line in Space … 107
98 Positive Normal Vector to a Plane … 109
99 Distance Between a Plane and a Point … 110
100 Distance Between a Line and a Point in Space … 111
101 Using Vectors on a Hip Roof … 112
102 Vectors Review … 114

Matrix Transformations and Fractal Figures

103 Matrix Multiplication … 116
104 Determinant of a Matrix … 117
105 Inverse of a 3×3 Matrix … 118
106 Matrix Images and Transformations … 119
107 Iterated Transformations* … 120
108 Combined Translation, Rotation, and Dilation … 121
109 Iterative Transformations and Fixed Points … 122
110 Rotation, Translation, and Dilation Practice … 124
111 Markov Chain Problem … 125
112 Multiple Transformations of the Same Figure … 126
113 Fractal Figures by Random Point Plotting* … 128
114 Hausdorff's Definition of (Fractal) Dimension … 130
115 Von Koch's Snowflake Curve … 131
116 Fractal Dimension of a River … 133
117 Foerster's Tree* … 135

Analytic Geometry of Conic Sections and Quadric Surfaces

118 Parametric Equations of Conic Sections … 137
119 Cartesian Equations of Conic Sections* … 138
120 Quadric Surfaces … 139
121 Cylinder in Nose Cone Problem … 140
122 Focus and Directrix of an Ellipse … 141
123 Focus, Directrix for Parabola and Hyperbola … 142

124 Two-Foci Property for an Ellipse   143
125 Pythagorean Property for Hyperbolas   144
126 More Analytic Properties of Conics   145
127 The Discriminant of a Conic Section   146
128 A Connection Between Conic Sections and Marketing   148
129 Reflecting Property of Ellipses   149
130 Computation of a Tangent Line   150
131 Equation from a Geometric Definition   152
132 A Quadratic-Quadratic System*   153

Polar Coordinates, Complex Numbers, and Moving Objects

133 Limaçon in Polar Coordinates   154
134 Roses and Circles in Polar Coordinates   155
135 Intersections of Polar Curves   156
136 Products of Complex Numbers   157
137 Complex Number Product Proof   158
138 Projectile Motion Problem   159
139 Hyperbola Construction   160
140 The Witch of Maria Agnesi   161
141 The Conchoid of Nicomedes   162
142 Cardioid Problem   163
143 Parametric Equations of Hypocycloids   164
144 Springs and Moving Ellipses   166
145 Is a Hanging Chain a Parabola?   167

Sequences and Series

146 Introduction to Sequences   168
147 Patterns in Sequences   169
148 Arithmetic and Geometric Sequences   170
149 Introduction to Series   171
150 Partial Sums of Arithmetic Series   172
151 Partial Sums of Geometric Series   173
152 Binomial Series and the Binomial Formula   174
153 Arithmetic and Geometric Series Problems   175
154 A Power Series for a Familiar Function   177
155 Complex Numbers in Exponential Form   178

Polynomial and Rational Functions, Limits, and Derivatives

156 Cubic Function Graphs   179
157 Synthetic Substitution   180
158 Sum and Product of the Zeros of a Polynomial   181
159 Fitting a Polynomial Function to Points   182
160 Rational Functions and Discontinuities   183
161 Limits and Curved Asymptotes   184
162 Rate of Change of a Polynomial Function   186

Solutions to the Explorations   189

Appendix: Programs for Graphing Calculators   261

Precalculus and Trigonometry Correlation Index   265

* Explorations 107, 113, 117, 119, and 132 require access to programs for graphing calculators.
These programs appear in the appendix and can be downloaded from *www.keypress.com/pte*.

To the Instructor

The materials in this book will give students hands-on practice with the concepts and techniques of precalculus and trigonometry. Most of the Explorations are intended to let students discover a concept on their own or in cooperative groups before the concept is reinforced through classroom discussion or lecture.

You can use the Explorations with any precalculus or trigonometry textbook. They generally are intended for use both with and without graphing calculators so that students can learn by graphical and numerical methods as well as by traditional algebraic methods. Students should also be able to verbalize their conjectures and conclusions and write a paragraph describing what they learned as a result of completing an Exploration. The sequence of topics in this book follows that of Paul Foerster's *Precalculus with Trigonometry: Concepts and Applications,* published by Key Curriculum Press in 2003. An extensive index of major topics is included to guide you to Explorations suitable for your immediate needs. Several of the Explorations (107, 113, 117, 119, and 132) require access to programs for graphing calculators. These programs appear in the appendix and can be downloaded from *www.keypress.com/pte.*

Most of the Explorations consist of a single page, enabling students to do their work right on the sheet in the space provided. This aspect is particularly important when students need an accurate graph on which to draw.

There are two ways to use the Explorations: in class and outside of class. For classroom use, you can have students work in **cooperative groups** for 15 to 30 minutes, with minimal instructor guidance, then follow up with 5 or 10 minutes of instruction to make sure they have not "discovered" something wrong! Another way you can use the Explorations is as **directed discovery exercises** in which you have students work the problems one at a time, on their own or in groups, then follow up with answers and discussion after each problem. After you follow up on a problem, you may want to tell students that they have a certain amount of time to do the next problem. You can even use the material of an Exploration as the basis for your **lecture** on a particular topic.

For out-of-class work, you can assign an Exploration as homework, then follow up on it during the next class period. Students can complete the Exploration on their own or they can work cooperatively with other students outside of class.

For out-of-class or in-class work, students will find it helpful to get more information about a particular topic by researching material in their textbooks. Some students will think they are outsmarting you by reading the text first and then working the Exploration. Actually, you are outsmarting them by getting them to read the book!

If students purchase or receive copies of this book of Explorations, you save photocopying time and expense and students can look up answers in the back of the book—an advantage if they are working on their own. If you reproduce the Explorations and distribute them to students as needed, you avoid any resistance students might have to a workbook and students can concentrate on one Exploration at a time. The fact that answers are *not* available using this approach might better fit your objectives for the Explorations.

The last question of an Exploration is often "What did you learn as a result of doing this Exploration that you did not know before?" You will find students' answers to this question interesting. Sometimes you will learn surprising things about your students or about their insights into mathematics. Often students' answers reveal where they are having difficulty or what students are confused about. Occasionally, you may even learn something new about precalculus or trigonometry!

Taken as a whole, the Explorations reveal virtually all of precalculus and trigonometry. You should be careful, however, not to go overboard with using the Explorations as a means of instruction. Students learn by a variety of teaching methods, and teachers need to accommodate their students' different learning styles. Sometimes the traditional lecture is the appropriate choice for introducing a topic, both to be time efficient and to prepare students for future courses in which the lecture method might be used extensively.

The name *Exploration* has been chosen deliberately. Students are learning concepts by exploring them on their own, with or without having read about the topic in their textbooks. If you avoid calling the Explorations *worksheets* or *reviews,* you'll avoid the connotation of tedium associated with those terms. Some instructors call them *games,* as there is some sport involved if the assignments aren't graded.

To the Student

This book contains activities called Explorations. Each Exploration concerns a particular concept or technique of precalculus or trigonometry. The Explorations were designed so that you can complete them with minimal guidance from your instructor. Although you can do them on your own, the Explorations are particularly effective when you work through them with classmates in cooperative groups. That way you can share your ideas with others and get feedback about whether any conjectures you have made are correct. Explorations 107, 113, 117, 119, and 132 require access to programs for graphing calculators. These programs appear in the appendix and can be downloaded from *www.keypress.com/pte*.

There are at least four ways you can learn precalculus and trigonometry—algebraically, graphically, numerically, and verbally. For example, you can prove a trigonometric identity **algebraically** by transforming the expression on one side into the expression on the other side. You can check the identity **numerically** by comparing tables of values of the expressions on the two sides. You can also check it **graphically** by plotting each side as an argument of a function and seeing that the graphs coincide. At the end of many Explorations, you will be asked to present your conclusions **verbally** and to describe what you have learned from doing the Exploration.

In the back of the book, you will find solutions to the Explorations. The most effective way to use the solutions is to do the entire Exploration first, then make sure your answers agree. If your answers don't agree, you should not consider the Exploration finished until you understand why the given solution is correct and why your original thinking may have been in error.

Using the index in this book, you can find additional information about the topic of a given Exploration. You can read about the topic in your textbook first, then try working the Exploration. Or you can work the Exploration first and use your textbook to confirm that what you have discovered is correct. Try both of these approaches and see which works best for you.

Best wishes for your study of precalculus and trigonometry!

Exploration 1: Finding Functions: Paper Cup Analysis   Date: _______________

Objective: Find an equation for calculating the height of a stack of paper cups.

1. Obtain several paper cups of the same kind. Measure the height of stacks containing 5, 4, 3, 2, and just 1 cup. Record the heights to the nearest 0.1 cm. State what kind of cup you used.

   Kind: _________________________________

Number	cm
1	
2	
3	
4	
5	

2. Plot the points in the table on this graph paper. Show the scale you are using on the vertical axis.

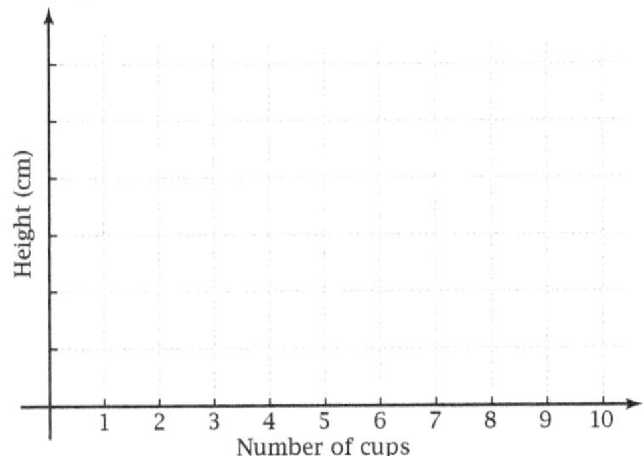

3. On average, by how much did the stack height increase for each cup you added? Show how you got your answer.

4. How tall would you expect a 10-cup stack to be? Show how you get your answer. Would this be twice as tall as a 5-cup stack?

5. Let x be the number of cups in a stack, and let y be the height of the stack, measured in centimeters. Write an equation for y as a function of x.

6. What is the name of the kind of function whose equation you wrote in Problem 5?

   ____

7. Show that your equation in Problem 5 gives a height close to the measured height for a stack of 3 cups.

8. Use your equation to predict the height of a stack of 35 cups. Round the answer to 1 decimal place.

9. What are the names of the processes of calculating a value *within* the range of the data, as in Problem 7, and *outside* the range of data, as in Problem 8?

   Within: ________________________________

   Outside: _______________________________

10. A cup manufacturer wants to package this kind of cup in boxes that are 45 cm long. What is the maximum number of cups the box could hold? Show how you get your answer.

11. What did you learn as a result of doing this Exploration that you did not know before?

Exploration 2: Names of Functions

Objective: Recall the names of certain kinds of functions.

1. $f(x) = 2x + 3$ is the equation for a **linear function.** Plot the graph and sketch the result here. Give a reason for the name *linear.*

2. $f(x) = x^2 - 6x + 10$ is the equation for a **quadratic function.** Plot the graph and sketch the result. Explain how the word *quadratic* is related to the word *quadrangle.*

3. $f(x) = 3x^{0.7}$ is the equation for a **power function.** Plot the graph and sketch the result. Why do you think it is called a *power* function?

4. $f(x) = 3 \times 0.7^x$ is the equation for an **exponential function.** Plot the graph and sketch the result. How does an exponential function differ from a power function algebraically? graphically?

5. $f(x) = \frac{24}{x}$ is the equation for an **inverse variation** power function. Plot the graph for $x > 0$ and sketch the result. Why do the words "y varies inversely with x" make sense for this function? Why can the function be called a *power* function?

6. $f(x) = x^4 - 4x^3 - 43x^2 + 130x + 168$ is the equation of this **quartic function.** Why do you think the name *quartic* is used for this function? Use your grapher to find the largest value of x at which the graph crosses the x-axis.

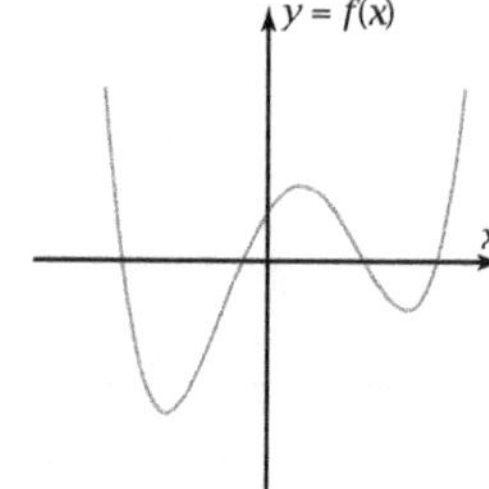

7. $f(x) = \frac{x-4}{x-3}$ is the equation of a **rational function.** Plot the graph and sketch the result. Why do you think it is called a *rational* function? What happens to the graph at $x = 3$?

8. What did you learn as a result of doing this Exploration that you did not know before?

Exploration 3: Restricted Domains and Boolean Variables

Objective: Use Boolean variables to plot graphs of functions in a restricted domain.

Weight Above and Below Earth's Surface: If there were a hole all the way through Earth and it were possible for you to go through it, you would be "weightless" at the center of Earth. This is because gravity would pull you with the same force in every direction. Between the center and the surface, your weight would vary directly with the distance from the center.

1. Kevin Vader (Darth's son) weighs 200 pounds on the surface of Earth. Earth's radius is about 4000 miles. Write the particular equation for Kevin's weight as a function of distance from the center when he is below the surface.

2. Above the surface, your weight varies inversely with the square of your distance from the center because the pull of gravity decreases as you recede from Earth. Write the particular equation for Kevin's weight as a function of distance from the center when he is above the surface.

3. Plot the graphs from Problems 1 and 2 as y_1 and y_2. Does your graph agree with this one? Where do the two graphs cross each other?

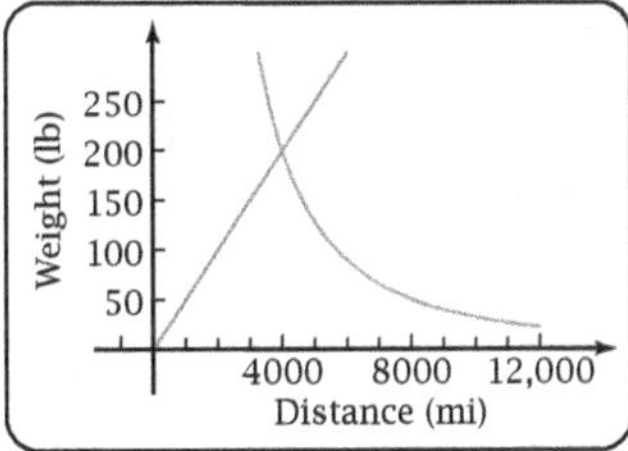

The linear graph should stop at $x = 4000$ miles, and the inverse square function graph should start at $x = 4000$ miles, the distance from the center to the surface.

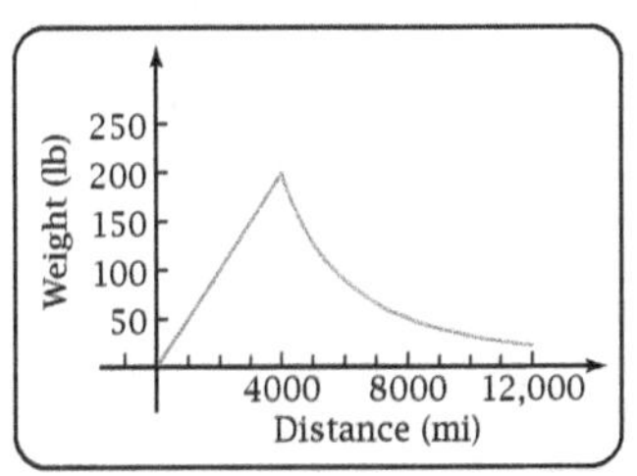

4. The **Boolean variable** $(x \geq 4000)$ equals 1 if x is greater than or equal to 4000 and 0 otherwise. Change your equation for y_2 by dividing it by this Boolean variable, then replot the graph. Explain why the grapher plots the same graph for y_2 when $x \geq 4000$ but plots nothing when $x < 4000$.

5. What Boolean variable could you divide y_1 by so that the grapher plots it only between 0 and 4000? Change the equation for y_1. Does the complete graph now match the one above Problem 4?

6. What word describes the set of x-values for which a particular function is defined? What word describes the corresponding set of y-values?

   x-values: ____

   y-values: ____

7. What did you learn as a result of doing this Exploration that you did not know before?

Exploration 4: Translations and Dilations, Numerically

Date: ___________

Objective: By calculating values and plotting points, discover the effect on a function graph of adding and multiplying by constants.

1. The table shows values of a **pre-image** function $y = f(x)$. The graph of f is a set of line segments connecting the points, shown dashed in the figure. Find values of the **image** function $g(x) = f(x) + 3$. For instance, $g(-2) = 2 + 3 = 5$. Plot the graph of this transformed function.

x	f(x)	g(x)
-2	2	
-1	3	
0	1	
1	-2	
2	0	
3	1	

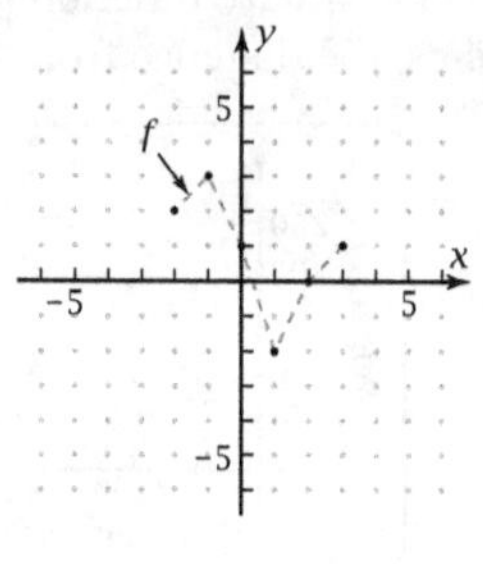

2. The transformation in Problem 1 is a **vertical translation** by 3 units. Give the meaning of a vertical translation.

3. Use the values of $f(x)$ in Problem 1 to make a table of values of a new image function, $g(x) = f(x - 3)$. For instance, $g(1) = f(1 - 3) = f(-2) = 2$. Plot the image of this transformed function.

x	g(x) = f(x - 3)
1	
2	
3	
4	
5	
6	

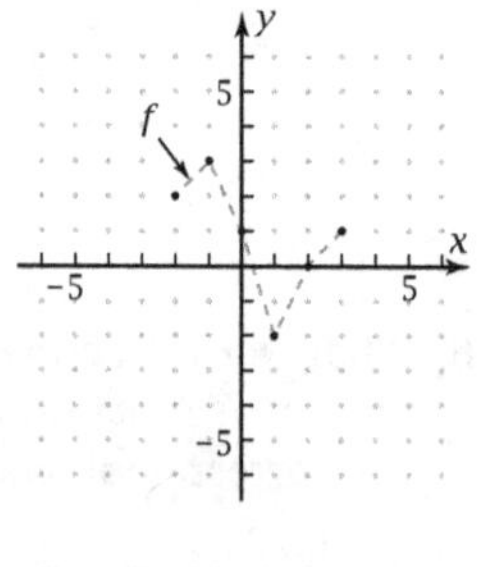

4. Describe the transformation in Problem 3.

5. Use the values of $f(x)$ in Problem 1 to make a table of values of a new image function, $g(x) = 2f(x)$. For instance, $g(-1) = 2f(-1) = 2 \cdot 3 = 6$. Plot the image of this transformed function.

x	g(x) = 2f(x)
-2	
-1	
0	
1	
2	
3	

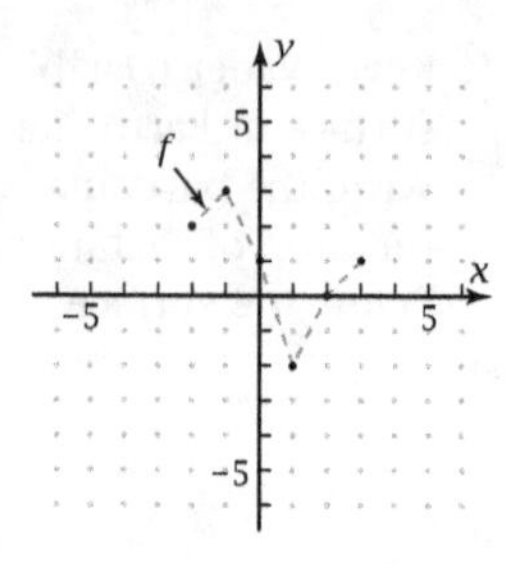

6. The transformation in Problem 5 is a **vertical dilation** by a factor of 2. Give the meaning of a vertical dilation, and explain how it differs from a vertical translation.

7. Use the values of $f(x)$ in Problem 1 to make a table of values of a new image function, $g(x) = f\left(\frac{1}{2}x\right)$. For instance,

$$g(-2) = f\left(\frac{1}{2} \cdot (-2)\right) = f(-1) = 3$$

Plot the image of this transformed function.

x	g(x) = f($\frac{1}{2}$x)
-4	
-2	
0	
2	
4	
6	

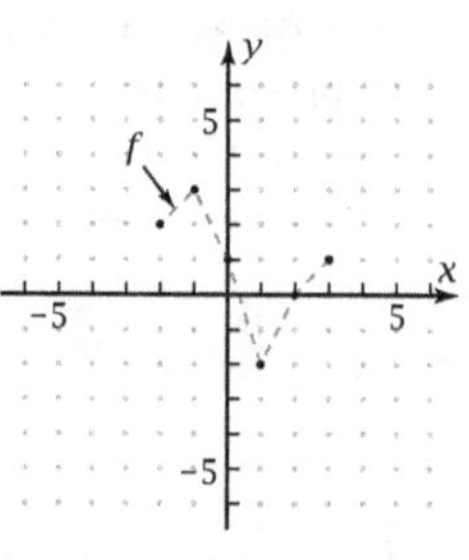

8. The transformation in Problem 7 is a **horizontal dilation.** By what factor is the graph dilated? How is that factor related to the $\frac{1}{2}$ in $f\left(\frac{1}{2}x\right)$?

9. What did you learn as a result of doing this Exploration that you did not know before?

Exploration 5: Translations and Dilations, Algebraically

Objective: Find the effect on a function graph of adding and multiplying by constants.

1. The graph below shows the **pre-image** function $f(x) = \frac{1}{1 + x^2}$. Plot this graph as y_1 on your grapher. Use the window shown, using GRID ON format.

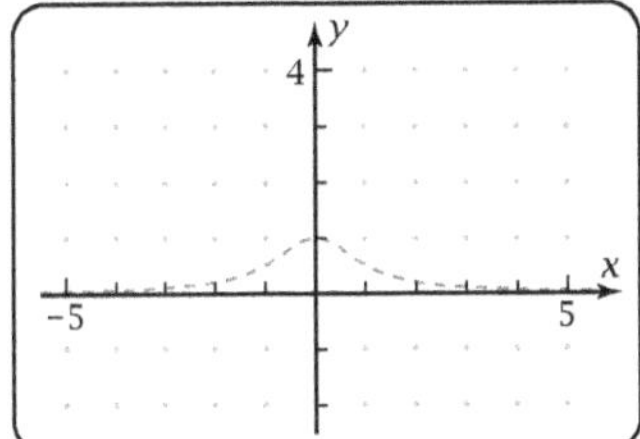

2. Plot the graph of $y_2 = f(x) + 3$. Sketch the result on the graph in Problem 1.

3. The transformation in Problem 2 is a **vertical translation** of 3. Give the meaning of a vertical translation.

4. Deactivate y_2 from Problem 2. Then plot $y_3 = f(x - 3)$. Sketch the result here.

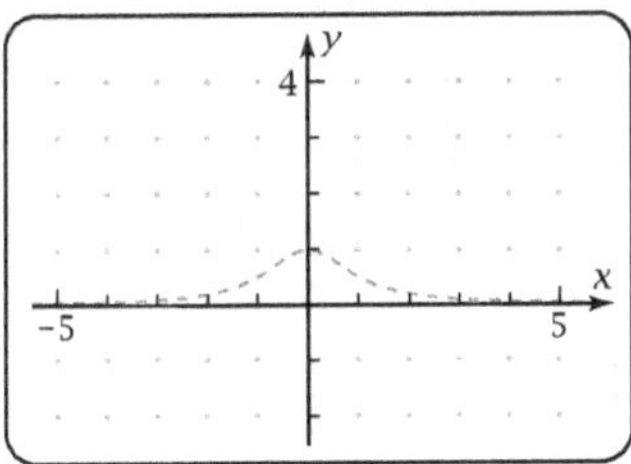

5. What words describe the transformation in Problem 4?

6. Deactivate y_3 from Problem 4. Then plot $y_4 = 3f(x)$. Sketch the result here.

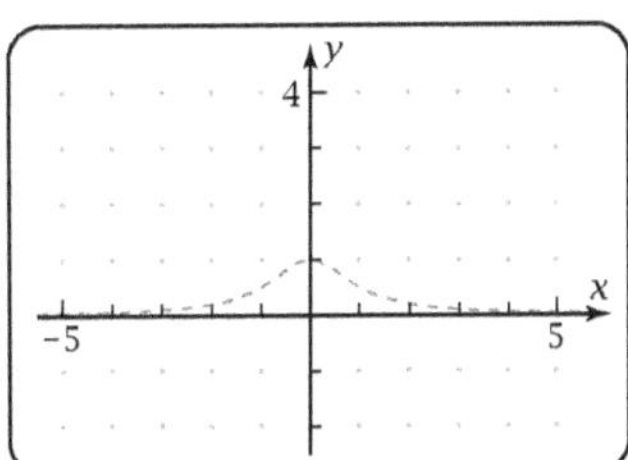

7. The transformation in Problem 6 is a **vertical dilation** by a factor of 3. Give the meaning of a vertical dilation, and explain how it differs from a vertical translation.

8. Deactivate y_4 from Problem 7. Then plot $y_5 = f(3x)$. Sketch the result here.

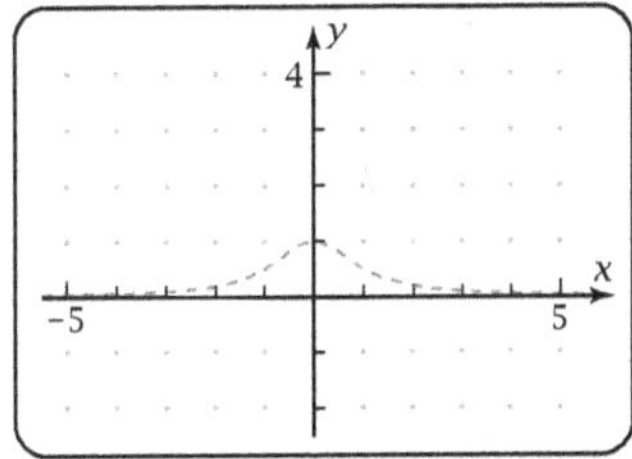

9. The transformation in Problem 8 is a **horizontal dilation.** By what factor is the graph dilated?

10. What did you learn as a result of doing this Exploration that you did not know before?

Exploration 6: Transformations from Graphs

Date: _________

Objective: Given the parent and transformed graphs, identify the transformation.

Identify the transformation of f (dotted) to get g (solid).

1. Verbally: _________________________________

   Equation: $g(x) =$ _________________________________

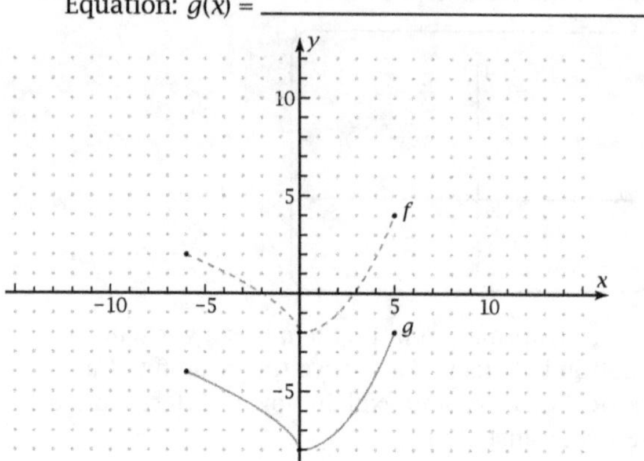

2. Verbally: _________________________________

   Equation: $g(x) =$ _________________________________

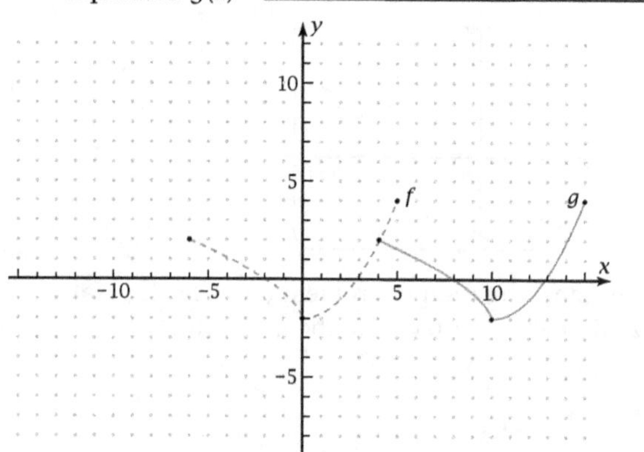

3. Verbally: _________________________________

   Equation: $g(x) =$ _________________________________

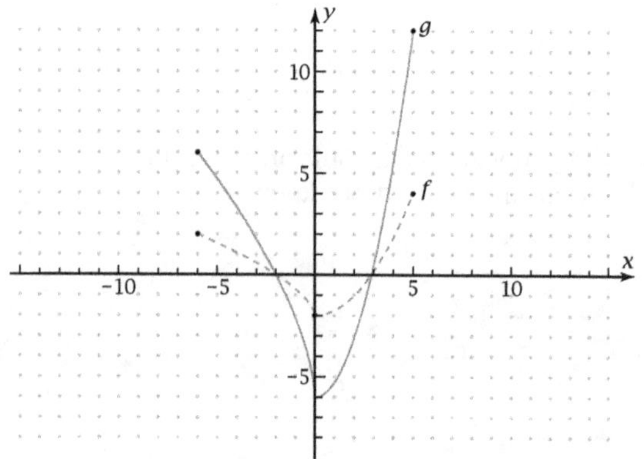

4. Verbally: _________________________________

   Equation: $g(x) =$ _________________________________

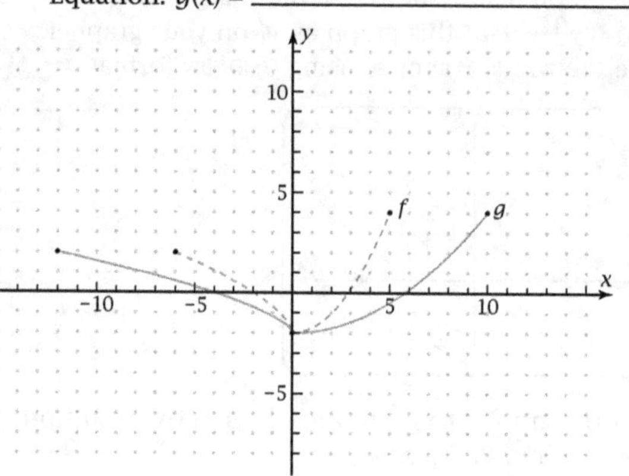

5. Verbally: _________________________________

   Equation: $g(x) =$ _________________________________

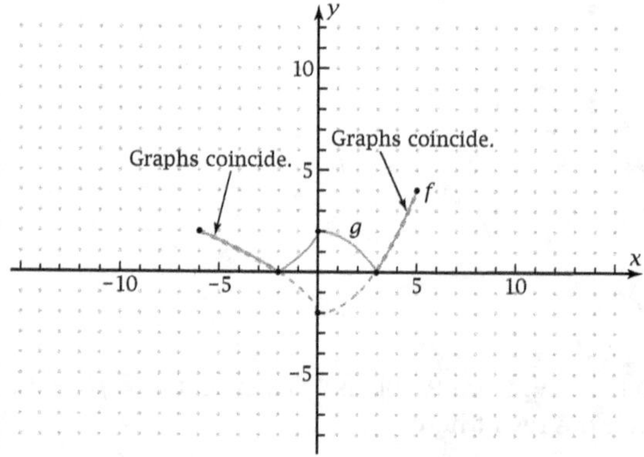

6. Verbally: _________________________________

   Equation: $g(x) =$ _________________________________

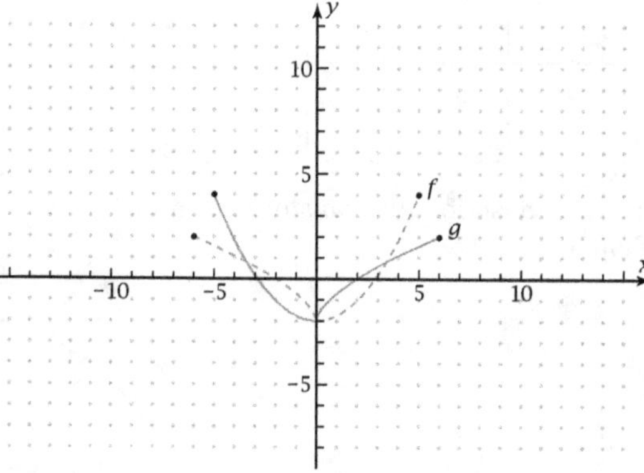

7. What did you learn as a result of doing this Exploration that you did not know before?

Exploration 7: Transformation Review

Date: _____________

Objective: Given the parent and transformed graphs, identify the transformation and confirm by grapher.

1. The figure shows the graph of $f(x) = -0.5x^2 + x + 3.5$ in the domain $-1 \le x \le 5$. Duplicate this graph on your grapher. Restrict the domain by dividing by the **Boolean variable** ($x \ge -1$ and $x \le 5$). Use GRID ON format to get the dots.

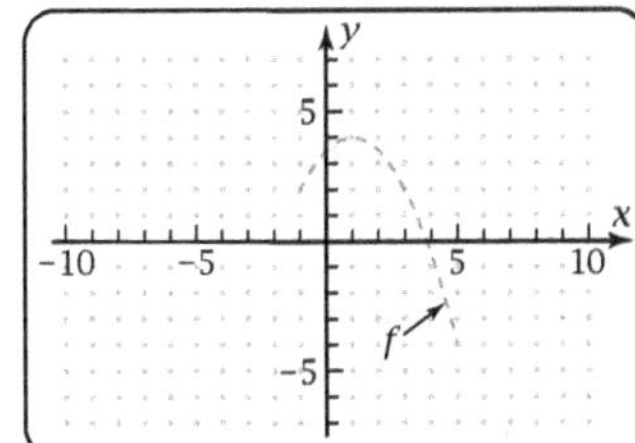

For Problems 2–6, identify the transformation of f (dotted) to get g (solid), and confirm by grapher.

2. Verbally: _________________________________

   Equation: $g(x) = $ ________________________

   Check: _______________

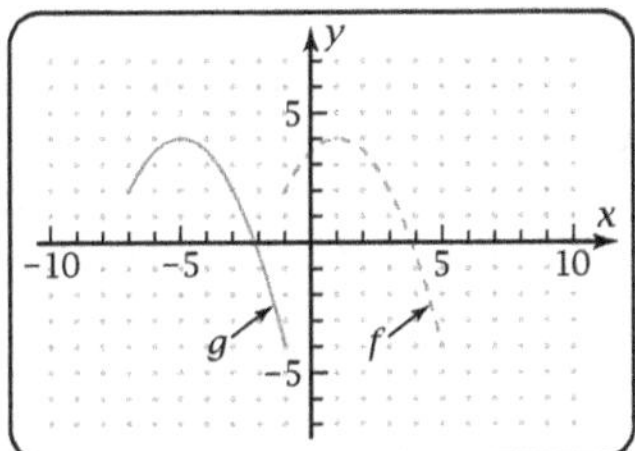

3. Verbally: _________________________________

   Equation: $g(x) = $ ________________________

   Check: _______________

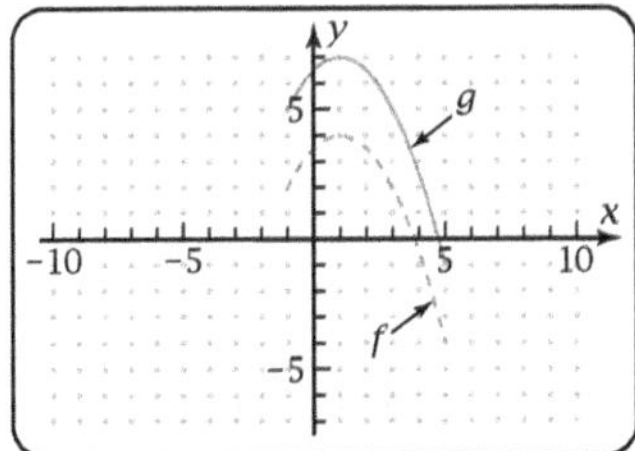

4. Verbally: _________________________________

   Equation: $g(x) = $ ________________________

   Check: _______________

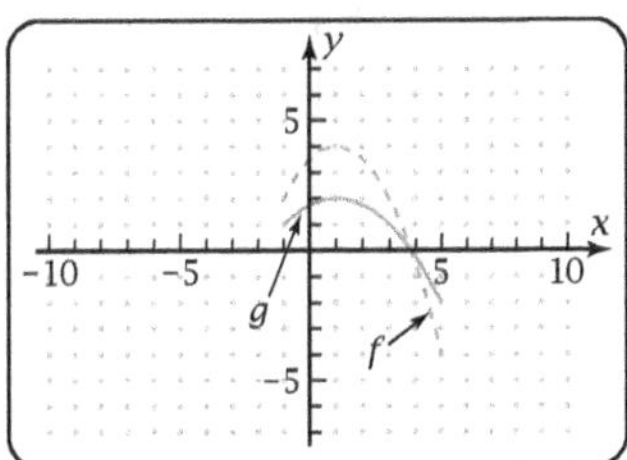

5. Verbally: _________________________________

   Equation: $g(x) = $ ________________________

   Check: _______________

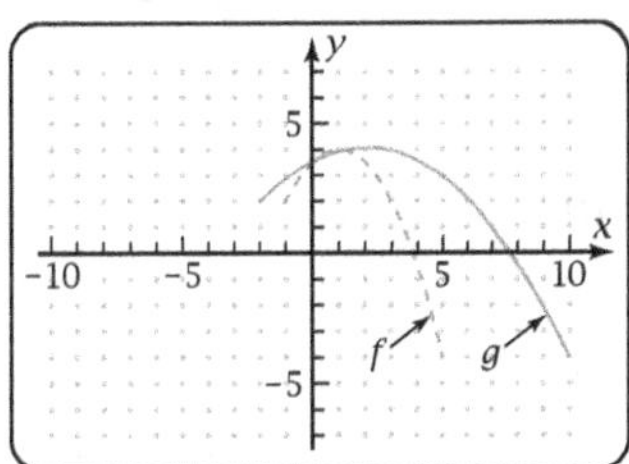

6. Verbally: _________________________________

   and _________________________________

   Equation: $g(x) = $ ________________________

   Check: _______________

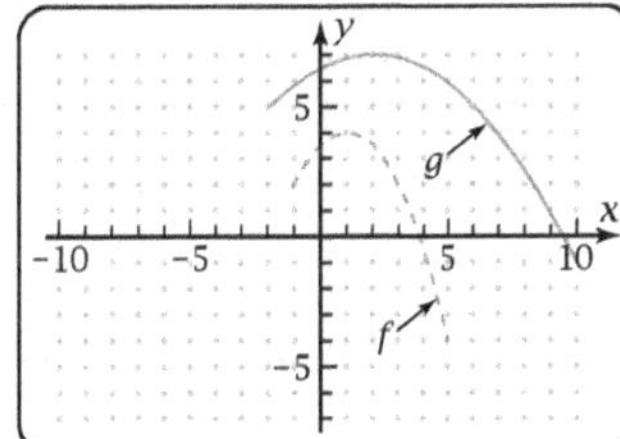

7. What did you learn as a result of doing this Exploration that you did not know before?

Exploration 8: Composition of Functions

Date: _____________

Objective: Find the composition of one function with another.

1. The figure shows two linear functions, *f* and *g*. Write the domain and range of each function.

   f: Domain: _______________ Range: _______________

   g: Domain: _______________ Range: _______________

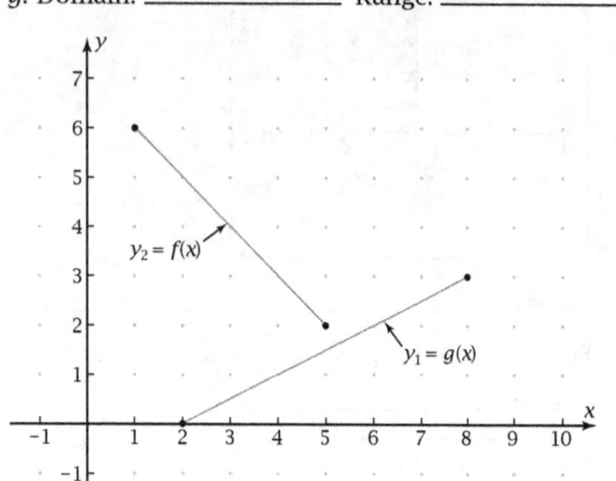

2. Read values of $g(x)$ from the graph and write them in this table. If the value of *x* is out of the domain, write "none."

x	g(x)
0	
1	
2	
3	
4	
5	
6	
7	
8	
9	

3. The symbol $f(g(x))$ is read "*f* of *g* of *x*." It means find the value of $g(x)$ first, and then find *f* of the answer. For instance, $g(5) = 1.5$. So $f(g(5)) = f(1.5) = 5.5$. Put another column into the table for values of $f(g(x))$. Write "none" where appropriate.

4. Show in the table an instance where $g(x)$ is defined but $f(g(x))$ is not defined.

5. Plot the values of $f(g(x))$ on the figure in Problem 1. If the points do not lie in a straight line, go back and check your work.

6. The function in Problem 5 is called the **composition** of *f* with *g*, which can be written $f \circ g$. What are the domain and range of $f \circ g$?

   Domain: _______________ Range: _______________

7. Find equations for functions *f* and *g*.

8. Enter in your grapher the *f* and *g* equations as y_1 and y_2, respectively. Use Boolean variables to make the functions have the proper domains. Then plot the graphs. Does the result agree with the given figure?

9. Enter $f \circ g$ in y_3 by entering $y_1(y_2(x))$. Plot this graph. Does it agree with the graph you drew in Problem 5?

10. By suitable algebraic operations on the equations in Problem 7, find an equation for $f(g(x))$. Simplify the equation as much as possible.

11. What did you learn as a result of doing this Exploration that you did not know before?

Exploration 9: Inverses of Functions

Objective: Find the inverse of a function graphically, numerically, or algebraically, and state whether or not the inverse is a function.

Problems 1–6 refer to the linear function $y = 2x - 5$.

1. Write the equation for the inverse relation by interchanging the variables. Then solve the resulting equation for y in terms of x.

2. The graph shows $y = 2x - 5$. Plot the graph of the inverse relation here.

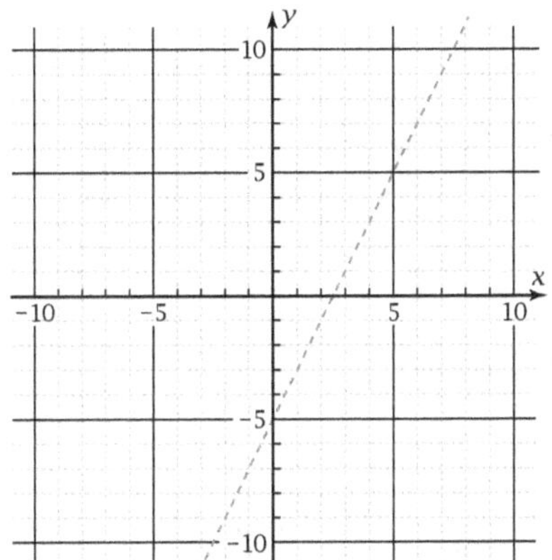

3. The inverse relation in Problems 1 and 2 is a function. How can you tell?

4. If the equation for the function is written as $f(x) = 2x - 5$, how could you write the equation for the inverse function using the $f(x)$ terminology?

5. Show that $f(3) = 1$ and $f^{-1}(1) = 3$. Explain why this is true, based on the definition of the inverse of a function.

6. Plot the line $y = x$. How are the graphs of f and f^{-1} related to this line?

Problems 7 and 8 refer to the quadratic function $y = x^2 - 4x + 7$, graphed here.

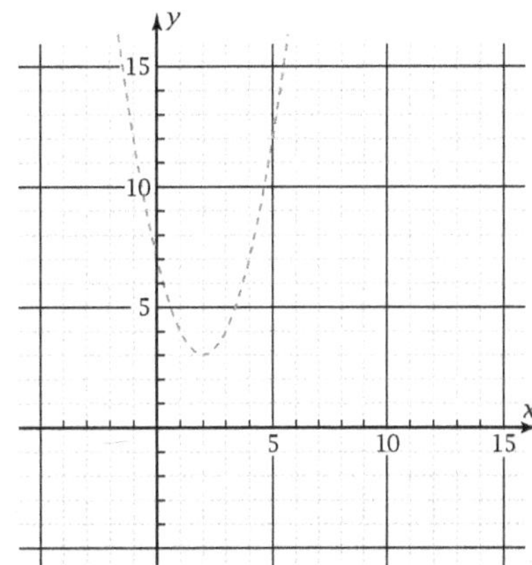

7. Plot the line $y = x$. Then plot the inverse of the function by reflecting the graph across this line.

8. Explain why the inverse of this function is not a function.

Problems 9 and 10 refer to the exponential function $f(x) = 2^x$, graphed here.

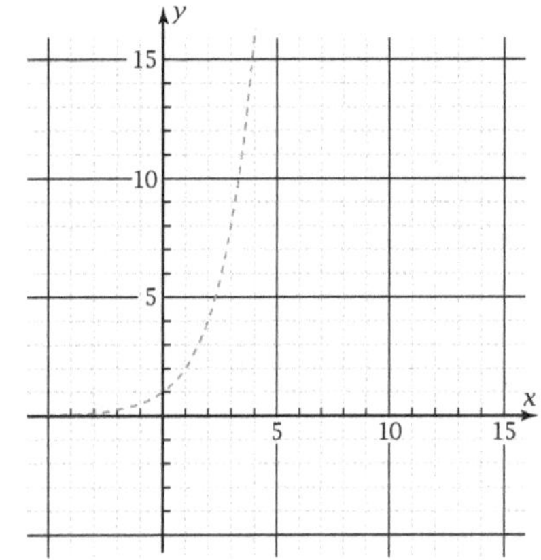

9. Find $f(0)$, $f(1)$, $f(2)$, and $f(3)$.

10. Find $f^{-1}(1)$, $f^{-1}(2)$, $f^{-1}(4)$, and $f^{-1}(8)$. Use these points to plot the graph of f^{-1}.

11. What did you learn as a result of doing this Exploration that you did not know before?

Exploration 10: Translation, Dilation, and Reflection   Date: __________

Objective: Show that you know the effects of various constants on the graph of a function.

Name the transformation and sketch the graph.

1. $y = \dfrac{1}{2} f(x)$

   Transformation: _____________________________

   ____

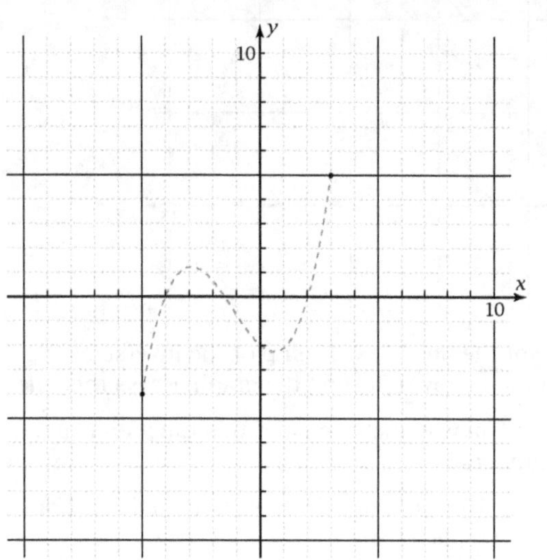

2. $y = f\!\left(\dfrac{1}{2}x\right)$

   Transformation: _____________________________

   ____

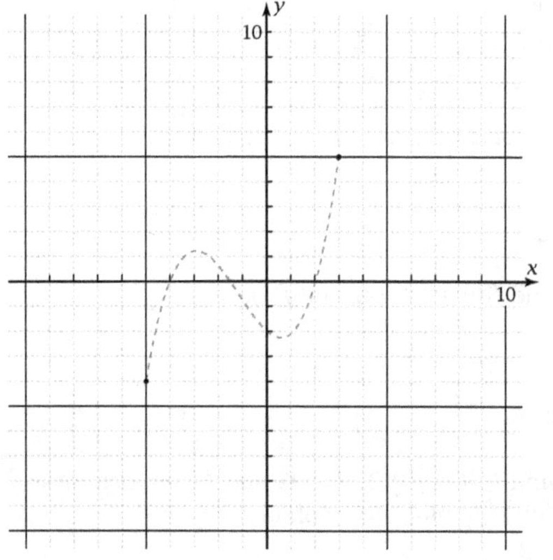

3. $y = f(x) - 4$

   Transformation: _____________________________

   ____

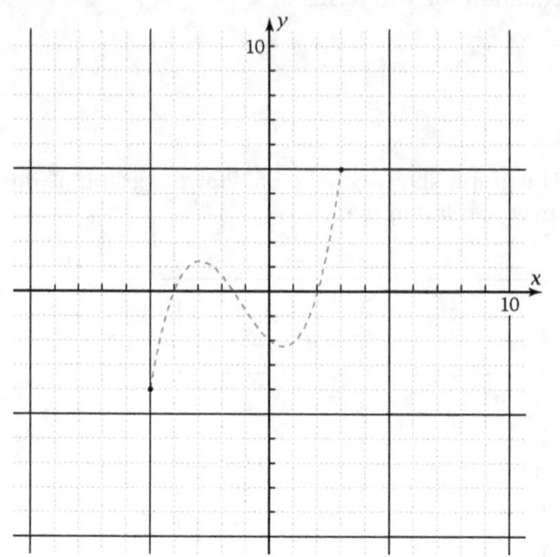

4. $y = f(x - 7)$

   Transformation: _____________________________

   ____

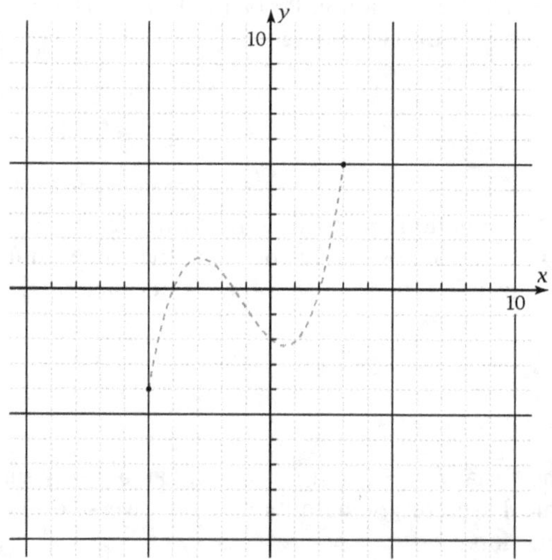

Precalculus and Trigonometry Explorations
©2004 Key Curriculum Press

Exploration 10: Translation, Dilation, and Reflection *continued*

5. $y = f(-x) = f(-1x)$

   Transformation (as a dilation): _________________

   ____

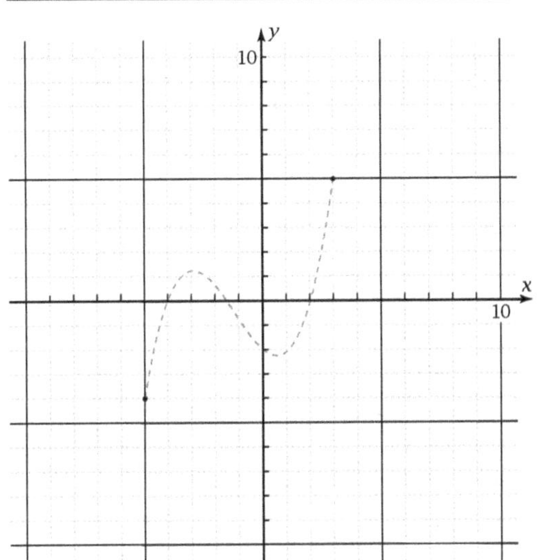

6. From the results of Problem 5, give another name for the transformation $y = f(-x)$.

   ____

7. $y = -f(x) = -1f(x)$

   Transformation (as a dilation): _________________

   ____

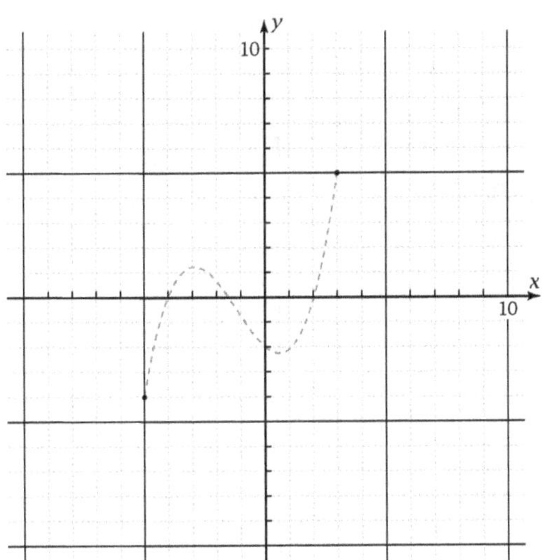

8. From the results of Problem 7, give another name for the transformation $y = -f(x)$.

   ____

9. $y = |f(x)|$ is an absolute value transformation. Is it a transformation in the x-direction or a transformation in the y-direction?

   ____

10. Sketch the graph of $y = |f(x)|$.

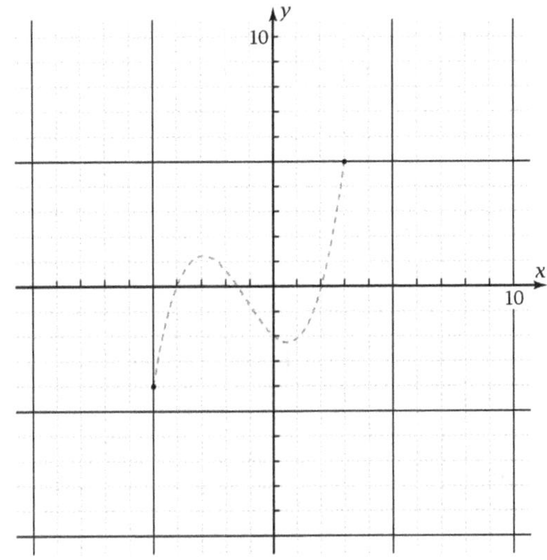

11. Sketch the graph of $y = f(|x|)$.

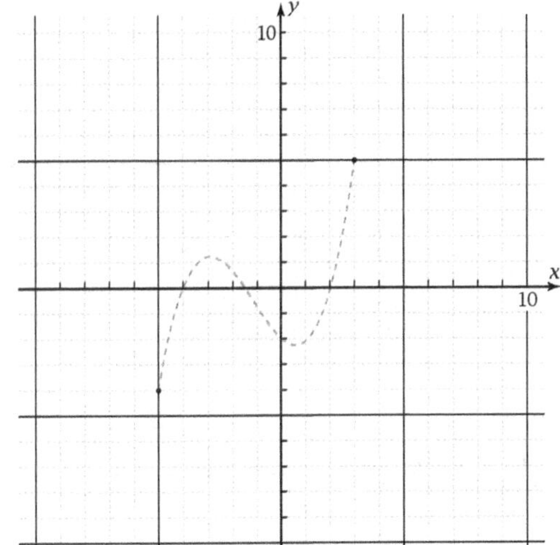

12. What did you learn as a result of doing this Exploration that you did not know before?

Exploration 11: Transformed Periodic Functions

Date: _________

Objective: Given a pre-image graph and a transformed graph of a periodic function, state the transformation(s).

Give the transformation applied to $f(x)$ (dashed) to get the solid graph, $y = g(x)$.

1. Verbally: _________________

   Equation: $y = g(x) = $ _________________

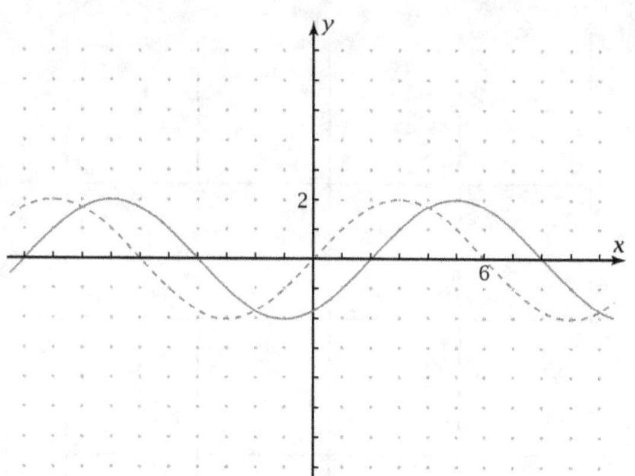

2. Verbally: _________________

   Equation: $y = g(x) = $ _________________

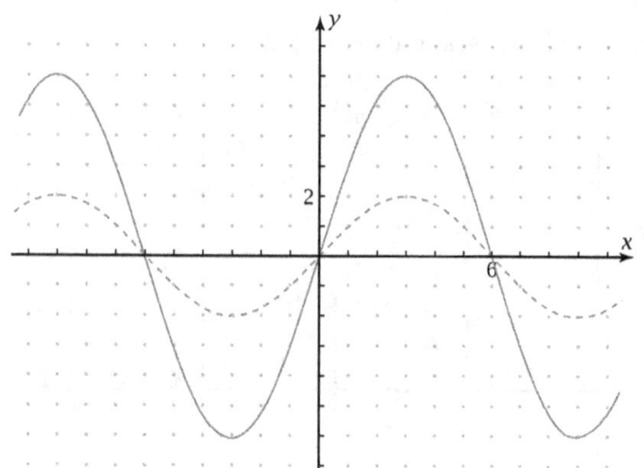

3. Verbally: _________________

   Equation: $y = g(x) = $ _________________

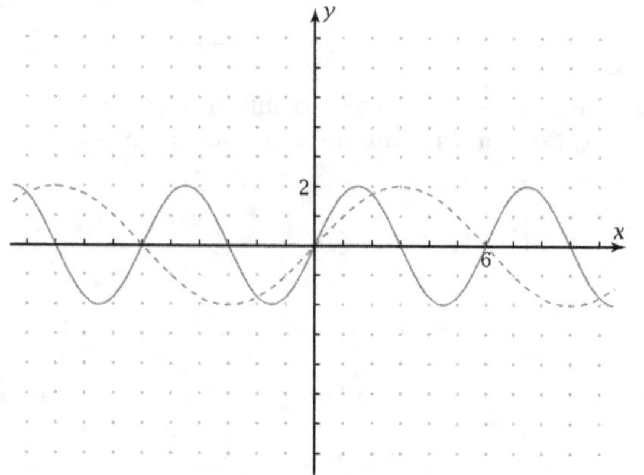

4. Verbally: _________________

   Equation: $y = g(x) = $ _________________

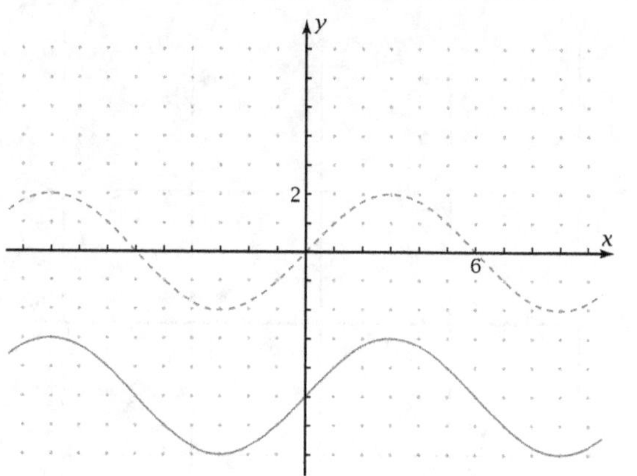

5. Verbally: _________________

   Equation: $y = g(x) = $ _________________

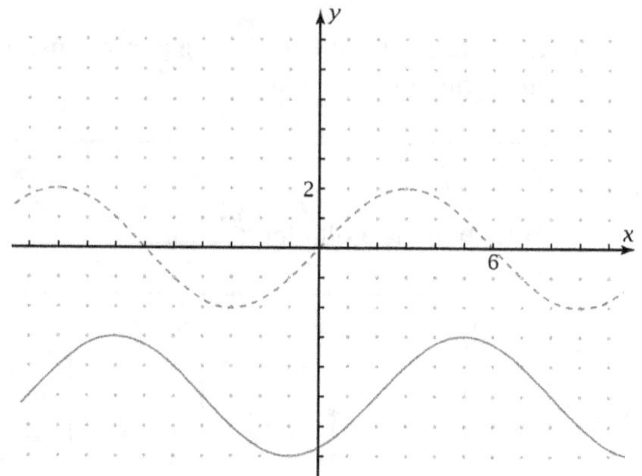

6. Verbally: _________________

   Equation: $y = g(x) = $ _________________

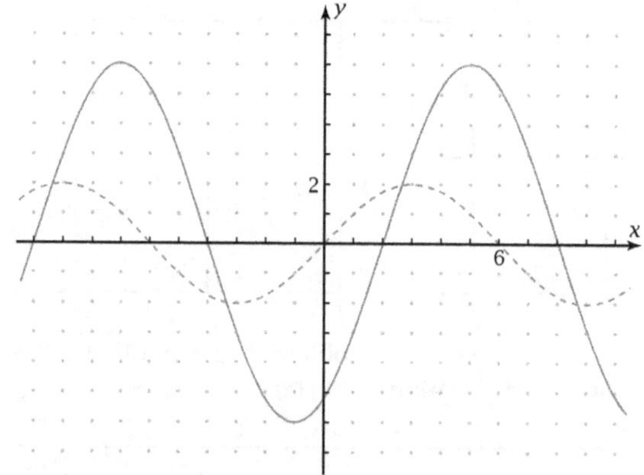

7. What did you learn as a result of doing this Exploration that you did not know before?

Exploration 12: Reference Angles

Objective: Learn about measures of angles in standard position and their reference angles.

1. The figure shows an angle, $\theta = 152°$, in **standard position**. The **reference angle**, θ_{ref}, is measured *counterclockwise* between the terminal side of θ and the nearest side of the horizontal axis. Show that you know what *reference angle* means by drawing θ_{ref} and calculating its measure.

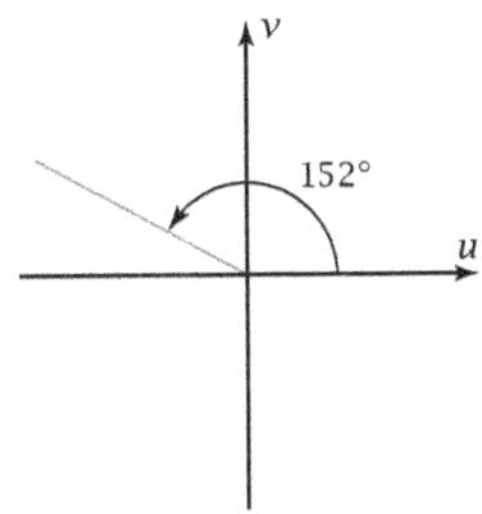

2. The figure shows $\theta = 250°$. Sketch the reference angle and calculate its measure.

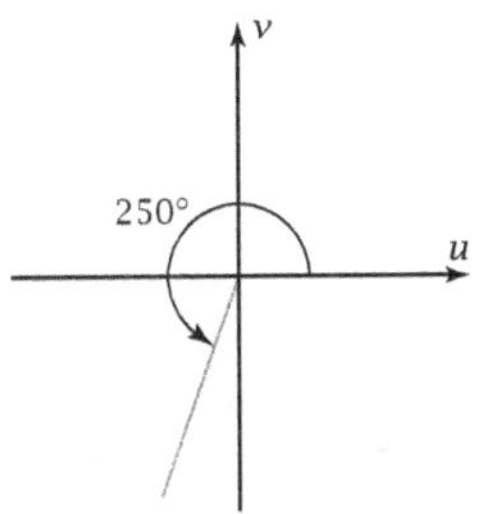

3. You should have drawn arrowheads on the arcs for the reference angles in Problems 1 and 2. If you haven't, draw them now. Explain why the arc for 152° goes from the terminal side to the u-axis but the arc for 250° goes from the u-axis to the terminal side.

4. Amos Take thinks the reference angle for 250° should go to the v-axis because the terminal side is closer to it than the u-axis. Tell Amos why his conclusion does not agree with the definition of *reference angle* in Problem 1.

5. Sketch an angle of 310° in standard position. Sketch its reference angle and find the measure of the reference angle.

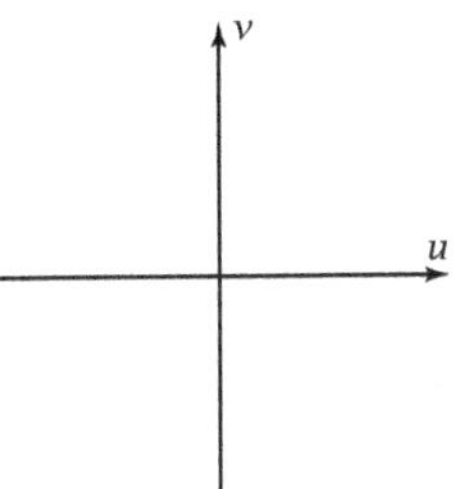

6. Sketch an angle whose measure is between 0° and 90°. What is the reference angle of this angle?

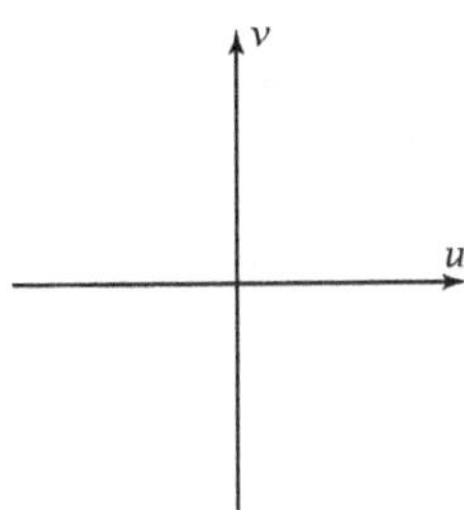

7. The figure shows an angle of −150°. Sketch the reference angle and find its measure.

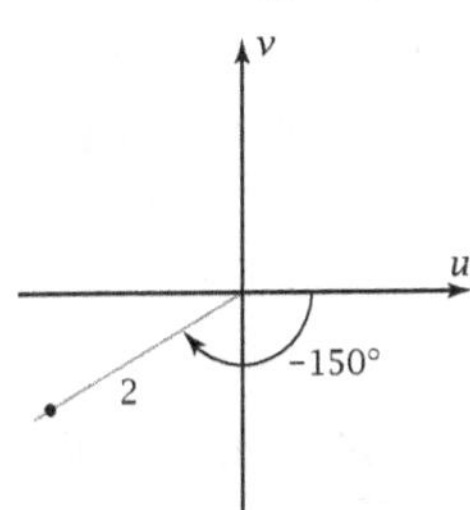

8. The figure in Problem 7 shows a point 2 units from the origin and on the terminal side of the angle. Draw a segment from this point perpendicular to the u-axis, thus forming a right triangle whose hypotenuse is 2 units long. Use what you recall from geometry to find the lengths of the two legs of the triangle.

9. What did you learn as a result of doing this Exploration that you did not know before?

Exploration 13: Definitions of Sine and Cosine

Date: ___________

Objective: Learn the formal definitions of sine and cosine functions.

1. The figure shows an angle of $\theta = 37°$ in standard position in a *uv*-coordinate system. Measure the angle with a protractor. Do you agree that it is 37°? ______________

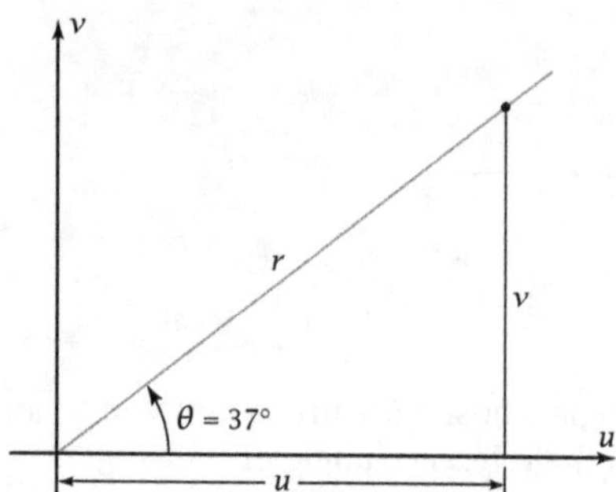

2. The figure shows a point on the terminal side of θ. The *u*- and *v*-coordinates of the point form a right triangle whose hypotenuse is the distance from the origin to the point. Measure the three distances, to the nearest 0.1 cm.

   Adjacent leg, $u =$ ______________

   Opposite leg, $v =$ ______________

   Hypotenuse, $r =$ ______________

3. You recall from previous courses that the **sine** and **cosine** of an angle in a right triangle are defined:

   $$\sin \theta = \frac{\text{opposite leg}}{\text{hypotenuse}} \qquad \cos \theta = \frac{\text{adjacent leg}}{\text{hypotenuse}}$$

   Use the answers in Problem 3 to calculate

   $\sin 37° \approx$ ______________ $\cos 37° \approx$ ______________

4. With your calculator in degree mode, find values of $\sin 37°$ and $\cos 37°$. Do your approximate values in Problem 3 agree with these precise values?

   $\sin 37° =$ ______________________

   $\cos 37° =$ ______________________

5. The definitions of sine and cosine can be extended to angles that measure rotation with the aid of the **reference angle.** Sketch an angle of $\theta = 125°$. Then mark and calculate the reference angle, θ_{ref}.

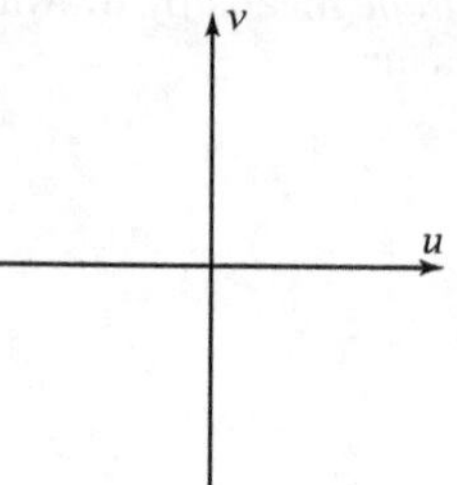

6. Use your calculator to find:

   $\sin \theta_{\text{ref}} =$ ______________ $\cos \theta_{\text{ref}} =$ ______________

7. The formal definitions of sine and cosine are:

   $$\sin \theta = \frac{\text{vertical coordinate}}{\text{radius}}$$

   $$\cos \theta = \frac{\text{horizontal coordinate}}{\text{radius}}$$

   Calculate $\sin 125°$ and $\cos 125°$. How are these numbers related to the sine and cosine of the reference angle in Problem 6? How do you explain that $\cos 125°$ is negative?

8. State what sign the sine and cosine will have for angles that terminate in:

   Quadrant I: sine ________   cosine ________

   Quadrant II: sine ________   cosine ________

   Quadrant III: sine ________   cosine ________

   Quadrant IV: sine ________   cosine ________

9. What did you learn as a result of doing this Exploration that you did not know before?

Precalculus and Trigonometry Explorations
©2004 Key Curriculum Press

Exploration 14: *uv*-Graphs and θy-Graphs of Sinusoids

Objective: Show a geometric relationship between angles plotted as angles and angles plotted along the θ-axis.

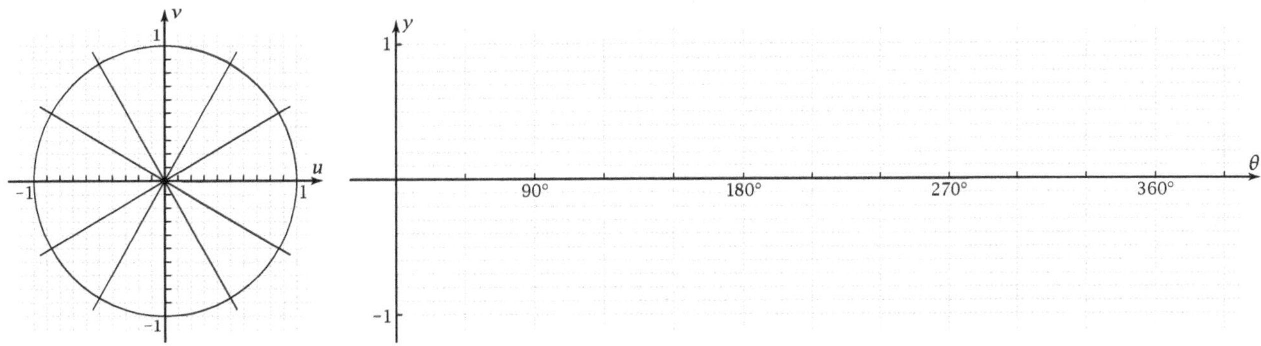

1. The left figure shows a unit circle in a *uv*-diagram with angles marked at every 30°. Read, to two decimal places, the coordinates (u, v) of the point where the ray at 60° cuts the unit circle.

2. Find cos 60° and sin 60° with your calculator. Explain how these numbers relate to the answers to Problem 1.

3. Plot the point $(\theta, y) = (60°, \sin 60°)$ on the θy-coordinate system on the right at the top of this Exploration. Draw a line segment showing how this point is related to the point you plotted in Problem 1.

4. Without actually calculating any more values, plot points on the graph of $y = \sin \theta$ for each 30° from 0° to 360°. Show segments connecting the appropriate points on the *uv*-diagram with points in the θy-diagram.

5. Connect the points in Problem 4 with a smooth curve. What geometrical figure is this curve?

6. Use your observation in Problem 2 to plot points on the graph of $y = \cos \theta$ for each 30° from $\theta = 0°$ to $\theta = 360°$. Connect the points with a smooth curve.

7. What transformation could you apply to the graph of $y = \sin \theta$ to get the graph of $y = \cos \theta$?

8. Explain the difference between the way the value of θ appears on the *uv*-diagram and the way it appears on the θy-diagram.

9. Why do you think the letters u and v, rather than the more common letters x and y, are used in the figure on the left at the top of this Exploration?

10. What did you learn as a result of doing this Exploration that you did not know before?

Exploration 15: Parent Sinusoids

Objective: Explore the graph of the parent function $y = \sin x$, and transform the graph.

1. The graph shows the function $y = \sin x$. Plot this graph as y_1 on your grapher. Use the window shown. Turn on the grid to get the dots. Does your graph agree with this figure? ___________

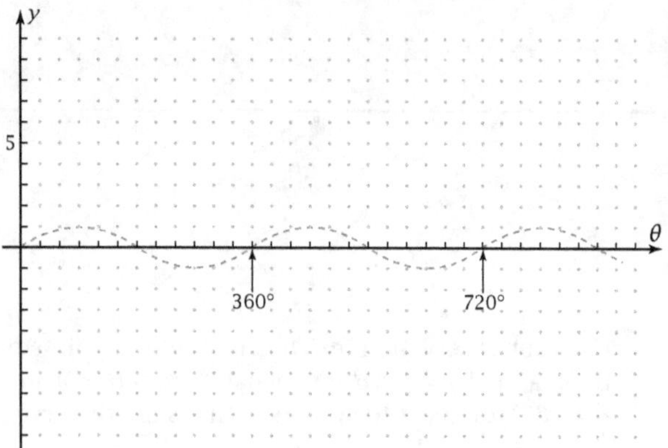

2. The **amplitude** of a periodic function is the vertical distance from the central axis to a high or low point. What is the amplitude of the sine function in Problem 1? Write the equation of the transformed function that would have an amplitude of 5.

3. Plot the transformed graph as y_2 on your grapher. Does the resulting graph really have an amplitude of 5? ___________

4. The solid graph shows a transformation of the sine function from Problem 1. Identify the transformation, and write the equation for the transformed graph. Confirm that your answer is correct by plotting your equation as y_3.

   Verbally: ________________________________

   Equation: ________________________________

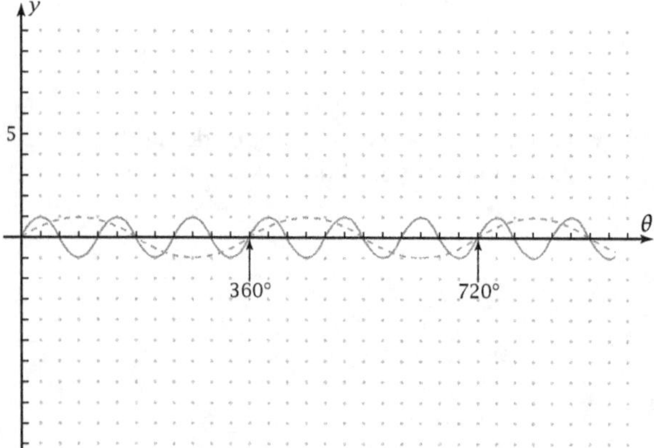

5. Write the equation for this transformed graph. Duplicate this graph on your grapher.

   Equation: ________________________________

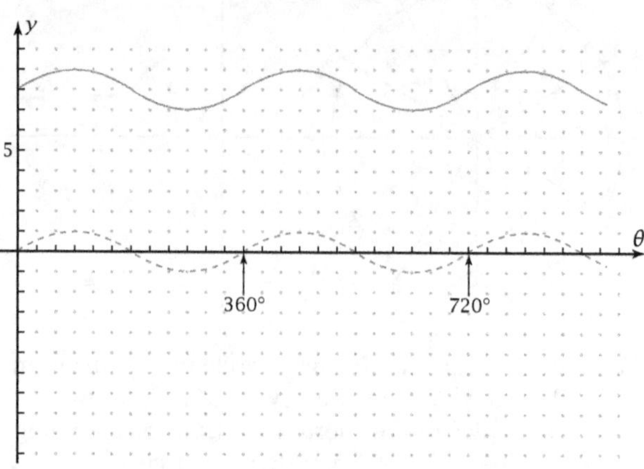

6. The dotted graph shows the result of three transformations. State each transformation, write the equation of the transformed graph, and duplicate the graph on your grapher.

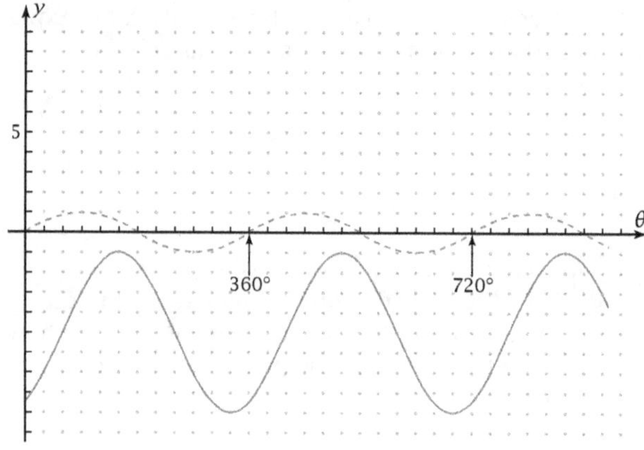

7. Degrees can be used to measure **rotation.** What do you think is the significance of the fact that the **period** of the sine function in Problem 1 is 360°?

8. What did you learn as a result of doing this Exploration that you did not know before?

Exploration 16: Values of the Six Trigonometric Functions

Objective: Find values of the six trigonometric functions, with or without a calculator.

1. Write the definitions of the six trigonometric functions of an angle in terms of the coordinates (u, v) of a point on the terminal side and the distance r from the origin to the point.

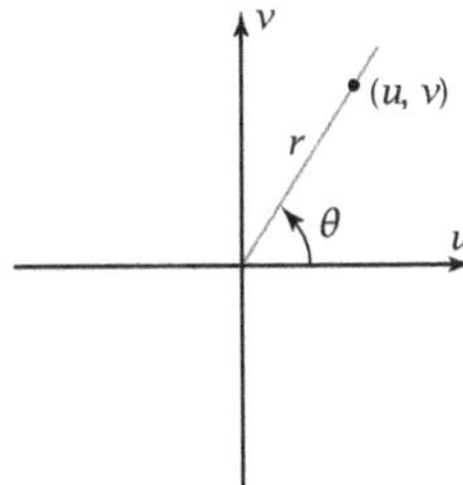

$\sin \theta =$

$\cos \theta =$

$\tan \theta =$

$\cot \theta =$

$\sec \theta =$

$\csc \theta =$

2. Sketch 123° in standard position. Then find the six trigonometric functions of 123°. Write the answers as decimals in ellipsis format.

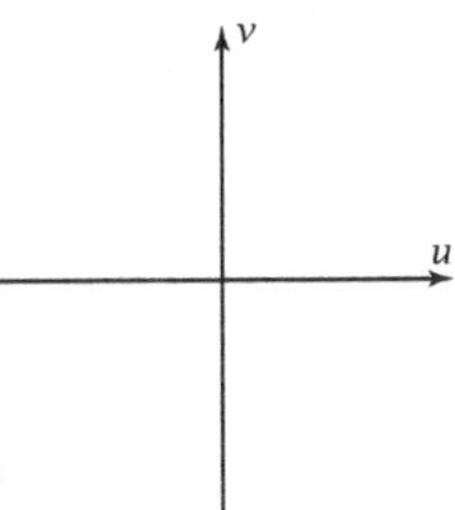

$\sin 123° =$

$\cos 123° =$

$\tan 123° =$

$\cot 123° =$

$\sec 123° =$

$\csc 123° =$

3. Explain why $\sin 123°$ is positive but $\tan 123°$ is negative.

4. The figure shows an angle, θ, in standard position. The terminal side contains the point $(-3, -7)$. Write the six trigonometric functions of θ exactly, as fractions involving radicals if necessary.

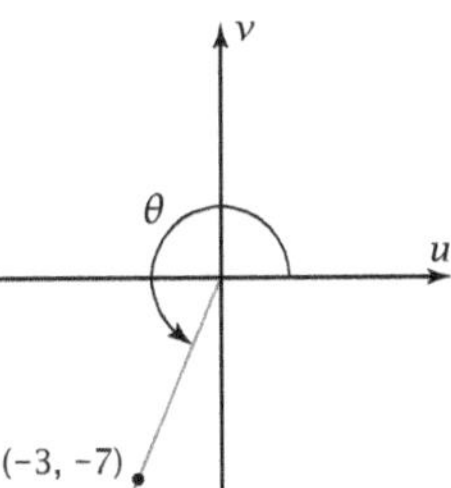

$\sin \theta =$

$\cos \theta =$

$\tan \theta =$

$\cot \theta =$

$\sec \theta =$

$\csc \theta =$

5. The figure shows an angle of 300° in standard position. Choose a convenient point on the terminal side, determine the values of u, v, and r, write them on the figure, and then find in *exact* form (no decimals) the six trigonometric functions of 300°.

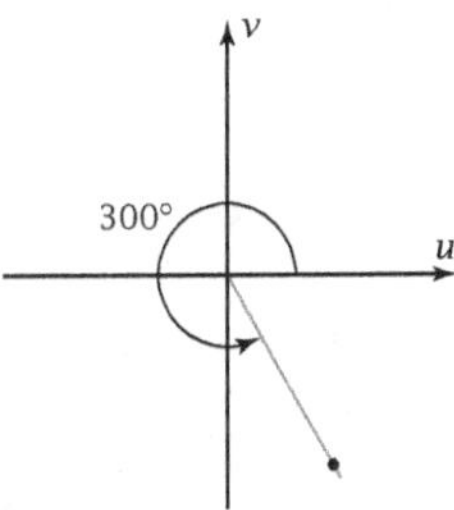

$\sin 300° =$

$\cos 300° =$

$\tan 300° =$

$\cot 300° =$

$\sec 300° =$

$\csc 300° =$

6. What did you learn as a result of doing this Exploration that you did not know before?

Exploration 17: Direct Measurement of Function Values

Objective: Use the definitions of sine, cosine, and tangent to calculate values from measurements on an accurate figure.

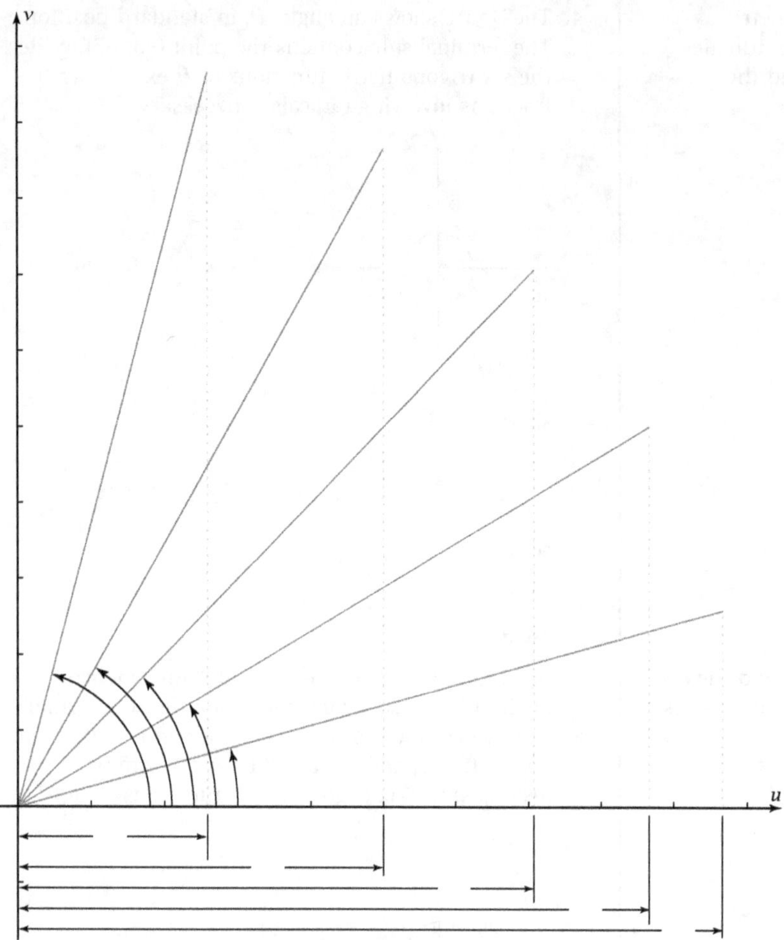

1. The figure shows line segments from the origin making angles with the u-axis of $\theta = 15°$, $30°$, $45°$, $60°$, and $75°$. Perpendiculars (dotted) are drawn from the ends of the segments to the u-axis, forming right triangles. For each triangle, measure the hypotenuse and the two legs, to the nearest 0.1 cm. Write the answers on the diagram.

2. Use the definitions of sine, cosine, and tangent and the lengths you measured to calculate the values of these functions for the five angles. Round the answers to two decimal places.

θ	$\sin \theta$	$\cos \theta$	$\tan \theta$
15°			
30°			
45°			
60°			
75°			

3. Use your grapher to make a table of values of sine, cosine, and tangent. Write the answers, rounded to two decimal places, in this table.

θ	$\sin \theta$	$\cos \theta$	$\tan \theta$
15°			
30°			
45°			
60°			
75°			

4. How well do your answers in Problem 2, found geometrically, compare with the answers found numerically in Problem 3?

5. What did you learn as a result of doing this Exploration that you did not know before?

Exploration 18: Measurement of Right Triangles

Objective: Given two pieces of information about a right triangle, find the other sides and angles.

1. The figure shows a right triangle with legs 4 cm and 3 cm. Do you agree that these measurements are correct? _________________

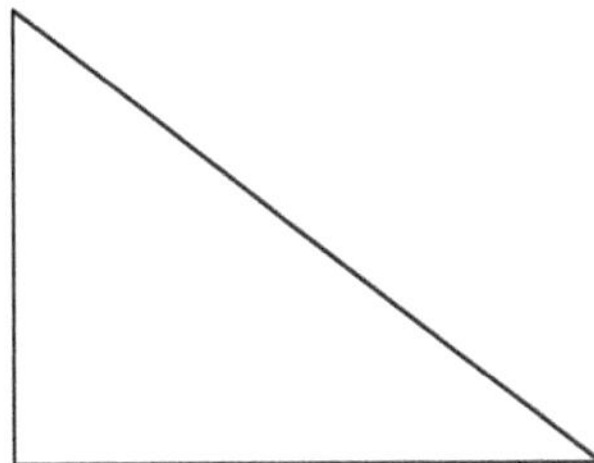

2. Mark the smaller acute angle as A. What number does tan A equal?

3. The measure of A is equal to the **inverse tangent** of the answer to Problem 2. This is found on your calculator as TAN^{-1}, and means "The angle whose tangent is. . . ." Calculate the measure of A. Store this answer as A in your calculator.

4. Use a protractor to measure A on the figure in Problem 1. Does your measured answer agree with the calculated answer?

5. Use the Pythagorean theorem to calculate the length of the hypotenuse.

6. Use the definition of cosine to calculate cos A.

7. Find cos A directly, using the value of A you stored in your calculator in Problem 3. Does the answer agree with your answer to Problem 6?

8. Sketch a figure representing a right triangle with hypotenuse 1066 ft and acute angle 28°. Label the longer leg as x.

9. For the triangle in Problem 8, $\frac{x}{1066}$ is one of the trigonometric functions of 28°. Which function?

10. Calculate the length of the longer leg, x.

11. There are two ways to calculate the length of the shorter leg of the triangle in Problem 8. Show that both ways give the same answer.

12. What did you learn as a result of doing this Exploration that you did not know before?

Exploration 19: Accurate Right Triangle Practice

Date: _________

Objective: Use trigonometric functions to calculate unknown side and angle measures for right triangles.

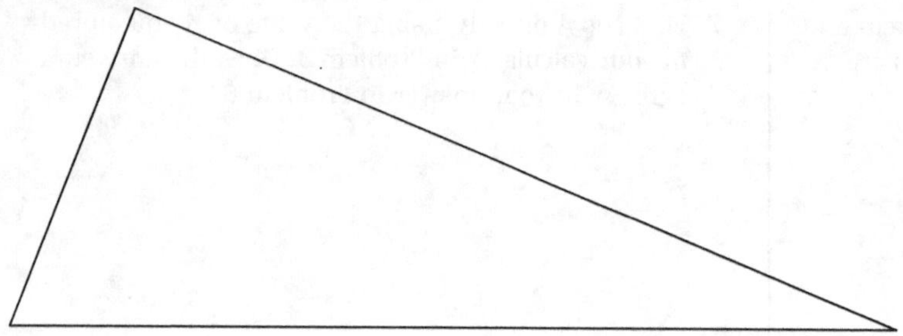

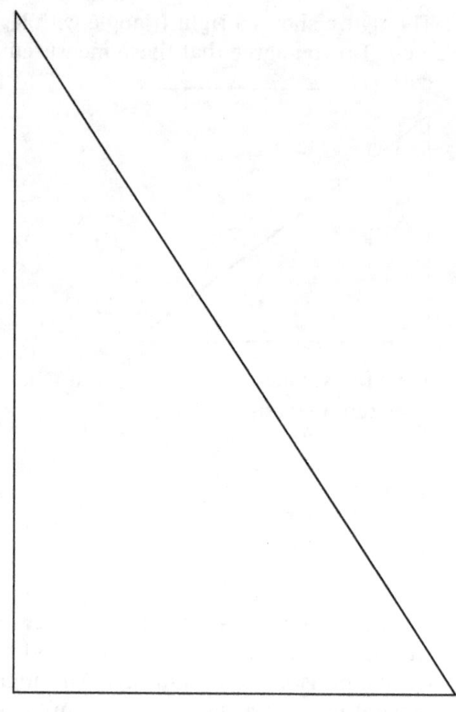

1. The figure above left shows a right triangle of hypotenuse 10 cm and larger acute angle 68°. Do you agree that these measurements are correct?

2. Calculate the length of the shorter leg. Show your work.

3. Measure the shorter leg. Does the measured value agree with your calculated value?

4. The figure to the right shows a right triangle with legs 6 cm and 9 cm. Do you find that these lengths are correct?

5. Calculate the measure of the smaller acute angle. Show your work.

6. Use the angle in Problem 5 to calculate the length of the hypotenuse. Show your work.

7. Calculate the hypotenuse again using the Pythagorean theorem. Does it agree with your answer to Problem 6?

8. What did you learn as a result of doing this Exploration that you did not know before?

Exploration 20: Right Triangles and the Empire State Building

Objective: Apply trigonometric functions to a right triangle problem from the real world.

The Empire State Building in New York was the tallest building in the world when it was built in 1931. To measure its height, a precalculus class finds that from a point on 5th Avenue leading to the building, the angle of elevation to the top of the building is 27°. They move 307 meters closer and find that the angle of elevation is now 38°.

1. Construct a figure showing the street and the two points where the angles were measured. Use a scale of 1 cm per 100 m. Construct the elevation angles from the two points. Where the terminal sides of these angles cross is the top of the building. Construct a perpendicular from this point representing the height of the building.

2. Let x be the distance from the closer point to the point where the perpendicular meets the ground. Let y be the height of the building. By accurate measurement on your figure, find estimates for x and y.

3. Write two equations involving trigonometric ratios with the two known angles, the known distance, 307 m, and the unknown distances, x and y.

4. By doing appropriate algebra on the two equations in Problem 3, calculate the values of x and y.

5. How well do the precise calculated values of x and y agree with your measured values of Problem 2?

6. Look on the Internet or in a reference book to find the actual height of the Empire State Building. State where you found the information.

7. What did you learn as a result of doing this Exploration that you did not know before?

Exploration 21: Periodic Daily Temperatures

Date: _____________

Objective: Transform the cosine function so that it fits, approximately, data on the average daily temperatures for a city.

Here are average daily high temperatures for San Antonio, by month, based on data collected over the past 100 years and published by NOAA, the National Oceanic and Atmospheric Administration. Such data are used, for example, in the design of heating and air conditioning systems.

Month	Temperature (°F)	Month	Temperature (°F)
Jan.	61.7	July	94.9
Feb.	66.3	Aug.	94.6
Mar.	73.7	Sept.	89.3
Apr.	80.3	Oct.	81.5
May	85.6	Nov.	70.7
June	91.8	Dec.	64.6

1. On the graph paper, plot the average daily high temperatures for two years. Assume that January is month 1 and so forth. Determine a time-efficient way for your group members to do the plotting. What should you plot for month zero? Connect the points with a smooth curve.

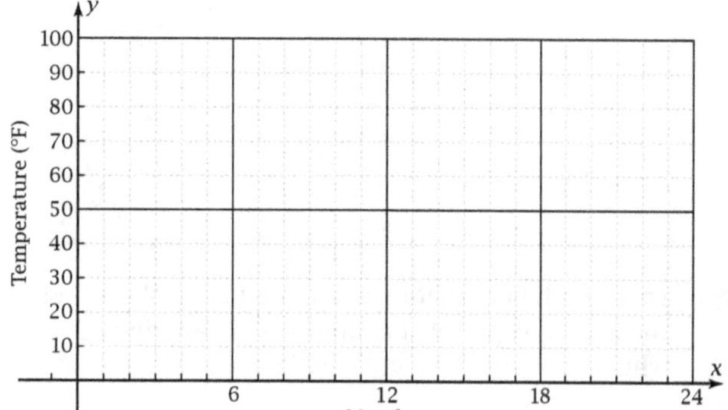

2. The graph of $y = \cos \theta$ completes a cycle each 360° (angle, not temperature). What horizontal dilation factor would make it complete a cycle each 12°, as shown? Write an equation for this transformed sinusoid and plot it on your grapher.

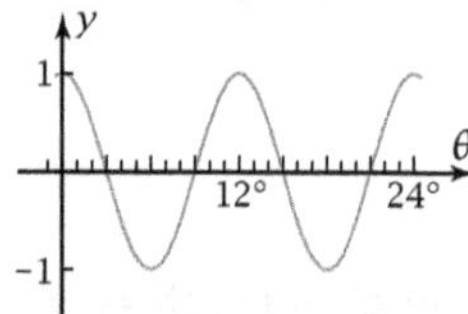

3. Earth rotates 360° around the Sun in 12 months. How do these numbers relate to the dilation factor you used in Problem 2?

4. The temperature graph in Problem 1 has a high point at $x = 7$ months. What transformation would you apply to the sinusoid in Problem 2 (dashed in the next figure) to make it have a high point at $\theta = 7°$ (solid) instead of at $\theta = 0°$? Write the equation and confirm it by plotting on your grapher.

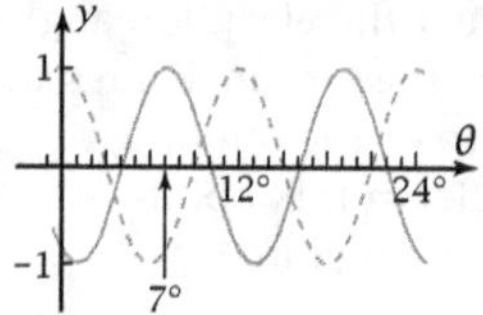

5. The average of the highest and lowest temperatures in the table is $\frac{94.9 + 61.7}{2} = 78.3$. Write an equation for the transformation that would translate the graph in Problem 4 upward by 78.3 units.

6. The 94.9 high point in Problem 1 is 16.6 units above 78.3, and the 61.7 low point is 16.6 units below 78.3. Write an equation for the transformation that would dilate the sinusoid in Problem 5 by a factor of 16.6 so that it looks like this graph. Confirm your answer by grapher.

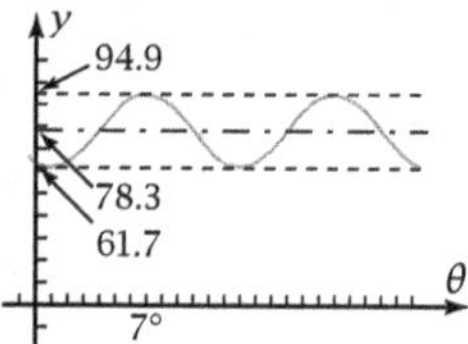

7. On your grapher, plot the points you plotted in Problem 1. How well does the sinusoidal equation in Problem 6 fit the points?

8. What did you learn as a result of doing this Exploration that you did not know before?

Exploration 22: Sine and Cosine Graphs, Manually   Date: __________

Objective: Find the shape of sine and cosine graphs by plotting them on graph paper.

1. On your grapher, make a table of values of $y = \sin \theta$ for each 10° from 0° to 90°. Set the mode to round to 2 decimal places. Plot the values on this graph paper. Also plot $y = \sin \theta$ for each 90° through 720°. Connect the points with a smooth curve, observing the shape you plotted for 0° to 90°.

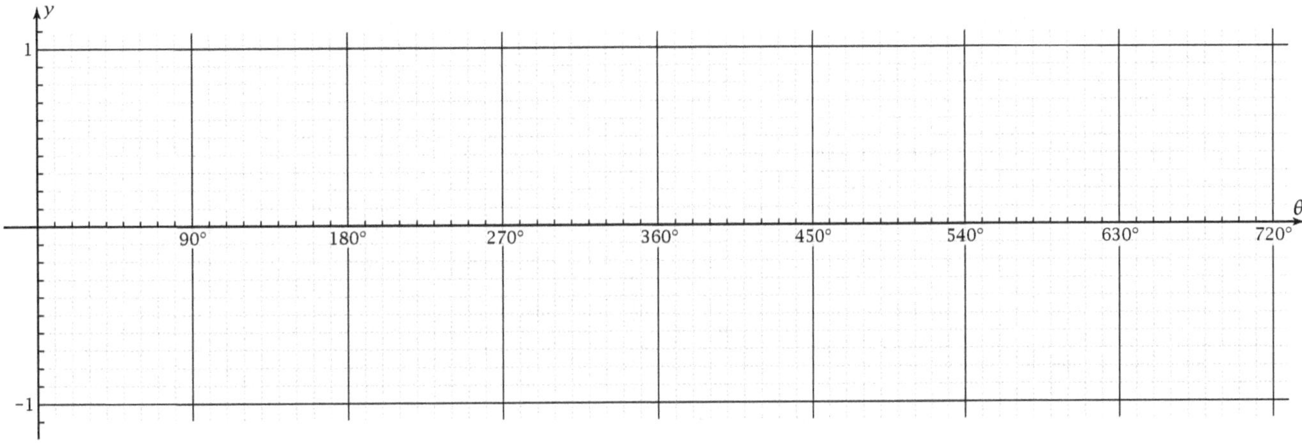

2. Plot the graph of $y = \cos \theta$ pointwise, the way you did for sine in Problem 1.

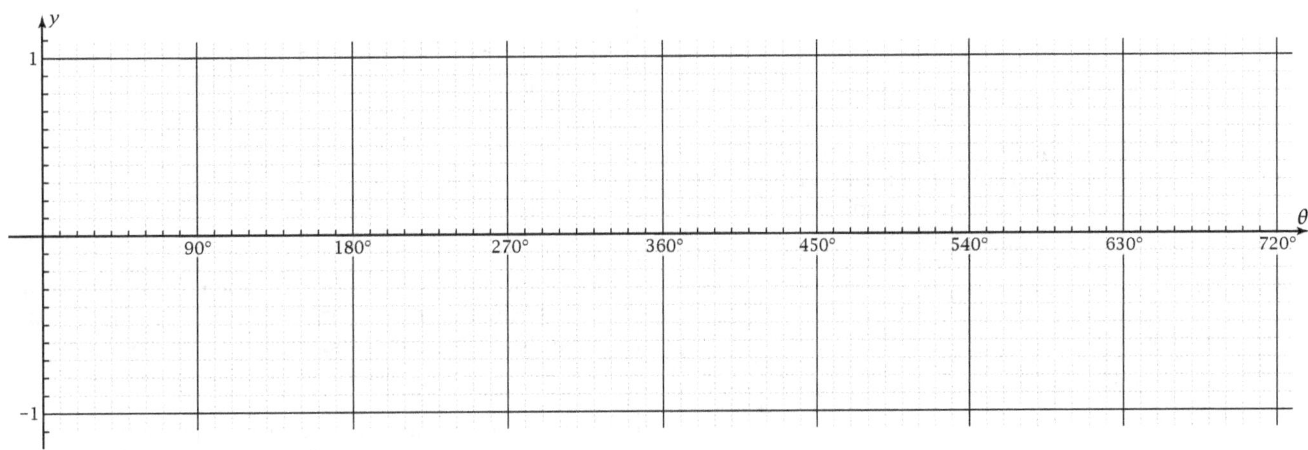

3. Find $\sin 45°$ and $\cos 65°$. Show that the corresponding points are on the graphs in Problems 1 and 2, respectively.

4. Find the inverse trigonometric functions $\theta = \sin^{-1} 0.4$ and $\theta = \cos^{-1} 0.8$. Show that the corresponding points are on the graphs in Problems 1 and 2, respectively.

5. What are the ranges of the sine and cosine functions?

6. Name a real-world situation where variables are related by a periodic graph like sine or cosine.

7. What did you learn as a result of doing this Exploration that you did not know before?

Exploration 23: Transformed Sinusoid Graphs

Date: ____________

Objective: Given the equation of a transformed sinusoid, sketch the graph, and vice versa.

1. Write the horizontal dilation factor, period, amplitude, phase displacement, and vertical displacement, and sketch the graph.

   $$y = 4 + 3\cos 2(\theta - 70°)$$

   Horizontal dilation factor: ____________

   Period: ____________

   Amplitude: ____________

   Phase displacement: ____________

   Vertical displacement: ____________

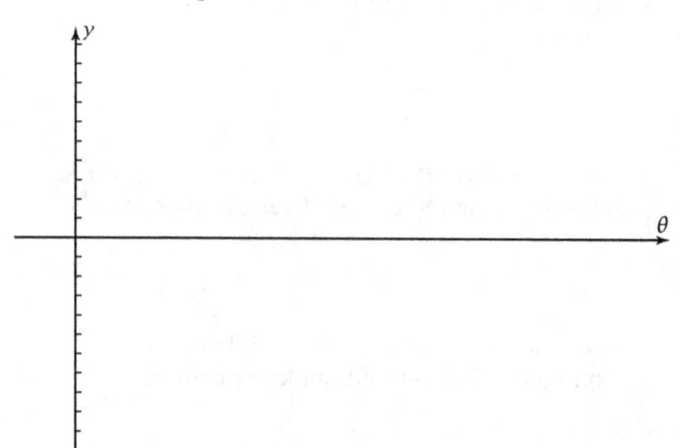

2. Write the horizontal dilation factor, period, amplitude, phase displacement, and vertical displacement, and sketch the graph.

   $$y = -2 + 4\sin 30(\theta + 1°)$$

   Horizontal dilation factor: ____________

   Period: ____________

   Amplitude: ____________

   Phase displacement: ____________

   Vertical displacement: ____________

3. Once you know the connection between the equation of a sinusoid and its graph, you can go backwards and write the equation from a given graph. For the following sinusoid, write the period, horizontal dilation factor, amplitude, phase displacement (for the cosine function), and vertical displacement. Then write the particular equation.

   Period: ____________

   Horizontal dilation factor: ____________

   Amplitude: ____________

   Phase displacement: ____________

   Vertical displacement: ____________

   Equation: ____________

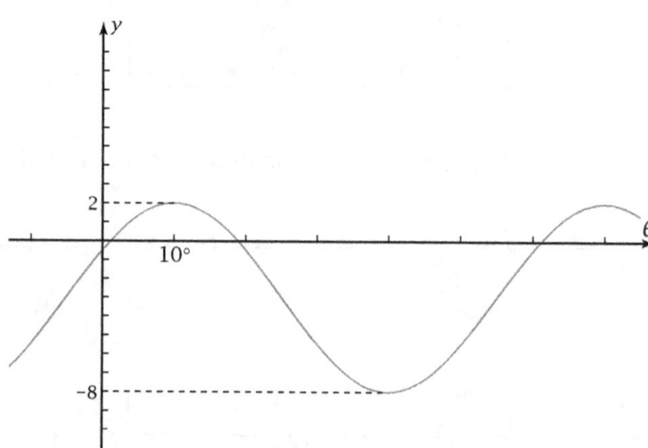

4. Confirm that your answer to Problem 3 is correct by entering the equation in the grapher and plotting the graph. Does your graph agree with the given figure?

   ____

5. By the most time-efficient method possible, find y for your equation in Problem 3 if $\theta = 35°$. Write the answer to as many decimal places as your grapher will give. Draw something on the given graph to show that your answer is reasonable.

6. What did you learn as a result of doing this Exploration that you did not know before?

Exploration 24: Sinusoidal Equations from Graphs   Date: _____________

Objective: Given the equation, sketch the sinusoid, and vice versa.

1. Sketch two cycles of this sinusoid:

   $$y = -3 + 5 \sin 4(\theta - 20°)$$

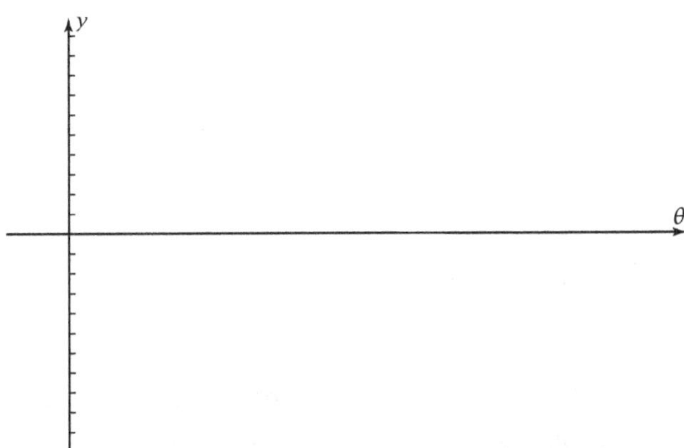

2. Write a particular equation (cos) for this sinusoid:

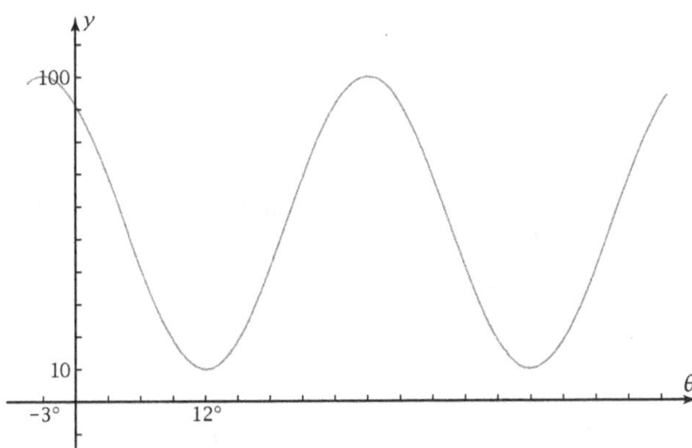

3. Write a particular equation for the sinusoid in Problem 2 using sine.

   ____

4. Plot your equation in Problem 2 as y_1 on your grapher. Plot the equation in Problem 3 as y_2. Use a different style for each graph. Do both graphs agree with the given graph?

5. This is a half-cycle of a sinusoid. Write a particular equation.

   ____

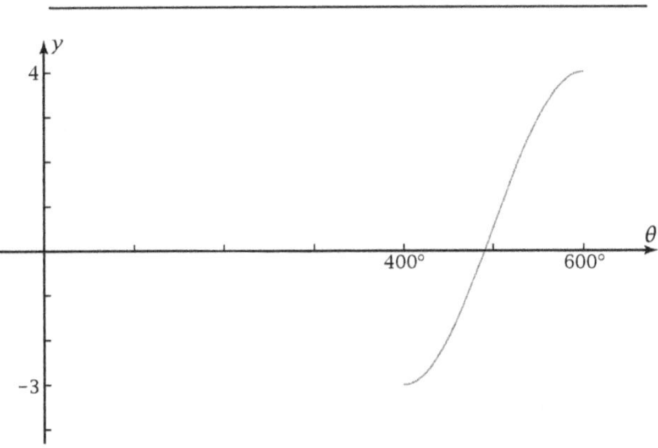

6. This is a quarter-cycle of a sinusoid. Write a particular equation.

   ____

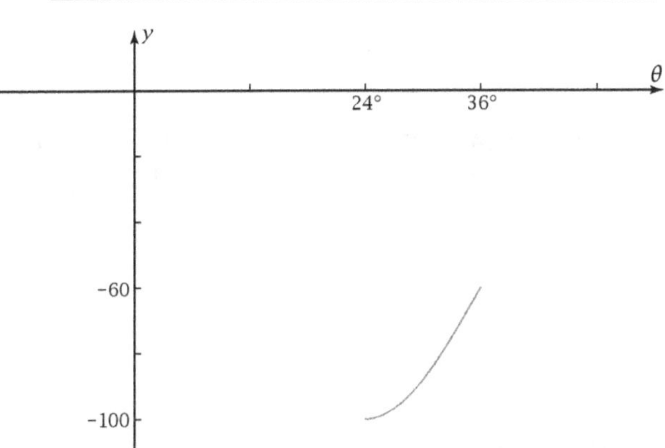

7. On the sinusoid in Problem 2, mark a point of inflection. Mark another point at which the graph is increasing but concave down.

8. What did you learn as a result of doing this Exploration that you did not know before?

Exploration 25: Tangent and Secant Graphs

Objective: Discover what the tangent and secant function graphs look like and how they relate to sine and cosine.

No graphers allowed for Problems 1–7.

1. The **reciprocal property** states that

$$\sec \theta = \frac{1}{\cos \theta}$$

   Without your grapher, use this property to sketch the graph of $y = \sec \theta$ on the same axes as the graph of the **parent function** $y = \cos \theta$. In particular, show what happens to the secant graph wherever $\cos \theta = 0$.

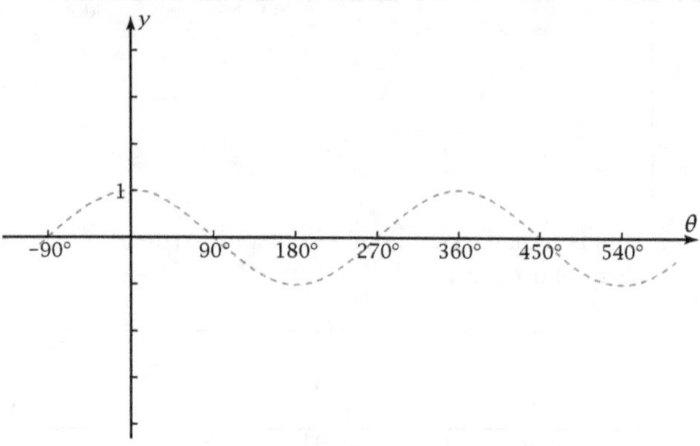

2. Write the **quotient property** expressing $\tan \theta$ as a quotient of two other trigonometric functions.

3. The next figure shows the parent functions $y = \sin \theta$ and $y = \cos \theta$. Based on the answer to Problem 2, determine where the asymptotes are for the graph of $y = \tan \theta$, and mark them on the figure.

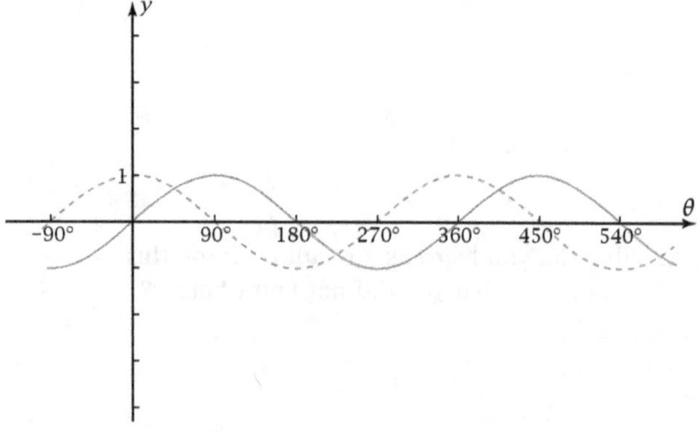

4. Based on the quotient property, find out where the θ-intercepts are for the graph of $y = \tan \theta$. Mark these intercepts on the figure in Problem 3.

5. At $\theta = 45°$, $\sin \theta$ and $\cos \theta$ are equal. Based on this fact, what does $\tan 45°$ equal? Mark this point on the graph in Problem 3. Mark all other points where $|\sin \theta| = |\cos \theta|$.

   $\tan 45° =$ __________________

6. Use the points and asymptotes you have marked to sketch the graph of $y = \tan \theta$ on the figure in Problem 3. (No graphers allowed!)

7. Check your graphs with your instructor. __________

Graphers allowed for the remaining problems.

8. On your grapher, plot the graph of $y = \csc \theta$. Sketch the result here.

9. On your grapher, plot the graph of $y = \cot \theta$. Sketch the result here.

10. At what values of θ are the points of inflection for $y = \tan \theta$? Explain why the tangent function has no critical points.

11. Explain why the graph of $y = \sec \theta$ has no points of inflection, even though the graph goes from concave up to concave down at various places.

12. What did you learn as a result of doing this Exploration that you did not know before?

Exploration 26: Transformed Tangent and Secant Graphs

Date: _____________

Objective: Sketch transformed tangent, cotangent, secant, and cosecant graphs, and find equations from given graphs.

1. For $y = 3 + \frac{1}{2}\tan 5(\theta - 7°)$, state

   The horizontal dilation: _______________

   The period: _______________

   The horizontal translation: _______________

   The vertical dilation: _______________

   The vertical translation: _______________

2. Sketch the graph of $y = 3 + \frac{1}{2}\tan 5(\theta - 7°)$, showing vertical asymptotes, horizontal axis, points of inflection, and other significant points.

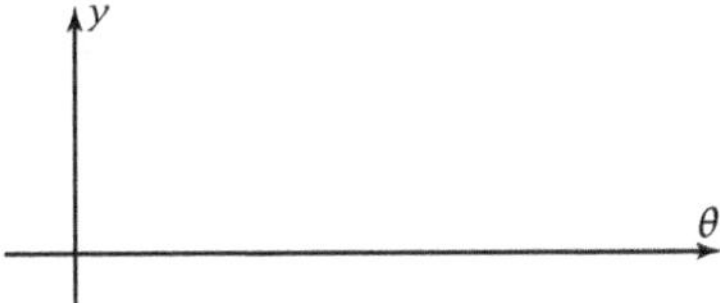

3. For the next graph, state

   The horizontal dilation: _______________

   The period: _______________

   The horizontal translation (for cotangent):

   ____

   The vertical dilation: _______________

   The vertical translation: _______________

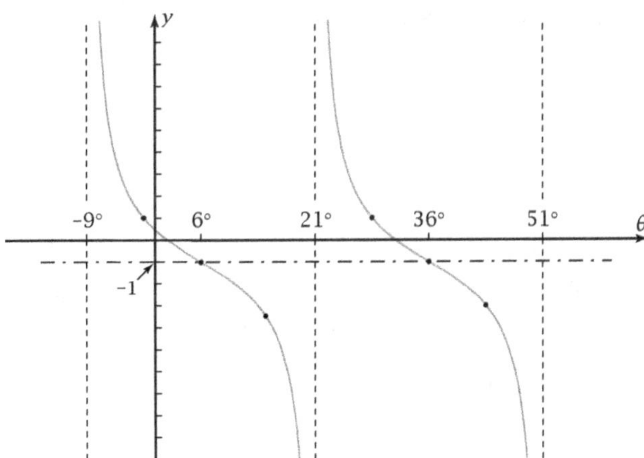

4. Write a particular equation for the graph in Problem 3. Check your answer by plotting on your grapher.

5. For $y = 1 + 3\csc 4(\theta + 10°)$, give

   The horizontal dilation: _______________

   The period: _______________

   The horizontal translation: _______________

   The vertical dilation: _______________

   The vertical translation: _______________

6. Sketch the graph of $y = 1 + 3\csc 4(\theta + 10°)$, showing vertical asymptotes, horizontal axis, and critical points.

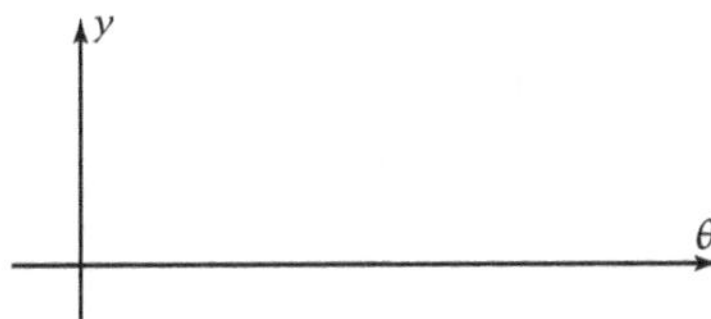

7. For the next graph, give

   The horizontal dilation: _______________

   The period: _______________

   The horizontal translation (for secant): _______________

   The vertical dilation: _______________

   The vertical translation: _______________

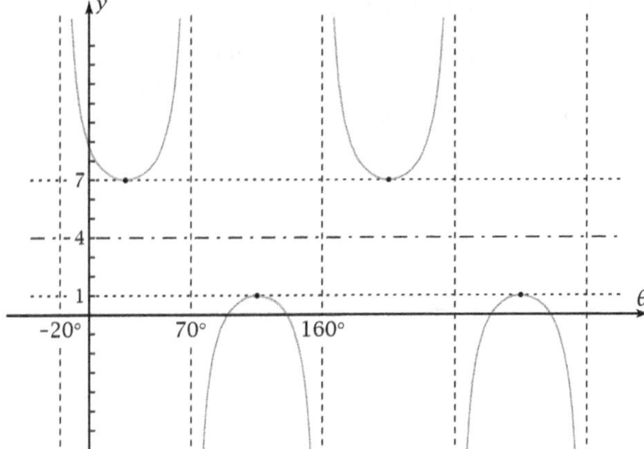

8. Write a particular equation for the graph in Problem 7. Check your answer by plotting on your grapher.

9. What did you learn as a result of doing this Exploration that you did not know before?

Exploration 27: Radian Measure of Angles 1

Date: ______________

Objective: Discover how angles are measured in radians by measuring around a circle with a flexible ruler.

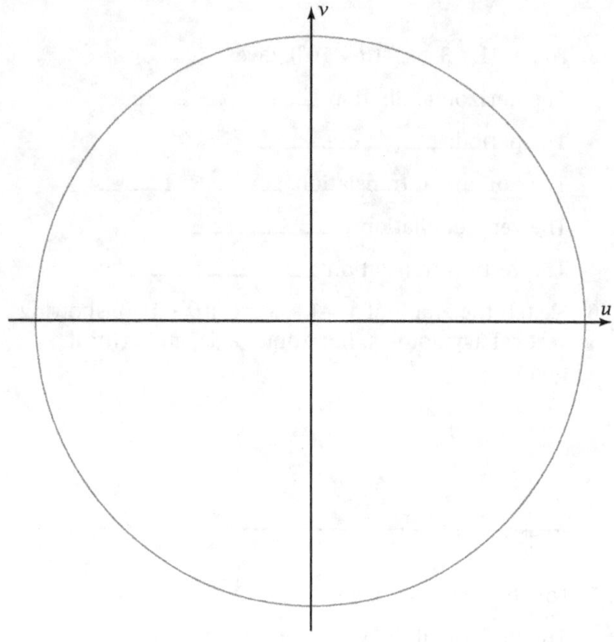

1. The figure shows a circle centered at the origin of a *uv*-coordinate system. On a flexible ruler (an index card will do), mark off a length equal to the radius of the circle. Start at the point where the circle intersects the positive *u*-axis and bend the ruler to mark off arcs of lengths 1, 2, and 3 units counterclockwise around the circle.

2. Draw a ray from the origin through the point corresponding to 1 radius length. Your drawing should look like this:

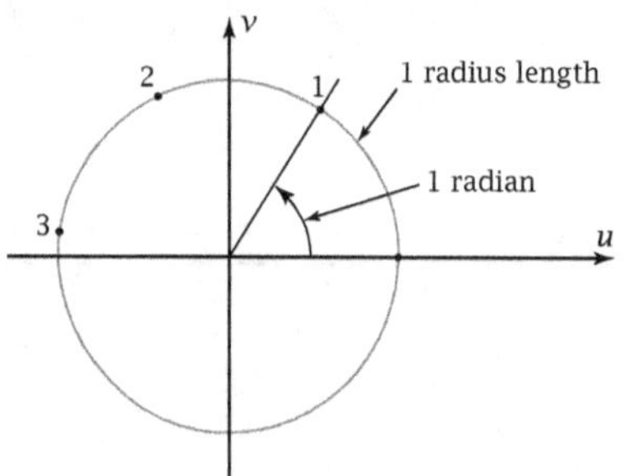

3. The resulting angle in standard position has measure 1 **radian.** Measure the number of degrees in this angle.

4. From geometry, recall that the circumference of a circle is $2\pi r$, where r is the radius. So there are 2π radians in a complete revolution. The fact that there are 360° in a complete revolution gives you a way to transform between degrees and radians. Calculate exactly the number of degrees in 1 radian. How does the measured value in Problem 3 compare with this exact answer?

5. Calculate the exact number of degrees in 3 radians. Draw a 3-radian angle on the previous figure. How close is the degree measure of your drawn angle to the exact value?

6. What did you learn as a result of doing this Exploration that you did not know before?

Exploration 28: Radian Measure of Angles 2

Date: ____________

Objective: Discover how angles are measured in radians by wrapping a string around a circle.

1. At the board, plot horizontal and vertical u- and v-axes. Obtain a roll of masking tape and place it with its center at the origin. Draw a circle on the board by tracing around the outside of the roll.

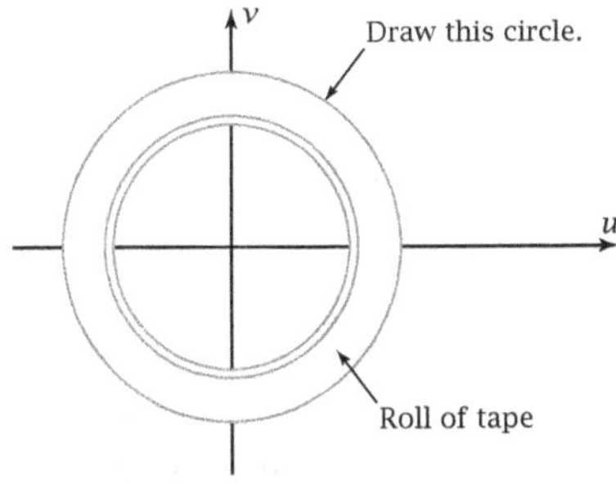

2. Remove the roll of tape from the board. Mark a "ruler" on a piece of string, with units equal to the radius of the circle you drew. Then attach the string to the roll of tape.

3. Put the roll back on the board in such a way that the starting point on the string is on the positive side of the u-axis. Wrap the string counterclockwise around the tape roll. Make marks on the board at the points 1, 2, 3, 4, 5, and 6 on the string. Then remove the tape roll again.

4. Draw rays through the points you marked on the board, like this:

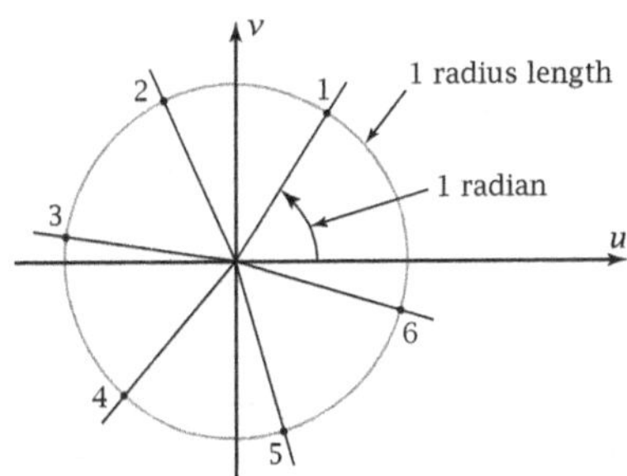

5. The central angles formed by the rays you drew have measures of 1, 2, 3, . . . **radians.** By measuring with a protractor, find out approximately how many degrees are in 1 radian.

6. An angle of 6 radians is not quite a complete revolution. How many radians would it take to make a complete revolution? Provide the exact value.

7. You should have answered "2π radians" for Problem 6. The fact that there are 360° in a complete revolution gives you a way to transform degrees to radians, and the other way around. Calculate exactly the number of degrees in 1 radian. How does the measured value in Problem 5 compare with this exact answer?

8. Calculate the exact number of degrees in 3 radians. Show a 3-radian angle on your board drawing. How close is the degree measure of your drawn angle to the exact value?

9. Explain why the size of a radian would be the same no matter what size circular object you use in place of the roll of tape in **Problem 1.**

10. What did you learn as a result of doing this Exploration that you did not know before?

Exploration 29: Circular Function Parent Graphs

Date: __________

Objective: Plot circular function sinusoids and tangent graphs.

1. Sketch the parent trigonometric function $y = \sin\theta$.

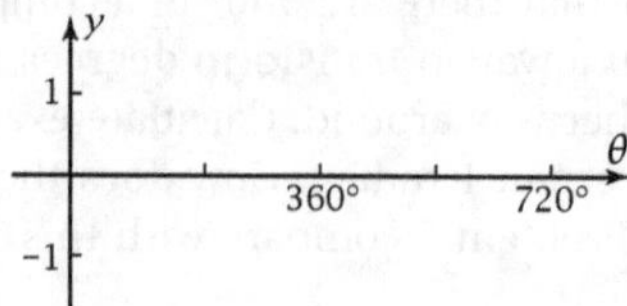

2. Sketch the parent trigonometric function $y = \cos\theta$.

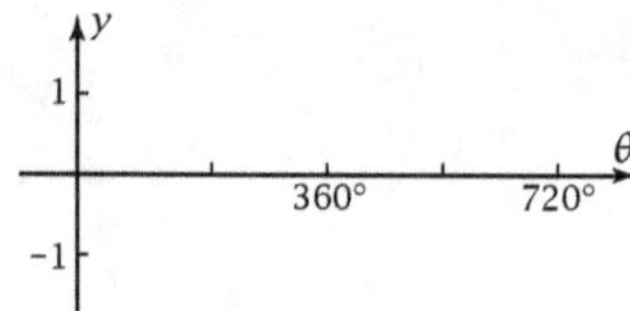

3. Sketch the parent trigonometric function $y = \tan\theta$.

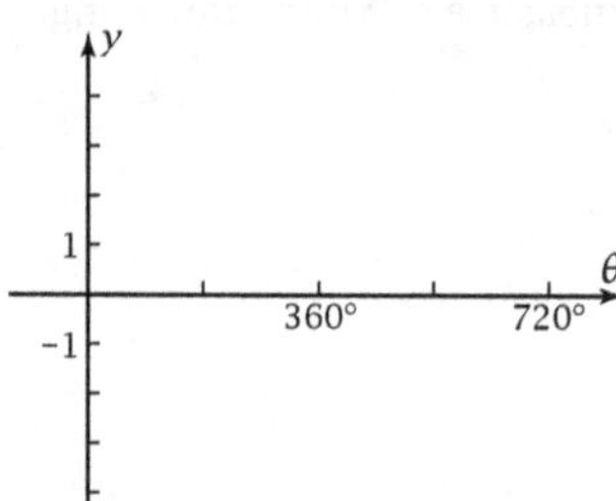

4. Set your grapher to radian mode. Set the window with an x-range of $[0, 4\pi]$ and the y-range as shown on the given graphs. Then plot the graph of the **circular function** $y = \sin x$. Sketch the result.

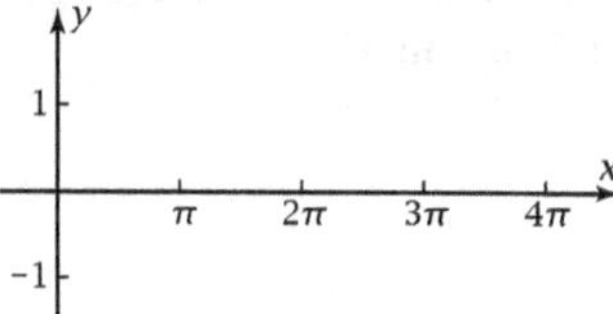

5. With your grapher still in radian mode, plot the graph of the circular function $y = \cos x$. Sketch the result.

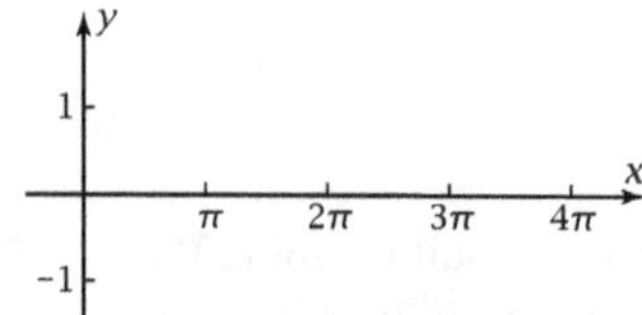

6. With your grapher still in radian mode, plot the graph of the circular function $y = \tan x$. Sketch the result.

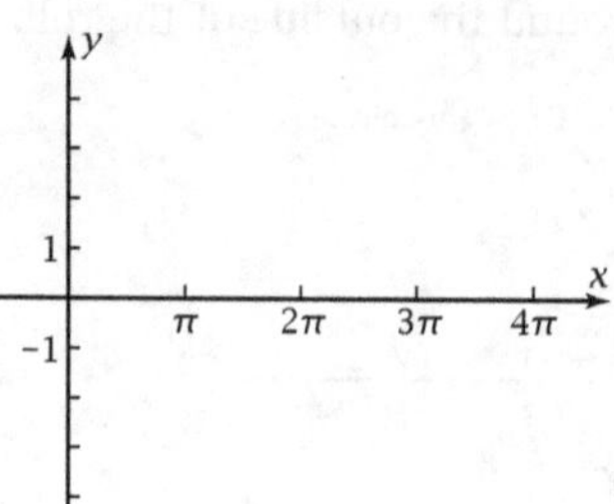

7. The only difference between the parent graphs for the circular function sinusoid and the ordinary trigonometric function sinusoid is the period. Explain how the periods of the two types of sinusoid relate to degrees and radians.

8. The graph here is a transformed circular function sinusoid. Using what you have learned about transformations, find a particular equation of this sinusoid. Confirm by grapher that your equation is correct.

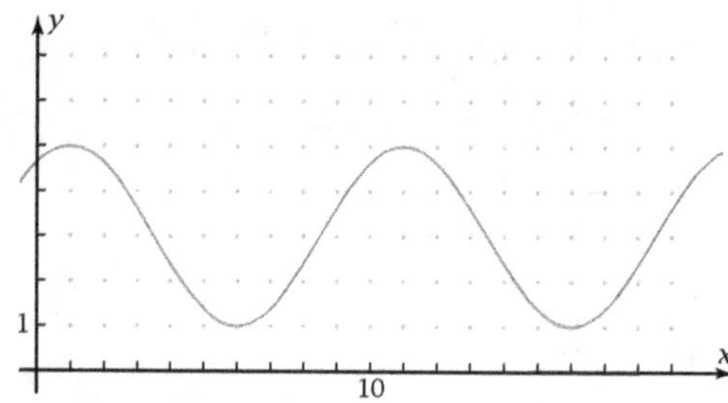

9. What did you learn as a result of doing this Exploration that you did not know before?

Exploration 30: Sinusoids, Given y, Find x Numerically

Objective: Find a particular equation for a given sinusoid and use it to graphically and numerically find x-values for a given y-value.

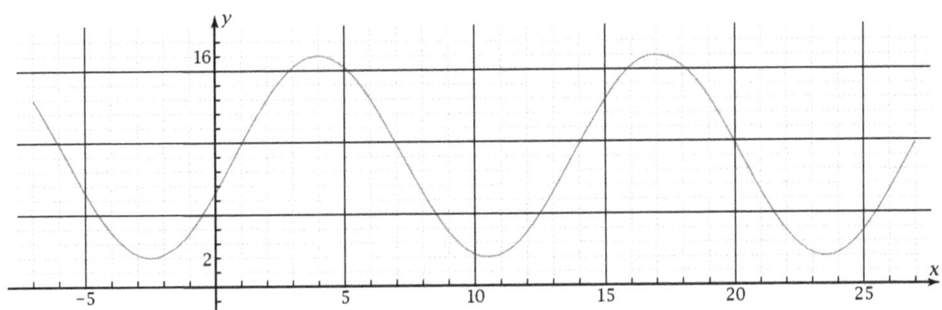

1. For the sinusoid shown, draw the line $y = 5$. Read from the graph the six values of x for which the line crosses the part of the graph shown. Write your answers to one decimal place.

   $x =$ _____________, _____________, _____________,

   _____________, _____________, _____________.

2. Write an equation for this sinusoid.

3. Plot the equation from Problem 2 on your grapher. Does it look like the given graph? _____________

4. Trace your graph in Problem 3 to $x = 17$. Does your graph have a high point there? _____________

5. Circle the leftmost point on the given graph at which $y = 5$. Plot the line $y = 5$, and use the intersect feature to find the value of x at this point.

   $x =$ _____________

6. Other values of x for which you can find $y = 5$ by adding multiples of the period to the value of x in Problem 5. Let n be the number of periods you add. Find two more values of x for which $y = 5$. Circle the three x-values in Problem 1 that are also answers to Problem 5 and this problem.

   Multiple, $n = 1$: $x =$ _____________

   Multiple, $n = 2$: $x =$ _____________

7. Put a box on the figure at a point whose x-value is not an answer to Problem 5 or 6. Use the intersect feature to find one of these x-values.

   $x =$ _____________

8. Add multiples of the period to the x-value in Problem 5 or 7 to find the other two x-values that are also on the graph. State what multiple of the period you added.

   Multiple, $n =$ _____________: $x =$ _____________

   Multiple, $n =$ _____________: $x =$ _____________

9. By adding an appropriate multiple of the period to the answer to Problem 5 or 7, find the first value of x greater than 1000 for which $y = 5$. At this value of x, will y be increasing or decreasing? How can you tell?

   Multiple, $n =$ _____________: $x =$ _____________

10. What did you learn as a result of doing this Exploration that you did not know before?

Exploration 31: Sinusoids, Given y, Find x Algebraically

Objective: Given the particular equation for a sinusoid and a value of y, calculate the corresponding x-values algebraically.

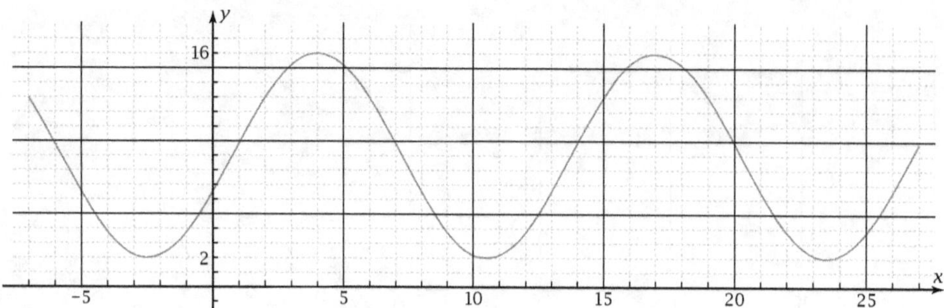

1. The sinusoid has equation

$$y = 9 + 7\cos\frac{2\pi}{13}(x - 4)$$

   Confirm that this equation gives the correct value of y when $x = 15$.

2. Your objective is to find algebraically the values of x given $y = 5$. Substitute 5 for y. Then do the algebra necessary to get x using an arccosine. Write the **general solution** in the form

   $x = $ (number) + (period)n or (number) + (period)n

3. Write the two values of x from the general solution in the $n = 0$ row of this table. By adding and subtracting multiples of the period, fill in the other rows in the table with more possible values of x.

n	x_1	x_2
−1		
0		
1		
2		

4. Circle the points on the given graph where the line $y = 5$ cuts the graph. For each point, tell the value of n at that point.

5. Find the two values of x if $n = 100$.

6. Find the first value of x greater than 1000 for which $y = 5$. What does n equal there?

7. What did you learn as a result of doing this Exploration that you did not know before?

Exploration 32: Using Sinusoids to Model Chemotherapy

Objective: Use sinusoids to predict events in the real world.

Chemotherapy Problem: Ima Patient has cancer. She must have a chemotherapy treatment once every three weeks. One side effect is that her red blood cell count goes down and then comes back up between treatments. On January 13 (day 13 of the year), she gets a treatment. At that time, her red blood cell count is at a high of 800. Halfway between treatments, the count drops to a low of 200. Assume that the red blood cell count varies sinusoidally with the day of the year, x.

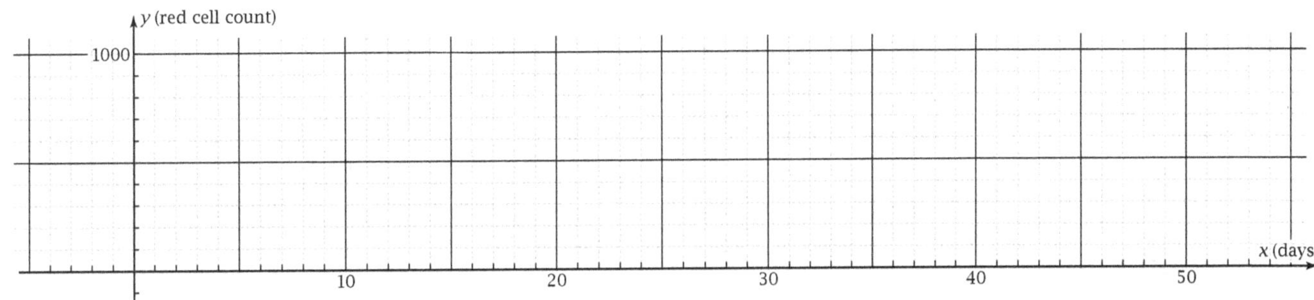

1. Draw the graph of the sinusoid on the given axes. Show enough cycles to fill the graph paper.

2. Write a particular equation for the (circular) sinusoid in Problem 1. It is recommended that you use the cosine function.

   ____

3. Enter your equation in your grapher. Plot the graph using the window shown. Explain how the graph verifies that your equation is correct.

4. Ima feels "good" if the red blood cell count is 700 or more, "bad" if the count is 300 or less, and "so-so" if the count is between 300 and 700. How will she be feeling on her birthday, March 19? Explain how you arrived at your answer.

5. Show on your graph the interval of dates between which Ima will feel "good" as she comes back from the low point after the January 13 treatment.

6. Find precisely the values of x at the beginning and end of the interval in Problem 5 by setting $y = 700$ and using appropriate numeric or graphical methods. Describe what you did.

   $x =$ ________________ and $x =$ ________________

7. What did you learn as a result of doing this Exploration that you did not know before?

Exploration 33: Modeling Geological Formations with Sinusoids

Date: __________

Objective: Use sinusoids to predict events in the real world.

The figure shows a vertical cross section through a piece of land. The y-axis is drawn coming out of the ground at the fence bordering land owned by your boss, Earl Wells. Earl owns the land to the left of the fence and is interested in buying land on the other side to drill a new oil well. Geologists have found an oil-bearing formation, which they believe to be sinusoidal in shape, beneath Earl's land. At $x = -100$ feet, the top surface of the formation is at its deepest, $y = -2500$ feet. A quarter-cycle closer to the fence, at $x = -65$ feet, the top surface is only 2000 feet deep. The first 700 feet of land beyond the fence is inaccessible. Earl wants to drill at the first convenient site beyond $x = 700$ ft.

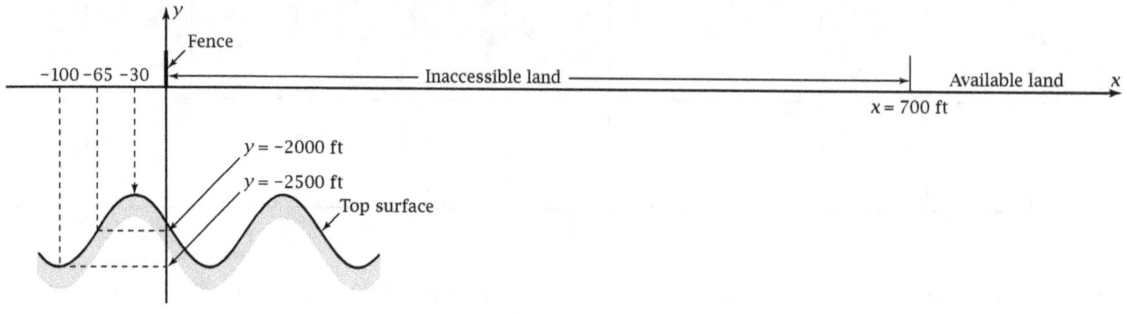

1. Find a particular equation for y as a function of x.

2. Plot the graph on your grapher. Use a window with an x-range of $[-100, 900]$. Describe how the graph confirms that your equation is correct.

3. Find graphically the first interval of x-values in the available land for which the top surface of the formation is no more than 1600 feet deep. Draw a sketch showing what you did.

4. Find algebraically the values of x at the ends of the interval in Problem 3.

5. Suppose that the original measurements were slightly inaccurate and that the value of x shown at -65 feet was at $x = -64$ instead. Would this fact make much difference in the answer to Problem 3? Use a time-efficient method to reach your answer. Explain what you did.

6. What did you learn as a result of doing this Exploration that you did not know before?

Exploration 34: Properties of Trigonometric Functions

Date: _______________

Objective: Use the properties of trigonometric functions to transform an expression to another given form.

1. Write the four trigonometric functions $\tan x$, $\cot x$, $\sec x$, and $\csc x$ in terms of $\sin x$ and $\cos x$.

2. The property $\sec x = \frac{1}{\cos x}$ is called a **reciprocal property.** Why do you think this is the property's name?

3. What is the name of the property $\tan x = \frac{\sin x}{\cos x}$?

4. Evaluate $\cos^2 0.6 + \sin^2 0.6$. Sketch a 0.6-radian angle in standard position, and use the drawing to explain the significance of your answer.

5. What is the name of the property $\cos^2 x + \sin^2 x = 1$?

6. Divide both sides of the equation in Problem 5 by $\cos^2 x$. Simplify the result to get a **Pythagorean property** relating $\sec x$ and $\tan x$.

7. Derive a Pythagorean property relating $\csc x$ and $\cot x$.

8. Derive another **quotient property** expressing $\tan x$ in terms of $\sec x$ and $\csc x$.

(Over)

Exploration 34: Properties of Trigonometric Functions *continued*

9. Transform the expression csc x tan x to sec x. Start by writing the given expression. Then substitute using appropriate properties and simplify.

11. Transform csc x tan x cos x to 1.

10. The expression csc x tan x in Problem 9 involves *two* functions. The answer, sec x, involves only *one* function. What could be your thought process in deciding how to start this problem?

12. What did you learn as a result of doing this Exploration that you did not know before?

Exploration 35: Transforming an Expression

Objective: Use the trigonometric properties to transform an expression to another given form.

1. Write the three reciprocal properties.

2. Write the two quotient properties.

3. Write the three Pythagorean properties.

4. Transform $\sec x - \cos x$ to $\sin x \tan x$. Beside each step, write the technique you used from your "list of things to try."

5. What did you learn as a result of doing this Exploration that you did not know before?

Exploration 36: Trigonometric Transformations

Date: _________

Objective: Use the properties of functions of one argument to transform trigonometric expressions.

1. Without consulting text, notes, or other students, write

   • The three reciprocal properties:

   • The two quotient properties:

   • The three Pythagorean properties:

2. Transform $(1 + \cos A)(1 - \cos A)$ to $\sin^2 A$.

3. Transform $\cot B + \tan B$ to $\csc B \sec B$.

4. Transform $\csc P \cos^2 P + \sin P$ to $\csc P$. (Try factoring out the $\csc P$ you want, first.)

5. On your grapher, plot $y_1 = \tan^2 x$ and $y_2 = \sec^2 x$. Use radian mode and a window with an x-range of $[0, \pi]$ and a y-range of $[0, 10]$. By appropriate tracing, find out how the two graphs are related to each other. How does this relationship correspond to the trigonometric properties? Check your graph with your instructor. _________

6. What did you learn as a result of doing this Exploration that you did not know before?

Exploration 37: Trigonometric Identities

Objective: Use the trigonometric properties to transform an expression to another given form.

No books or notes!

1. Write the three reciprocal properties.

2. Write the two quotient properties.

3. Write the three Pythagorean properties.

4. Prove the alternate form of the quotient property,

$$\tan x = \frac{\sec x}{\csc x}$$

5. Quick: Why does $\cot x = \frac{\csc x}{\sec x}$?

6. Prove that $\frac{\sec A}{\sin A} - \frac{\sin A}{\cos A} = \cot A$ is an identity.

7. Prove: $\frac{1}{1 - \cos B} + \frac{1}{1 + \cos B} = 2 \csc^2 B$

8. What did you learn as a result of doing this Exploration that you did not know before?

Exploration 38: Arccosine, Arcsine, and Arctangent   Date: ____________

Objective: Find values of arccosine, arcsine, and arctangent by calculator.

1. Find the degree measure of arccos 0.4 that equals $\cos^{-1}$ 0.4. Sketch the angle in a *uv*-coordinate system (draw the arc and the arrow). Show the reference triangle and label two of its sides.

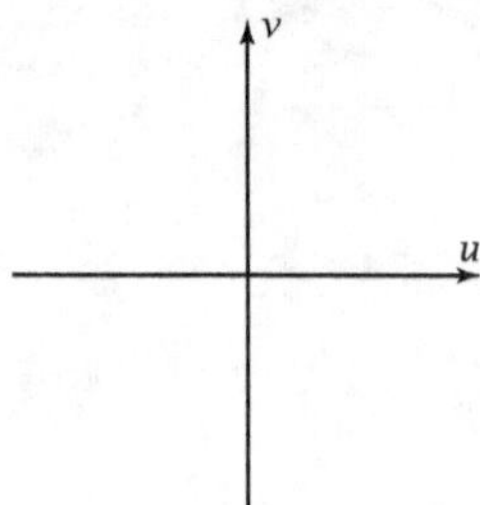

2. On the same axes, sketch another angle arccos 0.4 in a different quadrant. Sketch and label the reference triangle. Find the degree measure of the angle.

3. Write the general solution for arccos 0.4.

4. Find the value of arcsin 0.3 that equals $\sin^{-1}$ 0.3. Sketch the angle. Label two appropriate sides of the reference triangle.

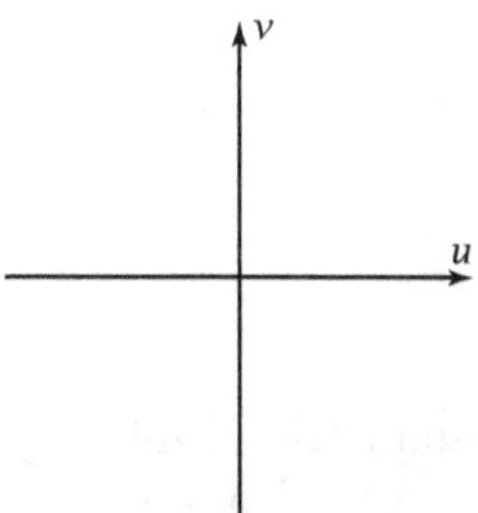

5. Sketch another angle arcsin 0.3 in a different quadrant. Sketch and label the reference triangle. Find the measure of the angle.

6. Write the general solution for arcsin 0.3.

7. Find $\sin^{-1}$ (−0.8) and a value of arcsin (−0.8) terminating in a different quadrant. Sketch the angles and label each reference triangle.

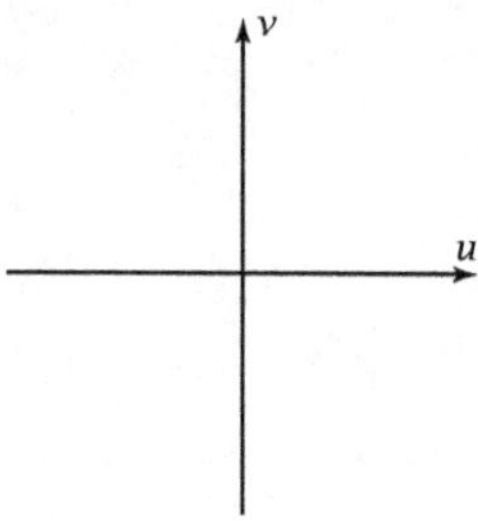

8. Find $\tan^{-1}$ 5 and a value of arctan 5 terminating in a different quadrant. Sketch the angles and label each reference triangle.

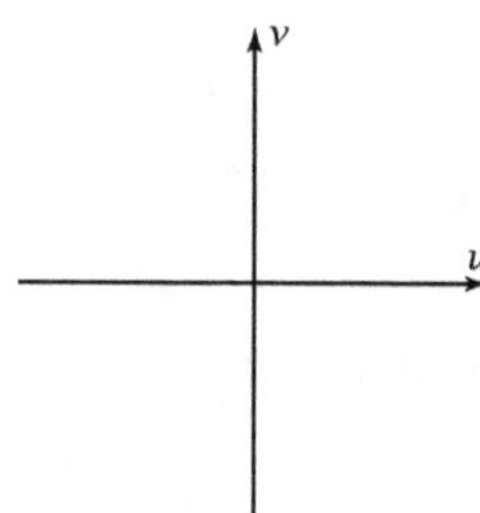

9. Write the general solution for arctan 5.

10. Find two values of arctan (−0.6) terminating in two different quadrants. Sketch the angles and label each reference triangle.

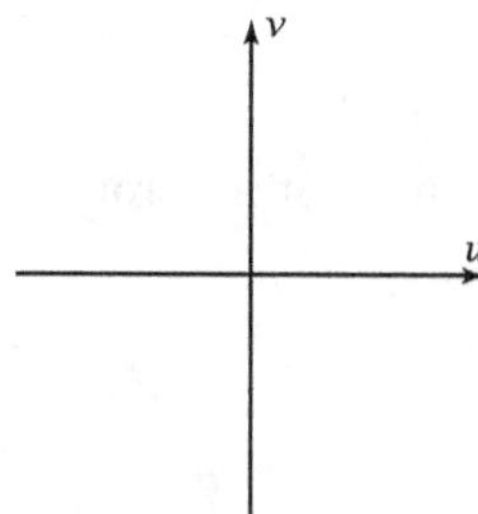

11. What did you learn as a result of doing this Exploration that you did not know before?

Exploration 39: Trigonometric Equations

Objective: Solve equations in which trigonometric functions appear.

1. Solve: $2 \cos(\theta - 17°) = 1$, $\theta \in [0°, 720°]$

2. Solve: $\tan^2 \theta - 2 \tan \theta - 3 = 0$, $\theta \in [-360°, 360°]$

3. Solve: $-1 - 5 \sin \theta = 2 \cos^2 \theta$, $\theta \in [-180°, 720°]$

4. The figure shows the graphs of
$$y_1 = -1 - 5 \sin \theta \qquad y_2 = 2 \cos^2 \theta$$

Show on the graph that all of your answers in Problem 3 are correct.

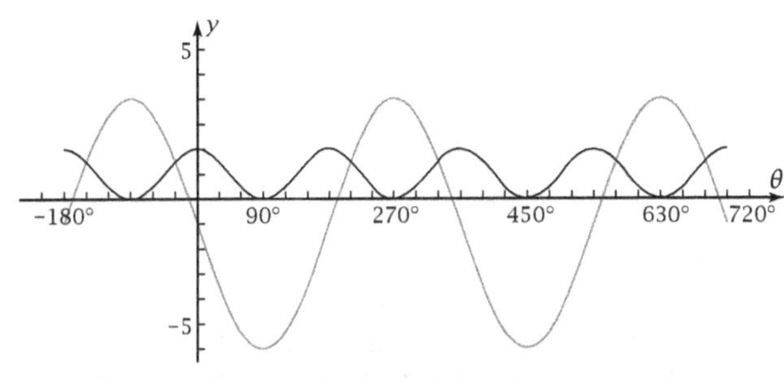

5. Explain how what you have been studying about transformation of trigonometric expressions allows you to solve trigonometric equations.

Exploration 40: Parametric Function Pendulum Problem

Date: _____________

Objective: Given a pendulum moving in both the *x*- and *y*-directions, predict its position, (*x*, *y*), at any given time, *t*.

A pendulum hangs from the ceiling. The point on the floor above which the pendulum bob is situated when it is at rest is the origin, (0, 0), of a Cartesian coordinate system. The *x*- and *y*-axes run parallel to the walls of the room and the *x-y* plane is horizontal.

1. Give the pendulum bob a displacement of 30 cm in the positive *x*-direction. Let it go and time its period. Sketch the graph of *x* as a function of time, *t* seconds, since it was released.

2. Starting with the pendulum bob at (0, 0), give it a push in the *y*-direction just hard enough to make it swing with an amplitude of 20 cm. Does the pendulum have the same period this time? Sketch the graph of *y* as a function of *t*.

3. Starting with the pendulum bob at (30, 0), give it a push in the *y*-direction just hard enough to make it have an amplitude of 20 cm in the *y*-direction, as in Problem 2. Sketch. Describe the path followed by the pendulum bob.

4. Assume that the graphs in Problems 1 and 2 are sinusoids. Write particular equations for *x* and *y* as functions of *t*.

    x = _________________________________

    y = _________________________________

5. Using your equations in Problem 4, make a table of values of *x* and *y* for various values of *t* from 0 through 1 complete orbit of the pendulum.

t	*x*	*y*

6. Plot the points (*x*, *y*) and connect them.

7. What geometrical figure describes the graph in Problem 6? _________________________________

8. What special name is given to a variable, such as *t* in this problem, upon which two or more other variables depend? _________________________________

9. What special name is given to functions in which two or more variables depend on the same independent variable? _________________________________

10. Put your grapher in parametric mode. Then enter the equations for *x* and *y* from Problem 4. Pick suitable ranges for *x*, *y*, and *t*. Have the calculator plot the graph. (Use ZOOM SQUARE to get equal scales on the axes.) Does the graph look like the graph you plotted in Problem 6? _______________

11. Just for fun, see if you can transform the two equations from Problem 4 into one equation in only *x* and *y* by eliminating the parameter *t*.

Exploration 41: Parametric Equations for Ellipses

Objective: Given a figure involving ellipses, write parametric equations and draw the graph on your grapher.

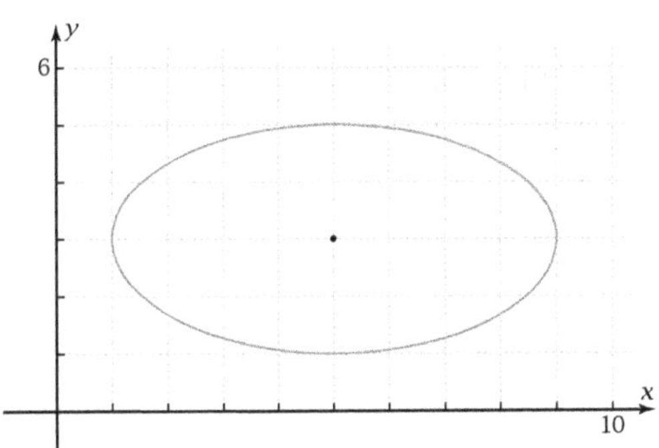

1. The figure shows an ellipse with center at (5, 3), x-radius = 4, and y-radius = 2. Write what you *think* the parametric equations of this ellipse are.

   $x =$ _________________________________

   $y =$ _________________________________

2. Plot these parametric equations on your grapher. Use a window with equal scales on both axes. Does the graph agree with the figure? ___________

For Problems 3–5, write parametric equations for each ellipse shown. Plot the ellipses using degree mode with $t \in [0°, 360]$ and a t-step of 10°. Use appropriate Boolean variables and dot style for the parts of the ellipses that are hidden. Then use the DRAW command to draw line segments and dots in appropriate places.

3. Cone:

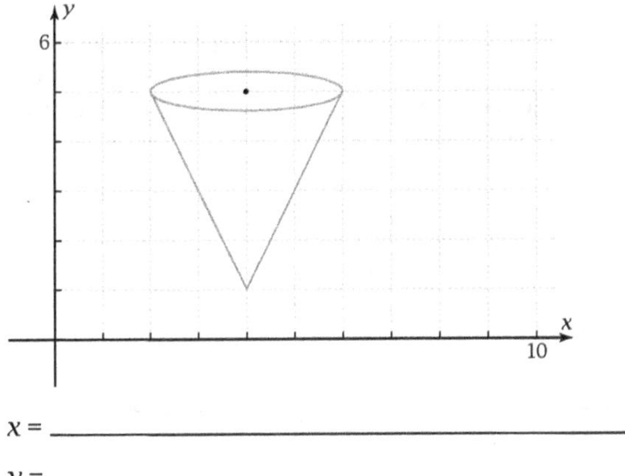

   $x =$ _________________________________

   $y =$ _________________________________

4. Frustum of a cone ("truncated" cone):

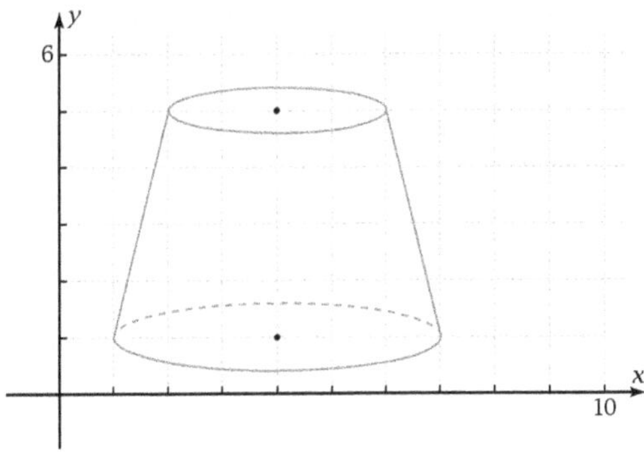

   Top:

   $x_1 =$ _________________________________

   $y_1 =$ _________________________________

   Bottom, solid (use a Boolean variable)

   $x_2 =$ _________________________________

   $y_2 =$ _________________________________

   Bottom, dashed (use a Boolean variable)

   $x_3 =$ _________________________________

   $y_3 =$ _________________________________

5. Cone oriented the other way:

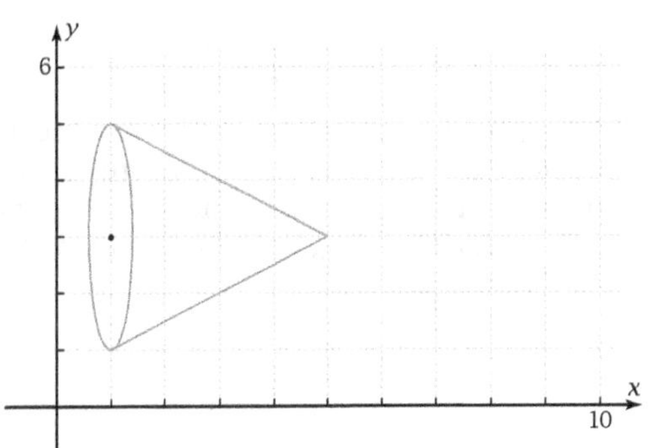

   Solid (use a Boolean variable)

   $x_1 =$ _________________________________

   $y_1 =$ _________________________________

   Dashed (use a Boolean variable)

   $x_2 =$ _________________________________

   $y_2 =$ _________________________________

6. What did you learn as a result of doing this Exploration that you did not know before?

Exploration 42: Graphs of Inverse Trigonometric Relations

Date: _____________

Objective: Investigate the graphs of inverse circular functions and relations by plotting them on your grapher.

This graph shows the inverse circular relation

$$y = \arcsin x$$

In this Exploration, you will learn how to plot this graph on your grapher.

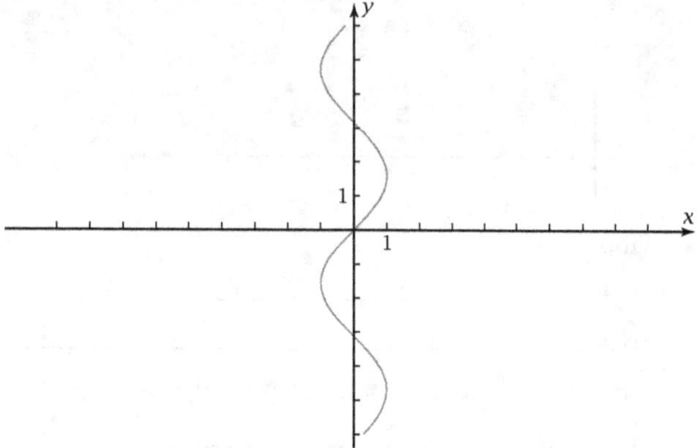

1. Recall that $y = \arcsin x$ means that $x = \sin y$. Calculate x for $y = 4$.

2. Write the general solution for $y = \arcsin x$. Use it to find the four values of arcsin 0.4 that are on the portion of the graph shown. Round to one decimal place. Plot points on the graph corresponding to the four answers.

3. Reproduce the given graph on your grapher. A clever way to do this is to use parametric mode. Enter the equation as

$$x_{1T} = \sin t$$

$$y_{1T} = t$$

Use a t-range of −9 to 9. Use [−9, 9] for the x-range and equal scales for the two axes. Check the graph with your instructor. _________________

4. Come back to function mode. Plot the graph of the **function** $y = \sin^{-1} x$. Because the calculator gives only the principal value of the inverse sine, the graph is called the **principal branch.** Darken the part of the given graph that corresponds to the principal branch.

5. Return to the parametric mode and enter the parent sine function graph. Enter

$$x_{2T} = t$$

$$y_{2T} = \sin t$$

Keep the x_{1T}, y_{1T} graph active. Sketch the resulting graph on the given figure.

6. What relationship do the sine and inverse sine graphs have to each other? How do they relate to the line $y = x$?

7. Plot the graphs of $y = \cos x$ and $y = \arccos x$ on your grapher. Use the same window as in the previous problems. Sketch the result here, along with the line $y = x$.

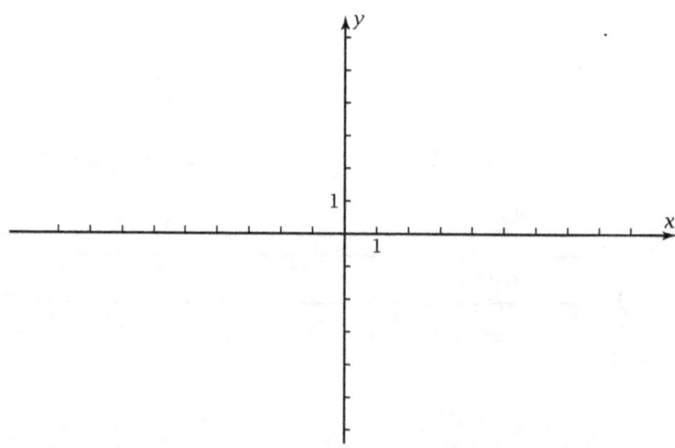

8. What did you learn as a result of doing this Exploration that you did not know before?

Exploration 43: Principal Branches of Inverse Trigonometric Relations

Objective: Figure out the principal branches of each of the six inverse circular functions.

1. On your grapher, plot the inverse circular function $y = \sin^{-1} x$. Use equal scales on the two axes. On the graph of $y = \arcsin x$ shown here, darken the **principal branch** of the inverse sine relation on the part of the graph that is the inverse sine function. Give the range of the principal branch.

   Range: _________________________________

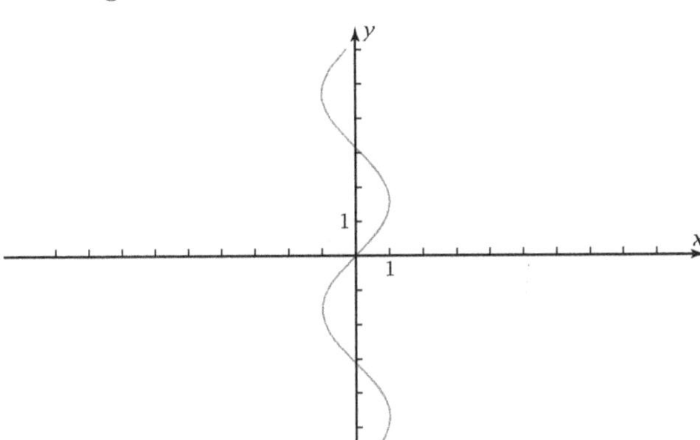

2. Use the technique in Problem 1 to find the range of $y = \cos^{-1} x$. Darken the principal branch on the graph of $y = \arccos x$, shown next.

   Range: _________________________________

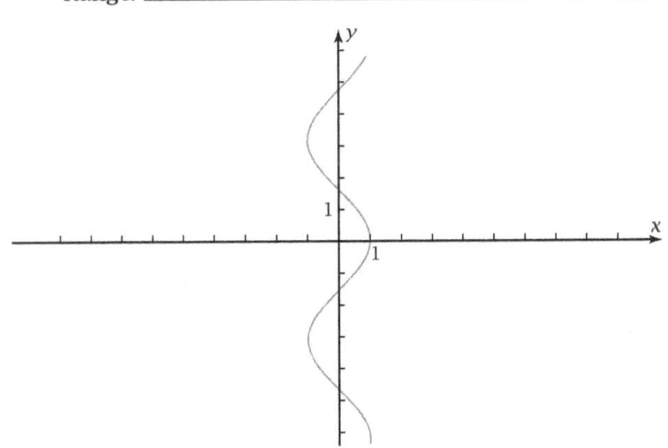

3. Why can't the range of the inverse cosine function $(y = \cos^{-1} x)$ be the same as the range of the inverse sine function $(y = \sin^{-1} x)$?

4. Use the technique in Problems 1 and 2 to find the range of $y = \tan^{-1} x$. Darken the principal branch on this graph of $y = \arctan x$.

   Range: _________________________________

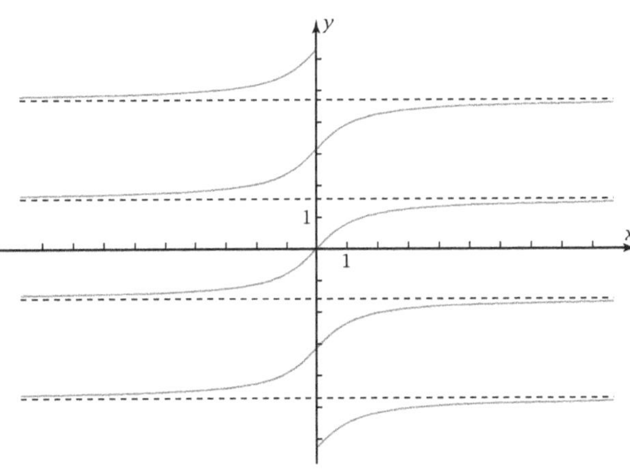

5. The range of $y = \sin^{-1} x$ is the closed interval $\left[-\frac{\pi}{2}, \frac{\pi}{2}\right]$. Explain why the range of $y = \tan^{-1} x$ cannot include the endpoints.

6. If the range of the inverse tangent function $(y = \tan^{-1} x)$ were $[0, \pi]$ (excluding $\frac{\pi}{2}$), like the range of $y = \cos^{-1} x$, then $y = \tan^{-1} x$ would still be a function. What disadvantage would there be to defining the range of $y = \tan^{-1} x$ this way?

(Over)

Exploration 43: Principal Branches of Inverse Trigonometric Relations *continued*

7. Duplicate the previous graph of $y = \arctan x$ on your grapher. Give the parametric equations you used. Check your graph with your instructor. _________

8. The graph here shows $y = \text{arccot}\, x$. How can you define the range of the function $y = \cot^{-1} x$ in such a way that the function is **continuous**? Darken this principal branch of $y = \text{arccot}\, x$.

   Range: _________________

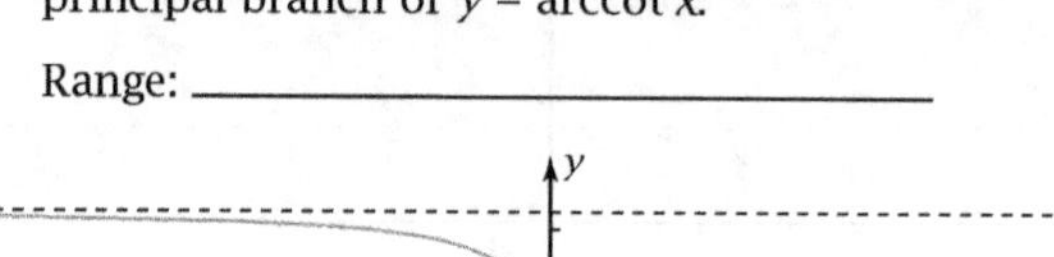

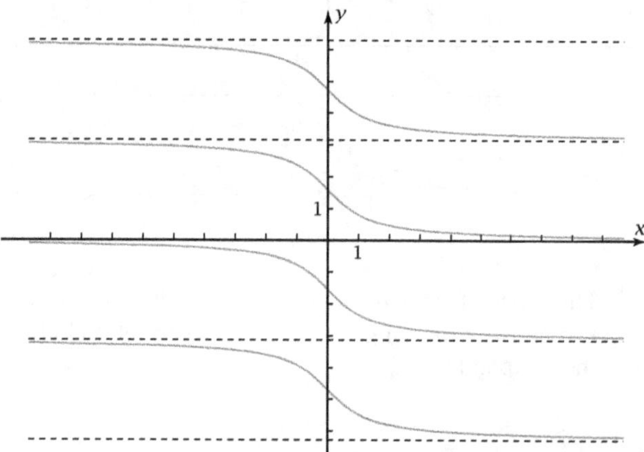

9. This next graph shows $y = \text{arcsec}\, x$. There is no way to restrict the range to make a continuous function $y = \sec^{-1} x$ and still use all of the domain. Darken what you think would be the best choice for the principal branch. Write the range.

   Range: _________________

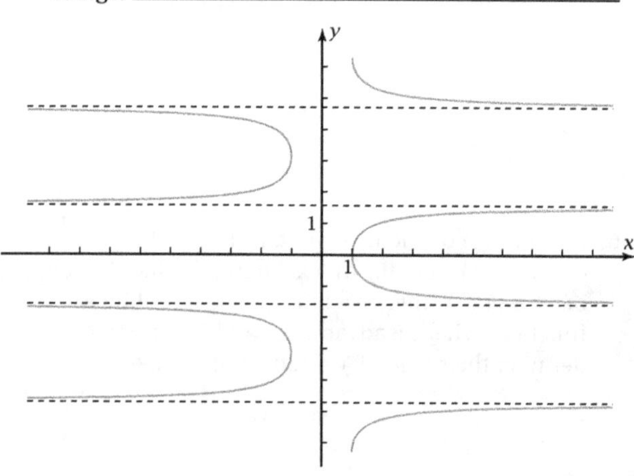

10. Look in the text to find out if the principal branch you chose in Problem 9 is the commonly accepted one.

11. This next graph shows $y = \text{arccsc}\, x$. Shade what you think the principal branch is. Write the range of the function you shaded.

    Range: _________________

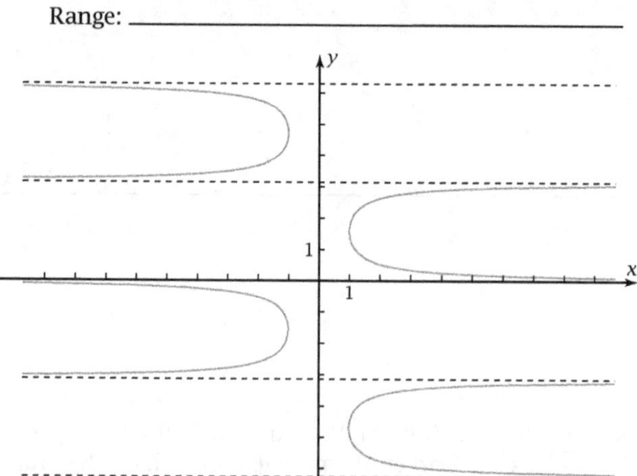

12. Does your answer to Problem 11 agree with the range listed in the text?

13. What are the criteria for picking the ranges of the principal branches of the six inverse circular functions?

14. What did you learn as a result of doing this Exploration that you did not know before?

Exploration 44: Inverse Trigonometric Relation Values

Objective: Calculate multiple values of inverse circular relations and confirm them graphically.

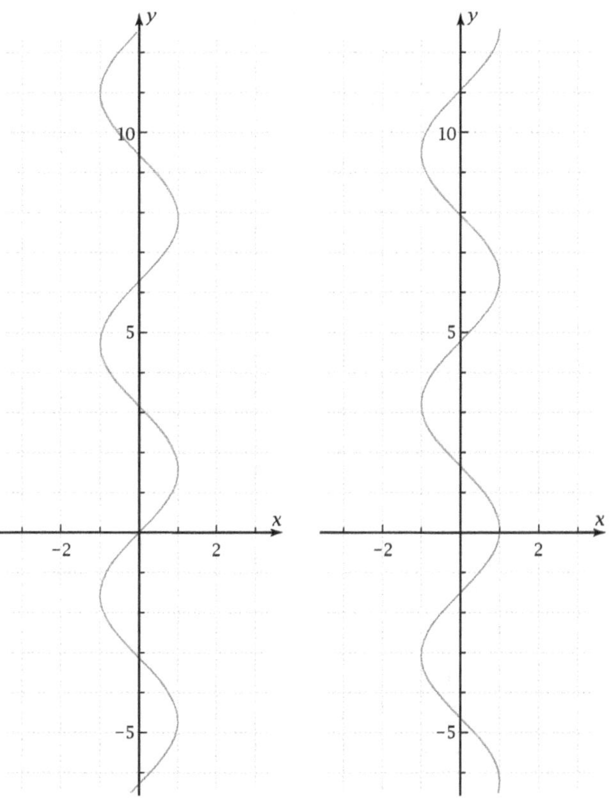

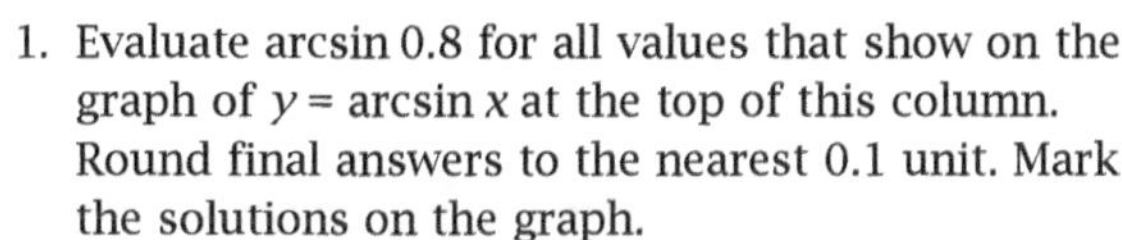

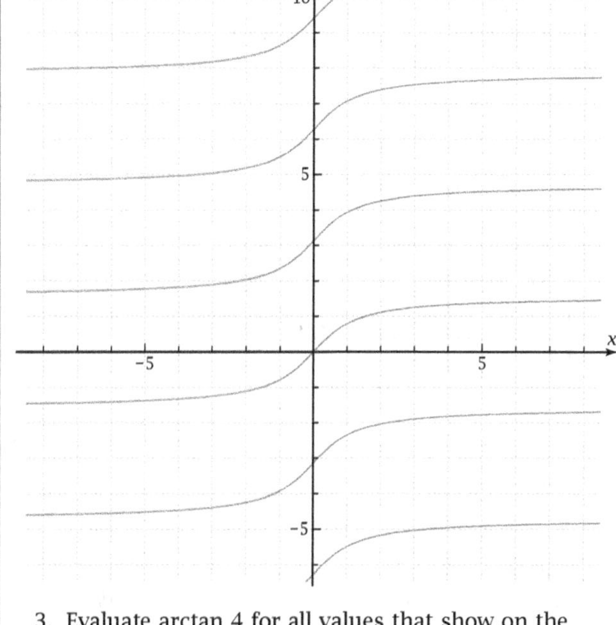

1. Evaluate arcsin 0.8 for all values that show on the graph of $y = \arcsin x$ at the top of this column. Round final answers to the nearest 0.1 unit. Mark the solutions on the graph.

2. Evaluate arccos (−0.3) for all values that show on the graph of $y = \arccos x$ at column top. Round final answers to the nearest 0.1 unit. Mark the solutions on the graph.

3. Evaluate arctan 4 for all values that show on the graph at the top of this column, of $y = \arctan x$. Round final answers to the nearest 0.1 unit. Mark the solutions on the graph.

4. Darken the **principal branch** of each graph. Circle the point your calculator gives for $\sin^{-1} 0.8$, $\cos^{-1} (-0.3)$, and $\tan^{-1} 4$ on the respective graphs, showing in each case that the value is on the principal branch.

5. What did you learn as a result of doing this Exploration that you did not know before?

Exploration 45: Linear Combination of Cosine and Sine

Objective: Write the linear combination $y = b \cos \theta + c \sin \theta$ as $y = A \cos(\theta - D)$, a sinusoid with a phase displacement.

The expression on the right in the equation

$$y = 3 \cos \theta + 4 \sin \theta$$

is called a **linear combination** of $\cos \theta$ and $\sin \theta$. That is, y equals a constant times cosine, plus a constant times sine. In this Exploration, you will learn how to express such a linear combination as a cosine with a phase displacement.

1. The graph shows

$$y_1 = 3 \cos \qquad \text{and} \qquad y_2 = 4 \sin \theta$$

Which graph is which?

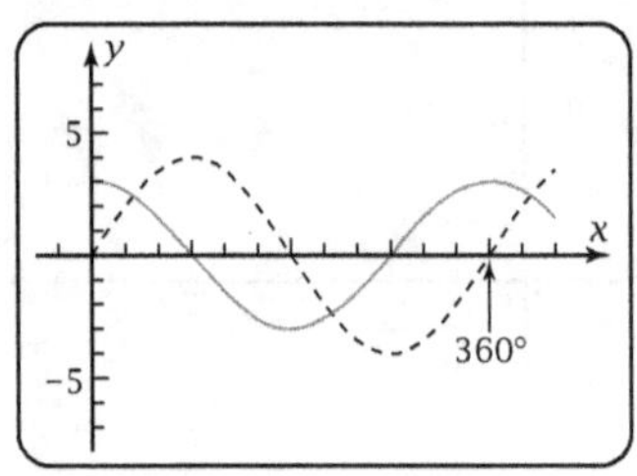

2. Plot y_3 and sketch it on the figure.

$$y_3 = 3 \cos \theta + 4 \sin \theta$$

3. The graph of y_3 is a sinusoid. Find the amplitude A and the phase displacement D using the MAXIMUM feature of your grapher.

$A =$ _________________ $D =$ _________________

4. Plot $y_4 = A \cos(\theta - D)$ using Problem 3 results. Does the graph coincide with y_3? _________________

5. The uv-diagram here shows an angle with $u = 3$, the coefficient of cosine in y_3, and $v = 4$, the coefficient of sine. Show that the hypotenuse equals A from Problem 3.

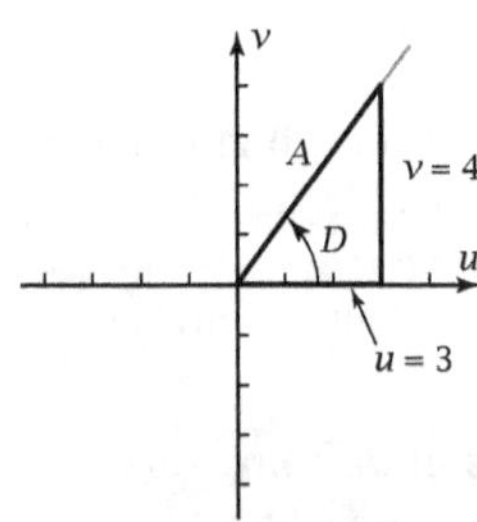

6. Show that the angle D in Problem 3 is a value of $\arctan \frac{4}{3}$, as shown in the figure in Problem 5.

7. Express as a cosine with a phase displacement:

$$y = -12 \cos \theta + 5 \sin \theta$$

Use the next uv-diagram to find the amplitude A and the phase displacement D. Show that D is a value of $\arctan \frac{5}{-12}$ but not the value of $\tan^{-1} \frac{5}{-12}$ that your calculator gives you.

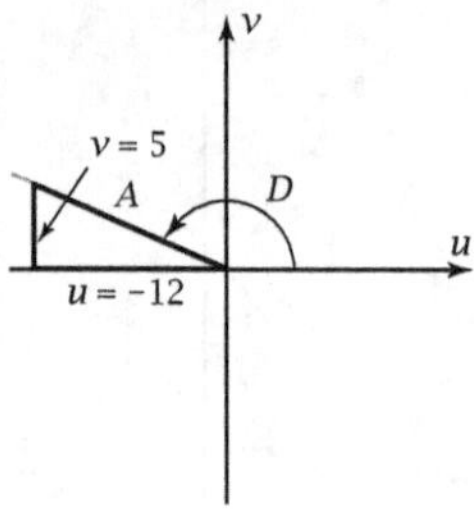

$y =$ _________________________________

8. Express as a cosine with a phase displacement:

$$y = -6 \cos \theta - 11 \sin \theta$$

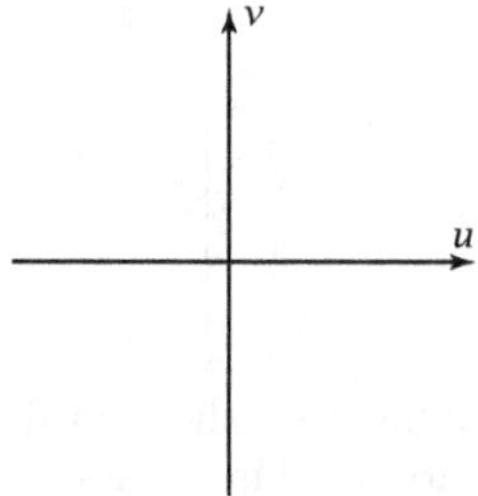

$y =$ _________________________________

9. Express as a cosine with a phase displacement:

$$y = 9 \cos \theta - 7 \sin \theta$$

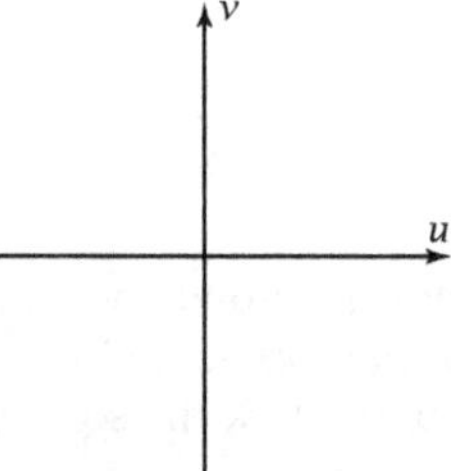

$y =$ _________________________________

10. What did you learn as a result of doing this Exploration that you did not know before?

Exploration 46: Cosine of a Difference

Objective: Write $\cos(A - B)$ in terms of $\cos A$, $\cos B$, $\sin A$, and $\sin B$.

1. Cosine does *not* distribute over subtraction or addition. For example,

   $$\cos(58° - 20°) \neq \cos 58° - \cos 20°$$

   Show by direct calculation that this is true. Explain your calculations.

2. It is possible to write $\cos(58° - 20°)$ exactly in terms of $\cos 58°$, $\cos 20°$, $\sin 58°$, and $\sin 20°$. By experimenting numerically with these four values, find out what this relationship is.

3. Make a conjecture:

   $\cos(A - B) =$ ______________________________

4. Have each group member pick another, different pair of values of A and B, and find out numerically whether or not your conjecture in Problem 3 works. If it doesn't, go back and modify the conjecture.

5. When you feel reasonably sure that your conjecture is correct, write a sentence explaining how you can write the cosine of the difference of two angles in terms of the cosines and sines of the angles. Start by writing

   "cosine (first angle − second angle) = . . ."

6. The equation in Problem 5 is called a **composite argument property.** Why do you think this is its name?

7. Apply the property in Problem 5 to

   $$y_1 = 6 \cos(\theta - 70°)$$

   The result should be a **linear combination** of $\cos \theta$ and $\sin \theta$. That is,

   $$y_2 = b \cos \theta + c \sin \theta$$

   where b and c are constants.

8. Plot the graphs of y_1 and y_2 on the same screen. Use a window with an x-range of $[0°, 720°]$. Is y_2 really a sinusoid with amplitude 6 and phase displacement 70° (for cosine)? How can you tell?

9. What did you learn as a result of doing this Exploration that you did not know before?

Exploration 47: Composite Argument Property Proof   Date: ____________

Objective: Prove algebraically that cos $(A - B)$ = cos A cos B + sin A sin B.

The figure shows angles A and B in standard position and angle $(A - B)$ between them. A unit circle cuts the terminal sides of A and B as shown.

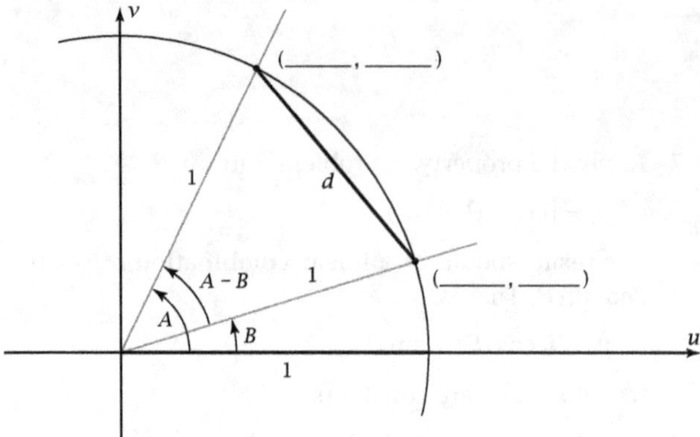

1. On the figure, write the coordinates (u, v) for the two points where the unit circle cuts the terminal side.

2. Recall that the **distance formula** says that if d is the distance between points (u_1, v_1) and (u_2, v_2), then

   $$d^2 = (u_2 - u_1)^2 + (v_2 - v_1)^2$$

   Use the distance formula to express d^2 in the figure in terms of cos A, cos B, sin A, and sin B. Expand the squares, and then use the Pythagorean property to simplify the answer.

The figure shows angle $(A - B)$ from before, rotated so that it is in standard position.

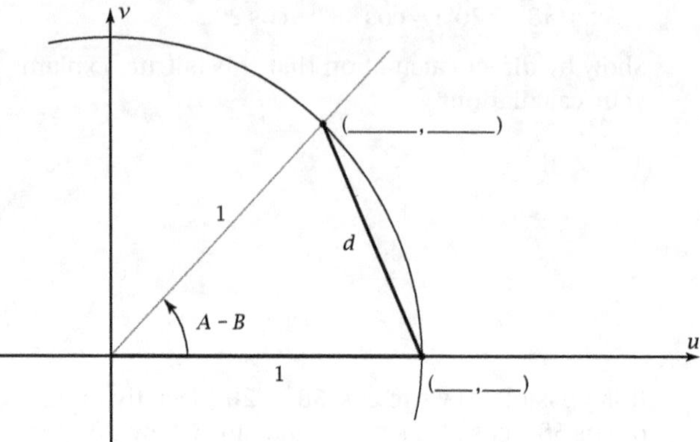

3. Use the distance formula to write the distance d^2 in terms of cos $(A - B)$ and sin $(A - B)$. Expand the squares and simplify.

4. Equate the two values of d^2 from Problems 2 and 3. Transform the resulting equation to show that

   $$\cos (A - B) = \cos A \cos B + \sin A \sin B$$

5. What did you learn as a result of doing this Exploration that you did not know before?

Exploration 48: Sum or Product of Sinusoids with Unequal Periods

Objective: Draw conclusions about graphs composed of a product or a sum of two sinusoids with unequal periods.

1. The graphs here are of

   $$y_1 = 9 \sin \theta \qquad \text{and} \qquad y_2 = \cos 9\theta$$

   Indicate which is which.

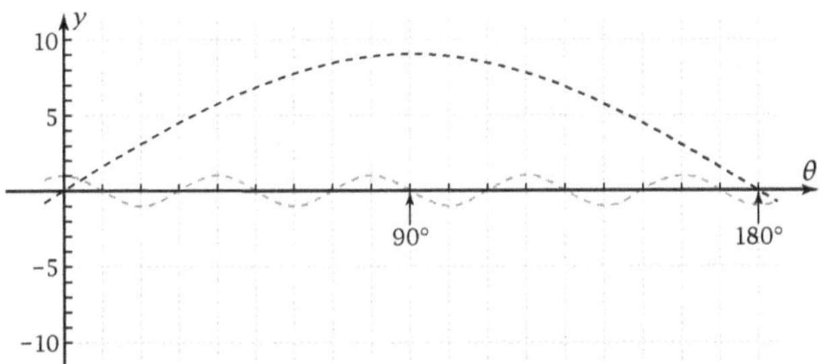

2. Without first plotting on your grapher, draw by **composition of ordinates** the graph of

   $$y = 9 \sin \theta \cdot \cos 9\theta$$

3. Plot y_1 and y_2 on your grapher. Also plot

   $$y_3 = y_1 * y_2$$

   Does your composed graph in Problem 2 come close to this actual graph? _________________

4. Recall that if $y = A \cos 9\theta$, then A is the vertical dilation (the amplitude) of the graph. From your work in Problems 1–3, write a few sentences describing the effect on the graph of making the dilation a *variable*, $A = 9 \sin \theta$, instead of just a constant.

5. These graphs duplicate the ones in Problem 1. By composition of ordinates, draw the graph of

   $$y = 9 \sin \theta + \cos 9\theta$$

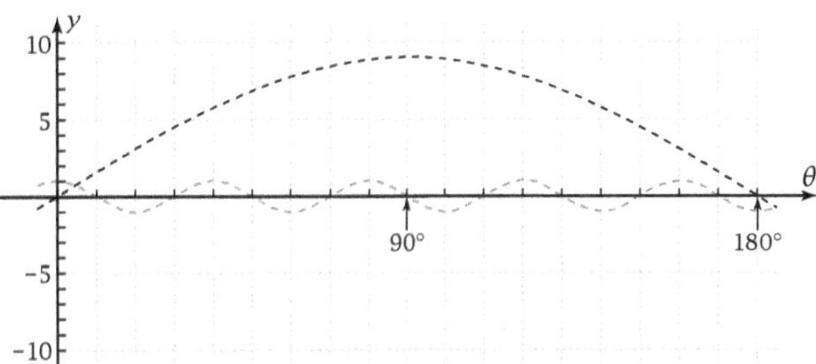

6. Plot $y = 9 \sin \theta + \cos 9\theta$ on your grapher. How closely does your graph in Problem 5 resemble the actual one? _________________

7. Recall that if $y = C + \cos 9\theta$, then C is the vertical translation of the sinusoid. From your work in this Exploration, write a few sentences describing the effect of making C a *variable*, $9 \sin \theta$, instead of just a constant.

8. Write conclusions:

   a. *Adding* sinusoids with much different periods produces . . .

   b. *Multiplying* sinusoids with much different periods produces . . .

9. What did you learn as a result of doing this Exploration that you did not know before?

Exploration 49: Harmonic Analysis

Date: _________

Objective: Given a graph that is a sum or product of sinusoids, find the equation.

This graph is composed of two sinusoids.

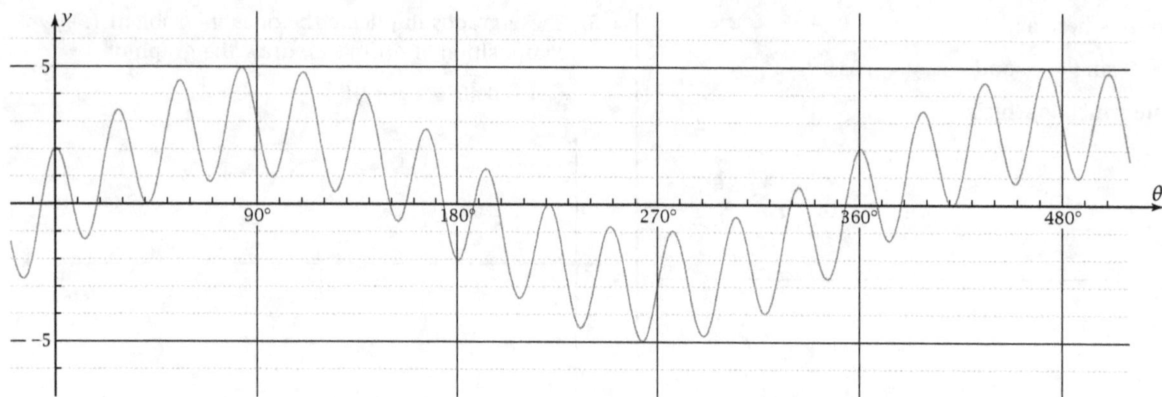

1. Were they added, or were they multiplied? _______________________

2. Sketch the sinusoid with the larger period. Use the result to write the particular equation of the given graph.

   Equation: _________________________________

3. Confirm your equation by plotting on your grapher. A window with an x-range of $[-20°, 200°]$ will allow you to see reasonable detail.

   Check? _______________

This graph is composed of two sinusoids.

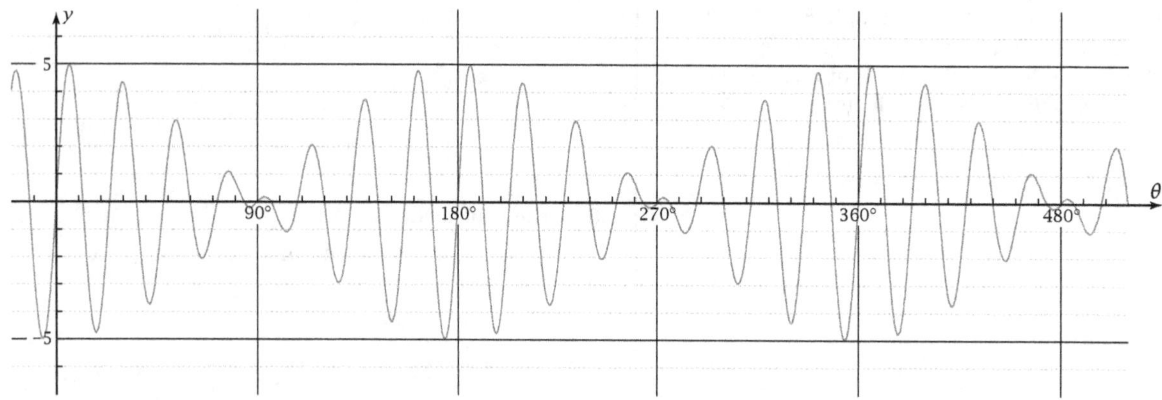

4. Were they added, or were they multiplied? _______________________

5. Sketch the sinusoid with the larger period. Use the result to write the particular equation of the given graph.

   Equation: _________________________________

6. Confirm your equation by plotting on your grapher. Use a window with an x-range of $[-20°, 200°]$.

   Check? _______________

7. What did you learn as a result of doing this Exploration that you did not know before?

Exploration 50: Harmonic Analysis Practice

Objective: Given a graph that is a sum or product of sinusoids, find the equation.

Write a particular equation for each graph, and check by plotting on your grapher.

1. Equation: ____ Grapher check: _________

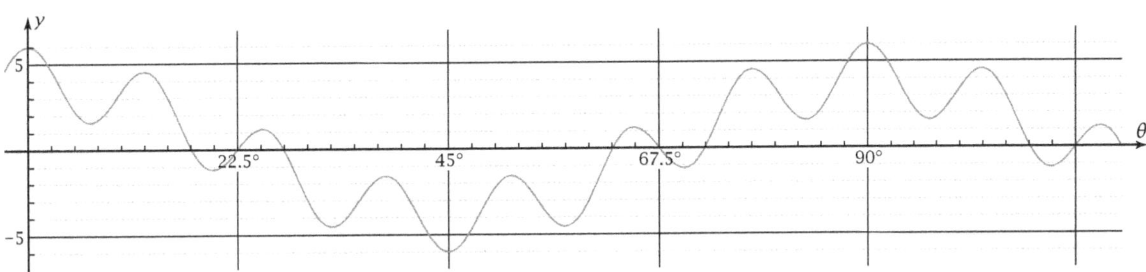

2. Equation: ____ Grapher check: _________

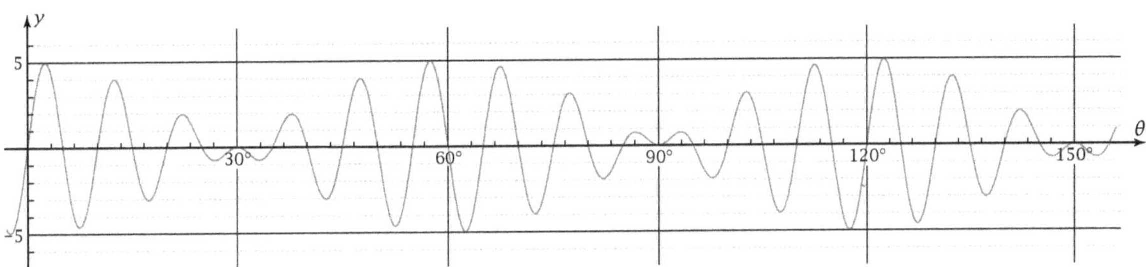

3. Equation: ____ Grapher check: _________

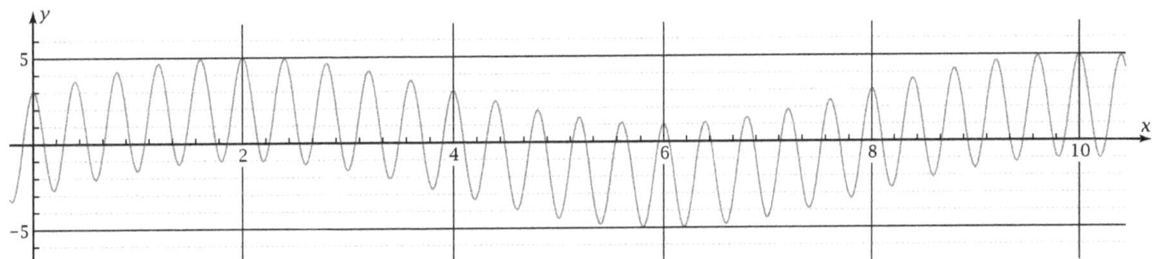

4. Equation: ____ Grapher check: _________

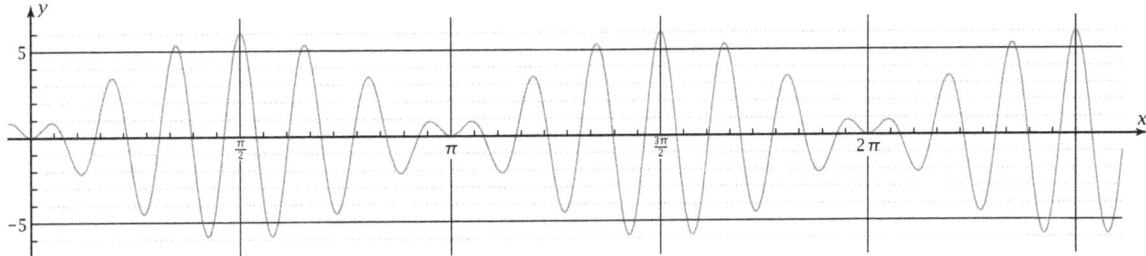

5. What did you learn as a result of doing this Exploration that you did not know before?

Exploration 51: More Harmonic Analysis Practice

Date: ____________

Objective: Given a graph that is a sum or product of sinusoids, find the equation.

Write a particular equation for each graph, and check by plotting on your grapher.

1. Equation: ____   Grapher check: _________

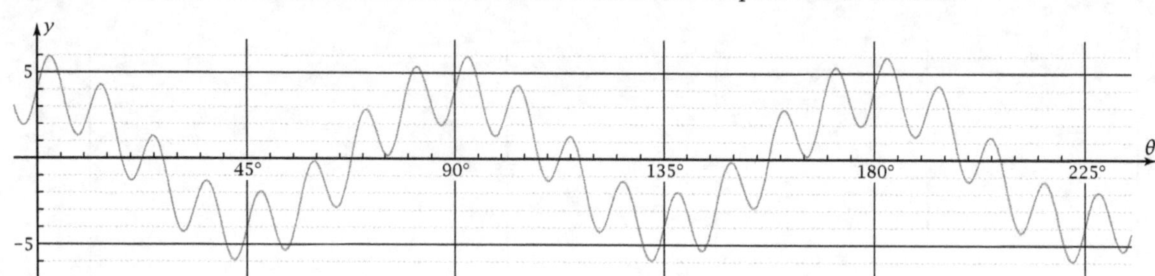

2. Equation: ____   Grapher check: _________

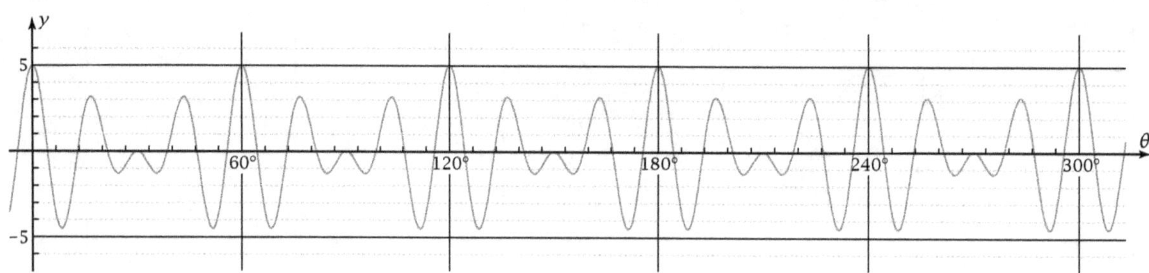

3. Equation: ____   Grapher check: _________

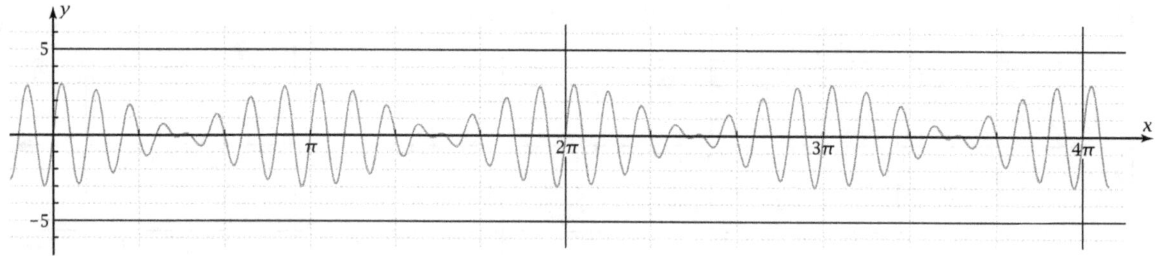

4. Equation: ____   Grapher check: _________

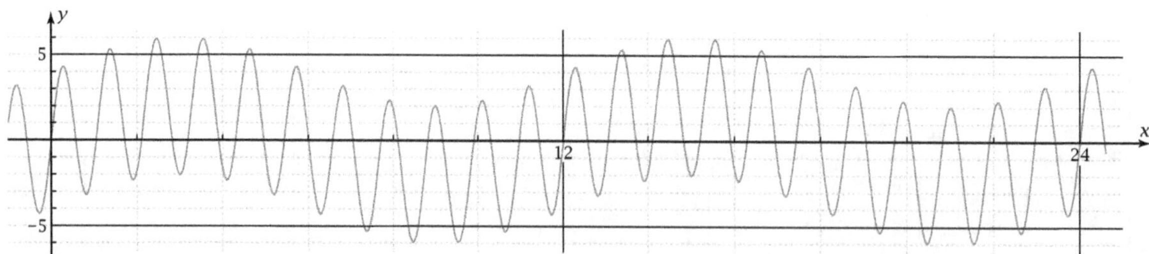

5. What did you learn as a result of doing this Exploration that you did not know before?

Exploration 52: Equivalence of Sinusoid Sums and Products

Objective: Show that a sum of two sinusoids with nearly equal periods is equivalent to a product of two sinusoids with very different periods.

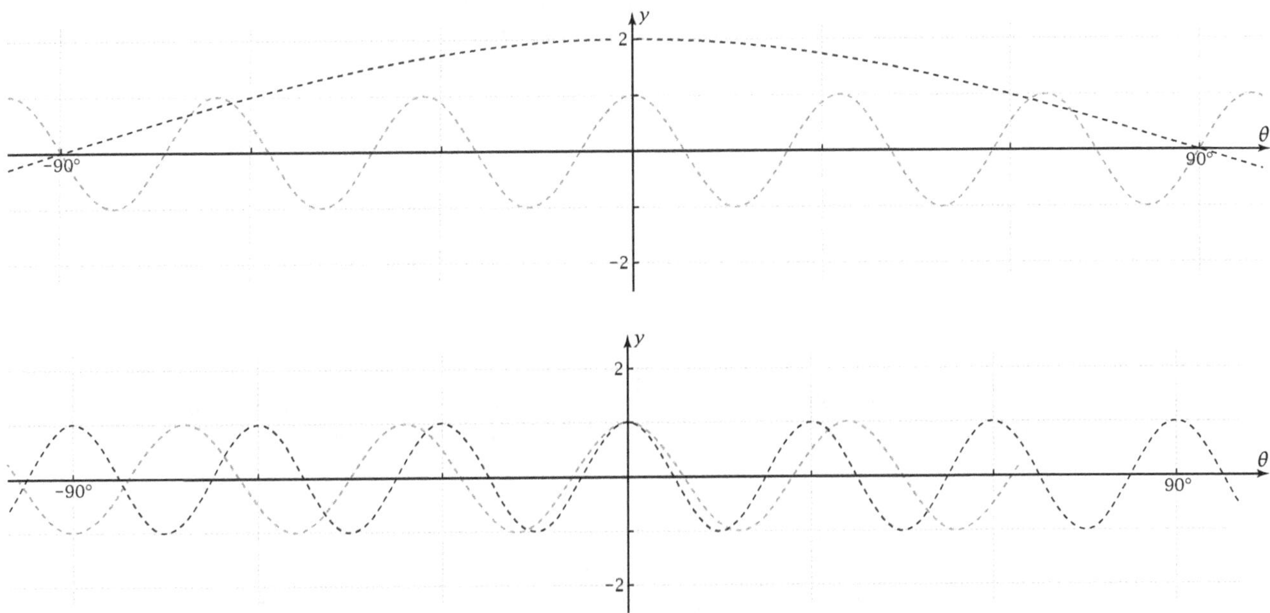

1. The first figure shows two graphs,

   $$y_1 = 2 \cos \theta$$
   $$y_2 = \cos 11\theta$$

   By composition of ordinates, sketch the graph of

   $$y = 2 \cos \theta \cos 11\theta$$

2. The second figure shows two more graphs,

   $$y_1 = \cos 12\theta$$
   $$y_2 = \cos 10\theta$$

   By composition of ordinates, sketch the graph of

   $$y = \cos 12\theta + \cos 10\theta$$

3. The two composed graphs in Problems 1 and 2 should be identical. This fact can be shown algebraically. Let

   $$\cos 12\theta = \cos (11\theta + \theta)$$

   and let

   $$\cos 10\theta = \cos (11\theta - \theta)$$

   By appropriate use of the composite argument properties, prove that

   $$\cos 12\theta + \cos 10\theta = 2 \cos 11\theta \cos \theta$$

4. The equation in Problem 3 is an example of one of the **sum and product properties.** These properties say that a sum of two sinusoids can be written as a product of two other sinusoids, and vice versa. Suppose that

   $$y = 2 \cos 20\theta \cos 2\theta$$

   Write y as a *sum* of two cosines.

5. Verify by plotting on your grapher that the given equation for y and your new equation for y give the same graph. Use a window with an x-range of [−45, 45]. Check your graph with your instructor.

   ____

6. What did you learn as a result of doing this Exploration that you did not know before?

Exploration 53: Double Argument Properties

Date: _____________

Objective: Express $\cos 2\theta$ and $\sin 2\theta$ in terms of functions of θ. Use the results to solve an equation containing $\cos 2\theta$.

1. Write the composite argument properties for $\cos(A + B)$ and for $\sin(A + B)$.

2. Write $\sin 2\theta$ as $\sin(\theta + \theta)$. Expand $\sin(\theta + \theta)$ using the composite argument for sine. The result should be an equation for $\sin 2\theta$ in terms of $\sin \theta$ and $\cos \theta$.

3. Use the technique of Problem 2 to find an equation for $\cos 2\theta$ in terms of $\cos \theta$ and $\sin \theta$.

4. Transform your answer to Problem 3 so that $\cos 2\theta$ is expressed in terms of $\cos \theta$ alone.

5. Transform your answer to Problem 3 so that $\cos 2\theta$ is expressed in terms of $\sin \theta$ alone.

6. Plot on your grapher the right side of your equation for $\sin 2\theta$ in Problem 2. Sketch here.

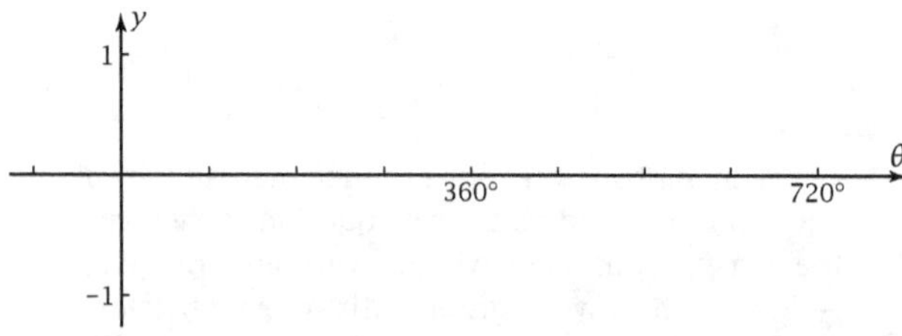

7. How does your graph in Problem 6 verify that your equation in Problem 2 is correct?

8. Solve $\cos 2\theta + \cos \theta = 1$ algebraically for $\theta \in [-100°, 850°]$. It is recommended that you first transform $\cos 2\theta$ so that it involves only $\cos \theta$. Don't forget the quadratic formula!

9. Plot the left and right members of the equation in Problem 8 on your grapher. Sketch here.

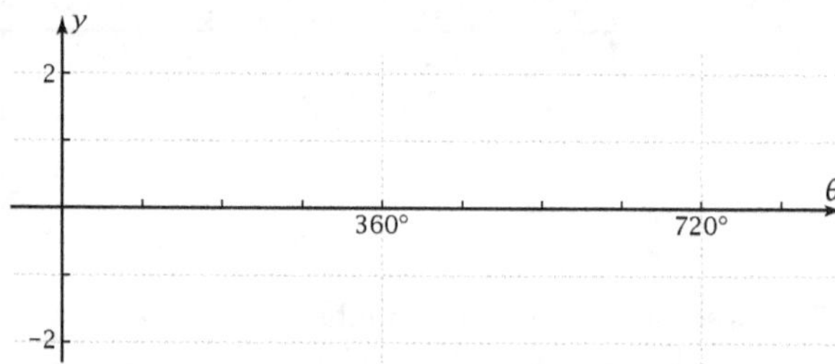

10. How does the graph in Problem 9 show that your answers to Problem 8 are correct?

11. What have you learned as a result of doing this Exploration that you did not know before?

Exploration 54: Algebraic Transformations

Date: _____________

Objective: Given a trigonometric expression, transform it to a specific trigonometric or algebraic form.

For Problems 1 to 26, write the given expression in the form specified.

1. $\sin 2x$, in terms of $\sin x$ *and* $\cos x$

2. $\sin^2 x$, in terms of $\cos 2x$

3. $\sin^2 x$, in terms of $\cos x$

4. $\cos^2 x$, in terms of $\sin x$

5. $\cos^2 x$, in terms of $\cos 2x$

6. $\cos 2x$, in terms of $\cos x$

7. $\cos 2x$, in terms of $\sin x$

8. $\cos 2x$, in terms of $\cos x$ *and* $\sin x$

9. $\tan 2x$, in terms of $\tan x$

10. $\sin\frac{1}{2}x$, in terms of $\cos x$

11. $\cos\frac{1}{2}x$, in terms of $\cos x$

12. $\tan\frac{1}{2}x$, in terms of $\cos x$

13. $\tan\frac{1}{2}x$, in terms of $\cos x$ *and* $\sin x$

14. $\sin 3x$, in terms of $\sin x$

15. $\cos 4x$, in terms of $\cos 2x$

16. $\cos 6x$, in terms of $\cos 3x$

17. $\sin x \cos x$, in terms of $\sin 2x$

18. $\sin x \cos y$, as a sum of sines or cosines

(Over)

Exploration 54: Algebraic Transformations *continued* Date: _____________

19. $\cos x \cos y$, as a sum of sines or cosines

20. $\cos x + \cos y$, as a product of sines and/or cosines

21. $\sin x + \sin y$, as a product of sines and/or cosines

22. $\sin x + \cos x$, as a single cosine with a phase displacement

23. $\sin 3x \sin 7x$, as a sum of sines or cosines of positive multiples of x

24. $\sin 3x + \sin 7x$, as a product of sines and/or cosines of positive multiples of x

25. $\sqrt{3} \cos x - \sin x$, as a cosine with a phase displacement

26. $4 \cos x - 4 \sin x$, as a cosine with a phase displacement

27. What have you learned as a result of doing this Exploration that you did not know before?

Exploration 55: Introduction to Oblique Triangles

Date: _____________

Objective: Given the lengths of two sides of a triangle and the measure of the included angle, measure the third side.

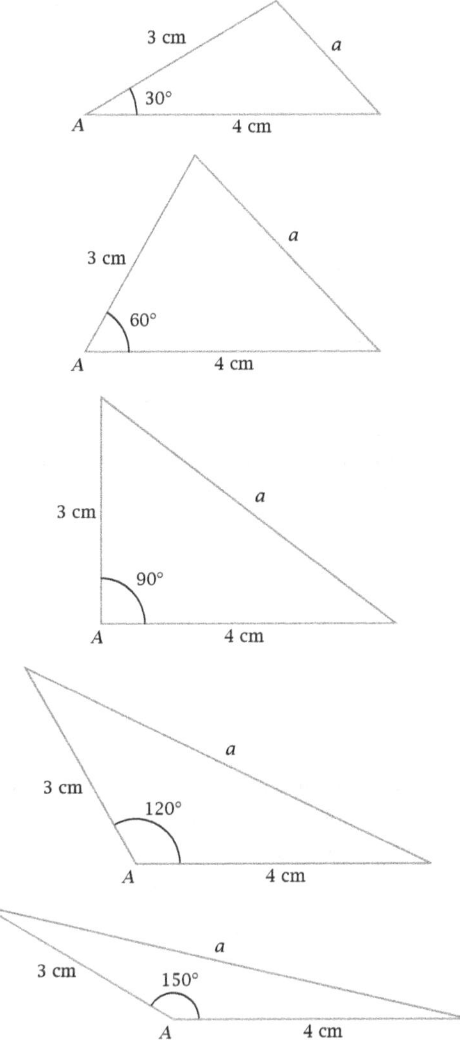

1. The figures show five triangles. Each has sides of 3 cm and 4 cm. They differ in the size of the angle included between the two sides. Are the measurements of the lengths of the sides correct as marked on each figure? _____________

2. Are the degree measurements correct as shown in each figure? _____________

3. Measure the third side, a, of each triangle. Record the results here, correct to one decimal place.

A (degrees)	a (cm)
30°	
60°	
90°	
120°	
150°	

4. What would a equal if $\angle A$ were:

   180°: $a =$ _____________   0°: $a =$ _____________

5. Plot the measured values of a as a function of the angle measure, A, from 0° to 180°. Connect the points with a smooth curve.

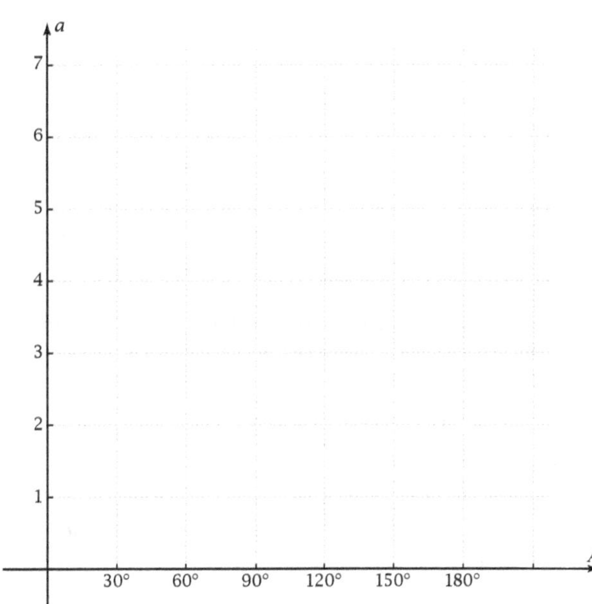

6. By the Pythagorean theorem, $a^2 = 3^2 + 4^2$ when A is 90°. What do you have to subtract to get the value of a^2 if A is less than 90°?

7. The answer to Problem 6 is part of the **law of cosines.** Use the law of cosines to calculate a for each value of angle A in the table of Problem 3.

8. What did you learn as a result of doing this Exploration that you did not know before?

Exploration 56: Derivation of the Law of Cosines
Date: ___________

Objective: Derive the law of cosines for predicting the third side of a triangle from two sides and the included angle.

The figure shows triangle ABC. Angle A has been placed in standard position in a uv-coordinate system.

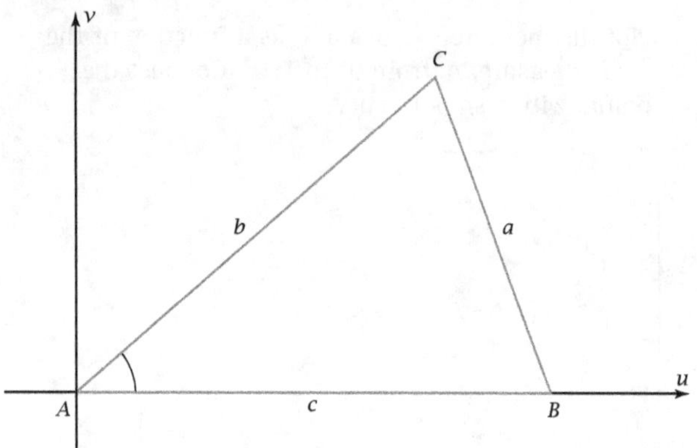

1. The sides that include angle A have lengths b and c. Write the coordinates of points B and C using b, c, and functions of angle A.

   B: $(u, v) =$ (____________, ____________)

   C: $(u, v) =$ (____________, ____________)

2. Use the **distance formula** to write the square of the length of the third side, a^2, in terms of b, c, and functions of angle A.

3. Simplify the answer to Problem 2 by expanding the square. Use the **Pythagorean property** for cosine and sine to simplify the terms containing $\cos^2 A$ and $\sin^2 A$.

4. The answer to Problem 3 is called the **law of cosines.** Show that you understand what the law of cosines says by using it to calculate the third side of this triangle.

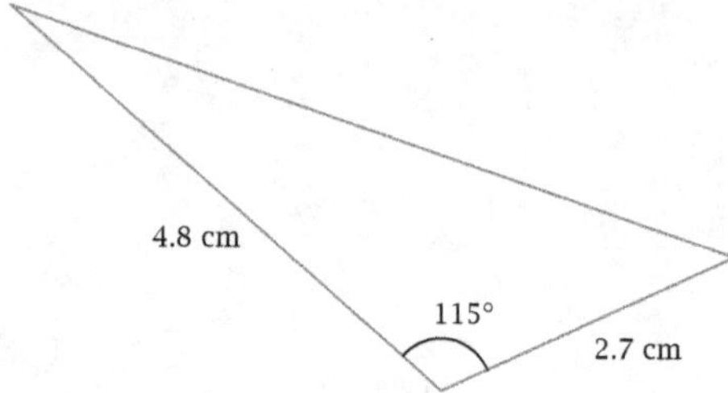

5. Measure the given sides and the angle of the triangle in Problem 4. Do you agree with the given measurements? Measure the third side. Does it agree with your calculated value?

6. Describe how the unknown side in the law of cosines is related to the given angle and how the given angle is related to the two given sides, using terms you studied in geometry.

7. What have you learned as a result of doing this Exploration that you did not know before?

Exploration 57: Angles by the Law of Cosines

Objective: Use the law of cosines to calculate the measure of an angle of a triangle if three sides are known.

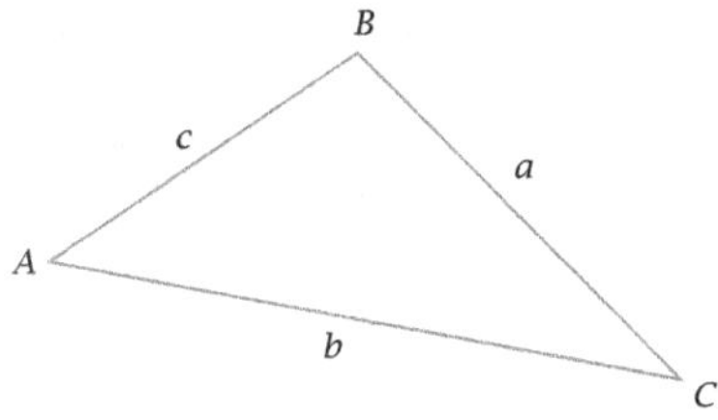

1. For $\triangle ABC$, write the law of cosines in the form involving $\cos B$.

____

2. Transform the equation in Problem 1 so that $\cos B$ is expressed explicitly in terms of a, b, and c.

3. Measure the sides of $\triangle ABC$ to the nearest 0.1 cm.

   $a \approx$ _________ $b \approx$ _________ $c \approx$ _________

4. Use the results of Problems 2 and 3 to calculate the measure of angle B. Round to the nearest degree.

5. Measure angle B with a protractor. _________________

6. How close did your calculated angle value come to the measured value? _________________

7. Calculate the measure of angle A using the law of cosines. Round to the nearest degree.

8. Calculate the measure of angle C using the law of cosines. Round to the nearest degree.

9. Do a *quick* calculation that checks your answers for the three angle measures.

10. Measure angles A and C.

    $A \approx$ _________________ $C \approx$ _________________

11. How close do your calculated angle values come to the measured values?

12. Just for fun, see if you can calculate the *area* of $\triangle ABC$ using only the sides and angles that have been measured so far.

13. What did you learn as a result of doing this Exploration that you did not know before?

Exploration 58: Area of a Triangle

Objective: Derive a *quick* method to calculate the area of a triangle from two sides and the included angle.

1. Sketch a rectangle circumscribing this triangle. Explain why the area of a triangle is *half* the area of the circumscribed rectangle.

2. By making appropriate measurements, find the area of $\triangle XYZ$.

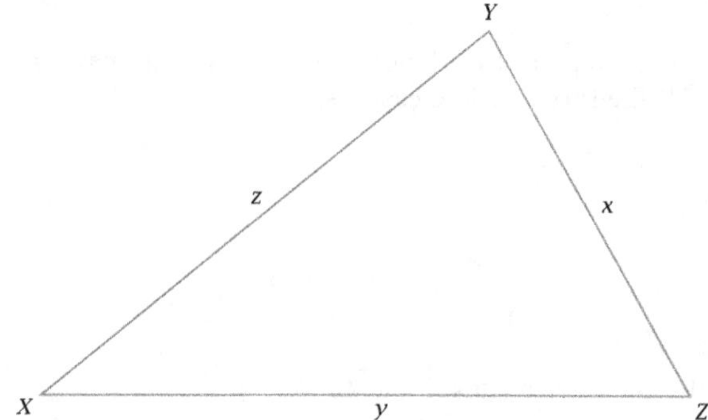

3. Draw u- and v-axes around $\triangle XYZ$ in such a way that angle X is in standard position. Then write an equation for the altitude, h, from Y to XZ as a function of the length z and angle X.

4. Use the result of Problem 3 to write an equation for the area of $\triangle XYZ$ in terms of y, z, and angle X.

5. Measure angle X and side z. Use the answers and your equation from Problem 4 to calculate the area of $\triangle XYZ$. How does the result compare with the answer you got in Problem 2?

6. Calculate the area of $\triangle XYZ$ again, this time using x, y, and angle Z.

7. Calculate the area of $\triangle XYZ$ once more, this time using x, z, and angle Y.

8. If you were to construct a triangle with sides 500 feet and 700 feet and included angle 70°, what would its area be?

9. If you increase the angle in Problem 8 to 140°, how much larger would the area of the triangle be? Explain your answer.

10. What did you learn as a result of doing this Exploration that you did not know before?

Exploration 59: Hero's Formula

Objective: Derive Hero's formula for calculating the area of a triangle given the measures of the three sides.

1. The figure shows $\triangle ABC$ with sides $a = 8$, $b = 7$, and $c = 11$. Use the law of cosines to calculate $\cos A$.

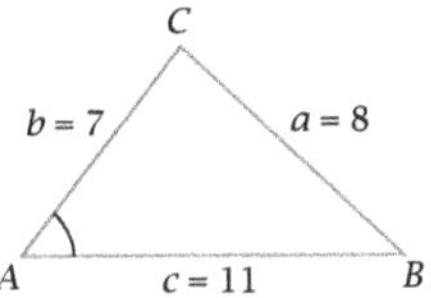

2. Use $\cos A$, without round-off, to calculate the measure of A and the area of $\triangle ABC$.

3. Calculate the area of $\triangle ABC$ again using **Hero's Formula:**

$$\text{Area} = \sqrt{s(s - a)(s - b)(s - c)}$$

where s is the **semiperimeter** of the triangle (half the perimeter). Do you get the same answer as in Problem 2?

4. The area formula from Section 6-3 is

$$\text{Area} = \frac{1}{2}bc \sin A$$

Explain why this can be written as

$$\text{Area} = \sqrt{\frac{1}{4}\, b^2 c^2 (1 - \cos^2 A)}$$

5. Explain why the result of Problem 4 can be written as

$$\text{Area} = \sqrt{\frac{1}{2}bc(1 + \cos A) \cdot \frac{1}{2}bc(1 - \cos A)}$$

6. Use the law of cosines to write $\cos A$ in terms of a, b, and c *without* substituting the values.

7. Show that $\frac{1}{2}bc(1 + \cos A)$ can be written as

$$\frac{(b + c)^2 - a^2}{4} = \frac{b + c + a}{2} \cdot \frac{b + c - a}{2}$$

(Over)

8. By substitutions such as you did in Problem 7, it is possible to transform the other factor under the radical in Problem 5 this way. (It is not necessary for you to *do* the transformations.)

$$\frac{1}{2}bc(1 - \cos A) = \frac{a - b + c}{2} \cdot \frac{a + b - c}{2}$$

Using this information and the result of Problem 7, the area can be written as

$$\text{Area} = \sqrt{\frac{b + c + a}{2} \cdot \frac{b + c - a}{2} \cdot \frac{a - b + c}{2} \cdot \frac{a + b - c}{2}}$$

The first fraction under the radical is the semiperimeter, $s = \frac{1}{2}(a + b + c)$. Show that the other three fractions can be written as $(s - a)$, $(s - b)$, and $(s - c)$, respectively. From the results, derive Hero's formula,

$$\text{Area} = \sqrt{s(s - a)(s - b)(s - c)}$$

9. Use Hero's formula to find the area of $\triangle XYZ$ if $x = 50$ cm, $y = 60$ cm, and $z = 80$ cm.

10. Show that Hero's formula indicates that there is no possible $\triangle BAD$ if $b = 10$ ft, $a = 12$ ft, and $d = 26$ ft.

11. What did you learn as a result of doing this Exploration that you did not know before?

Exploration 60: The Law of Sines

Objective: Use the ratio of a side length to the sine of the opposite angle to find other parts of a triangle.

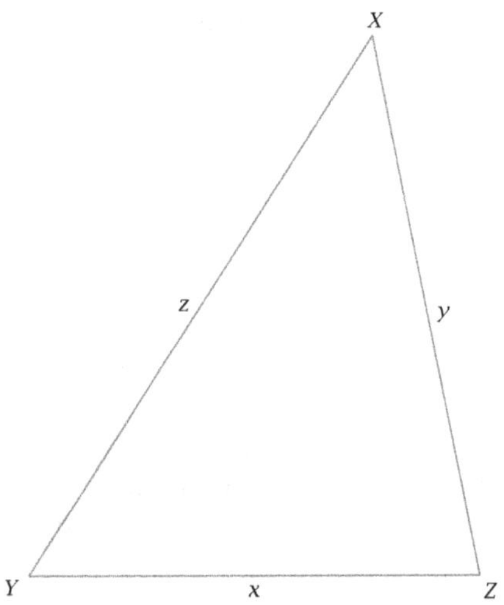

1. In $\triangle XYZ$, are the following measurements correct?

   $y = 6.0$ cm _____________ $z = 7.0$ cm _____________

   $Y = 57°$ _____________ $Z = 78°$ _____________

2. Assuming that the measurements in Problem 1 are correct, calculate these ratios:

   $$\frac{y}{\sin Y} = \text{_________________}$$

   $$\frac{z}{\sin Z} = \text{_________________}$$

3. The **law of sines** states that within a triangle, the ratio of the length of a side to the sine of the opposite angle is constant. Do the calculations in Problem 2 seem to confirm this property? _________

4. Measure angle X. _________________________________

5. Assuming that the law of sines is correct,

   $$\frac{x}{\sin X} = \frac{y}{\sin Y}$$

   Use this information and the measured value of X to calculate the length x.

6. Measure side x. Does your measurement agree with the calculated value in Problem 5? _______________

The law of sines can be derived **algebraically.**

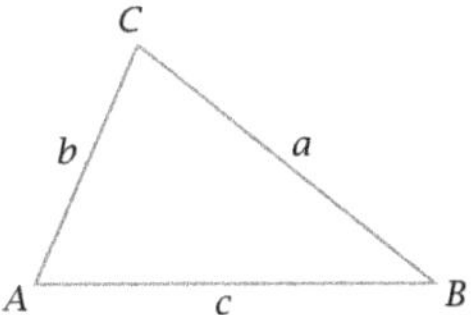

7. For $\triangle ABC$, use the area formula to write the area *three* ways: one involving angle A, one involving angle B, and one involving angle C.

   ____

   ____

   ____

8. The area of a triangle is *independent* of the way you *measure* that area, so all three area expressions in Problem 7 are equal to each other. Write a *three*-part equation expressing this fact.

9. Divide all three "sides" of the equation in Problem 8 by whatever is necessary to leave only the sines of the angles in the numerators. Simplify.

10. The equation you should have gotten in Problem 9 is the **law of sines.** Explain why it is equivalent to the law of sines as written in Problem 5.

11. What did you learn as a result of doing this Exploration that you did not know before?

Exploration 61: The Law of Sines for Angles

Objective: Discover the hazards of using the law of sines to find an angle of a triangle.

The figure shows $\triangle ABC$ (not to scale) with $a = 7$, $b = 4$, and $c = 10$.

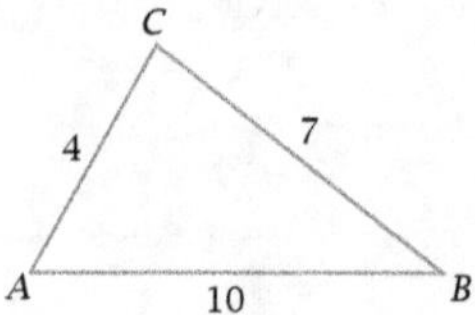

1. Use the law of cosines to find the measure of angle A. Store the answer in your calculator without round-off.

2. Use the answer to Problem 1 and the law of sines to calculate the measure of angle C.

3. Calculate the measure of angle C again, directly from the three side lengths, using the law of cosines.

4. Do your answers to Problems 2 and 3 agree? If not, describe how you can use the *general* solution for inverse sine (arcsin instead of $\sin^{-1}$) to get the correct answer by the law of sines.

5. Triangle XYZ has $x = 11$ in., $y = 8$ in., and angle $Y = 40°$. Use the law of sines to find *two* possible values for the measure of angle X. Sketch both triangles.

6. What did you learn as a result of doing this Exploration that you did not know before?

Exploration 62: The Ambiguous Case, SSA

Objective: Investigate what can happen if SSA (side, side, angle) is given in a triangle.

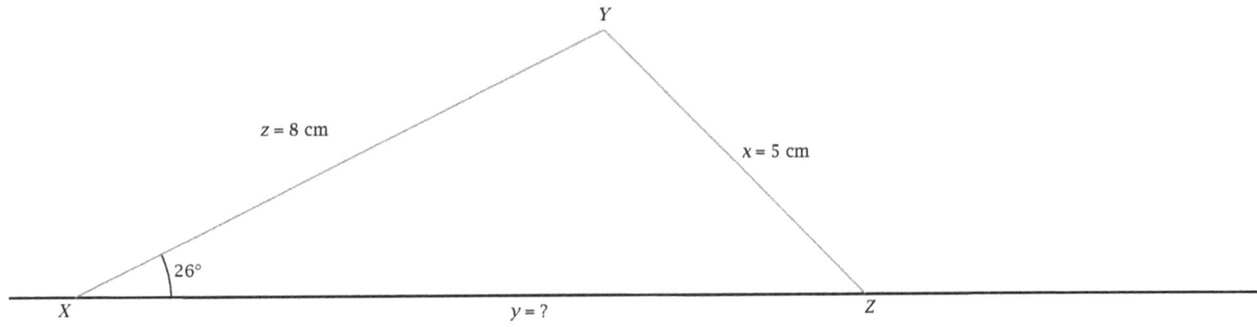

1. In $\triangle XYZ$, side $x = 5$ cm, side $z = 8$ cm, and angle $X = 26°$. Do you agree with these measurements?

   ____

2. Draw another possibility for $\triangle XYZ$ with the same values of x, z, and X but a different value of y. For both triangles, measure side y.

   $y =$ _______________ or $y =$ _______________

3. What does the word **ambiguous** mean? Why is case SSA called *ambiguous*?

4. If side z and angle X remain fixed and side x is increased to 9 cm, will there still be two possible triangles? Draw the result on the given figure.

5. If side z and angle X remain fixed and side x is decreased to 3 cm, how many possible triangles will there be? Show your conclusion on the given figure.

6. When side x is perpendicular to side y, there is exactly one triangle. Calculate the value of x in this case. Confirm that it is correct by taking its measurement on the figure.

7. By clever application of the law of cosines and the quadratic formula, it is possible to *calculate* both values of y in Problem 2. Do this calculation.

8. Show how the technique of Problem 7 justifies the answer to Problem 4.

9. Show how the technique of Problem 7 justifies the answer to Problem 5.

10. What did you learn as a result of doing this Exploration that you did not know before?

Exploration 63: Golf and the Ambiguous Case (Inspired by Chris Sollars)

Objective: Analyze a real-world problem involving the ambiguous case.

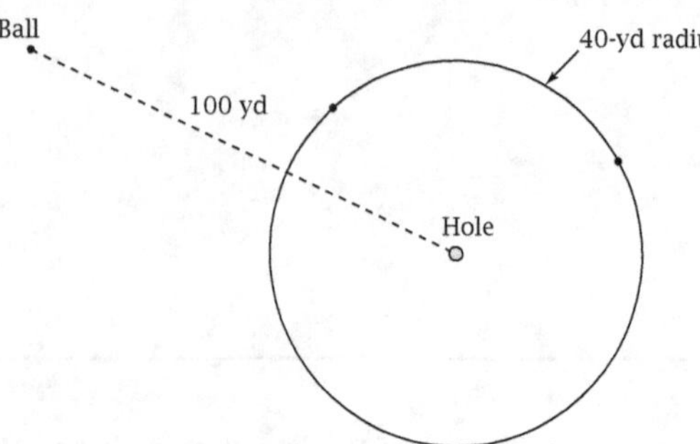

Golf Ball Problem: Dolph Ball hits his tee shot. The ball winds up exactly 100 yd from the hole. On his second shot, the ball winds up exactly 40 yd from the hole, somewhere on the circle shown in the figure.

1. If Dolph's second shot went on a line 15° to the right of the line to the hole, plot on the figure the path the ball took. Show one point where the path crosses the circle. Then draw a triangle with the distance the ball traveled to get to this point as one of its sides and the 40 yd and 100 yd as the other two sides.

2. Use the law of cosines to calculate the *two* possible distances the shot in Problem 1 could have gone. Show both distances on the figure.

3. If Dolph's second shot had gone on a line 30° to the right of the line to the hole, plot the path of the ball on this copy of the figure. Show by calculation that the ball could not have come within 40 yd of the hole.

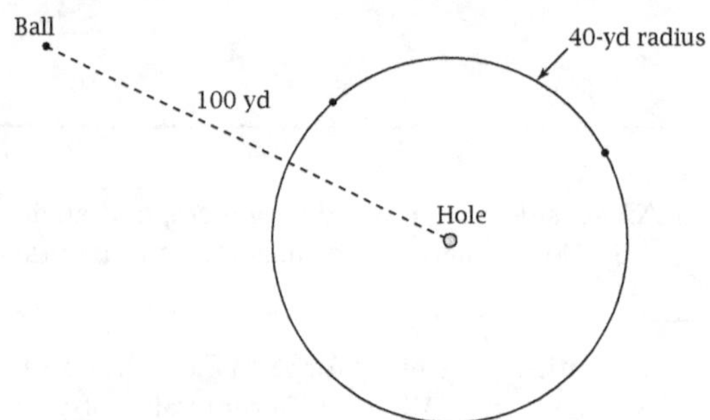

4. Show on this copy of the figure the path of the ball at the maximum angle Dolph's second shot could have made and still have come to rest 40 yd from the hole. Calculate the measure of this angle. Does the angle you drew have this measure? _______________

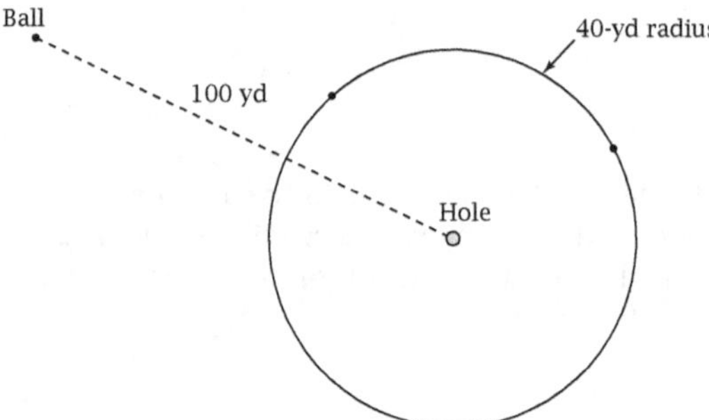

5. What did you learn as a result of working this Exploration that you did not know before?

Exploration 64: Introduction to Vectors

Objective: Use the properties of triangles to add vectors.

1. The figure shows two vectors starting from the origin. One ends at the point (4, 7), and the other ends at (5, 3). Translate one of the two vectors so that the vectors are in position to be added. Then draw the resultant vector—the sum of the two vectors.

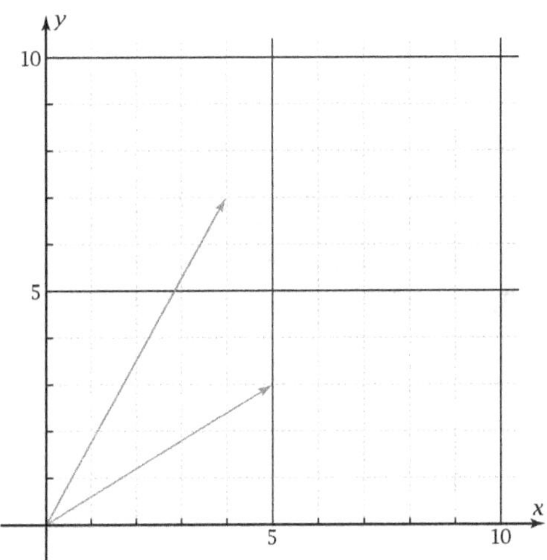

2. Calculate the length of the resultant vector in Problem 1 and the angle it makes with the *x*-axis.

3. The two given vectors and the resultant vector form a triangle. Calculate the measure of the largest angle in this triangle.

4. Calculate the measure of the angle between the two vectors when they are placed tail-to-tail, as they were given in Problem 1.

5. In Problem 1, you translated one of the vectors. Show on the figure that you would have gotten the *same* resultant vector if you had translated the *other* vector. Use a different color than you used in Problem 1.

6. The vectors in Problem 1 have **components** in the *x*-direction and in the *y*-direction. These components are vectors that can be added together to equal the given vector. On this copy of the figure, show how the components of the longer vector can be added to give that vector.

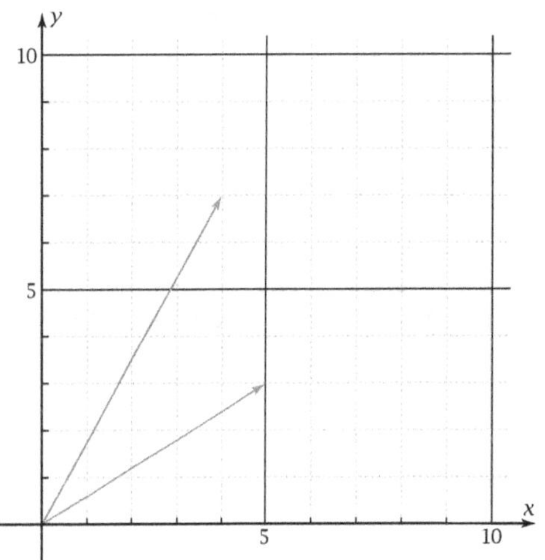

7. Give an *easy* way to get the components of the sum of the two given vectors in Problem 1.

8. What did you learn as a result of doing this Exploration that you did not know before?

Exploration 65: Navigation Vectors

Objective: Work navigation problems using components of vectors.

A ship sails for 20 miles along a bearing of $\beta_1 = 325°$ and then turns and sails on a bearing of $\beta_2 = 250°$ for 7 more miles, as shown in the figure.

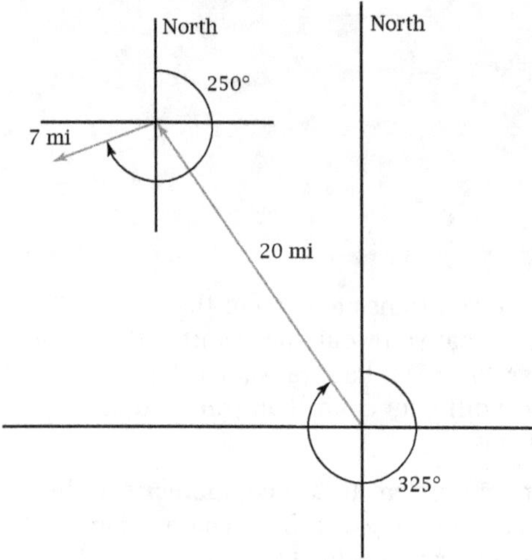

1. Bearings are measured clockwise from North. Show bearings of 0°, 90°, 180°, and 270° on the figure. What bearing is equivalent to 360°?

2. Assume that the unit vectors $\vec{i}$ and $\vec{j}$ point along bearings of 0° and 90°, the same as they would for angles in standard position. Use this fact to write components for the two given vectors.

3. Find the $\vec{i}$ and $\vec{j}$ components for the resultant displacement vector.

4. Find bearing β for the resultant vector.

5. Find the magnitude of the resultant vector. Then write the vector as a magnitude and bearing.

A ship sails with a velocity of 20 knots (nautical miles per hour) on a bearing of 325°. The water has a current of velocity 7 knots on a bearing of 250°.

6. Draw a diagram showing the two vectors coming from the origin.

7. The ship's resultant velocity is the vector sum of the two velocity vectors. Find this resultant velocity vector.

8. What did you learn as a result of doing this Exploration that you did not know before?

Exploration 66: The Ship's Path Problem

Objective: Work a real-world triangle problem given only the descriptive words.

Suppose that you are on a rescue mission in the Gulf of Suez. Your ship is steaming east when its sonar detects the disabled submarine 6 (nautical) miles away, at an angle of 28° to the south of the easterly path.

1. Construct a diagram showing the rescue ship, its easterly path, and the submarine southeast of the ship. Use 1 cm for 1 nautical mile.

2. The submarine can transmit distress signals that can be detected at a 4-mile radius. Calculate how far your rescue ship would have had to go on its easterly path before the distress signals would have reached it had it not detected the submarine.

3. Plot the distance you calculated in Problem 2 on the diagram from Problem 1. Is the point really 4 (scale) miles from the submarine? _________________

4. Assume that your rescue ship missed the distress signal the first time it could have been detected and continues on its easterly path. Calculate the farthest your ship can be from the initial point in Problem 2 while still able to hear the distress signal.

5. Calculate the angle at the submarine between the rescue ship's initial position in Problem 1 and its position in Problem 4.

6. Find the area of the triangular region of ocean with vertices at the submarine, the ship's initial Problem 1 position, and its Problem 4 position.

7. The submarine also carries short-range radio transmitters that will transmit radio signals only 2 miles. Was your ship ever within range of these radio signals? Show how you reach your conclusion.

8. What did you learn as a result of doing this Exploration that you did not know before?

Exploration 67: Area of a Regular Polygon

Date: __________

Objective: Apply your knowledge of triangle trigonometry to a new situation.

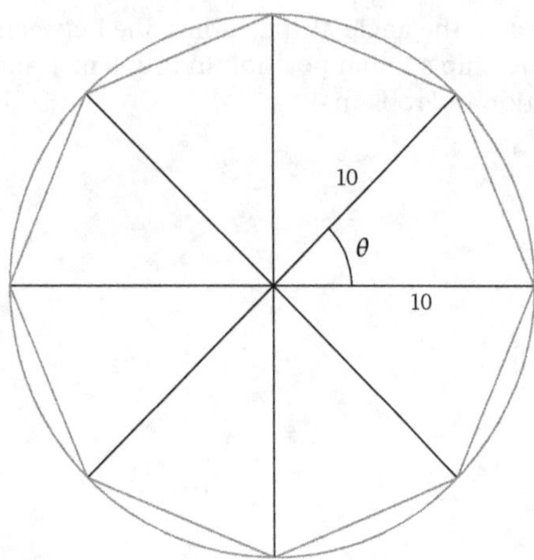

1. The figure shows a regular octagon inscribed in a circle of radius 10 units. Diagonals connecting the opposite vertices divide the octagon into eight triangles meeting at the center. Calculate the measure of central angle θ. Use the result and the appropriate triangle property to calculate the area of one triangle. Then find the area of the octagon.

2. Suppose a regular n-gon (an n-sided polygon) is inscribed in the circle shown. Write an equation expressing the area of an n-gon as a function of n.

3. Enter the equation of Problem 2 in the y= menu of your grapher. Make a table of values of area for n-gons, starting with a triangle.

4. There are two of these regular polygons that have *integer* areas. Which ones? What are the areas?

5. Write a short program to calculate and display the areas of each n-gon, starting at $n = 3$ and going to at least $n = 600$. Set the mode so that your grapher displays a fixed number of digits, at least nine. Then run the program.

6. As the program is running, what do you notice is happening to more and more digits in the areas?

7. The areas are approaching a **limit** as n gets larger and larger. What do you think it means for a quantity to approach a limit? What number does that limit equal? Explain geometrically why that number is reasonable.

8. Name one new thing you learned as a result of doing this Exploration.

Exploration 68: Graphical Patterns in Functions

Date: _____________

Objective: Find the particular equation of a linear, quadratic, power, or exponential function from a given graph.

1. Identify what kind of function is graphed, and find its particular equation.

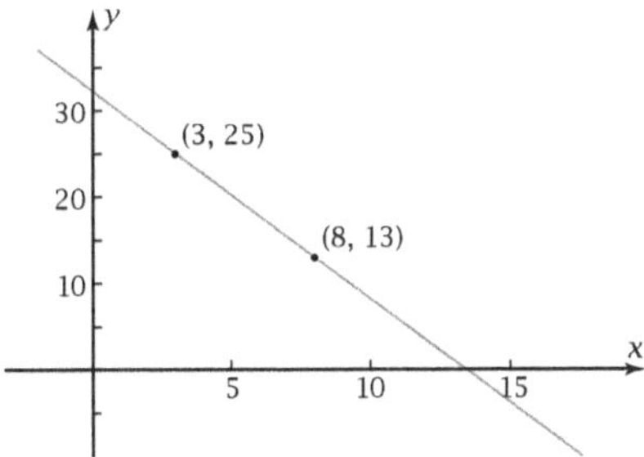

4. What graphical evidence do you have that the function graphed is an exponential function, not a power function? Find its particular equation.

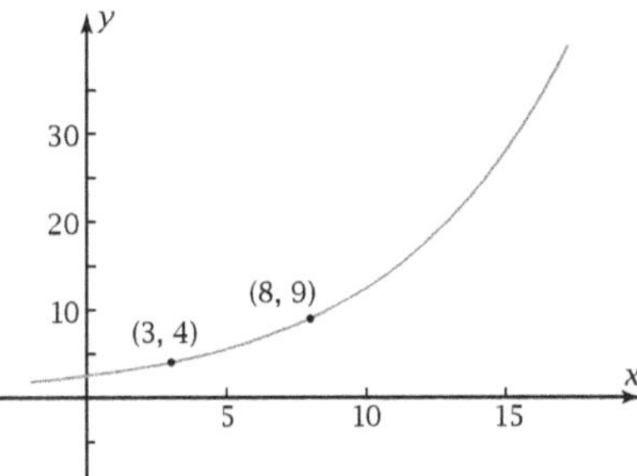

2. Check your answer to Problem 1 graphically. Does your graph agree with the given one?

3. Is the graph in Problem 1 concave up, concave down, or neither?

5. Check your answer to Problem 4 graphically. Does your graph agree with the given one?

6. Is the graph in Problem 4 concave up, concave down, or neither?

(Over)

Exploration 68: Graphical Patterns in Functions *continued*

7. What graphical evidence do you have that this function is a power function, not an exponential function? Find its particular equation.

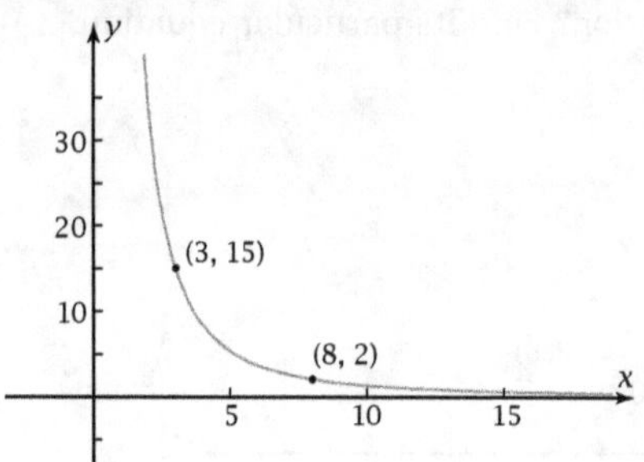

10. Identify what kind of function is graphed, and find its particular equation.

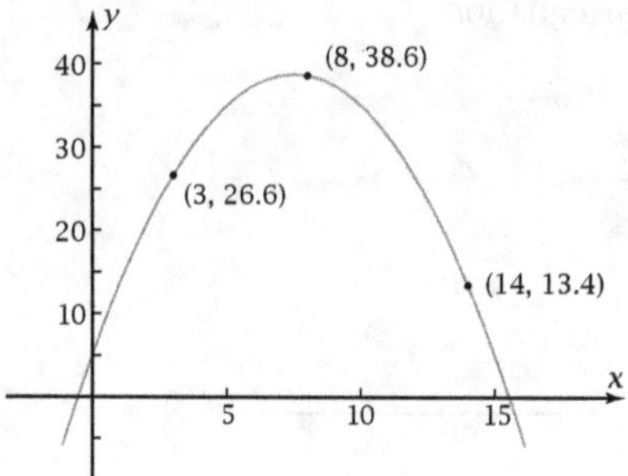

8. Check your answer to Problem 7 graphically. Does your graph agree with the given one?

11. Check your answer to Problem 10 graphically. Does your graph agree with the given one?

9. Is the graph in Problem 7 concave up, concave down, or neither?

12. Is the graph in Problem 10 concave up, concave down, or neither?

13. What did you learn as a result of doing this Exploration that you did not know before?

Exploration 69: Numerical Patterns in Function Values

Date: _____________

Objective: For exponential, power, or linear functions, find a pattern in the y-values for regularly spaced x-values.

For the exponential function $y = 0.2 \cdot 3^x$:

1. Find y for the given values of x.

x	y
2	
4	
6	
8	
10	
12	

2. If $x_1 = 4$ and $x_2 = 6$, how does y_2 compare to y_1?

3. If $x_1 = 8$ and $x_2 = 10$, how does y_2 compare to y_1?

4. Show in the table that anytime 2 is added to x, y is multiplied by 9.

5. The property in Problem 4 is called the

____

property of exponential functions.

For the power function $y = 5x^3$:

6. Find y for the given values of x.

x	y
2	
4	
6	
8	
10	
12	

7. Show in the table all instances you can find where $x_2 = 2x_1$. Show how y_2 compares to y_1.

8. Complete the sentence: "Multiplying x by 2 __________

_________________________________."

9. The property you observe in Problem 8 is called the **multiply–multiply property of power functions.** Write a sentence stating this property.

For the linear function $y = 7x + 17$:

10. Find y for the given values of x.

x	y
2	
4	
6	
8	
10	
12	

11. Show in the table three instances in which $x_2 = x_1 + 4$. Show how y_2 compares to y_1.

12. Complete the sentence: "Adding 4 to x __________

_________________________________."

13. Use your findings in Problem 10 to complete:

"The _________________________________

property of linear functions states that

____

____

____

____

_________________________________."

14. On the back of this sheet, prove algebraically that if $y = a \cdot x^n$ and if $x_2 = cx_1$ for some constant c, then $y_2 = k \cdot y_1$ for some constant k.

Exploration 70: Patterns for Quadratic Functions

Objective: Find patterns for the y-values in quadratic functions similar to the add–add property for linear functions.

1. The points in the table are not fit by a linear, power, or exponential function because they decrease and then increase. Show by table that they are fit by the **quadratic function**

$$q(x) = 0.2x^2 - 1.3x + 14$$

x	$q(x)$
2	12.2
4	12.0
6	13.4
8	16.4
10	21.0

2. Find the differences between consecutive y-values. Then find the **second differences,** i.e., the differences between the consecutive differences. What do you notice?

3. Recall that the general equation of a quadratic function is $y = ax^2 + bx + c$, where a, b, and c stand for constants. Substitute the first three ordered pairs from the table in Problem 1 to get three linear equations involving a, b, and c. Solve this system of equations using matrices. Write the particular equation, and state whether it agrees with the one in Problem 1.

4. On the same screen, plot the graph of $q(x)$ and the five data points. You may use the STAT feature on your grapher. Sketch the result here.

5. Trace the graph of function q to each value of x in the table. Do the five points lie on the graph?

6. Show that a quadratic function fits the data in this table by finding second differences. Find the particular equation, and show that these values satisfy the equation.

x	$f(x)$
1	12.3
4	34.8
7	44.7
10	42.0
13	26.7

7. What did you learn as a result of doing this Exploration that you did not know before?

Exploration 71: Equations from Given Values Practice

Objective: Given a set of regularly spaced x- and y-values, find a particular equation and use it to find other values.

For each set of points, state what pattern the points follow and what kind of function has that pattern. Substitute given points into the general equation to calculate the constants, and give the particular equation. Then calculate the indicated function value.

1.

x	$f(x)$
2	2940
4	1440.6
6	705.894
8	345.88806
10	169.4851494

Find $f(13.7)$.

3.

x	$h(x)$
2	87.5
4	122.1
6	156.7
8	191.3
10	225.9

Find $h(100)$.

2.

x	$g(x)$
2	5.6
4	44.8
6	151.2
8	358.4
10	700

Find $g(22.3)$.

4.

x	$q(x)$
2	9.2
4	12.6
6	19.2
8	29.0
10	42.0

Find $q(-7.9)$.

5. What did you learn as a result of doing this Exploration that you did not know before?

Exploration 72: Introduction to Logarithmic Functions

Objective: Show that the function $y = \log x$ has the multiply–add property, and use this fact to recall the definition of logarithm.

1. Enter $y_1 = \log x$ (for **logarithm**) in your grapher. Then make a table of values.

x	$y_1 = \log x$	x as Power of 10
0.1		
1		
10		
100		
1000		

2. In the table in Problem 1, write each x-value in Problem 1 as a power of 10. From the result, state the relationship between the x-values and the y-values.

3. Show that the values in Problem 1 have the **multiply–add** property. Explain verbally what this property means.

4. Make another table of values of $y_1 = \log x$. Show that these values also have the multiply–add property.

x	$y_1 = \log x$
1	
5	
25	
125	
625	

5. On the same screen, plot $y_1 = \log x$ and $y_2 = 10^x$. Use a window with an x-range of $[-3, 3]$, and use equal scales on the two axes. Sketch the result.

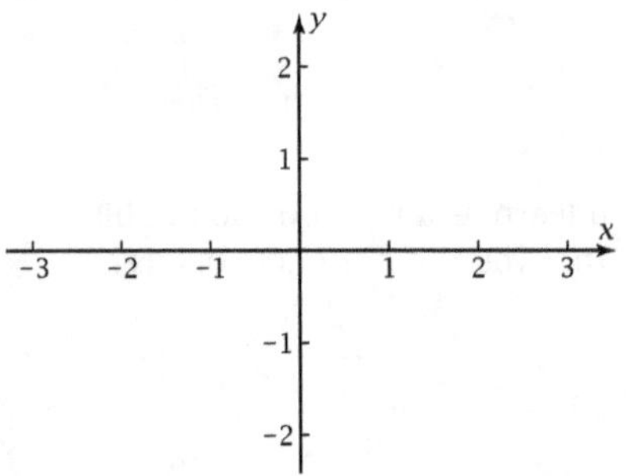

6. Find log 1776. Then, without round-off, raise 10 to the power of the answer you got.

log 1776 = _________________________

$10^{\log 1776} =$ _________________________

What important thing do you notice?

7. Find $10^{1.8}$. Then, without round-off, find the logarithm of the answer.

$10^{1.8} =$ _________________________

$\log 10^{1.8} =$ _________________________

What do you notice?

8. Write the definition of logarithm. From the results of your work on this Exploration, what is the base of the function "log" on your grapher?

9. Show that the values in this table have the multiply–add property. Explain what this means.

x	y
9	2
81	4
729	6
6561	8
59049	10

10. In Problem 9, y is a logarithm of x. From the definition of logarithm, state what the base of the logarithm is.

11. Explain why this statement is true: **"A logarithm is an exponent."**

12. What did you learn as a result of doing this Exploration that you did not know before?

Exploration 73: The Logistic Function for Population Growth

Date: _______________

Objective: Fit the **logistic function** to restricted population growth.

Suppose that what the table lists are populations of a small community, in thousands of people. The figure shows a scatter plot of the data.

x (years)	y (thousands of people)
1	2
2	3
3	5
4	9
5	13
6	19
7	27
8	32
9	36
10	39

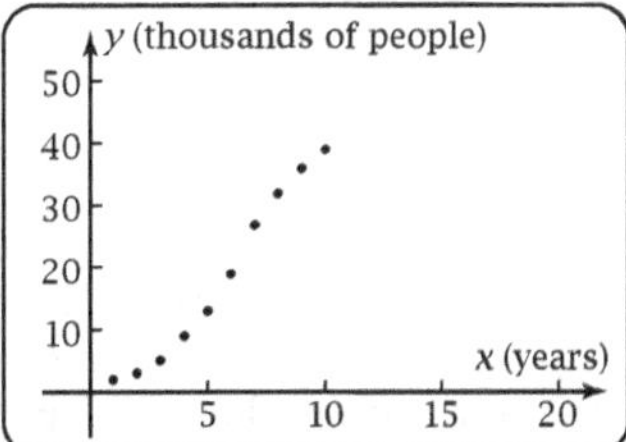

1. Plot the data on your grapher.

2. At first the population seems to be increasing exponentially with time. On the given figure, sketch the graph of an exponential function that would fit the first six data points reasonably well.

3. Toward the end of the 10-year period, the function seems to be leveling off. A function that models such population growth is the **logistic function.** Its general equation is

$$y = \frac{c}{1 + ae^{-bx}}$$

where x and y are the variables, e is the base of natural logarithm, and $a, b,$ and c stand for constants. The community has room for about 43 thousand people, meaning $c = 43$. Calculate a and b using the first and the tenth points. Write the particular equation, and plot it on the same screen as the data. Sketch the result on the given figure.

4. What does the logistic function indicate the population was at time $x = 0$ years?

5. What graphical evidence do you have that the maximum population in the community is 43,000?

6. Look up the word *logistic* in a dictionary, and find the origin of the word.

7. What did you learn as a result of doing this Exploration that you did not know before?

Exploration 74: Introduction to Linear Regression

Date: _______________

Objective: Find the sum of the squares of the residuals for a function found by linear regression.

Turkey Problem: Tom raises turkeys. He records the weight, y, measured in pounds, of one of his turkeys over several months, x.

x	y
2	7
5	9
8	18
11	17
14	24

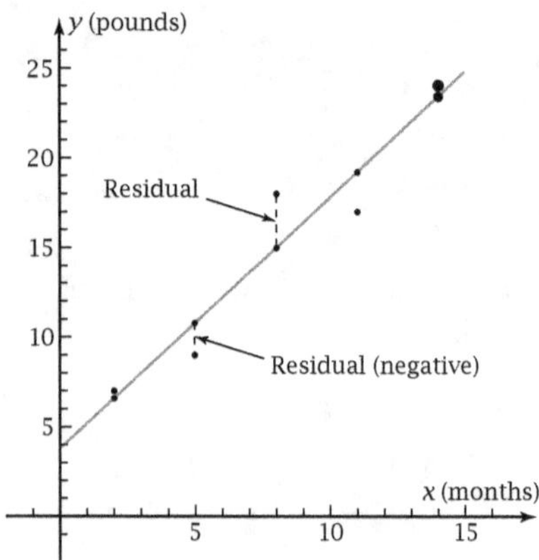

1. The graph is a scatter plot of the data, along with the linear regression line. Run linear regression on the data to find the particular equation of this line. Use $\hat{y}$ (pronounced "y hat") to distinguish the y-values for the regression equation from the y-values in the data. Store this equation as y_1 in your grapher.

   $\hat{y} =$ _______________________

2. Calculate the value of $\hat{y}$ for each value of x in the table. Record the results in a new, third column in the table.

3. The dotted lines on the graph show by how much each data point deviates from the regression line. This is called the *residual deviation,* or simply the **residual.** Calculate the residual, $y - \hat{y}$, for each x-value, and record the results in a fourth column.

4. Show that the sum of the residuals is zero.

5. Square each residual and record the results in a fifth column. Then find the **sum of the squares of the residuals.** This number is abbreviated SS_{res}. It is a measure of how well the equation fits the data. The smaller the SS_{res}, the better the fit.

   $SS_{res} =$ _______________________

6. In Problem 1, you found $\hat{y} = 1.4x + 3.8$. The value of SS_{res} for this equation is the lowest possible. Demonstrate that this is correct by using the function $y_2 = 1.4x + 3.9$, which increases the y-intercept by 0.1, to calculate SS_{res} again. Show that the answer is greater than in Problem 5. What does this fact tell you about the new function?

x	y
2	7
5	9
8	18
11	17
14	24

$SS_{res} =$ _______________________

7. Use the function $y_3 = 1.5x + 3.8$, which increases the slope of the regression equation by 0.1, to calculate SS_{res} a third time. What do the results indicate about how well $\hat{y}$, y_2, and y_3 fit the given data?

x	y
2	7
5	9
8	18
11	17
14	24

$SS_{res} =$ _______________________

8. What reason can you think of to explain why the turkey's weight *decreased* between the 8th and the 11th month?

9. What did you learn as a result of doing this Exploration that you did not know before?

Exploration 75: Sums of Squares of Residuals   Date: _____________

Objective: Find the sum of the squares of the residuals for a function found by linear regression.

Suppose that these data have been measured for the related variables x and y.

x	y
3	41
5	37
7	29
9	28
11	23
13	26
15	19
17	9
19	8
21	4

1. Show by linear regression that the best-fitting linear function is $\hat{y} = -2x + 46.4$. How do you interpret the fact that the correlation coefficient is negative?

2. Plot the given data on a scatter plot. On the same screen, plot the regression line. How well does the linear function fit the data?

3. On your grapher, make three lists, one showing $\hat{y}$ for each point, one showing the residual $y - \hat{y}$, and a third showing the squares of the residuals, $(y - \hat{y})^2$. Copy the results into the table.

4. Calculate SS_{res} from the squares of the residuals. See if you can discover a time-efficient way to find the sum using built-in features on your grapher.

5. This table has the same data as before. Find SS_{res} again using $y_2 = -2x + 46$. How do you interpret the fact that the answer is *greater* than SS_{res} using the regression equation $\hat{y} = -2x + 46.4$?

x	y
3	41
5	37
7	29
9	28
11	23
13	26
15	19
17	9
19	8
21	4

6. Calculate $\bar{x}$ and $\bar{y}$, the averages of x and y. Show algebraically that the point $(\bar{x}, \bar{y})$ is on the regression line $\hat{y} = -2x + 46.4$.

7. The line $y_3 = -2.1x + 47.6$ also contains the "average-average" point $(\bar{x}, \bar{y})$, but it has a slope of -2.1 instead of -2. Plot the line on the same screen as in Problem 2. Can you tell from the graphs which line fits the data better?

8. Find SS_{res} using the equation in Problem 7. Based on your answer, how can you tell that this line does *not* fit the data as well as does the regression line?

9. What did you learn as a result of doing this Exploration that you did not know before?

Exploration 76: The Correlation Coefficient

Date: _____________

Objective: Learn the formula by which your grapher calculates the regression equation from a set of data.

x	y
3	7
6	9
9	18
12	17
15	24

The table shows the weight, y, of a turkey at times x months after it was hatched. The graph shows a scatter plot of these data.

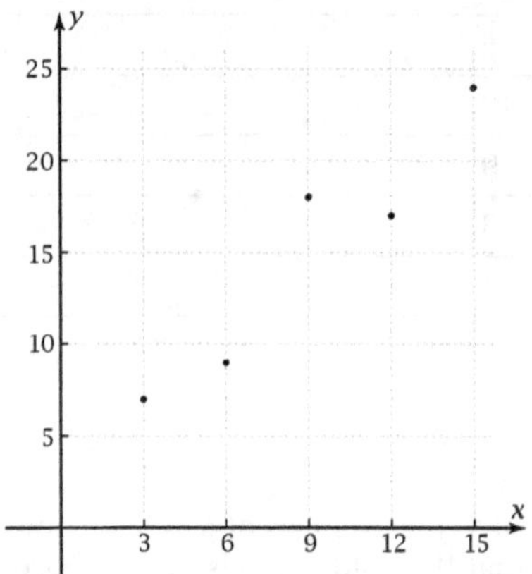

1. Find the average of the y-values, $\bar{y}$. Plot a horizontal line on the graph at $y = \bar{y}$.

2. The **deviation** of each point is its directed distance from the line $y = \bar{y}$ to the point. Show that you understand the meaning of *deviation* by drawing the deviation of each point on the graph.

3. Calculate the deviation and the square of the deviation for each data point. Record these in the table at the top of this column. Then calculate the **sum of the squares of the deviations, SS_{dev}**.

4. Run linear regression on the data. Write the regression equation, and record the value of r^2 and the value of the correlation coefficient r.

5. On this scatter plot, plot the graph of the regression equation.

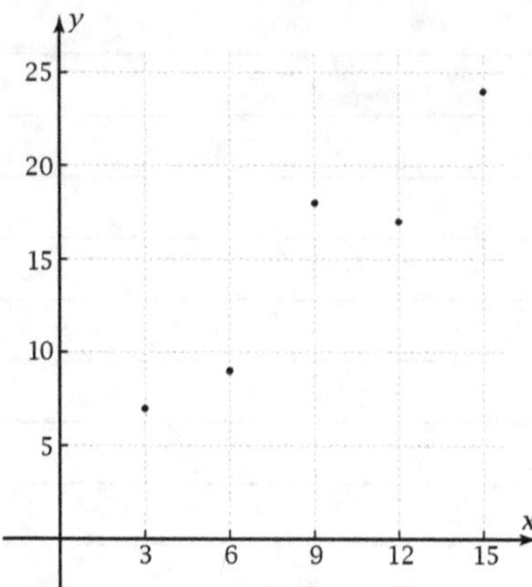

6. Recall that the **residual** of each data point is the directed vertical distance from the regression line to the point. Draw on the graph the residual for each point.

7. On this copy of the data table, calculate $\hat{y}$ for each point. Then calculate the square of each residual. Use the result to calculate SS_{res}.

x	y	$\hat{y}$	$(y - \hat{y})^2$
3	7		
6	9		
9	18		
12	17		
15	24		

8. SS_{res} is the amount of SS_{dev} that remains after regression has taken out as much of it as possible. The fraction of SS_{dev} that is removed by the linear regression is called the coefficient of determination. Calculate

$$\frac{SS_{dev} - SS_{res}}{SS_{dev}}$$

9. On the back of this sheet, write a paragraph stating where the coefficient of determination shows up in the linear regression by grapher. Explain how the correlation coefficient r is calculated from the coefficient of determination.

Exploration 77: Formula for the Correlation Coefficient

Objective: Derive a formula for calculating the correlation coefficient from sums of deviations without finding the regression equation.

Gasoline Consumption Problem: These data show the amount of gas left in your tank at various times after you start a long highway trip.

x (hr)	y (gal)
2	20
3	17
7	8

1. Find $\bar{x}$ and $\bar{y}$, the averages of x and y. Use the answer to put columns in the table for the deviations $(x - \bar{x})$ and $(y - \bar{y})$. You may do this in a time-efficient way using the TABLE feature of your grapher.

2. Put in columns for $(x - \bar{x})^2$, $(y - \bar{y})^2$, and $(x - \bar{x})(y - \bar{y})$. Then find these sums:

   $\Sigma(x - \bar{x})^2 =$ ______________ $\Sigma(y - \bar{y})^2 =$ ______________ $\Sigma(x - \bar{x})(y - \bar{y}) =$ ______________

3. Using the numbers from Problem 2, evaluate the fraction

   $$\frac{\left[\Sigma(x - \bar{x})(y - \bar{y})\right]^2}{\Sigma(x - \bar{x})^2\,\Sigma(y - \bar{y})^2}$$

4. Run a linear regression on the given data. Does r^2, the coefficient of determination, equal the answer to Problem 3? If not, check your work.

In the rest of this Exploration, you will do algebraic calculations to show that your observation in Problem 4 is always true.

5. By definition, $SS_{res} = \Sigma(y - \hat{y})^2$ for any point $(x, \hat{y})$ on the regression line $\hat{y} = mx + b$. Also, the average-average point $(\bar{x}, \bar{y})$ is on the regression line, so $\bar{y} = m\bar{x} + b$. Use this information to show that

   $$SS_{res} = \Sigma[(y - \bar{y}) - m(x - \bar{x})]^2 = \Sigma(y - \bar{y})^2 - 2m\,\Sigma(x - \bar{x})(y - \bar{y}) + m^2\,\Sigma(x - \bar{x})^2$$

6. The expression $\Sigma(x - \bar{x})(y - \bar{y})$ that appears in Problem 3 can be written as $\Sigma xy - \frac{1}{n}\Sigma x\,\Sigma y$, where n is the number of data points. Show the algebraic steps in doing this. For this purpose, it helps to realize, for example, that a constant multiplier such as $\bar{x}$ can be factored out of $\Sigma \bar{x}y$ to give $\bar{x}\Sigma y$ and that $\bar{x} = \frac{1}{n}\Sigma x$.

(Over)

Exploration 77: Formula for the
Correlation Coefficient *continued*

7. The regression formula for m is

$$m = \frac{n\sum xy - \sum x \sum y}{n\sum x^2 - (\sum x)^2}$$

Use this fact and the result of Problem 6 to show that

$$m = \frac{\sum(x-\bar{x})(y-\bar{y})}{\sum(x-\bar{x})^2}$$

8. Substitute the latter fraction for m in the latter expression for SS_{res} in Problem 5, thus showing that

$$SS_{res} = \sum(y-\bar{y})^2 - \frac{\left[\sum(x-\bar{x})(y-\bar{y})\right]^2}{\sum(x-\bar{x})^2}$$

(Don't be afraid of the algebraic manipulations!)

9. By definition, the coefficient of determination is

$$r^2 = \frac{SS_{dev} - SS_{res}}{SS_{dev}}$$

Recall that $SS_{dev} = \sum(y-\bar{y})^2$. Substitute this value of SS_{dev} and the value of SS_{res} from Problem 8 to show that the following is true. (Don't be afraid of the algebraic manipulations!)

$$r^2 = \frac{\left[\sum(x-\bar{x})(y-\bar{y})\right]^2}{\sum(x-\bar{x})^2 \, \sum(y-\bar{y})^2}$$

10. Finding r involves taking the square root of r^2. Evaluate r for the data in Problem 1 using the formula

$$r = \frac{\sum(x-\bar{x})(y-\bar{y})}{\sqrt{\sum(x-\bar{x})^2} \, \sqrt{\sum(y-\bar{y})^2}}$$

What do you notice about the sign of the answer?

11. What did you learn as a result of doing this Exploration that you did not know before?

Exploration 78: Violent Crimes and Regression Inaccuracies

Objective: Make conclusions about interpolation and extrapolation using a function derived by linear regression.

An aspiring reporter is attempting to draw some conclusions about the crime rate in the United States. Here are actual numbers of arrests for violent crimes reported for the years from 1971 to 1997. Let x be the number of years that have elapsed since 1970, and let y be the number of violent crime arrests reported in each year x.

Year	Total	Year	Total
1971	323060	1985	497560
1972	350410	1986	553900
1973	380560	1987	546300
1974	429350	1988	625900
1975	451310	1989	685500
1976	411630	1990	705500
1977	435720	1991	718890
1978	469700	1992	742130
1979	467700	1993	754110
1980	475160	1994	778730
1981	490460	1995	796250
1982	526200	1996	729900
1983	499390	1997	717750
1984	493960		

1. To make things simpler, the reporter does linear regression with the eight years 1974 and 1989 through 1995. The year 1974 was included to give a wider range of data and therefore, perhaps, a more accurate mathematical model. Show that the correlation is very good and that interpolating to 1975 gives a highly accurate result.

2. Show numerically that interpolating to 1985 and extrapolating to 1997 both produce quite inaccurate results.

3. Enter all of the data into your grapher. Then make a scatter plot. Does the result look like the one given here? _______________

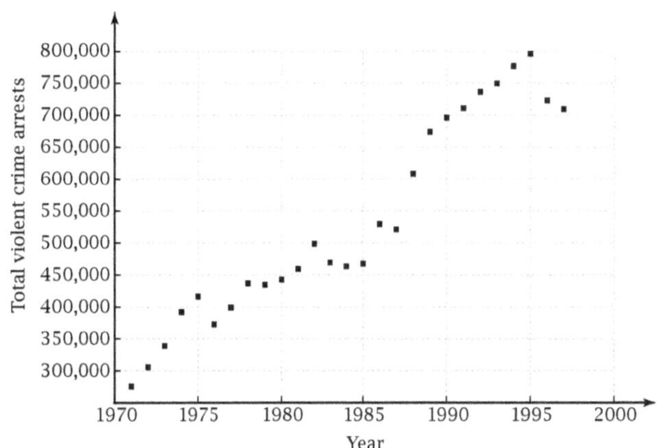

4. Find the particular equation of the linear function that best fits the entire data set. Record the correlation coefficient. Plot this function as y_2 and the linear function from Problem 1 as y_1 on your grapher. Sketch the results on the figure.

5. A reporter looking only at the data from 1971 through 1985 would observe that the number of crime arrests was leveling off. Suppose the reporter assumed that a quadratic function was a reasonable mathematical model. What quadratic function would best fit this range of data? Plot this quadratic function on your grapher, and sketch the result on the previous figure. How far off would the quadratic function's prediction be for the number of crime arrests in 1997?

6. What did you learn as a result of doing this Exploration that you did not know before?

Exploration 79: Coffee Data Residual Plot

Date: _________

Objective: Use a residual plot to make conclusions about a cooling cup of coffee.

You pour a cup of coffee. Rather than drinking it right away, you let it cool, measuring its temperature at 1-minute intervals. This table and graph show y, the number of degrees above room temperature, as a function of x minutes since you poured the coffee.

x (min)	y (°C)
2	51.5
3	45.5
4	40.2
5	35.5
6	31.3
7	27.4
8	23.9
9	20.6
10	17.5
11	14.5

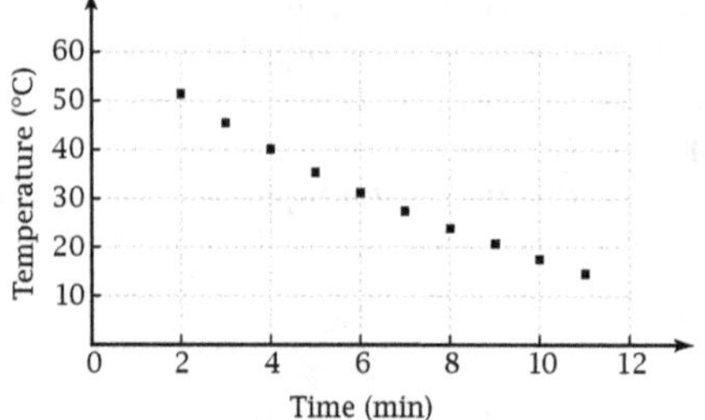

1. Based on the shape and concavity of the graph, explain why both an exponential function and a power function would be reasonable mathematical models for temperature as a function of time.

2. Explain why both exponential and power functions have reasonable endpoint behavior for large values of x but only the exponential function behaves correctly at the left end of the domain.

3. Confirm by exponential and power regression that an exponential function fits the data better than a power function. How do you decide?

4. Plot the exponential function of Problem 3 and the data on the same screen. Sketch the result on the given figure.

5. Calculate the residuals for each point using the exponential function from Problem 3. Make a residual plot using a window with an x-range of [0, 12] and a suitable range for the residuals. Show your graph to your instructor. _________________

6. After you have checked your residual plot with your instructor, sketch it here.

7. What does the fact that the residual plot shows a pattern indicate about how well the exponential function fits the data? What could you conclude if a residual plot showed no discernible pattern?

8. What did you learn as a result of doing this Exploration that you did not know before?

Exploration 80: Airplane Fuel and Non-Linear Regression

Objective: Mathematically model the fuel consumption of an airplane.

The table and scatter plot show the fuel consumption rate, speed, and passenger capacity of various models of commercial airplanes.

Plane	Speed	Fuel (gal/hr)	Seats
DC-9-50	374	915	122
DC-9-10	381	743	71
F-100	383	646	97
DC-9-30	385	810	101
B737-100/200	389	824	113
B737-400	414	792	144
B737-300	417	776	131
MD-80	432	933	141
B727-200	437	1287	150
A320-100/200	460	820	148
B757-200	465	1050	186
A300-600	473	1678	249
B767-200ER	488	1409	181
B767-300ER	495	1602	214
L-1011-100/200	495	2428	310
DC-10-10	500	2287	289
DC-10-30	520	2667	265
B747-100	520	3638	447
B777	521	2117	292
MD-11	524	2462	253
B747-400	538	3410	396

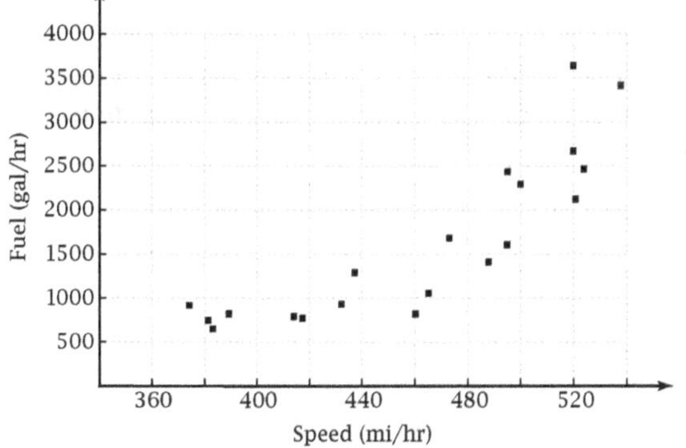

1. Based on the shape and concavity of the graph, explain why both an exponential and a power function would be reasonable mathematical models for fuel consumption rate as a function of speed.

2. Find the best-fitting exponential function and the best-fitting power function for the data. Write the equation for each function and the correlation coefficient. Plot both equations on the same screen as the scatter plot. Sketch each on the given figure. Based on the numerical and graphical evidence, which kind of function seems to fit better?

3. Make a residual plot for the exponential function. Sketch the result. Based on the residual plot, are the residuals caused by some systematic error in the assumption of an exponential function, or are they just the result of random differences?

4. Make a residual plot for the power function. Sketch the result. Based on the two residual plots, does one function seem to explain the variation of fuel consumption with speed better than the other, or are they about the same? How can you tell?

5. What did you learn as a result of doing this Exploration that you did not know before?

Exploration 81: Carbon Dioxide and Regression on Residuals

Date: _______________

Objective: Eliminate a pattern in a residual plot by finding a function that fits the residuals and adding it to the regression equation.

This data and table show the concentration of carbon dioxide in the atmosphere over a two-year period.

x (months)	y (ppm)	x (months)	y (ppm)
Jan. 1	323.8	Jan. 13	343.5
Feb. 2	324.8	Feb. 14	344.9
Mar. 3	325.9	Mar. 15	346.1
Apr. 4	327.2	Apr. 16	347.4
May 5	328.1	May 17	348.4
June 6	329.6	June 18	349.9
July 7	331.5	July 19	352.0
Aug. 8	333.5	Aug. 20	354.5
Sept. 9	335.5	Sept. 21	365.5
Oct. 10	338.0	Oct. 22	358.9
Nov. 11	340.2	Nov. 23	360.8
Dec. 12	342.1	Dec. 24	362.6

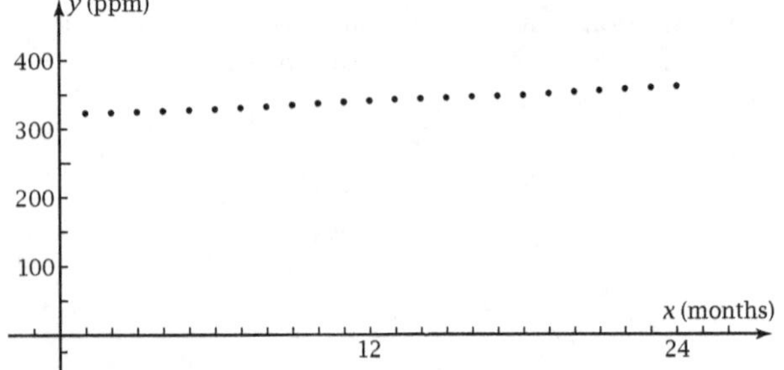

1. Show that an exponential function fits the data with a high correlation. Write the particular equation of this exponential function and store it as y_1.

2. Plot the exponential function and the scatter plot of the data on the same screen. Does the function seem to fit the points well?

3. Make a residual plot for the data and the exponential function. What pattern do you notice?

4. Do sinusoidal regression on the residuals. Write the particular equation and store it as y_2. Plot y_2 on the same screen as the residuals. Sketch the result.

5. Let y_3 be the sum of the exponential function and the residuals. That is, $y_3 = y_1 + y_2$. Plot y_3 on the same screen as the scatter plot of the original data, as you did in Problem 2. Can you tell that the y_3 function fits the data any better than y_1?

6. Make a residual plot for y_3 and the original data. Sketch the result here. What can you conclude from the resulting pattern?

7. What did you learn as a result of doing this Exploration that you did not know before?

Exploration 82: Counting Principles for "And" or "Or"

Objective: Calculate the number of elements in an event described by "and" or "or."

At a particular elementary school, there are 22 first-graders—12 girls and 10 boys. Suppose that one child is chosen at random from this class. Let A be the event "The child is a girl," and let B be the event "The child is a boy."

1. What do $n(A)$ and $n(B)$ equal, the number of ways of choosing a girl and the number of ways of choosing a boy? _____________________

2. What is $n(A \text{ or } B)$, the number of ways the child chosen could be a girl or a boy? _______________ How does this number relate to $n(A)$ and $n(B)$?

3. Events A and B are said to be **mutually exclusive.** Explain what this means.

Suppose that two children are selected at random, one after the other, from this class.

4. For each way of choosing a girl, there are ten ways of choosing a boy. Find $n(A \text{ and } B)$, the number of ways of choosing a girl and a boy. How does this number relate to $n(A)$ and $n(B)$?

5. Events A and B are also said to be **independent.** Why do you think they are called independent?

Three of the girls and four of the boys are redheads. The figure shows Event A, the person chosen is a girl, and Event C, the person chosen is a redhead.

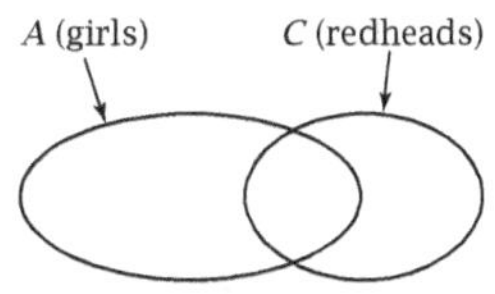

6. Write $n(C)$, the number of ways the one person chosen could be a redhead?

7. What does $n(A \text{ and } C)$ equal?

8. Events A and C are said to be **overlapping events.** Explain why, in this case, $n(A \text{ or } C)$ is *not* equal to $n(A) + n(C)$. How can $n(A \text{ or } C)$ be calculated from $n(A)$ and $n(C)$ for overlapping events?

9. Complete this sentence: "If X and Y are not mutually exclusive, then $n(X \text{ or } Y) = \text{—?—}$."

10. Complete this sentence: "If X and Y are independent, then $n(X \text{ and } Y) = \text{—?—}$."

11. What did you learn as a result of doing this Exploration that you did not know before?

Exploration 83: Probability of Various Permutations   Date: _________

Objective: Calculate the probability of various arrangements that can be made from the elements in a given set.

1. What is meant by a **permutation** of a set of objects?

2. How many different permutations can be made from the letters in SEQUOIA?

3. How many of the permutations of the letters in SEQUOIA begin with a vowel and end with a consonant?

4. What is the probability that a permutation of SEQUOIA picked at random begins with a vowel and ends with a consonant?

5. What is the probability that a permutation of SEQUOIA picked at random begins with a vowel and ends with another vowel?

6. If you pick three different letters at random from SEQUOIA, how many different three-letter "words" could be formed?

7. How many of the permutations in Problem 6 begin with the letter *S*?

8. How many different "words" can be made from the six letters in MESSES, considering that the three *S*s are the same and the two *E*s are the same?

9. In how many different ways could the letters in SEQUOIA be arranged in a circle if you are concerned only with which letters are to the left and right of which other letters, not where the letters appear on the circle?

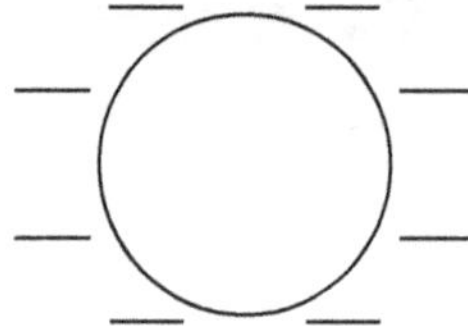

10. In how many different ways could seven children be arranged around a merry-go-round if one child sits in the middle and the other six sit around the outside?

11. What did you learn as a result of doing this Exploration that you did not know before?

Exploration 84: Properties of Probability

Objective: Calculate probabilities of various events from other probabilities.

Grades Problem: Assume that your probability of getting an "A" in English is $P(E) = 0.8$ and that your probability of getting an "A" in History is $P(H) = 0.7$.

1. How do you calculate $P(E \text{ and } H)$, the probability of getting an "A" in both English and History, assuming that the two events are independent?

2. Show that you *cannot* find $P(E \text{ or } H)$, the probability of getting an "A" in English or getting an "A" in History, just by adding the two probabilities.

3. How do you calculate $P(E \text{ or } H)$?

4. What is $P(\text{not } E)$, the probability that you do *not* get an "A" in English? What is $P(\text{not } H)$?

5. What is $P(\text{not } E \text{ and not } H)$, the probability that you do not get an "A" in English and do not get an "A" in History?

6. Find $P(E \text{ and not } H)$.

7. Find $P(H \text{ and not } E)$.

8. These four events are mutually exclusive. By adding their probabilities, show that there are no other possible events for this grades problem.

   E and H

   E and not H

   H and not E

   not E and not H

9. What is meant by the symbol $P(H|E)$? What can you tell about events E and H if $P(H|E) = P(H)$?

10. What is meant by the symbol $P(E \cup H)$? What is meant by the symbol $P(E \cap H)$? How can you calculate $P(E \cup H)$ from $P(E)$, $P(H)$, and $P(E \cap H)$?

11. What did you learn as a result of doing this Exploration that you did not know before?

Exploration 85: The Binomial Distribution

Date: _____________

Objective: Calculate and plot probabilities in a binomial probability distribution.

1. Have one member of your group flip a thumbtack 10 times. Based on the results, what does the probability seem to be that the tack lands "point up" on any one flip?

2. Based on your answer to Problem 1, if you were to flip the tack 10 more times, what would be the probability that it lands point up all 10 of those times?

3. What is the probability that the tack in Problem 2 lands point down all 10 times?

4. What is the probability that the tack in Problem 2 lands point up exactly 1 time?

5. What is the probability that the tack in Problem 2 lands point up exactly 0 or exactly 1 time?

6. Let $P(x)$ be the probability that the tack in Problem 2 lands point up an exact number of times x. Calculate these values:

   $P(0)$ _________________________________

   $P(1)$ _________________________________

   $P(2)$ _________________________________

   $P(3)$ _________________________________

   $P(4)$ _________________________________

   $P(5)$ _________________________________

   $P(6)$ _________________________________

   $P(7)$ _________________________________

   $P(8)$ _________________________________

   $P(9)$ _________________________________

   $P(10)$ ________________________________

7. Plot the graph of this probability distribution.

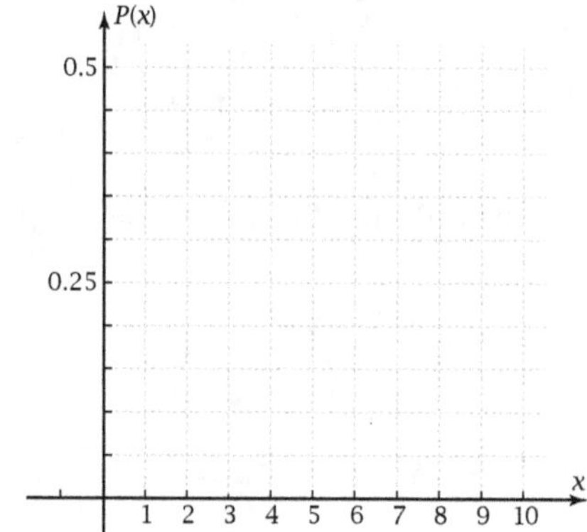

8. What is the probability that the number of point-downs is between 4 and 7, inclusive?

9. What did you learn as a result of doing this Exploration that you did not know before?

Exploration 86: Mathematical Expectation

Date: _____________

Objective: Calculate the "payoff" expected for a random experiment.

Grades Problem (Problems 1–3): Ernie is good in Mathematics and Languages but not as good in History and Chemistry. He figures that these are his probabilities of earning various grades in the five courses he is taking.

	A	B	C
Mathematics	95%	5%	
English	90%	10%	
Spanish	80%	15%	5%
Chemistry	70%	20%	10%
History	55%	30%	15%

1. Ernie gets 4 grade points for an A in any subject, 3 for a B, and 2 for a C. What is Ernie's mathematically expected number of grade points for the mathematics course?

2. Calculate Ernie's mathematically expected number of grade points in each of the other courses.

3. Ernie's GPA (grade point average) is the total number of grade points made divided by the total number of courses. What is his mathematically expected GPA?

Bent Coin Problem (Problems 4 and 5): You "pay" 10 tokens for a game. You flip a bent coin 5 times. The probability of heads on any one flip is 40%.

4. Calculate your probabilities of getting heads exactly 0, 1, 2, 3, 4, and 5 times in the five flips.

   $P(0) = $ ______________________________________

   $P(1) = $ ______________________________________

   $P(2) = $ ______________________________________

   $P(3) = $ ______________________________________

   $P(4) = $ ______________________________________

   $P(5) = $ ______________________________________

5. Your payoff for five heads is 100 tokens (less the 10 tokens you paid to play the game). Your payoff for four heads is 50 tokens and for three heads is 20 tokens (in both instances, less the 10 tokens you "paid" to play). For any other number of heads, you win nothing (and lose your 10 tokens). Calculate the mathematical expectation for this random experiment.

6. What did you learn as a result of doing this Exploration that you did not know before?

Exploration 87: Introduction to Three-Dimensional Vectors

Objective: Plot and do operations with three-dimensional vectors by using their components.

1. Draw vector $\vec{v} = 5\vec{i} + 7\vec{j} + 10\vec{k}$ as a position vector in three-dimensional coordinates. Do this by plotting the three components head to tail, starting with $5\vec{i}$. Show the "box" surrounding the vector that makes the vector look three-dimensional.

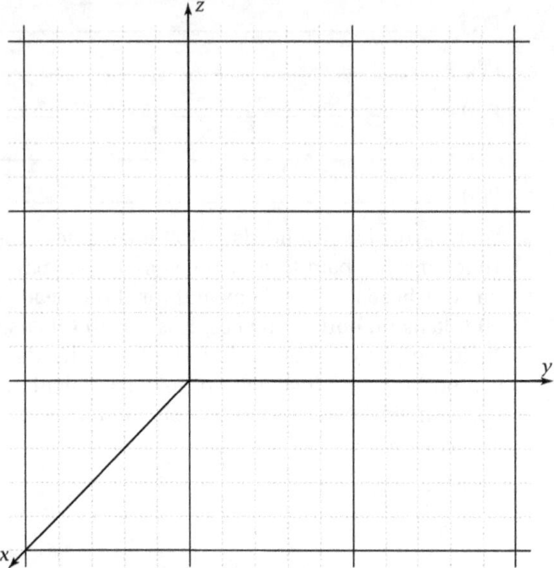

2. Show that you know what the three unit vectors $\vec{i}, \vec{j},$ and $\vec{k}$ mean by sketching them starting from the origin in Problem 1.

3. Find the length of $\vec{v}$ in Problem 1 using the three-dimensional Pythagorean theorem.

4. Write the vector $3\vec{v}$ as a sum of components.

5. Show by means of the three-dimensional Pythagorean theorem that $3\vec{v}$ is really three times as long as $\vec{v}$.

6. Write as a sum of components a *unit* vector in the direction of $\vec{v}$.

7. Let $\vec{a} = 4\vec{i} - 3\vec{j} + 8\vec{k}$. Quickly find $\vec{v} + \vec{a}$, where $\vec{v} = 5\vec{i} + 7\vec{j} + 10\vec{k}$, as in Problem 1.

8. Find displacement vector $\vec{b}$ from the point (1, 5, 2) to the point (7, 3, 11).

9. Find the position vector of the point 0.3 of the way from the point (1, 5, 2) to the point (7, 3, 11).

10. What did you learn as a result of doing this Exploration that you did not know before?

Exploration 88: Introduction to the Scalar Product of Two Vectors

Objective: With guidance from your instructor or text, find out the meaning of the **inner product** of two vectors.

1. The figure shows two vectors pointing in the same direction, $\vec{a}$ 8 units long and $\vec{b}$ 6 units long. What is the **dot product,** $\vec{a} \cdot \vec{b}$, of these two vectors?

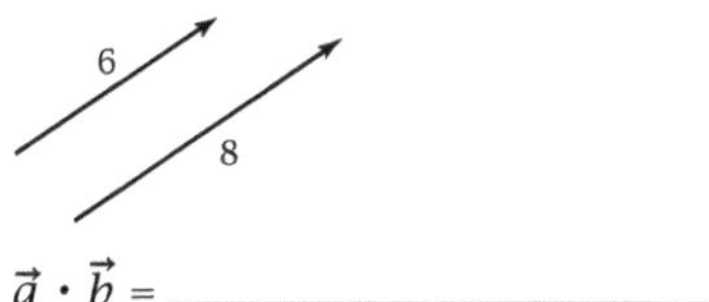

$\vec{a} \cdot \vec{b} =$ _______________

2. The figure shows two vectors as in Problem 1 but with $\vec{b}$ pointing in the direction *opposite* $\vec{a}$. Find the dot product.

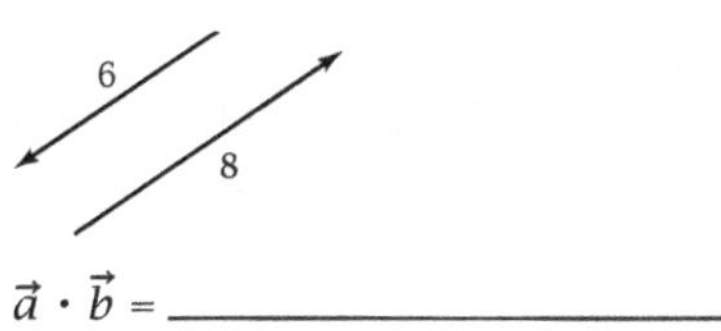

$\vec{a} \cdot \vec{b} =$ _______________

3. The figure shows two vectors, $\vec{a}$ and $\vec{b}$, at right angles to each other. Find the dot product.

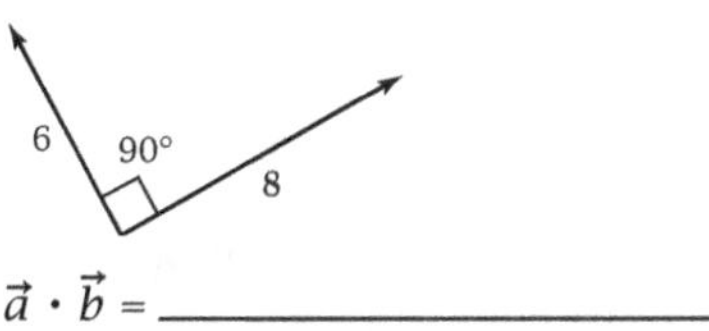

$\vec{a} \cdot \vec{b} =$ _______________

4. The figure shows two vectors, $\vec{a}$ and $\vec{b}$, placed tail-to-tail, with an angle of 50° between them. Find the dot product.

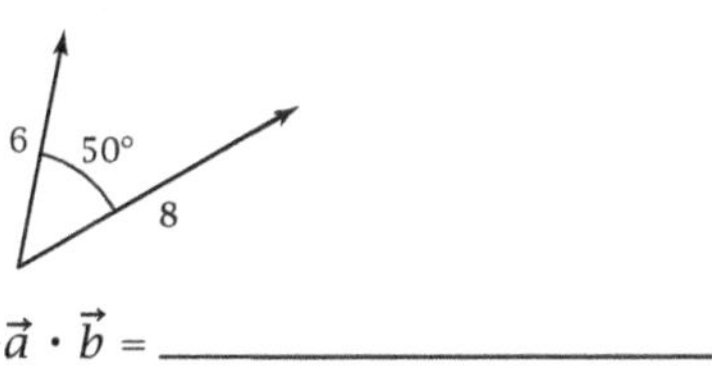

$\vec{a} \cdot \vec{b} =$ _______________

5. The figure shows two vectors with an angle of 110° between them when placed tail-to-tail. Find the dot product.

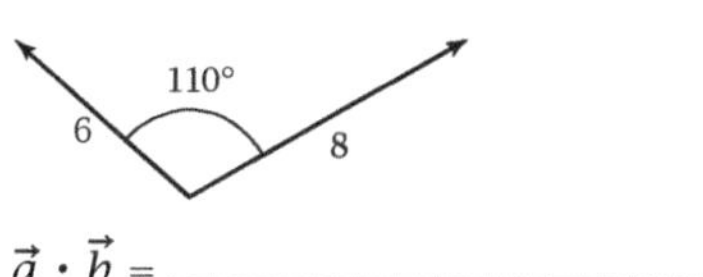

$\vec{a} \cdot \vec{b} =$ _______________

6. State the definition of *dot product*.

7. Find:

$\vec{i} \cdot \vec{i} =$ _________, $\vec{j} \cdot \vec{j} =$ _________, $\vec{k} \cdot \vec{k} =$ _________

$\vec{i} \cdot \vec{j} =$ _________, $\vec{j} \cdot \vec{k} =$ _________, $\vec{i} \cdot \vec{k} =$ _________

8. Let $\vec{a} = 2\vec{i} + 5\vec{j} + 7\vec{k}$.

Let $\vec{b} = 9\vec{i} + 3\vec{j} + 4\vec{k}$.

Find $\vec{a} \cdot \vec{b}$.

9. Give two other names for *dot product*.

10. From what you found in Problem 8, there is an easy way to find the dot product of two vectors. Show that you have discovered this way by finding quickly the dot product $\vec{c} \cdot \vec{d}$.

$$\vec{c} = 4\vec{i} - 6\vec{j} + 9\vec{k}$$
$$\vec{d} = 2\vec{i} + 5\vec{j} - 3\vec{k}$$

11. On the other side of this page, find $|\vec{c}|$ and $|\vec{d}|$. Use the result and the definition of *dot product* to calculate the angle between $\vec{c}$ and $\vec{d}$.

Exploration 89: Projections of Vectors

Objective: Use the dot product to find the **scalar projection** and the **vector projection** of one vector on another.

The figure shows $\vec{p}$, the **vector projection of $\vec{a}$ on $\vec{b}$.** It is called a *projection* because it is the same as the shadow that $\vec{a}$ would cast on $\vec{b}$ if a light were shined perpendicular to $\vec{b}$ in the plane of vectors $\vec{a}$ and $\vec{b}$.

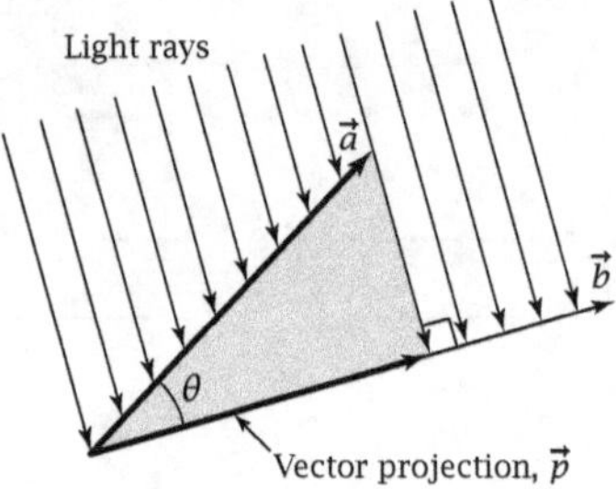

1. If $\vec{a}$ is 5 units long and forms an angle $\theta = 33°$ with $\vec{b}$, how long would $\vec{p}$ be? Use trigonometry to find this **scalar projection** of $\vec{a}$ on $\vec{b}$. Store the answer as p.

2. If $\vec{b} = 9\vec{i} + 3\vec{j} + 4\vec{k}$, how long is $\vec{b}$? Write a unit vector $\vec{u}$ in the direction of $\vec{b}$. Show $\vec{u}$ on the given figure.

3. The vector projection $\vec{p}$ is p times the unit vector $\vec{u}$. Write $\vec{p}$ in terms of its components.

4. Let $\vec{b} = 9\vec{i} + 3\vec{j} + 4\vec{k}$ (as in Problem 2).

   Let $\vec{c} = 2\vec{i} + 5\vec{j} + 7\vec{k}$.

   Find the angle θ between $\vec{b}$ and $\vec{c}$ when they are placed tail-to-tail. Store the answer.

5. Let $\vec{p}$ be the vector projection of $\vec{c}$ on $\vec{b}$. Use the length of $\vec{c}$ and the angle θ from Problem 4 to find p, the scalar projection of $\vec{c}$ on $\vec{b}$. Store it.

6. Find $\vec{p}$, the vector projection of $\vec{c}$ on $\vec{b}$. Write the answer in terms of its components.

7. The figure shows vector $\vec{v}$ 6 units long making an angle of 140° with $\vec{b}$. Draw $\vec{p}$, the vector projection of $\vec{v}$ on $\vec{b}$.

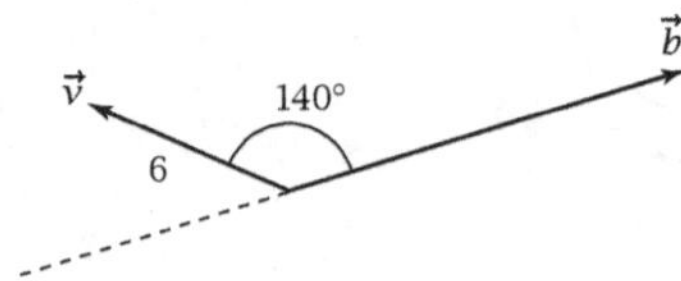

8. Let $\vec{b} = 9\vec{i} + 3\vec{j} + 4\vec{k}$, as in Problems 2 and 4. From Problems 1 and 5 you learned that the scalar projection of $\vec{v}$ on $\vec{b}$ will be

   $$p = |\vec{v}| \cos \theta$$

   Calculate p. Explain the significance of the fact that p is *negative* in this case.

9. Calculate the vector projection of $\vec{v}$ on $\vec{b}$. Write $\vec{p}$ in terms of its components.

10. Explain why the scalar projection of one vector on another is not always equal to the length of the vector being projected and why, therefore, the symbol $|\vec{p}|$ cannot be used for scalar projection.

11. Let $\vec{r} = 3\vec{i} - 5\vec{j} + 2\vec{k}$.

    Let $\vec{s} = 4\vec{i} + 8\vec{j} - 7\vec{k}$.

    Find the vector projection of $\vec{r}$ on $\vec{s}$.

12. What is a relatively quick way to calculate scalar and vector projections using the dot product and the length of the vector that you're projecting onto? Show that you get the same answer as in Problem 11.

13. What did you learn as a result of doing this Exploration that you did not know before?

Exploration 90: Summary of Vector Properties

Date: _____________

Objective: Find the sum, difference, and dot product of 3-D vectors and apply them.

1. Draw a sketch showing the geometrical meaning of the sum of two vectors, $\vec{a} + \vec{b}$.

2. Draw a sketch showing the geometrical meaning of the difference of two vectors, $\vec{a} - \vec{b}$.

3. Let $\vec{a} = 3\vec{i} + 8\vec{j} + 4\vec{k}$.

   Let $\vec{b} = 2\vec{i} - 7\vec{j} + 9\vec{k}$.

   Draw $\vec{a}$ on the given axes. Show the "box" that makes it look three-dimensional.

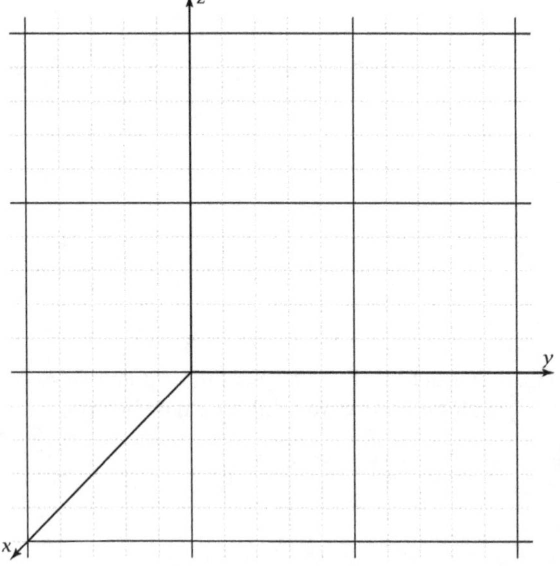

4. Find $|\vec{a}|$ and $|\vec{b}|$.

5. Find $\vec{a} + \vec{b}$ and $\vec{a} - \vec{b}$.

6. If the tail of $\vec{b}$ is placed at the head of $\vec{a}$, how far is it from the tail of $\vec{a}$ to the head of $\vec{b}$?

7. If $\vec{a}$ and $\vec{b}$ are placed tail-to-tail, write the vector that goes from the head of $\vec{a}$ to the head of $\vec{b}$.

8. Find the dot product, $\vec{a} \cdot \vec{b}$.

9. Give two other names for *dot product*.

10. Is the angle between $\vec{a}$ and $\vec{b}$ acute or obtuse? How can you tell?

11. Find the degree measure of the angle between $\vec{a}$ and $\vec{b}$ when they are placed tail-to-tail.

12. Draw a sketch illustrating $\vec{p}$, the vector projection of $\vec{a}$ on $\vec{b}$.

13. Find p, the scalar projection of $\vec{a}$ on $\vec{b}$. Explain why p is negative in this case.

14. Find a unit vector, $\vec{u}$, in the direction of $\vec{b}$.

15. Find the vector projection of $\vec{a}$ on $\vec{b}$.

16. Sketch the position vector, $\vec{v}$, to the point that is 0.6 of the way from the head of $\vec{a}$ to the head of $\vec{b}$ when $\vec{a}$ and $\vec{b}$ are placed tail-to-tail. Calculate the components of $\vec{v}$.

17. What did you learn as a result of doing this Exploration that you did not know before?

Exploration 91: Equation of a Plane Normal to a Vector

Objective: Given a point on a plane and a vector normal to the plane, find a Cartesian equation of the plane.

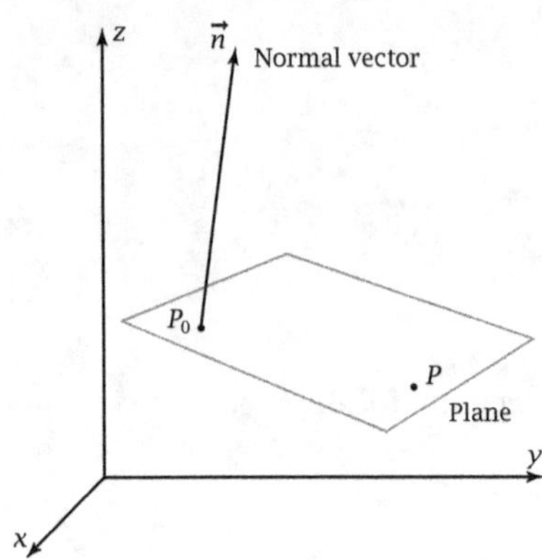

The figure shows, in general, a plane and its normal vector (vector perpendicular to the plane). Assume that the normal vector is $\vec{n} = 3\vec{i} + 7\vec{j} + 10\vec{k}$ and that it starts at point $P_0(8, 9, 4)$ on the plane.

1. Let $P(x, y, z)$ be a (variable) point on the plane. Draw displacement vector $\overrightarrow{P_0P}$.

2. Explain why the dot product $\vec{n} \cdot \overrightarrow{P_0P}$ equals 0.

3. Write $\overrightarrow{P_0P}$ in terms of the coordinates $P_0(8, 9, 4)$ and $P(x, y, z)$.

4. Substitute the answer to Problem 3 and the components of $\vec{n}$ into the equation $\vec{n} \cdot \overrightarrow{P_0P} = 0$. Simplify the result.

5. The equation in Problem 4 is a Cartesian equation of the plane shown. By observing the pattern in the answer, show how you can get the equation *quickly*.

6. Use the equation of the plane to find the z-coordinate of the point for which $x = 5$ and $y = 2$.

7. Find the x-intercept of the plane, the value of x where the plane crosses the x-axis.

8. Explain what is meant by a *variable* point in the plane.

9. What did you learn as a result of doing this Exploration that you did not know before?

Exploration 92: Introduction to the Cross Product

Objective: Discover the meaning of and way of computing the cross product of two vectors.

Let $\vec{a} = 3\vec{i} + 5\vec{j} + 7\vec{k}$.

Let $\vec{b} = 11\vec{i} + 2\vec{j} + 13\vec{k}$.

The vector $\vec{c} = 51\vec{i} + 38\vec{j} - 49\vec{k}$ is the **cross product** of $\vec{a}$ and $\vec{b}$, written $\vec{a} \times \vec{b}$. In this Exploration, it is your objective to find out the meaning of *cross product* and how it is calculated.

1. Find $|\vec{a}|$, $|\vec{b}|$, and $|\vec{a} \times \vec{b}|$. Does the length of the cross product vector equal the product of the lengths of the two factors?

2. Find the angle θ between $\vec{a}$ and $\vec{b}$ when they are placed tail-to-tail. Show that
$$|\vec{a} \times \vec{b}| = |\vec{a}||\vec{b}| \sin \theta$$

3. Find $(\vec{a} \times \vec{b}) \cdot \vec{a}$ and $(\vec{a} \times \vec{b}) \cdot \vec{b}$. From the answers, what do you conclude about the direction of the cross product with respect to the direction of the two vectors being cross multiplied?

4. Look up the **right-hand rule** in your textbook. When all members of your group understand it, have one representative of your group demonstrate it to your instructor. _________

5. State the formal definition of *cross product*.

6. Explain why $\vec{i} \times \vec{i}$, $\vec{j} \times \vec{j}$, and $\vec{k} \times \vec{k}$ all equal zero.

7. Explain why $\vec{i} \times \vec{j} = \vec{k}$, $\vec{j} \times \vec{k} = \vec{i}$, and $\vec{k} \times \vec{i} = \vec{j}$.

8. Explain why $\vec{j} \times \vec{i}$ is the *opposite* of $\vec{i} \times \vec{j}$.

9. For $\vec{a}$ and $\vec{b}$ at the beginning of this Exploration, calculate $\vec{a} \times \vec{b}$ by distributing each term in the first factor to each term in the second factor. Write the nine terms in the answer on three lines, as you did the first time you calculated a dot product. Simplify the result by taking advantage of the results of Problems 6, 7, and 8, but leave the nine terms in the same three lines.

(Over)

Exploration 92: Introduction to the Cross Product *continued*

10. Simplify the answer to Problem 9 by combining like terms and thereby show that the answer really is the vector $\vec{c}$ shown before Problem 1.

11. Based on your work in Problem 9, why do you think that the cross product is also called the **outer product**?

12. Why is the cross product also called the **vector product**?

13. The coefficients of vectors $\vec{a}$ and $\vec{b}$ can be written as a **determinant** with the three unit vectors as the top row, like this:

$$\begin{vmatrix} \vec{i} & \vec{j} & \vec{k} \\ 3 & 5 & 7 \\ 11 & 2 & 13 \end{vmatrix}$$

Expand this determinant, thereby showing that the answer is $\vec{a} \times \vec{b}$.

14. Find $\left(4\vec{i} - 5\vec{j} + 8\vec{k}\right) \times \left(3\vec{i} + 6\vec{j} - 7\vec{k}\right)$ quickly.

15. What did you learn as a result of doing this Exploration that you did not know before?

Exploration 93: Dot and Cross Product, and Their Uses

Objective: Find dot and cross products of vectors, and use the results to find angles, equations of planes, and areas.

For this Exploration, let these be three points in space:

$A(5, 7, 3)$

$B(4, -2, 6)$

$C(2, -6, 1)$

1. Find vectors $\overrightarrow{AB}$ and $\overrightarrow{AC}$.

2. Find $\overrightarrow{AB} \cdot \overrightarrow{AC}$. Show your calculation.

3. Find the angle θ between $\overrightarrow{AB}$ and $\overrightarrow{AC}$ if they are placed tail-to-tail. Store the answer.

4. Find $\overrightarrow{AB} \times \overrightarrow{AC}$ using your program that calculates cross products.

5. Show that $|\overrightarrow{AB} \times \overrightarrow{AC}|$ really *does* equal $|\overrightarrow{AB}||\overrightarrow{AC}| \sin \theta$, as the definition tells you.

6. Show that $\overrightarrow{AB} \times \overrightarrow{AC}$ is really perpendicular to $\overrightarrow{AB}$ and $\overrightarrow{AC}$.

7. Use normal vector $\vec{n} = \overrightarrow{AB} \times \overrightarrow{AC}$ and point A to find the particular equation of the plane containing point A and vectors $\overrightarrow{AB}$ and $\overrightarrow{AC}$.

8. Show algebraically that the plane in Problem 7 really *does* contain point C.

9. Find the x-intercept of the plane in Problem 7.

10. Find the area of the triangle in space that has points A, B, and C as its vertices.

11. What did you learn as a result of doing this Exploration that you did not know before?

Exploration 94: Direction Angles and Direction Cosines

Objective: Find the angles a vector makes with the coordinate axes.

The figure shows vector $\vec{v} = 4\vec{i} + 5\vec{j} + 12\vec{k}$. Vector $\vec{v}$ makes angles of α, β, and γ (Greek letters alpha, beta, and gamma, respectively) with the x-, y-, and z-axes, respectively. These angles are called the **direction angles** of $\vec{v}$. In this exploration, you will learn about these angles and about their cosines, the **direction cosines** of $\vec{v}$.

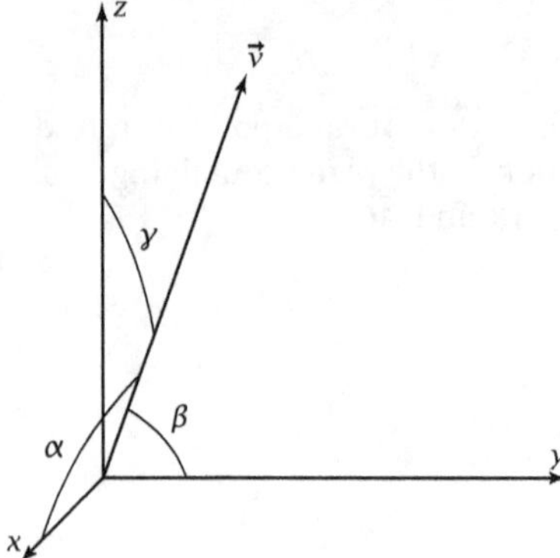

1. Use dot product to find the degree measures of α, β, and γ. Store these answers, without round-off.

2. Find $\cos^2 \alpha + \cos^2 \beta + \cos^2 \gamma$ using the stored direction angles from Problem 1. What interesting thing do you notice about the answer?

3. In Problem 2, you discovered the **Pythagorean property** of direction cosines. Write $\cos \alpha$, $\cos \beta$, and $\cos \gamma$ exactly in radical form. The work you did in Problem 1 should help. Then square the answers and add them, thereby showing again the Pythagorean property.

4. Find a unit vector, $\vec{u}$, in the direction of $\vec{v}$. Show that this vector is

$$\vec{u} = c_1\vec{i} + c_2\vec{j} + c_3\vec{k}$$

where c_1, c_2, and c_3 are the direction cosines of $\vec{v}$.

5. Show that $\vec{a} = \frac{9}{17}\vec{i} - \frac{8}{17}\vec{j} + \frac{12}{17}\vec{k}$ is a unit vector. Then use your observation in Problem 4 to find the direction angles quickly.

6. Vector $\vec{b}$ has direction cosines $c_1 = 0.3$ and $c_2 = -0.4$. Calculate the two possible values of c_3. Hold three pencils in space, erasers touching, to illustrate how there can be two values of the third direction angle when you know the other two.

7. What did you learn as a result of doing this Exploration that you did not know before?

Exploration 95: Vector Equation of a Line in Space

Date: _____________

Objective: Use vectors to write the equation of a line in space.

The figure shows a general line in space. Assume that the fixed point $P_0(4, 9, 11)$ is on the line. Position vector $\vec{r}$ goes to the variable point $P(x, y, z)$ on the line. P is at a directed distance d units from P_0. Your goal is to get an equation expressing $\vec{r}$ in terms of d so that you can calculate the coordinates of various points on the line.

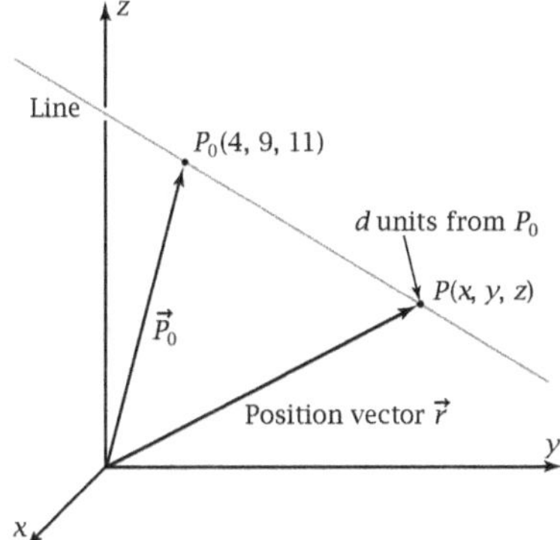

1. Let $\vec{u} = \frac{3}{7}\vec{i} + \frac{6}{7}\vec{j} + \frac{2}{7}\vec{k}$ be a vector in the direction of the line. Prove that $\vec{u}$ is a *unit* vector.

2. Write the direction cosines of vector $\vec{u}$.

3. Show that you understand the definitions of $\vec{u}$ and d by drawing the vector $d\vec{u}$ at an appropriate place on the figure.

4. Let vector $\vec{P_0}$ be the position vector to point P_0. Explain why vector $\vec{r}$ is given by

$$\vec{r} = \vec{P_0} + d\vec{u}$$

5. Substitute the coordinates for P_0 and for $\vec{u}$ into the equation from Problem 4. Combine like terms to get an equation for $\vec{r}$ in terms of its components.

6. Find the point on the line for which $d = 21$.

7. How far from P_0, and in which direction, is the point on the line for which $x = 50$?

8. Find the point on the line for which $x = 50$.

9. The line intersects the xy-plane at a point where $z = 0$. Find the value of d for which $z = 0$. Use the answer to find this intersection point.

10. What did you learn as a result of doing this Exploration that you did not know before?

Exploration 96: Intersection of a Line and a Plane in Space

Date: _______________

Objective: Find the point where a line given by a vector equation intersects a plane in space.

The figure shows, in general, a line through the point P_0 in the direction of the unit vector $\vec{u}$. Assume that the coordinates of point P_0 are (7, 11, 3) and that the unit vector is $\vec{u} = \frac{3}{7}\vec{i} + \frac{6}{7}\vec{j} + \frac{2}{7}\vec{k}$. The line intersects the plane at the point whose coordinates (x, y, z) you need to find.

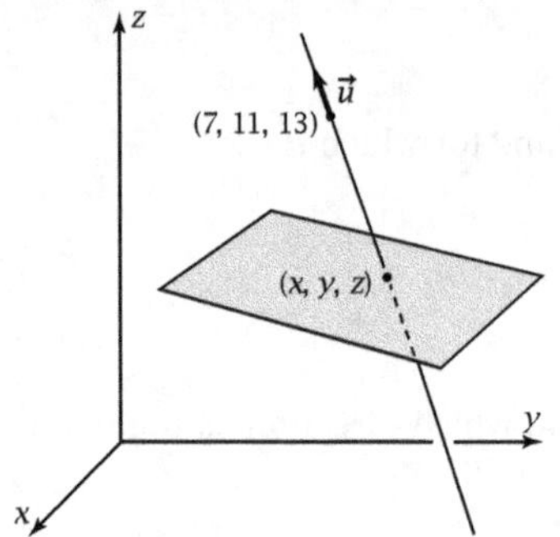

1. Prove that $\vec{u}$ *is* a unit vector.

2. Write the vector equation of the line.

3. Assume that points $A(5, 10, 1)$, $B(6, 3, 6)$, and $C(12, 1, 4)$ are on the plane shown in the figure.

   Find:

   $\overrightarrow{AB} =$ _______________________________

   $\overrightarrow{AC} =$ _______________________________

4. Find a vector normal to the plane.

5. Find an equation of the plane containing A, B, and C.

6. At the point where the line intersects the plane, the $\vec{i}$-, $\vec{j}$-, and $\vec{k}$-coefficients of the line equal the corresponding values of x, y, and z for the plane. By substituting these coefficients for x, y, and z in the equation for the plane, find the directed distance, d, from the known point (7, 11, 3) to the intersection point (x, y, z).

7. Find the x-, y-, and z-coordinates of the intersection point.

8. What did you learn as a result of doing this Exploration that you did not know before?

Exploration 97: Other Equations of a Line in Space

Objective: Learn about direction numbers, parametric equations, and standard equations of a line in space.

The figure shows a general line in space. Assume that the fixed point $P_0(4, 9, 11)$ is on the line and the line is parallel to the fixed **direction vector** $\vec{v} = 2\vec{i} + 5\vec{j} - 3\vec{k}$. Position vector $\vec{r}$ goes to variable point $P(x, y, z)$ on the line.

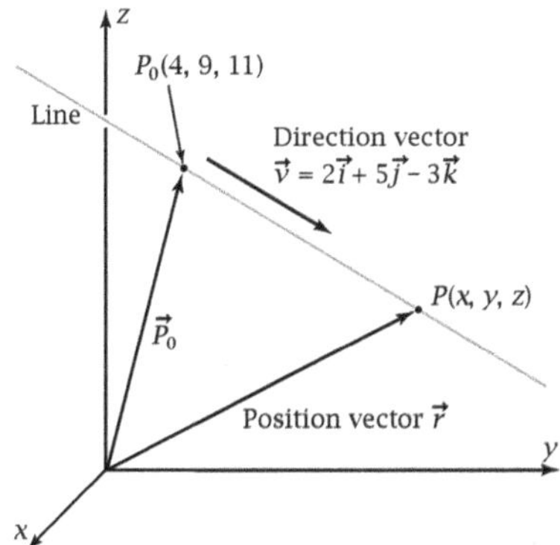

1. A vector equation for point P on the line can be written

   $$\vec{r} = \vec{P}_0 + t\vec{v}$$

   where t is a **parameter** equal to the number of vector $\vec{v}$ lengths that can be placed along the segment from P_0 to P. Find position vector $\vec{r}$ if $t = 2$ and if $t = -1.7$.

2. How far is P from P_0 if $t = 2$, as in Problem 1?

3. The numbers 2, 5, and -3 in $\vec{v} = 2\vec{i} + 5\vec{j} - 3\vec{k}$ are called **direction numbers** for $\vec{v}$. What is the relationship between direction numbers for a vector and direction cosines?

4. Show that by using direction numbers, you can write the equation for variable point P on the line as

   $$\vec{r} = (4 + 2t)\vec{i} + (9 + 5t)\vec{j} + (11 - 3t)\vec{k}$$

5. Because $|\vec{v}| = \sqrt{38}$, another vector equation for the line is

   $$\vec{r} = \left(4 + \frac{2}{\sqrt{38}}d\right)\vec{i} + \left(4 + \frac{5}{\sqrt{38}}d\right)\vec{j} + \left(4 + \frac{-3}{\sqrt{38}}d\right)\vec{k}$$

   What advantage does the equation in Problem 4 have over this equation? What advantage does this equation have over the one in Problem 4?

6. Because $\vec{r} = x\vec{i} + y\vec{j} + z\vec{k}$, the equation for $\vec{r}$ in Problem 4 allows you to write **parametric equations** for the line in space.

   $$x = 4 + 2t$$
   $$y = 9 + 5t$$
   $$z = 11 - 3t$$

   Show that the parametric equations give an answer equivalent to Problem 1 if $t = 2$.

7. Show that the parametric equations in Problem 6 can be used to get the **symmetric equations**

   $$\frac{x - 4}{2} = \frac{y - 9}{5} = \frac{z - 11}{-3}$$

(Over)

8. Use the symmetric equations in Problem 7 to find the z-coordinate of a point on the line if $x = 7$.

10. State one advantage of using the symmetric equations of a line instead of the vector or parametric equations.

9. State one advantage of using parametric equations for a line instead of the vector equation.

11. What did you learn as a result of doing this Exploration that you did not know before?

Exploration 98: Positive Normal Vector to a Plane

Date: _____________

Objective: Find out whether a given point is on the same side or on the opposite side of a given plane as the origin.

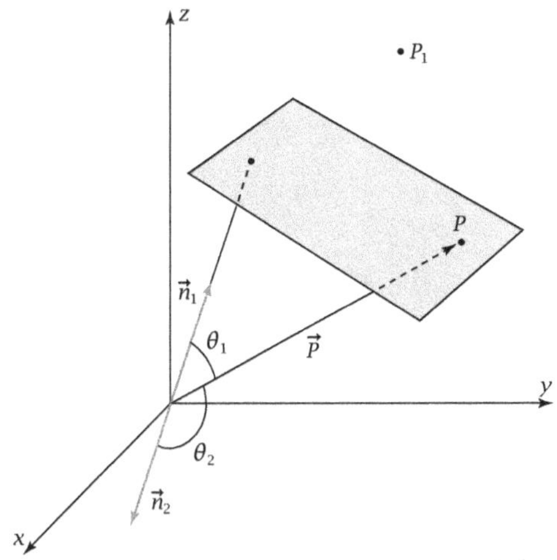

The figure shows a general plane in space. Vector $\vec{n}_1$ is a **positive normal vector** (that is, it points from the origin toward the plane). Vector $\vec{n}_2$ is a **negative normal vector.** Point $P(x, y, z)$ is on the plane.

1. If the equation of the plane is

   $$6x + 5y - 4z = -12$$

   find the coordinates of a point P on the plane. You could do this by substituting 1 for x and 2 for y and then calculating z.

2. Write a vector $\vec{n}$ normal to the plane. Write the position vector $\vec{P}$ to point P. Then find the scalar product $\vec{n} \cdot \vec{P}$.

3. Based on your answer to Problem 2, explain why the normal vector you wrote is a *negative* normal vector. Write a positive normal vector to the plane.

4. Transform the equation in Problem 1 so that you use the coefficients of the *positive* normal vector in it. What do you notice about the sign of the constant term, D, in the transformed equation?

5. Point $P_1(x_1, y_1, z_1)$ is a fixed point not on the plane. Draw vector $\overrightarrow{PP_1}$.

6. Suppose that P_1 has coordinates (10, 20, 30). Find $\overrightarrow{PP_1} \cdot \vec{n}$, where $\vec{n}$ is the positive normal vector to the plane.

7. Is the angle between $\overrightarrow{PP_1}$ and $\vec{n}$ acute or obtuse? What does this information tell you about whether P_1 lies on the same side of the plane as the origin or on the opposite side?

8. Given the plane $3x + 6y + 2z = 36$, state whether the point $P_1(1, 2, 3)$ lies on the same side of the origin as the plane or on the opposite side.

9. What did you learn as a result of doing this exercise that you did not know before?

Exploration 99: Distance Between a Plane and a Point

Objective: Use vectors to find the distance between a given plane and a given point in space.

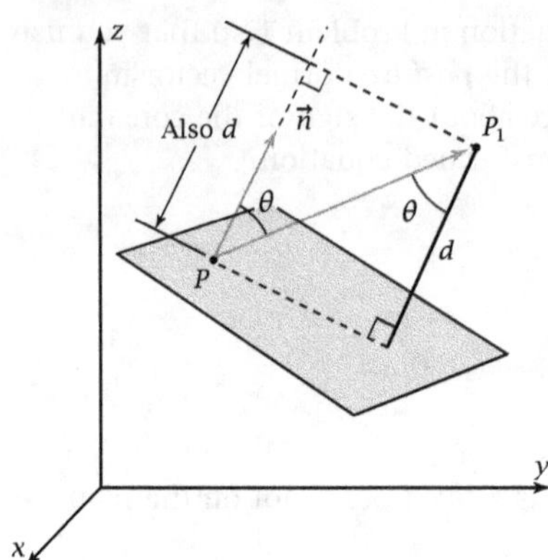

The figure shows a general plane in space with its normal vector $\vec{n}$. You need to find the (perpendicular) distance d between the plane and point P_1 (not on the plane).

1. If the equation of the plane is

$$6x + 5y + 4z = 28$$

find the coordinates of a point P on the plane. You could do this by substituting 1 for x and 2 for y and then calculating z.

2. If P_1 is (11, 22, 14), find vector $\overrightarrow{PP_1}$.

3. Quick: Write vector $\vec{n}$.

4. Find angle θ between $\vec{n}$ and $\overrightarrow{PP_1}$.

5. Explain why the angle at P_1 between $\overrightarrow{PP_1}$ and the perpendicular to the plane is also equal to θ.

6. Find d using trigonometry on the right triangle shown in the figure.

7. If you go through the computations in Problems 4–6 without actually doing the arithmetic, many parts cancel out. Show how to calculate d using a time-efficient method.

8. Demonstrate that you understand the significance of what you have learned by calculating quickly the distance between plane $7x + 3y + 10z - 30 = 0$ and the point (−9, −2, 17). (You may substitute $x = 1$ and $y = 1$ to find z for point P.)

9. What did you learn as a result of doing this Exploration that you did not know before?

Exploration 100: Distance Between a Line and a Point in Space

Date: _____________

Objective: Use vectors to find the distance between a given line and a given point in space.

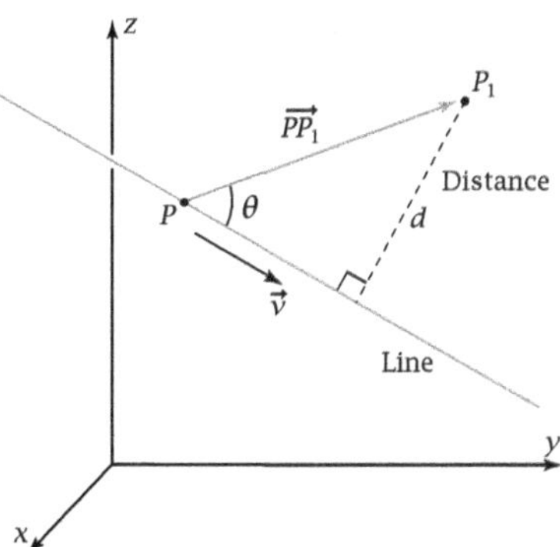

The figure shows a general line in space in the direction of vector $\vec{v}$. You need to find the (perpendicular) distance d between the line and point P_1 not on the line.

1. Vector $\overrightarrow{PP_1}$ goes from point P on the line to point P_1. Using right triangle trigonometry, find d in terms of the hypotenuse $\left|\overrightarrow{PP_1}\right|$ and the angle θ.

2. From the definition of *cross product*, recall

$$\left|\vec{v} \times \overrightarrow{PP_1}\right| = \left|\vec{v}\right|\left|\overrightarrow{PP_1}\right| \sin \theta$$

Use this information and the answer to Problem 1 to write an equation for d in terms of $\vec{v}$ and $\overrightarrow{PP_1}$.

3. Suppose that a line has the vector equation

$$\vec{r} = \left(5 + \frac{6}{11}t\right)\vec{i} + \left(3 - \frac{2}{11}t\right)\vec{j} + \left(-1 + \frac{9}{11}t\right)\vec{k}$$

Why do you think the variable t is used in the equation instead of the more familiar variable d?

4. For the line in Problem 3, give a point P on the line and a vector $\vec{v}$ in the direction of the line.

5. If point P_1 (not on the line) has coordinates $(-3, 2, 5)$, write $\overrightarrow{PP_1}$.

6. Find $\vec{v} \times \overrightarrow{PP_1}$ and $\left|\vec{v} \times \overrightarrow{PP_1}\right|$.

7. Show that $\vec{v}$ in Problem 4 is a *unit* vector. Use this fact and the results of previous problems to find out how far point P_1 is from the line.

8. What did you learn as a result of doing this Exploration that you did not know before?

Exploration 101: Using Vectors on a Hip Roof

Date: __________

Objective: Use vectors to find angles, distances, and areas of surfaces of a hip roof.

The figure shows a "hip roof" covering a building that measures 12 feet wide and 20 feet long. The roof rises to 4 feet above the ceiling of the building, which is in the *xy*-plane of a three-dimensional coordinate system. The two triangular faces of the roof have upper vertices that are 5 feet horizontally from the respective ends of the building.

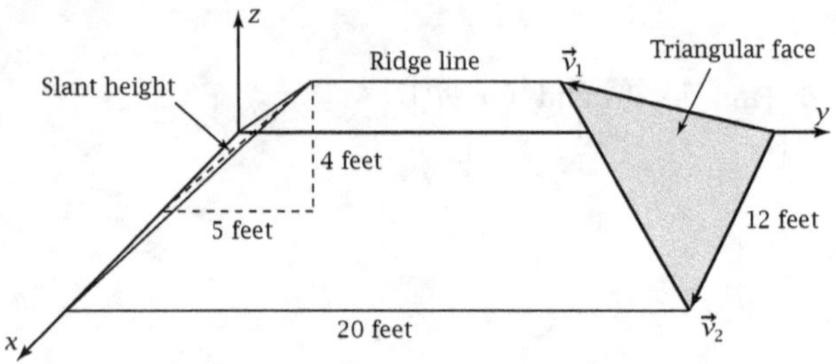

1. Vector $\vec{v}_1$ goes from one corner of the building to the end of the ridge line (see the figure). What are the coordinates of the beginning (the tail) of $\vec{v}_1$? What are the coordinates of the end (the head) of $\vec{v}_1$? Write $\vec{v}_1$ in terms of the unit vectors $\vec{i}, \vec{j}$, and $\vec{k}$.

2. Find the length of the rafters that will go along the joint between the triangular faces and the trapezoidal faces of the roof.

3. Vector $\vec{v}_2$ goes from one corner of the building to the next corner along the width of the building. Write $\vec{v}_2$ in terms of $\vec{i}, \vec{j}$, and $\vec{k}$.

4. Find the angle between $\vec{v}_1$ and $\vec{v}_2$.

5. Find $\vec{v}_1 \times \vec{v}_2$.

6. How many square feet of roofing will be needed to cover each triangular face of the roof?

7. Find the slant height of the triangular faces by finding the hypotenuse of the dotted triangle shown in the figure.

8. Will the slant height of the trapezoidal faces be equal to the slant height of the triangular faces?

9. Find a unit vector in the direction of $\vec{v}_1$.

10. Write a vector equation of the line in space that goes along vector $\vec{v}_1$.

11. A wire will run through the attic under the hip roof from the origin to a point on the line in Problem 10 that is 3 units from (0, 20, 0). What is the shortest length the wire can be?

Exploration 101: Using Vectors on a Hip Roof *continued*   Date: _______________

12. Construct the roof by drawing on graph paper the two triangular pieces and the two trapezoidal pieces. Tape the pieces together.

13. Do the base angles of the triangular pieces really equal the value you calculated in this Exploration? _________

14. Will the lengths of the rafters in Problem 2 really equal the length you calculated? _________

15. Is the ridge line of the roof really 4 feet above the xy-plane? _________

16. What did you learn as a result of doing this Exploration that you did not know before?

Exploration 102: Vectors Review

Objective: Show that you understand dot and cross products of vectors applied to lines and planes in space.

Problems 1–14 refer to the line in space that has vector equation

$$\vec{r} = \left(5 + \frac{12}{17}d\right)\vec{i} + \left(7 + \frac{8}{17}d\right)\vec{j} + \left(3 + \frac{9}{17}d\right)\vec{k}$$

1. Sketch the line in space.

2. The coordinates of point P_0 on the line are in the equation of the line. Give the coordinates of P_0.

3. Find a point on the line 34 units from P_0.

4. The components of unit vector $\vec{u}$ along the line are also in the equation. Give these components of $\vec{u}$.

5. Write the direction cosines $\cos \alpha$, $\cos \beta$, and $\cos \gamma$ of $\vec{u}$. Show that

$$\cos^2 \alpha + \cos^2 \beta + \cos^2 \gamma = 1$$

6. Write the direction angles of $\vec{u}$.

7. Find the point on the line where $y = 23$.

8. Point P_1 has coordinates (20, 23, 13). Explain quickly how you know that P_1 is *not* on the line.

9. Draw the displacement vector $\overrightarrow{P_0P_1}$ from P_0 to P_1 on your figure from Problem 1. Find this vector in terms of its components.

10. Find the angle between $\overrightarrow{P_0P_1}$ and $\vec{u}$.

11. Using trigonometry, find the distance between P_1 and the line.

12. Find $\overrightarrow{P_0P_1} \times \vec{u}$.

13. Show that the formula

$$\text{distance} = \left|\overrightarrow{P_0P_1} \times \vec{u}\right|$$

gives the same answer as in Problem 11.

14. Write the part of the definition of $\vec{a} \times \vec{b}$ that tells the length of the cross product. Based on this definition, explain why the formula in Problem 13 gives the distance in Problem 11.

A plane in space has the Cartesian equation
$2x + 4y + 3z = 36$.

15. Show that $P_0(4, 1, 8)$ is on the plane.

16. Show that $P_1(9, 7, 17)$ is *not* on the plane.

17. Write a vector $\vec{n}$ normal to the plane.

18. Draw a sketch that shows clearly that you know the meaning of *positive* normal vector to a plane.

19. Is vector $\vec{n}$ in Problem 17 a positive normal vector or a negative normal vector? How can you tell?

20. On the sketch in Problem 18, show $\vec{n}$ starting at P_0 and coming out normal to the plane and point P_1 somewhere above the plane.

21. Find vector $\overrightarrow{P_0P_1}$.

22. Find the angle between $\vec{n}$ and $\overrightarrow{P_0P_1}$.

23. Using trigonometry, find the distance from P_1 to the plane.

24. One form of the general equation of a plane in space is $Ax + By + Cz = D$. Show that the formula

$$\text{distance} = \frac{|Ax_1 + By_1 + Cz_1 - D|}{\sqrt{A^2 + B^2 + C^2}}$$

gives the same answer as in Problem 23, where x_1, y_1, and z_1 are coordinates of point P_1.

25. Is P_1 on the same side of the plane as the origin or on the opposite side? How can you tell?

26. The line in Problems 1–14 intersects the plane in Problems 15–25 at some point (x, y, z) in space. At this point, the $\vec{i}$-, $\vec{j}$-, and $\vec{k}$-coefficients for the line equal the x-, y-, and z-coordinates of the plane. By substituting these coefficients for the line into the equation of the plane, find d for the point (x, y, z).

27. Find the coordinates of the point where the line intersects the plane.

28. What did you learn as a result of doing this Exploration that you did not know before?

Exploration 103: Matrix Multiplication

Date: _____________

Objective: Find out how to multiply two matrices.

1. This matrix is called a _________ × _________ matrix. What numbers go in the blanks?

$$\begin{bmatrix} 3 & 5 & 2 & 1 \\ 8 & 9 & 4 & 5 \\ 0 & 6 & 7 & 4 \end{bmatrix}$$

2. Multiply these matrices without using a calculator.

$$\begin{bmatrix} 3 & 2 \\ 4 & 5 \\ 1 & 8 \end{bmatrix}\begin{bmatrix} 1 & 3 & 7 & 2 \\ 4 & 5 & 4 & 1 \end{bmatrix}$$

3. Explain why these matrices *cannot* be multiplied.

$$\begin{bmatrix} 1 & 3 & 7 & 2 \\ 4 & 5 & 4 & 1 \end{bmatrix}\begin{bmatrix} 3 & 2 \\ 4 & 5 \\ 1 & 8 \end{bmatrix}$$

4. Multiply these matrices without a calculator.

$$\begin{bmatrix} 3 & 7 \\ 4 & 2 \end{bmatrix}\begin{bmatrix} 5 & 6 \\ 1 & 8 \end{bmatrix}$$

5. Multiply these matrices without a calculator.

$$\begin{bmatrix} 5 & 6 \\ 1 & 8 \end{bmatrix}\begin{bmatrix} 3 & 7 \\ 4 & 2 \end{bmatrix}$$

6. Give two reasons why matrix multiplication does not commute.

7. Multiply these matrices without a calculator.

$$\begin{bmatrix} 9 & 4 & 4 \\ 5 & 2 & 1 \\ 4 & 3 & 6 \end{bmatrix}\begin{bmatrix} 1 & 0 & 0 \\ 0 & 1 & 0 \\ 0 & 0 & 1 \end{bmatrix} =$$

8. Multiply these matrices without a calculator.

$$\begin{bmatrix} 1 & 0 & 0 \\ 0 & 1 & 0 \\ 0 & 0 & 1 \end{bmatrix}\begin{bmatrix} 9 & 4 & 4 \\ 5 & 2 & 1 \\ 4 & 3 & 6 \end{bmatrix} =$$

9. The matrix $\begin{bmatrix} 1 & 0 & 0 \\ 0 & 1 & 0 \\ 0 & 0 & 1 \end{bmatrix}$ is called the **identity matrix for multiplication.** Why do you suppose this name is given to it?

10. Multiply these matrices using your grapher.

$$\begin{bmatrix} 9 & 4 & 4 \\ 5 & 2 & 1 \\ 4 & 3 & 6 \end{bmatrix}\begin{bmatrix} 1.8 & -2.4 & -0.8 \\ -5.2 & 7.6 & 2.2 \\ 1.4 & -2.2 & -0.4 \end{bmatrix} =$$

11. Multiply these matrices using your grapher.

$$\begin{bmatrix} 1.8 & -2.4 & -0.8 \\ -5.2 & 7.6 & 2.2 \\ 1.4 & -2.2 & -0.4 \end{bmatrix}\begin{bmatrix} 9 & 4 & 4 \\ 5 & 2 & 1 \\ 4 & 3 & 6 \end{bmatrix} =$$

12. The left matrix in Problem 11 is called the **multiplicative inverse** of the other matrix. Why do you think this is what it's called?

13. Your grapher has been programmed to find the inverse of a matrix. Enter the right matrix in Problem 11 as [A]. Then press the keys needed to find $[A]^{-1}$. What do you get for the answer?

14. What did you learn as a result of doing this Exploration that you did not know before?

Exploration 104: Determinant of a Matrix

Objective: Find the determinant of a 2×2 matrix and a 3×3 matrix.

Here is a 2×2 **matrix** and its corresponding **determinant**.

$$[M] = \begin{bmatrix} 3 & 4 \\ 2 & -5 \end{bmatrix} \qquad \det [M] = \begin{vmatrix} 3 & 4 \\ 2 & -5 \end{vmatrix}$$

The vertical bars are used to distinguish a determinant from its parent matrix. To evaluate a determinant, you first **expand** it. Write down the 3 and the 4 from the top row, with a negative sign between the terms. Then multiply the 3 and 4 by what is left when you cover up their row and column.

Think:

$$\begin{vmatrix} 3 & 4 \\ 2 & -5 \end{vmatrix} \qquad \begin{vmatrix} 3 & 4 \\ 2 & -5 \end{vmatrix}$$

Write: $\quad 3(-5) \quad - \quad 4(2) \quad = -23 \leftarrow$ Ans.

1. Evaluate this determinant. $\begin{vmatrix} 7 & 9 \\ -6 & 8 \end{vmatrix}$

A 3×3 determinant is expanded the same way. This time, when you cover up the row and column of an element, "what is left" is a 2×2 determinant. The signs between the terms alternate.

Think:

$$\begin{vmatrix} 2 & 3 & 4 \\ -1 & 5 & 7 \\ 9 & -8 & -6 \end{vmatrix} \quad \begin{vmatrix} 2 & 3 & 4 \\ -1 & 5 & 7 \\ 9 & -8 & -6 \end{vmatrix} \quad \begin{vmatrix} 2 & 3 & 4 \\ -1 & 5 & 7 \\ 9 & -8 & -6 \end{vmatrix}$$

Write:

$$\begin{vmatrix} 2 & 3 & -4 \\ -1 & 5 & 7 \\ 9 & -8 & -6 \end{vmatrix}$$

$$= 2 \begin{vmatrix} 5 & 7 \\ -8 & -6 \end{vmatrix} - 3 \begin{vmatrix} -1 & 7 \\ 9 & -6 \end{vmatrix} + (-4) \begin{vmatrix} -1 & 5 \\ 9 & -8 \end{vmatrix}$$

2. By expanding these three **minor** determinants, show that the 3×3 determinant equals 371.

3. Evaluate by expanding by minors:

$$\begin{vmatrix} 7 & -3 & -4 \\ 3 & -8 & 6 \\ -2 & 1 & -5 \end{vmatrix}$$

4. Enter into your grapher the matrix from which the determinant in Problem 3 comes. Then evaluate the determinant of the matrix using the built-in feature on your grapher. Show that you get the same answer as in Problem 3.

5. What did you learn as a result of doing this Exploration that you did not know before?

Exploration 105: Inverse of a 3 × 3 Matrix

Objective: Find out how to invert a 3 × 3 matrix "by hand," and confirm your answer by calculator.

The inverse of matrix [M] is given by

$$[M]^{-1} = \frac{1}{\det\,[M]} \cdot \text{adj}\,[M]$$

where det [M] is the determinant of [M] and adj [M] is the adjoint of [M]. In this Exploration, you will learn how to find the adjoint of a 3 × 3 matrix and use it to find the inverse.

1. Let

$$[M] = \begin{bmatrix} 2 & 3 & 4 \\ 5 & 1 & 2 \\ 6 & 8 & 7 \end{bmatrix}$$

   The **transpose** of [M] is

$$[M]^{T} = \begin{bmatrix} 2 & 5 & 6 \\ 3 & 1 & 8 \\ 4 & 2 & 7 \end{bmatrix}$$

   Explain what you must do to [M] to get $[M]^{T}$.

2. Each element of $[M]^{T}$ has a **minor** determinant that is found by covering up the row and column of that element. For instance, the minor of 2 in the upper left corner is

$$\begin{vmatrix} 1 & 8 \\ 2 & 7 \end{vmatrix} = -9$$

   Show that the minor for 5 in the first row, second column of $[M]^{T}$ is equal to −11.

3. Each element of $[M]^{T}$ also has a **cofactor,** equal to the minor determinant with this alternating sign applied:

$$\begin{matrix} + & - & + \\ - & + & - \\ + & - & + \end{matrix}$$

   Write the cofactor of 2 from Problem 1 and the cofactor of 5 from Problem 2.

4. The **adjoint** of [M], written adj [M], is the matrix of the cofactors of $[M]^{T}$. By evaluating the other seven cofactors, show that

$$\text{adj}\,[M] = \begin{bmatrix} -9 & 11 & 2 \\ -23 & -10 & 16 \\ 34 & 2 & -13 \end{bmatrix}$$

5. Show that the determinant of [M] is det [M] = 49.

6. The inverse of [M] is

$$[M]^{-1} = \frac{1}{49}\begin{bmatrix} -9 & 11 & 2 \\ -23 & -10 & 16 \\ 34 & 2 & -13 \end{bmatrix} = \begin{bmatrix} \frac{9}{49} & \frac{11}{49} & \frac{2}{49} \\ \frac{23}{49} & \frac{10}{49} & \frac{16}{49} \\ \frac{34}{49} & \frac{2}{49} & \frac{13}{49} \end{bmatrix}$$

   Check this by finding $[M]^{-1}$ on your calculator and then multiplying the answer by 49. Does the result equal adj [M]?

7. What did you learn as a result of doing this Exploration that you did not know before?

Exploration 106: Matrix Images and Transformations

Date: _____________

Objective: Transform an image by multiplying by a matrix.

1. Matrix [M] describes the triangle shown here. What do the columns represent?

$$[M] = \begin{bmatrix} 1 & 4 & 1 \\ 1 & 2 & 3 \end{bmatrix}$$

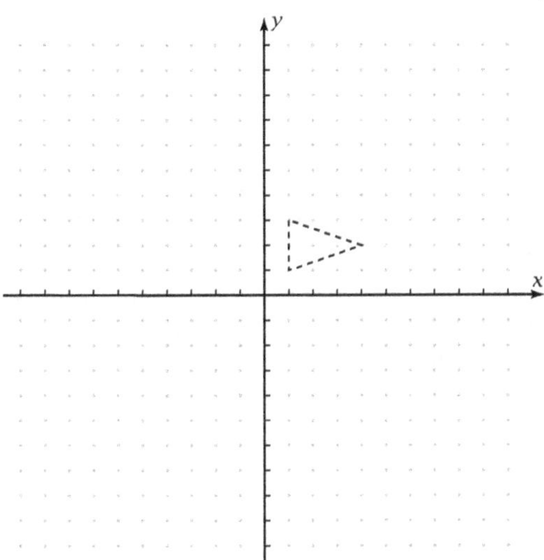

2. Matrix [A] is a **transformation matrix.** Multiply [A][M] and write the answer here.

$$[A] = \begin{bmatrix} 2 & 0 \\ 0 & 2 \end{bmatrix}$$

3. Explain why you can't multiply [M][A].

4. The figure corresponding to the answer in Problem 2 is called the **image** of [M] for the transformation [A]. Plot the image on the given figure.

5. How would you describe the transformation defined by matrix [A]?

6. The original triangle whose matrix is [M] is called the **pre-image.** Why is this its name?

7. Matrix [B] is another transformation matrix.

$$[B] = \begin{bmatrix} \cos 90° & \cos 180° \\ \sin 90° & \sin 180° \end{bmatrix}$$

Apply the transformation by calculating [B][M]. Plot the result on the given figure.

8. Matrix [B] in Problem 7 **rotates** the pre-image by 90° counterclockwise. Write a matrix [C] that you think will rotate the pre-image by 40° counterclockwise. To do this, you should realize that the angle in the second column is 90° more than the angle in the first column. Apply the transformation by calculating [C][M]. Write the answer with elements rounded to one decimal place.

9. Plot the image from Problem 8 on the given figure. Extend one side of the pre-image and the corresponding side of the image and measure the angle with a protractor. Did the transformation really rotate the pre-image by 40°?

10. Write a transformation matrix [D] that both **dilates** by a factor of 2 and rotates by 240°. Calculate [D][M]. Write the image matrix with elements rounded to one decimal place, and plot the image on the given figure.

11. What did you learn as a result of doing this Exploration that you did not know before?

Exploration 107: Iterated Transformations

Date: __________

Objective: Find out what happens to a geometrical figure if you apply a transformation iteratively (over and over again).

The figure shows the rectangle with matrix [D].

$$[D] = \begin{bmatrix} 12 & 12 & 6 & 6 \\ 5 & 2 & 2 & 5 \end{bmatrix}$$

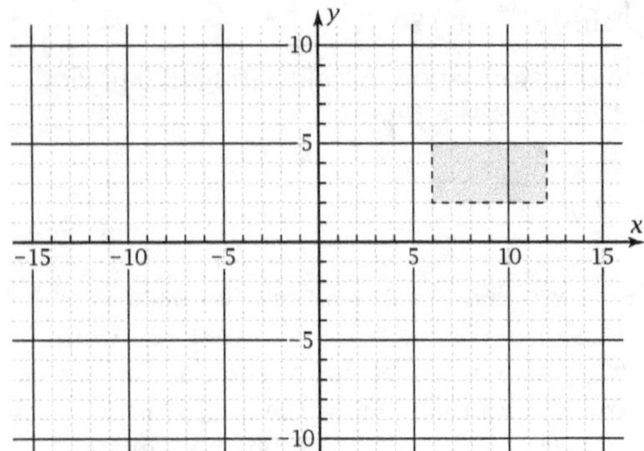

Transform this rectangle by reducing it by a factor of 0.9 and rotating it 30° counterclockwise.

1. Write a 2 × 2 matrix that does the rotation and dilation. Store it as [A] in your grapher.

2. Store the pre-image matrix for the given rectangle as [D]. Set your grapher's mode so that it shows exactly one decimal place. Then calculate [A][D]. Write the image matrix here, and plot it on the given figure.

3. Store the image matrix as matrix [E]. Then apply the transformation [A] to the image matrix by calculating [A][E]. Write the second image matrix here, and plot it on the given figure.

4. Store the second image as [E], and then get a third image. Plot it on the given figure.

5. Describe what is happening to the images as you apply the transformation repeatedly.

6. Write or obtain a program to plot the pre-image and the image rectangles. Before you run the program, the transformation matrix should be stored as [A] and the pre-image matrix as [D]. When you execute the program, the grapher should store the pre-image as [E] and then draw the pre-image on the screen. Each time you press ENTER, it should apply the transformation by multiplying [A][E], store the image in [E], draw the image, and then pause. The line command is useful for this program to connect the vertices of the rectangles.

7. Run three iterations of the program. Do the resulting images agree with the ones you drew in Problems 2, 3, and 4? __________

8. Perform the iterations until the rectangles get so small you can't distinguish one image from the next. Describe the resulting figure. To what point do the images converge? (Display [E].)

9. Apply iteratively a 0.95 reduction and 20° rotation on the dart-shaped pre-image given by matrix *D*. Draw the pre-image and the first few iterations on the given axes. Sketch the pattern, followed by the rest of them.

$$[D] = \begin{bmatrix} 6 & 8 & 10 & 8 \\ 1 & 2 & 1 & 6 \end{bmatrix}$$

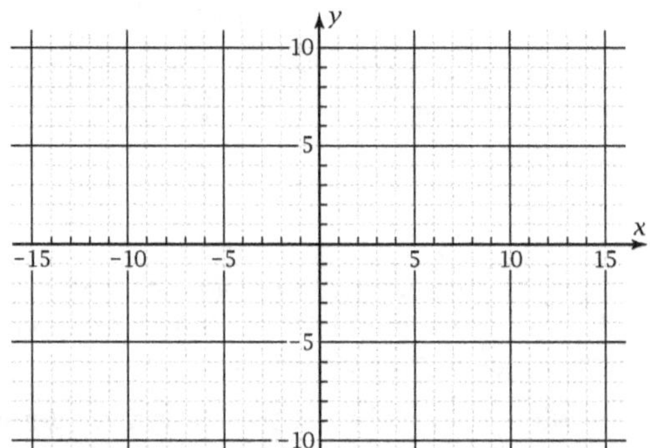

10. What did you learn as a result of doing this Exploration that you did not know before?

Exploration 108: Combined Translation, Rotation, and Dilation

Date: _______________

Objective: Perform matrix operations that **translate** a given figure as well as rotate and dilate it.

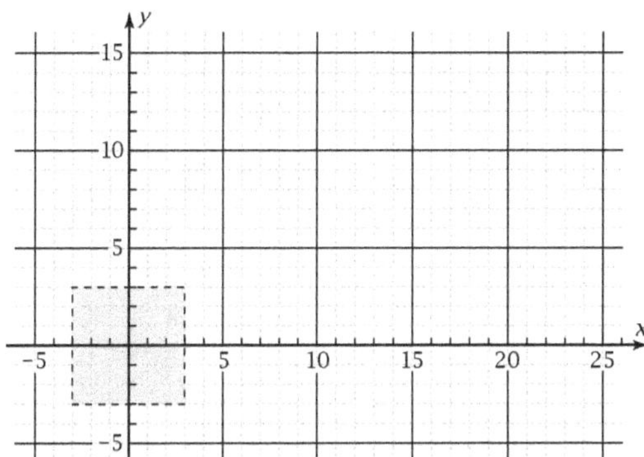

1. The figure shows the square specified by matrix

$$[D] = \begin{bmatrix} 3 & 3 & -3 & -3 \\ 3 & -3 & -3 & 3 \end{bmatrix}$$

   Draw the image formed by translating this square by 5 units in the x-direction and 4 units in the y-direction. Write the matrix for the image.

2. Insert a row of 1s in [D]. Multiply it by the transformation matrix [A], as shown.

$$[A][D] = \begin{bmatrix} 1 & 0 & 5 \\ 0 & 1 & 4 \\ 0 & 0 & 1 \end{bmatrix} \begin{bmatrix} 3 & 3 & -3 & -3 \\ 3 & -3 & -3 & 3 \\ 1 & 1 & 1 & 1 \end{bmatrix}$$

3. What do you notice about the first two rows of the image matrix in Problem 2?

4. Explain how the 5 and the 4 in the third column of [A], along with the 1s in the bottom row of [D], accomplish the translations. (Try multiplying by hand to see!)

5. Describe what the row 0, 0, 1 at the bottom of [A] does.

6. What special 2 × 2 matrix is in the upper left corner of [A] in Problem 2? How would you modify [A] so that it accomplishes a 70% reduction and a 30° counterclockwise rotation as well as the translations?

7. Run your iterative transformation program with [A] as modified in Problem 6 and [D] as in Problem 2. Sketch the first image and the path of the centers of the images on the given figure.

8. If your work in Problem 7 is correct, you should have found that the squares are being **attracted** to a **fixed point.** Find *graphically* the approximate coordinates of this "attractor." _______________

9. Find the fixed point *numerically* by finding $[A]^{50}[D]$. _______________

10. Change the pre-image matrix to

$$[D] = \begin{bmatrix} 16 & 16 & 10 & 10 \\ 2 & 8 & 8 & 2 \\ 1 & 1 & 1 & 1 \end{bmatrix}$$

   Run the program again. Iterate at least 20 times. Sketch the first image and the path of the centers of the images on the given graph. Do the images converge to the same fixed point as in Problem 8, to a different fixed point, or to no point?

11. What did you learn as a result of doing this Exploration that you did not know before?

Exploration 109: Iterative Transformations and Fixed Points

Objective: Draw images of iterative transformations of a figure, and find the fixed point.

The figure shows the dart specified by

$$[D] = \begin{bmatrix} 6 & 8 & 10 & 8 \\ 1 & 2 & 1 & 6 \\ 1 & 1 & 1 & 1 \end{bmatrix}$$

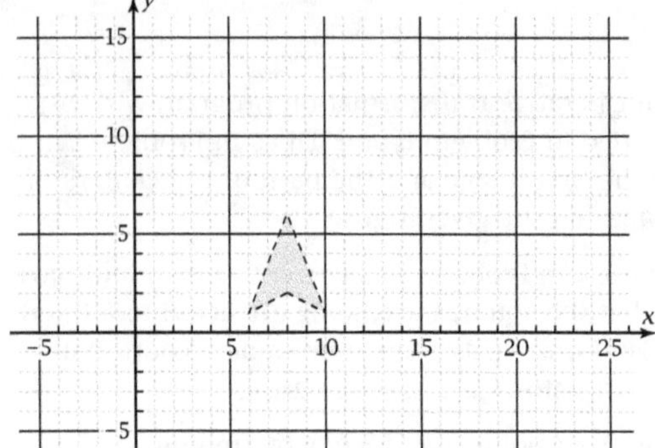

In this Exploration, you will perform various transformations on this dart iteratively (repeatedly) and reach some conclusions about the location of the fixed point.

1. Write a 3 × 3 transformation matrix, [A], for:
   - 80% reduction ("80% of the original size")
   - Rotation of 30° counterclockwise
   - Translation of 4 in the *x*-direction
   - Translation of 2 in the *y*-direction

   [A] =

2. Apply the transformation in Problem 1 iteratively to the pre-image [D]. Sketch the images on the given figure.

3. To what fixed point are the images attracted? How can you tell? _________________

4. Change the transformation matrix in Problem 1 so that the rotation is 60° counterclockwise. Apply the transformation iteratively. Sketch the results.

   [A] =

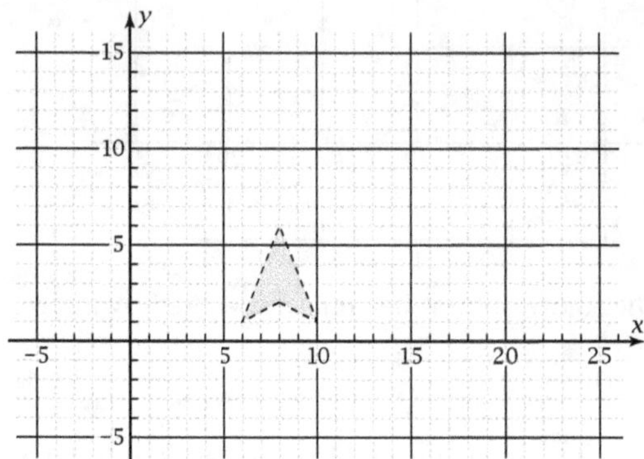

5. Do the images in Problem 4 seem to be attracted to the same fixed point as in Problem 3? Show numbers to justify your answer.

6. Apply [A] in Problem 4 to this pre-image. Sketch.

$$[D] = \begin{bmatrix} 10 & 14 & 14 & 10 \\ 2 & 2 & 4 & 4 \\ 1 & 1 & 1 & 1 \end{bmatrix}$$

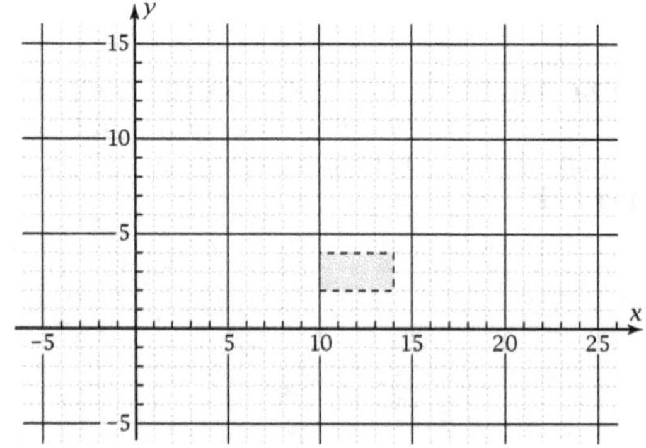

7. To what fixed point are the images in Problem 6 attracted? _________________

8. What affects the location of the fixed point—the transformation matrix or the pre-image matrix? Give two pieces of evidence to support your answer.

9. Calculate *algebraically* the location (x, y) of the fixed point in Problem 4.

10. In finding the image for the second iteration, you apply [A] to the result of [A][D]. That is,

    [A]([A][D])

    Show that matrix multiplication is associative by calculating

    ([A][A])[D]

    and comparing the answers.

11. You can write the fiftieth iteration $[A]^{50}[D]$. Find $[A]^{50}$. What limit does each of the four numbers in the rotation and translation part of the answer seem to be approaching? What limits do the two numbers in the translation part of the matrix seem to be approaching? Give a numerical method you can use to find the fixed point of a transformation *quickly*.

12. What did you learn as a result of doing this Exploration that you did not know before?

Exploration 110: Rotation, Translation, and Dilation Practice

Objective: Practice graphing iterated matrix transformations of plane figures.

1. Describe what transformations are applied by matrix

$$[T] = \begin{bmatrix} 0.7\cos 30° & 0.7\cos 120° & 5 \\ 0.7\sin 30° & 0.7\sin 120° & 4 \\ 0 & 0 & 1 \end{bmatrix}$$

____

____

____

____

2. Write a transformation matrix, [A], to dilate (reduce) by a factor of 0.95, rotate by 20° counterclockwise, and translate 3 spaces in the x-direction and 2 spaces in the y-direction. Enter [A] into your grapher.

3. Write a 3 × 4 pre-image matrix, [D], for the dart shown here. Enter it into your grapher.

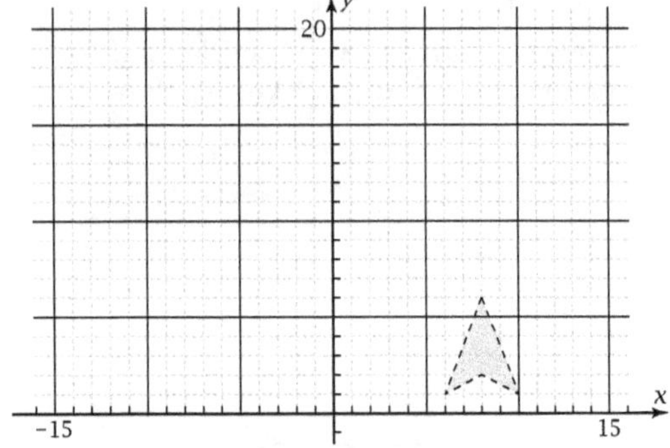

4. Set your grapher's mode so that it displays only one decimal place. Then multiply [A][D] using the matrices from Problems 2 and 3. Write the result here, and plot the image on the graph.

5. Apply the transformations in Problem 4 iteratively by using your program. Choose an appropriate window. Complete enough iterations so that the image does not change any more. Check your graph with your instructor. ______________

6. Find graphically the coordinates of the fixed point to which the darts in Problem 5 are attracted.

7. Find the fixed point numerically using $[A]^{100}$. Write the answer in ellipsis format with four decimal places.

8. On the back of this sheet, find the fixed point algebraically. Write the answer here, using ellipsis format with four decimal places.

9. Explain the purpose of the 1, 1, 1, 1 in the third row of the pre-image matrix [D] and the 0, 0, 1 in the third row of transformation matrix [A].

10. Explain why the pre-image matrix [D] was stored as [E] at the beginning of the program.

11. What did you learn as a result of doing this Exploration that you did not know before?

Exploration 111: Markov Chain Problem

Objective: Use iterative multiplication of matrices in a nongeometrical real-world problem.

Television Network Loyalty Problem: In a poll of television viewers, a research company found the percentages of viewers who change from the evening news on one network to that on another network one month later.

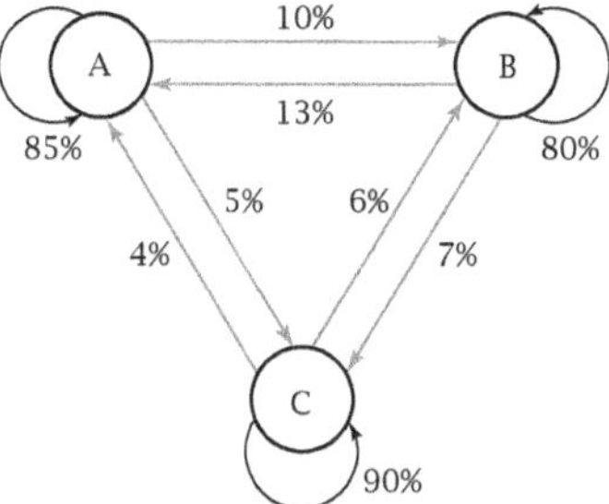

For instance, for network A, 85% stayed with A, 10% switched to network B, and 5% switched to network C. The company uses these numbers as probabilities of what will happen in subsequent months. They arrange the numbers in **transition matrix** [T].

$$[T] = \begin{bmatrix} 0.85 & 0.10 & 0.05 \\ 0.13 & 0.80 & 0.07 \\ 0.04 & 0.06 & 0.09 \end{bmatrix} \begin{matrix} A \\ B \\ C \end{matrix}$$
$$\quad\; A \quad\;\; B \quad\;\; C$$

Each element represents the probability that a viewer who watches the network in the *row* one month will be watching the network in the *column* the next month.

1. At the beginning of January, A has 35 million viewers, B has 20 million, and C has 43 million. These numbers are recorded in the "viewers" matrix, $[V_0]$.

   $$[V_0] = [35 \quad 20 \quad 43]$$
   $$\quad\;\; A \quad\; B \quad\;\; C$$

   Explain why the number of viewers at the beginning of February is given by

   $$[V_1] = [V_0][T]$$

2. Show that the number of viewers for A in $[V_1]$ is equal to the number who stayed with A plus the number that transferred from B and from C to A.

3. Show that the viewers matrix $[V_2]$ at the beginning of March can be found either as

   $$[V_2] = [V_1][T] \text{ or } [V_2] = [V_0][T]^2$$

4. In the most time-efficient way, find the number of viewers predicted for each network the following January, one full year later. Assume that the probabilities remain constant.

5. Assuming that the probabilities remain constant, the number of viewers for each network approach a fixed limit as the number of months becomes very large. Find approximations for these limits numerically.

6. The matrices $[V_0], [V_1], [V_2], \ldots$ form a **Markov chain.** On the Internet or in some other reference source, find out who Markov is or was. Write a paragraph summarizing your findings.

7. What did you learn as a result of doing this Exploration that you did not know before?

Exploration 112: Multiple Transformations of the Same Figure

Date: ____________

Objective: See what happens when you perform several different transformations iteratively on the same pre-image.

The graph shows the rectangle

$$[E] = \begin{bmatrix} 2 & 2 & -2 & -2 \\ 0 & 10 & 10 & 0 \\ 1 & 1 & 1 & 1 \end{bmatrix}$$

This rectangle is to be transformed using four different transformation matrices, individually and then combined.

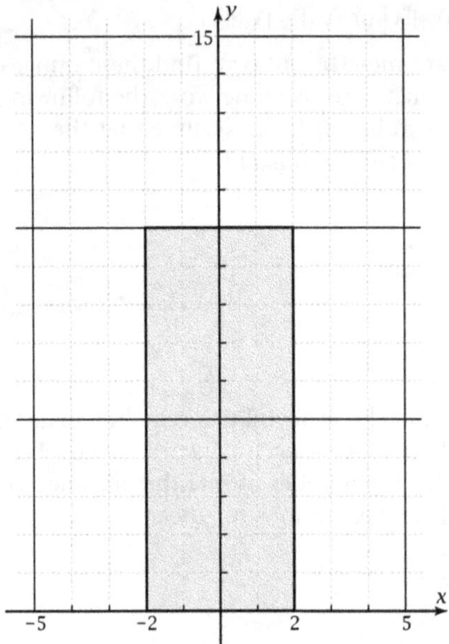

1. Enter the given pre-image matrix as [E].

2. Enter these four transformation matrices:

$$[A] = \begin{bmatrix} 0.8\cos(-3°) & 0.8\cos 87° & 0 \\ 0.8\sin(-3°) & 0.8\sin 87° & 3 \\ 0 & 0 & 1 \end{bmatrix}$$

$$[B] = \begin{bmatrix} 0.3\cos(-52°) & 0.3\cos 38° & 0 \\ 0.3\sin(-52°) & 0.3\sin 38° & 2 \\ 0 & 0 & 1 \end{bmatrix}$$

$$[C] = \begin{bmatrix} 0.3\cos 46° & 0.3\cos 136° & 0 \\ 0.3\sin 46° & 0.3\sin 136° & 3 \\ 0 & 0 & 1 \end{bmatrix}$$

$$[D] = \begin{bmatrix} 0 & 0 & 0 \\ 0 & 0.3 & 0 \\ 0 & 0 & 1 \end{bmatrix}$$

3. Perform these four transformations. Set the mode so that the grapher displays image matrices with elements rounded to one decimal place. Write down the four image matrices.

   [A][E] =

   [B][E] =

   [C][E] =

   [D][E] =

4. Plot these four images on the graph paper (left).

5. If you perform transformation [A] to the first image matrix, it is equivalent to the transformation [A]([A][E]). Calculate [A][E]; then calculate [A]Ans. Write the result here.

   [A]([A][E]) =

6. Calculate in *one* step the transformation [A][A][E].

   [A][A][E] =

7. The fact that the answers to Problems 5 and 6 are *equal* illustrates a property of matrix multiplication. Which property?

   ____

8. Calculate these transformations:

   [B][A][E] =

   [A][B][E] =

9. The fact that the two answers in Problem 8 are *not* equal illustrates that matrix multiplication does *not* have a certain property. What property?

   ____

Exploration 112: Multiple Transformations
of the Same Figure *continued*

10. Perform the following 16 transformations. Record the answers here.

   [A][A][E] =

   [A][B][E] =

   [A][C][E] =

   [A][D][E] =

   [B][A][E] =

   [B][B][E] =

   [B][C][E] =

   [B][D][E] =

   [C][A][E] =

   [C][B][E] =

   [C][C][E] =

   [C][D][E] =

   [D][A][E] =

   [D][B][E] =

   [D][C][E] =

   [D][D][E] =

11. Plot the 16 images on this graph paper.

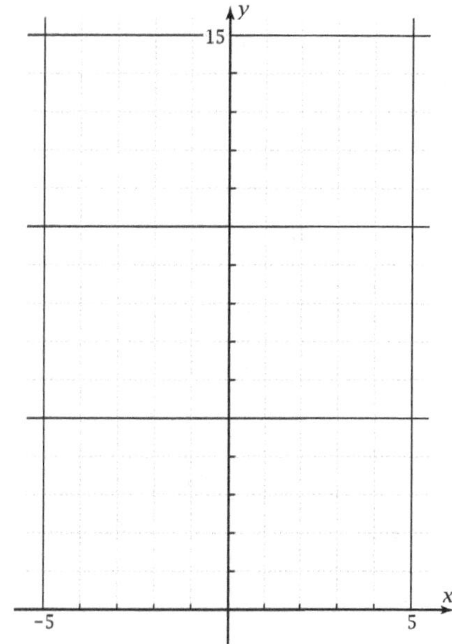

12. The four images in Problem 4 are called the results of the **first iteration.** The 16 images of the first four images (Problem 11) are the results of the **second iteration.** How many images result from the third iteration? the fourth iteration? the tenth iteration?

13. What have you learned as a result of doing this Exploration that you did not know before?

Exploration 113: Fractal Figures by Random Point Plotting

Objective: Generate a figure determined by multiple iterative transformations *one* point at a time.

In this Exploration, you will use these matrices to generate, one point at a time, the figure to which the images are attracted.

$$[A] = \begin{bmatrix} 0.8\cos(-3°) & 0.8\cos 87° & 0 \\ 0.8\sin(-3°) & 0.8\sin 87° & 3 \\ 0 & 0 & 1 \end{bmatrix}$$

$$[B] = \begin{bmatrix} 0.3\cos(-52°) & 0.3\cos 38° & 0 \\ 0.3\sin(-52°) & 0.3\sin 38° & 2 \\ 0 & 0 & 1 \end{bmatrix}$$

$$[C] = \begin{bmatrix} 0.3\cos 46° & 0.3\cos 136° & 0 \\ 0.3\sin 46° & 0.3\sin 136° & 3 \\ 0 & 0 & 1 \end{bmatrix}$$

$$[D] = \begin{bmatrix} 0 & 0 & 0 \\ 0 & 0.3 & 0 \\ 0 & 0 & 1 \end{bmatrix}$$

1. Enter the pre-image matrix $[E] = \begin{bmatrix} 2 \\ 5 \\ 1 \end{bmatrix}$. Then apply iteratively these transformations, picked at random, and record the ten image points.

   Transformation: [A][E]   Point: ________________

   Transformation: [C]Ans   Point: ________________

   Transformation: [B]Ans   Point: ________________

   Transformation: [A]Ans   Point: ________________

   Transformation: [D]Ans   Point: ________________

   Transformation: [C]Ans   Point: ________________

   Transformation: [C]Ans   Point: ________________

   Transformation: [B]Ans   Point: ________________

   Transformation: [A]Ans   Point: ________________

   Transformation: [D]Ans   Point: ________________

2. Plot the ten points in Problem 1. Do you see any patterns?

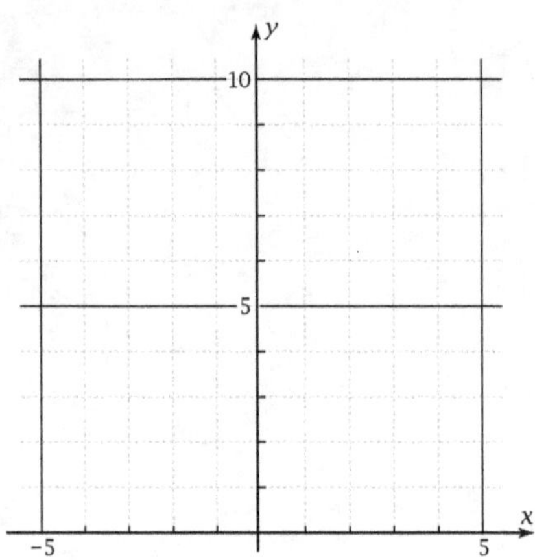

3. Download the program BARNSLEY (named for Michael Barnsley of Georgia Tech). Then set up your grapher to use the program as follows:

   - Enter the matrices [A], [B], [C], and [D] given in Problem 1.
   - Enter probabilities in List 1: 0.7, 0.13, 0.13, 0.04. These indicate the probabilities of picking matrix [A], [B], [C], or [D], respectively.
   - Set the window: x: [−11, 11], y: [0, 15]
   - Deactivate all functions and StatPlots.

4. Run the BARNSLEY program. When you are prompted to do so, input the following:

   - Pre-image: (2, 5)
   - Number of points: 100

5. Do you see any patterns in the image points? For instance, do points seem to be attracted to some regions but not to others?

6. Run the BARNSLEY program again using 1000 points. How does this **strange attractor** compare with this image?

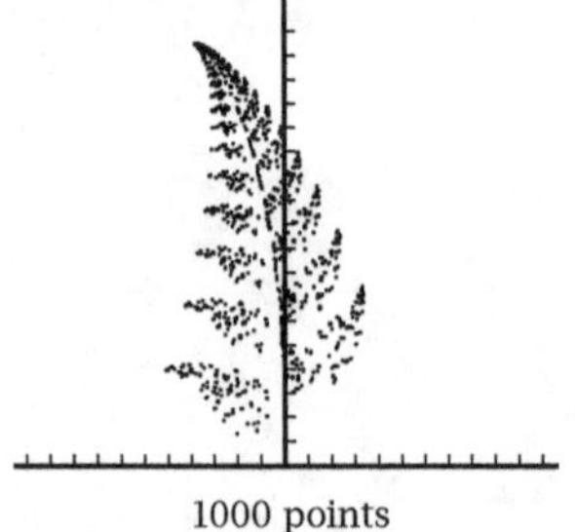

1000 points

Exploration 113: Fractal Figures by Random Point Plotting *continued*

7. Save the image on the screen. For TI-83, press

   DRAW

   STO

   StorePic

   VARS

   Picture

   Pic1

8. Run the BARNSLEY program with 100 points using several other starting pre-images. Is the form of the strange attractor influenced by the pre-image?

9. Explain why the Barnsley technique is more practical for generating the figure than are the rectangle transformations of Exploration 112.

10. Change the matrices as given here, to give the figure different horizontal translations (in boldface).

$$[A] = \begin{bmatrix} 0.8\cos(-3°) & 0.8\cos 87° & \mathbf{1} \\ 0.8\sin(-3°) & 0.8\sin 87° & 3 \\ 0 & 0 & 1 \end{bmatrix}$$

$$[B] = \begin{bmatrix} 0.3\cos(-52°) & 0.3\cos 38° & \mathbf{2} \\ 0.3\sin(-52°) & 0.3\sin 38° & 2 \\ 0 & 0 & 1 \end{bmatrix}$$

$$[C] = \begin{bmatrix} 0.3\cos 46° & 0.3\cos 136° & \mathbf{-4} \\ 0.3\sin 46° & 0.3\sin 136° & 3 \\ 0 & 0 & 1 \end{bmatrix}$$

$$[D] = \begin{bmatrix} 0 & 0 & 0 \\ 0 & \mathbf{0.5} & 0 \\ 0 & 0 & 1 \end{bmatrix}$$

   Then run the BARNSLEY program with these transformations. Use the same window and 1000 points. Describe the changes in the image.

11. When you apply a *single* transformation iteratively, the images are attracted to a *single* fixed point. When you apply *multiple* transformations iteratively, as in this Exploration, the attractor is the figure you see emerging from the dots. What is the name of an attractor such as this?

12. In this Exploration, you have been introduced to complicated figures generated by a rather simple code applied over and over thousands of times. It is beginning to be believed that this is the way complex living things develop. Some relatively simple information is stored in the genes. Then the computations are done iteratively, millions of times. A slight difference in the numbers stored in the genes makes the difference between getting a fern leaf and getting some other life form! Try another change in the transformation matrices and see what figure you get.

13. What did you learn as a result of doing this Exploration that you did not know before?

Exploration 114: Hausdorff's Definition of (Fractal) Dimension

Objective: Learn a definition of the concept of *dimension*, and apply it to self-similar figures generated iteratively.

Here is a (three-dimensional) self-similar cube of edge 1 unit. It is cut into smaller cubes of edge $r = \frac{1}{5}$ unit. Note that the smaller cubes are similar to the original cube.

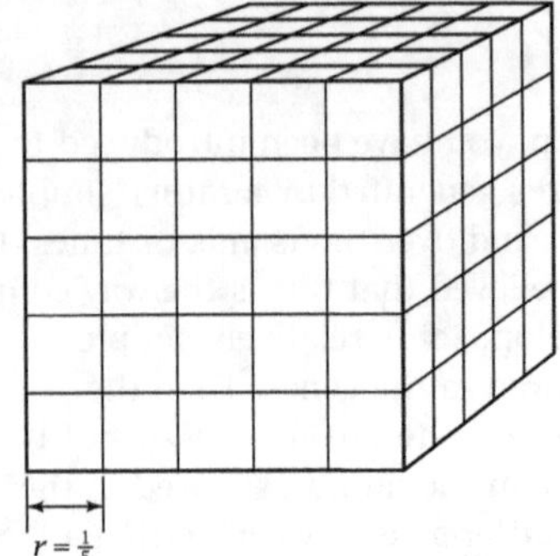

1. How many small cubes will there be? _______________

2. Write the answer to Problem 1 as a power of 5.

   ____

3. What physical quantity does the exponent in Problem 2 represent? _______________

4. Suppose that the cube in Problem 1 is divided into N smaller cubes, each of edge r units. Write N in terms of r. _______________

5. Transform the equation $\left(\frac{1}{r}\right)^3 = N$ so that 3 is by itself on the left and N and r are on the right. (Recall the properties of logarithms!)

The results of Problem 5 are the basis for the definition of dimension, D, of a self-similar object. It was proposed by Felix Hausdorff, who lived from 1868 to 1942.

$$D = \frac{\log N}{\log \left(\frac{1}{r}\right)}$$

where N is the number of self-similar pieces into which an object can be divided and r is the length of a given piece as a fraction of the length of the original pre-image. In order for this definition to apply, you must be able to subdivide the object infinitely.

6. Apply Hausdorff's definition to this square and thereby show that a square is (as you would expect) two-dimensional.

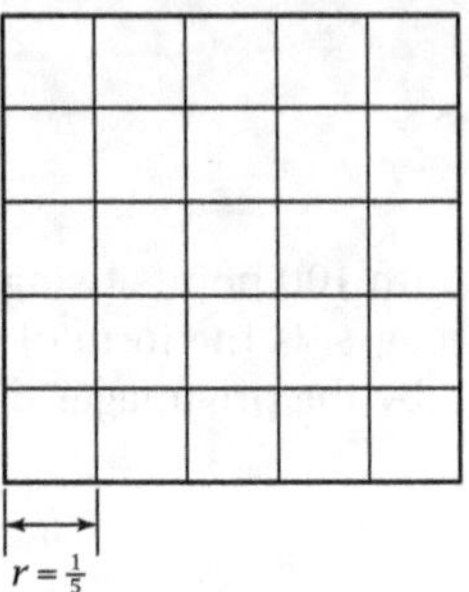

7. In a previous exercise, you generated Sierpiński's gasket, shown here.

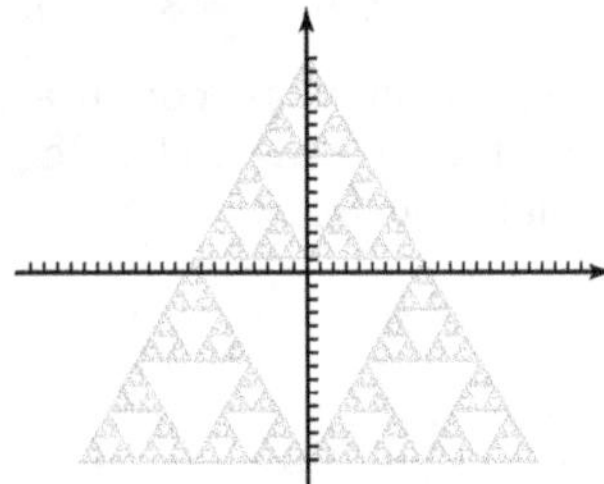

On the drawing, show that the gasket is self-similar by dividing it into three congruent pieces, each of which is similar to the whole figure. Then show how one of these pieces can be divided into three other pieces, each of which is also similar to the whole figure.

8. Because the subdividing into self-similar pieces can go on infinitely, Hausdorff's definition of dimension applies. Calculate the dimension of the Sierpiński gasket. Surprising?

9. The original pre-image in Problem 7 is a triangle of base 20 and altitude 30. What is its area? What is the total area of the first iteration? the second iteration? What is the total area of the tenth iteration? the 100th iteration? What limit do the total areas of the iterations approach as the number of iterations approaches infinity? Surprising?

10. What did you learn as a result of doing this Exploration that you did not know before?

Exploration 115: Von Koch's Snowflake Curve

Date: _______________

Objective: Perform four transformations iteratively, starting with a line segment, and generate a "fractured" curve.

The figure shows the line segment $\begin{bmatrix} 12 & 12 \\ 6 & -6 \\ 1 & 1 \end{bmatrix}$.

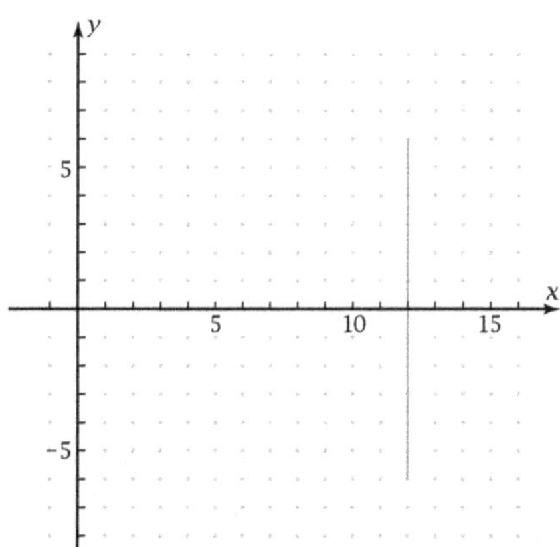

Using this segment as the pre-image, write 3×3 transformation matrices for Problems 1–4.

1. Reduce the pre-image to $\frac{1}{3}$ of its length and then translate so that its upper end is at the upper end of the pre-image, as shown here.

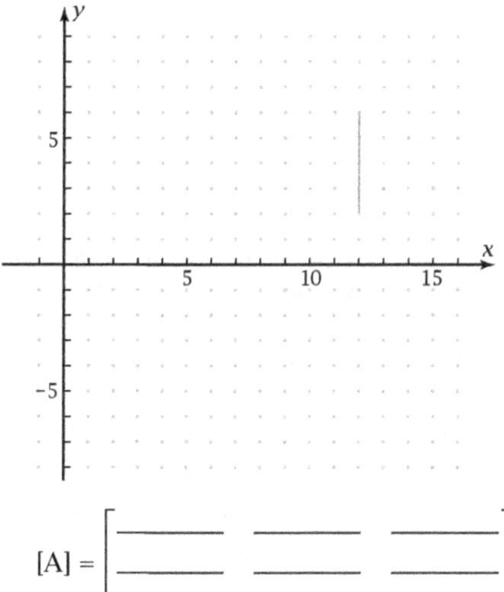

$$[A] = \begin{bmatrix} \underline{\quad} & \underline{\quad} & \underline{\quad} \\ \underline{\quad} & \underline{\quad} & \underline{\quad} \\ \underline{\quad} & \underline{\quad} & \underline{\quad} \end{bmatrix}$$

2. Reduce the pre-image to $\frac{1}{3}$ of its length and then translate so that its lower end is at the lower end of the pre-image.

$$[B] = \begin{bmatrix} \underline{\quad} & \underline{\quad} & \underline{\quad} \\ \underline{\quad} & \underline{\quad} & \underline{\quad} \\ \underline{\quad} & \underline{\quad} & \underline{\quad} \end{bmatrix}$$

3. Reduce the pre-image to $\frac{1}{3}$ of its length, rotate it 60° counterclockwise, and then translate so that the upper end is at the lower end of the image in Problem 1, as shown here. (The dashed line shows the pre-image rotated and dilated. You must figure out the necessary translations. Keep at least four decimal places.)

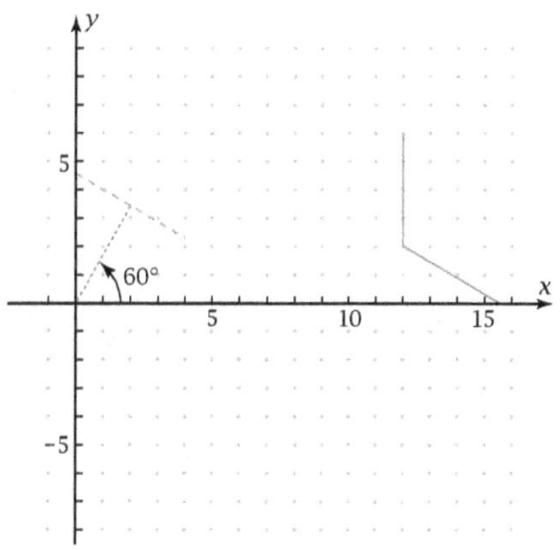

$$[C] = \begin{bmatrix} \underline{\quad} & \underline{\quad} & \underline{\quad} \\ \underline{\quad} & \underline{\quad} & \underline{\quad} \\ \underline{\quad} & \underline{\quad} & \underline{\quad} \end{bmatrix}$$

4. Reduce the pre-image to $\frac{1}{3}$ of its length, rotate it 60° clockwise, and then translate so that the lower end is at the upper end of the image in Problem 2. Keep at least four decimal places.

$$[D] = \begin{bmatrix} \underline{\quad} & \underline{\quad} & \underline{\quad} \\ \underline{\quad} & \underline{\quad} & \underline{\quad} \\ \underline{\quad} & \underline{\quad} & \underline{\quad} \end{bmatrix}$$

5. Apply the four transformations to the original pre-image. Write the results here. Round the elements of the image matrices to one decimal place.

$$[E_1] = \begin{bmatrix} \underline{\quad} & \underline{\quad} \\ \underline{\quad} & \underline{\quad} \\ \underline{\quad} & \underline{\quad} \end{bmatrix} \qquad [E_2] = \begin{bmatrix} \underline{\quad} & \underline{\quad} \\ \underline{\quad} & \underline{\quad} \\ \underline{\quad} & \underline{\quad} \end{bmatrix}$$

$$[E_3] = \begin{bmatrix} \underline{\quad} & \underline{\quad} \\ \underline{\quad} & \underline{\quad} \\ \underline{\quad} & \underline{\quad} \end{bmatrix} \qquad [E_4] = \begin{bmatrix} \underline{\quad} & \underline{\quad} \\ \underline{\quad} & \underline{\quad} \\ \underline{\quad} & \underline{\quad} \end{bmatrix}$$

(Over)

Exploration 115: Von Koch's Snowflake Curve *continued*

6. Plot the four images from Problem 5 here.

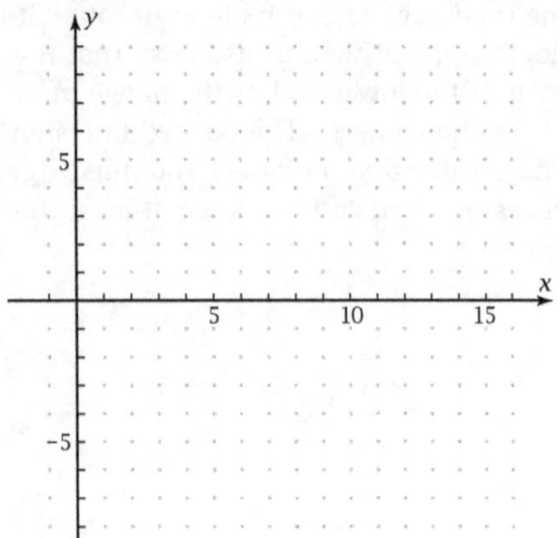

7. Apply each of the four transformations to each of the four images in Problem 5. (Be time-efficient!)

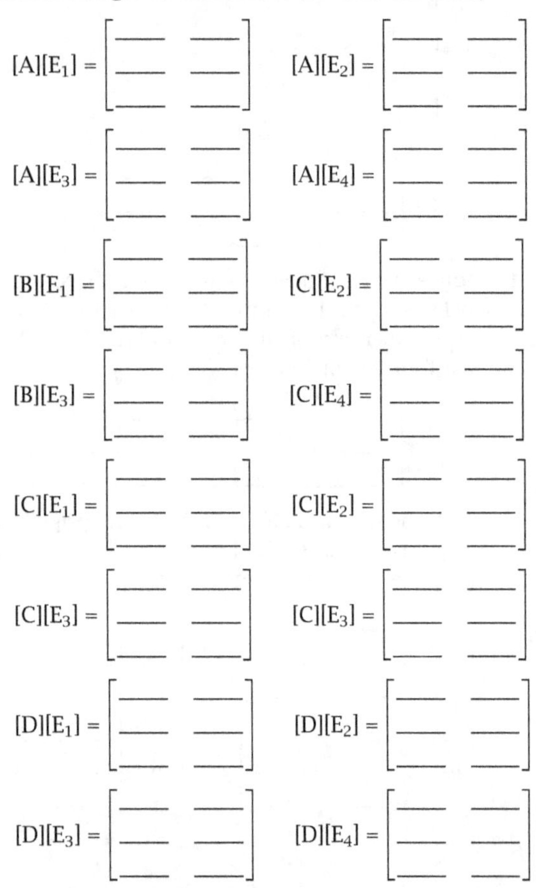

$$[A][E_1] = \begin{bmatrix} \underline{\quad} & \underline{\quad} \\ \underline{\quad} & \underline{\quad} \end{bmatrix} \qquad [A][E_2] = \begin{bmatrix} \underline{\quad} & \underline{\quad} \\ \underline{\quad} & \underline{\quad} \end{bmatrix}$$

$$[A][E_3] = \begin{bmatrix} \underline{\quad} & \underline{\quad} \\ \underline{\quad} & \underline{\quad} \end{bmatrix} \qquad [A][E_4] = \begin{bmatrix} \underline{\quad} & \underline{\quad} \\ \underline{\quad} & \underline{\quad} \end{bmatrix}$$

$$[B][E_1] = \begin{bmatrix} \underline{\quad} & \underline{\quad} \\ \underline{\quad} & \underline{\quad} \end{bmatrix} \qquad [C][E_2] = \begin{bmatrix} \underline{\quad} & \underline{\quad} \\ \underline{\quad} & \underline{\quad} \end{bmatrix}$$

$$[B][E_3] = \begin{bmatrix} \underline{\quad} & \underline{\quad} \\ \underline{\quad} & \underline{\quad} \end{bmatrix} \qquad [C][E_4] = \begin{bmatrix} \underline{\quad} & \underline{\quad} \\ \underline{\quad} & \underline{\quad} \end{bmatrix}$$

$$[C][E_1] = \begin{bmatrix} \underline{\quad} & \underline{\quad} \\ \underline{\quad} & \underline{\quad} \end{bmatrix} \qquad [C][E_2] = \begin{bmatrix} \underline{\quad} & \underline{\quad} \\ \underline{\quad} & \underline{\quad} \end{bmatrix}$$

$$[C][E_3] = \begin{bmatrix} \underline{\quad} & \underline{\quad} \\ \underline{\quad} & \underline{\quad} \end{bmatrix} \qquad [C][E_3] = \begin{bmatrix} \underline{\quad} & \underline{\quad} \\ \underline{\quad} & \underline{\quad} \end{bmatrix}$$

$$[D][E_1] = \begin{bmatrix} \underline{\quad} & \underline{\quad} \\ \underline{\quad} & \underline{\quad} \end{bmatrix} \qquad [D][E_2] = \begin{bmatrix} \underline{\quad} & \underline{\quad} \\ \underline{\quad} & \underline{\quad} \end{bmatrix}$$

$$[D][E_3] = \begin{bmatrix} \underline{\quad} & \underline{\quad} \\ \underline{\quad} & \underline{\quad} \end{bmatrix} \qquad [D][E_4] = \begin{bmatrix} \underline{\quad} & \underline{\quad} \\ \underline{\quad} & \underline{\quad} \end{bmatrix}$$

8. Plot the 16 images from Problem 7 on this dot paper.

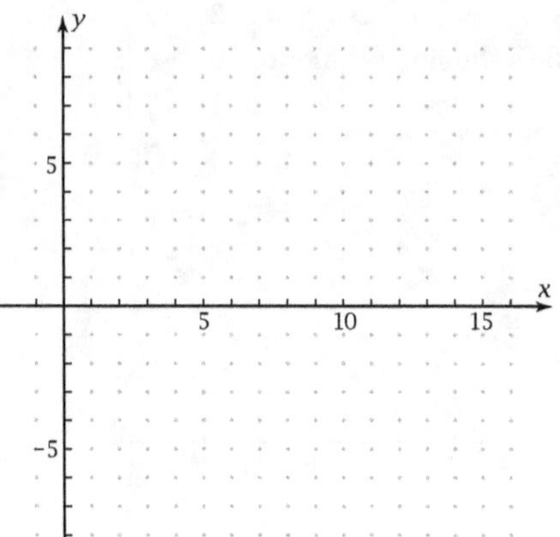

9. If the transformations are iterated infinitely many times, the result is part of Helge von Koch's **snowflake curve.** The second iteration of that curve is shown here, and the result of many iterations in Problem 7 is also shown (rotated 90° and enlarged). Explain why parts of this image can be said to be **self-similar.**

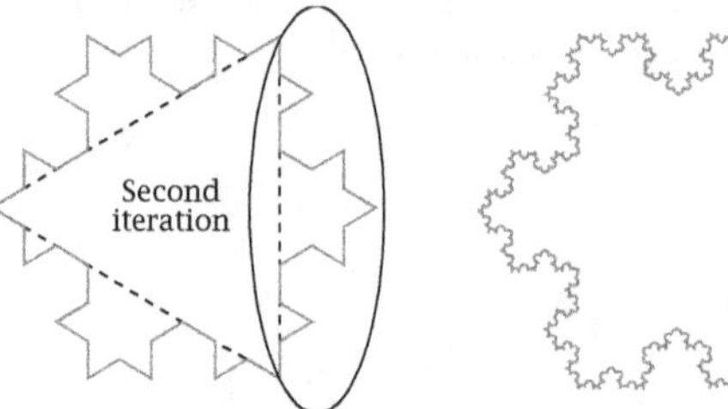

10. The length of the pre-image is 12 units. The first iteration you drew in Problem 6 has total length 16 units because it has four (self-similar) segments that are each 4 units long. What is the total length of the second iteration? the third iteration? the fourth iteration? Use the pattern you discover to find the length of the 100th iteration.

   What would be the total length of this part of the final snowflake curve? Surprising?

11. Use Barnsley's method to show that you get the same **strange attractor** when you start with *one* point as a pre-image, and do the four transformations iteratively, at random, on the resulting images.

12. What did you learn as a result of doing this Exploration that you did not know before?

Exploration 116: Fractal Dimension of a River

Date: _____________

Objective: Apply Hausdorff's definition of *dimension* to the Guadalupe River.

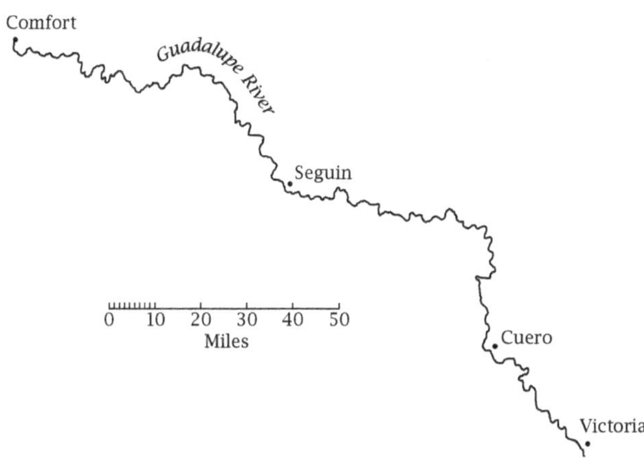

The figure shows the Guadalupe River between the towns of Comfort and Victoria. As the crow flies, the two towns are 150 miles apart. But the river is longer because it meanders. In this Exploration, you will find out something surprising about the actual length of the river.

1. Draw the bee-line distance from Comfort to Victoria. Do you agree that this distance is 150 miles?

____

2. If you use a ruler 150 miles long, the river would measure $N = 1$ ruler length. If you use a ruler only 50 miles long, then measuring the distance between points on the river, you will get *more* than three ruler lengths. Starting at Comfort, draw a line segment to a point on the river a direct distance of 50 miles from Comfort. Draw other 50-mile segments connecting points on the river until you reach Victoria. (The last segment will be a fraction of 50 miles.) How long does the river appear to be, using a 50-mile ruler? To two decimal places, how many ruler lengths is this?

3. This table shows the number of ruler lengths, N, as a function of the length of the ruler. Do you agree with N for a 20-mile ruler? _____________

Ruler (mi)	N Pieces	Ratio, r	$\frac{1}{r}$
150	1		
50	3.16		
20	8.2		
10	17.2		
5	35.0		
2	90.5		

4. In the table, fill in the columns for the ratio

$$r = \frac{\text{ruler length}}{150}$$

and for the reciprocal, $\frac{1}{r}$.

5. Enter lists for $\frac{1}{r}$ and N on your grapher. Use these to calculate other lists containing values of $\log \frac{1}{r}$ and $\log N$. Write the values here, rounded to two decimal places.

$\log \frac{1}{r}$	$\log N$

6. Plot $\log N$ versus $\log \frac{1}{r}$ on this graph paper. If any points do not lie on a straight line, go back and check your work.

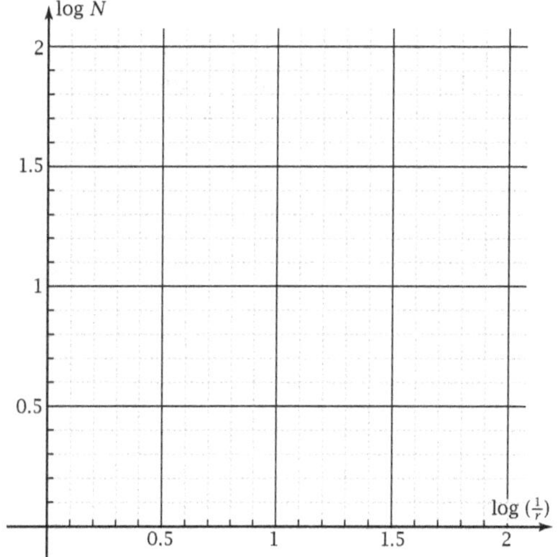

7. Perform linear regression to find the best-fitting linear function for $\log N$ as a function of $\log \frac{1}{r}$. Write the particular equation here.

(Over)

8. Use the equation in Problem 7 to calculate N if $\frac{1}{r} = 15$ (10-mile ruler). How close does the answer come to 17.2 in the table?

9. The graph in Problem 6 and the equation in Problem 7 show that the line goes almost through the origin. Assuming that it does, the equation is

$$\log N = m \log \frac{1}{r}$$

By dividing both sides by $\log \frac{1}{r}$, show that m satisfies Hausdorff's definition of *dimension*. What is the dimension of the Guadalupe River?

10. Based on your equation in Problem 7, what would N equal if you used a 1-inch-long ruler? How many miles long would the river be? Surprising?

11. What did you learn as a result of doing this Exploration that you did not know before?

Exploration 117: Foerster's Tree

Objective: Show that you understand the fundamentals of iterated transformations and the resulting fractal figures.

The figure on the left shows a vertical segment 10 units high, starting at the origin. This pre-image is to be transformed into a "tree" with three self-similar pieces, each 6 units long, as shown in the right figure.

- Left branch at +30° to the trunk, starting at $y = 5$
- Trunk, starting at the origin
- Right branch at −30° to the trunk, starting at $y = 4$

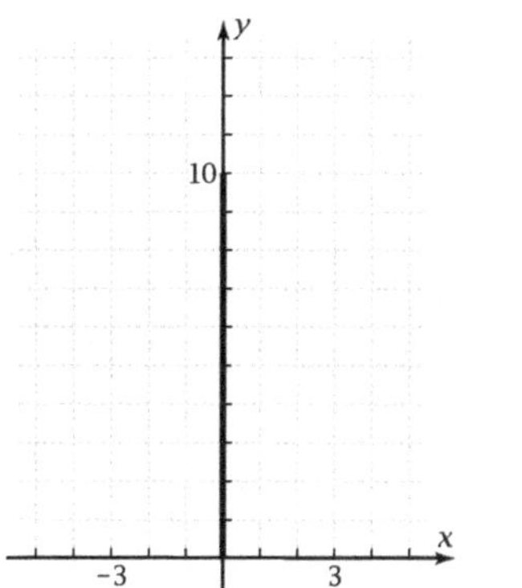
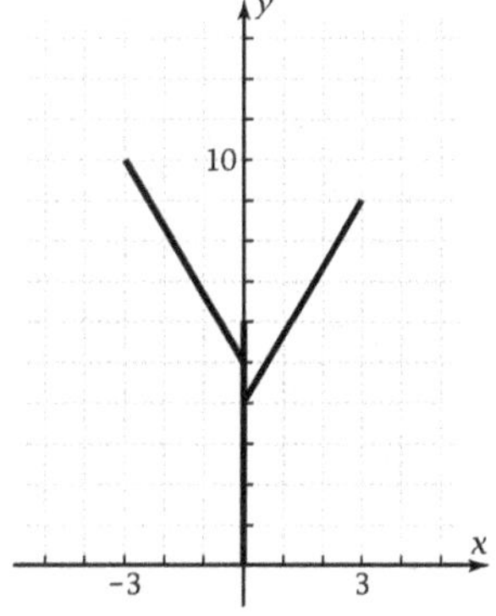

1. Write the pre-image on the left as a 3 × 2 matrix with 1 and 1 on the bottom row.

2. Write three 3 × 3 transformation matrices to do these things:

   [A] to transform the pre-image to the left branch

   [B] to transform the pre-image to the trunk

   [C] to transform the pre-image to the right branch

3. Apply transformation [A] iteratively using the ITRANS program and sketch. Use a window with an x-range of [−9, 9] and a y-range of [0, 12]. Because your grapher program is expecting a 3 × 4 pre-image, enter the pre-image this way:

$$[D] = \begin{bmatrix} 0 & 0 & 0 & 0 \\ 10 & 0 & 10 & 0 \\ 1 & 1 & 1 & 1 \end{bmatrix}$$

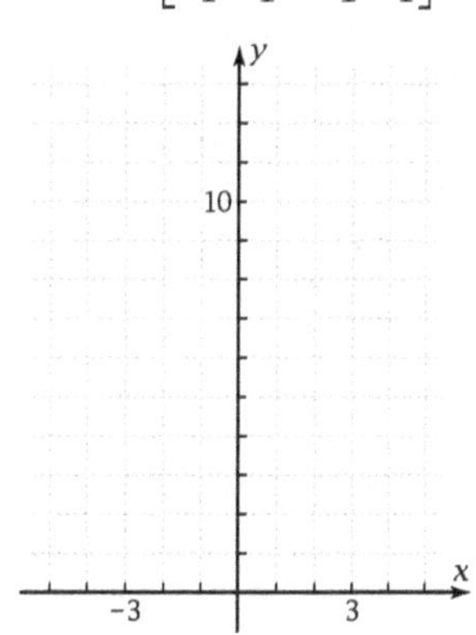

4. Calculate the fixed point numerically, and plot it on the graph paper in Problem 3. Do the images approach this fixed point?

5. The figure on the right preceding Problem 1 shows the three images in the first iteration. Store in your grapher the three transformation matrices from Problem 2. Then calculate (to one decimal place) the nine images in the second iteration.

   [A][A][D]   [A][B][D]

   [A][C][D]   [B][A][D]

   [B][B][D]   [B][C][D]

   [C][A][D]   [C][B][D]

   [C][C][D]

(Over)

6. Plot the nine images of the second iteration on the graph paper on the left here.

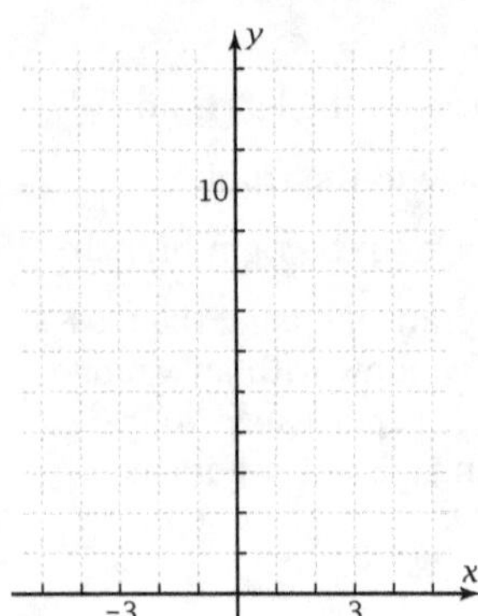 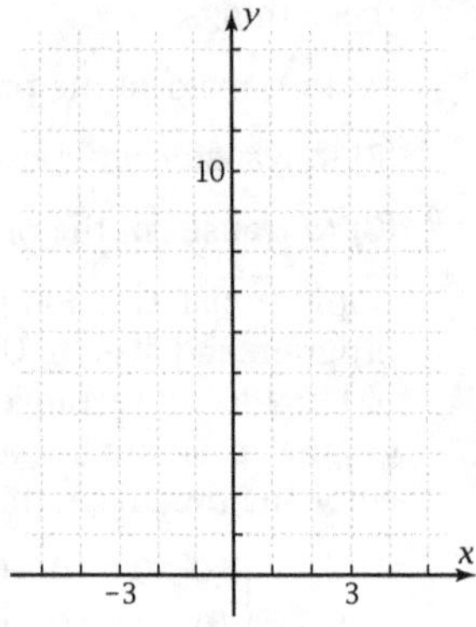

7. Sketch on the graph paper on the right what you think the tree might look like if the iterations were performed infinitely many times.

8. Use Barnsley's method to see what the tree would look like if the iterations were performed infinitely. Because your program is expecting four transformation matrices, enter the probabilities as

$$\frac{1}{3}, \frac{1}{3}, \frac{1}{3}, \text{ and } 0$$

in L_1 on the STAT menu. Does the result look like this figure?

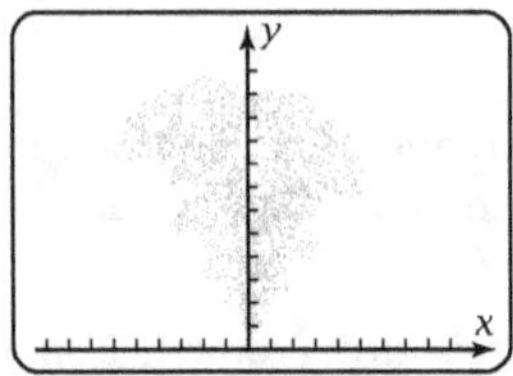

9. Describe Barnsley's method verbally.

10. What is the name of the kind of figure in Problem 8, to which points are attracted by iterating several different transformations?

11. What relationship do you notice between the fixed point in Problem 4 and the tree in Problem 8? (The image is made up of the fixed points for all possible sequences of the three transformations.)

12. For the first, second, third, and 100th iterations, calculate the sum of the lengths of the images. If the iterations were performed infinitely many times, what would the sum of the lengths of the images approach?

13. Each iteration divides each previous segment into three self-similar pieces, each of which is 0.6 times as long as before. Let N be the number of pieces, and let r be the ratio of the length of each piece to the length of the original pre-image. Complete this table of values:

Iteration	r	$\frac{1}{r}$	N
0			
1			
2			
3			
4			
5			

14. Put lists in your grapher for $\log \frac{1}{r}$ and $\log N$. By regression, show that $\log N$ is a linear function of $\log \frac{1}{r}$. Write the particular equation.

15. Write Hausdorff's definition of *dimension*. Use the result of Problem 14 to find quickly the dimension of the tree.

16. What is the name of a figure that has a fractional dimension?

17. What did you learn as a result of doing this Exploration that you did not know before?

Exploration 118: Parametric Equations of Conic Sections

Date: ____________

Objective: Plot graphs of conic sections in parametric form, and relate the result to the Cartesian equation.

1. Put your grapher in parametric mode and radian mode. Set a t-range of 0–2π and a window, with equal scales on the two axes, that has an x-range of $[-10, 10]$. Plot these **parametric equations** and sketch the result.

$$x = \cos t$$
$$x = \sin t$$

2. Square both sides of both equations in Problem 1. Then add the two equations, left side to left side and right side to right side. Use the Pythagorean property of circular functions to show that the result is equivalent to the **unit circle** $x^2 + y^2 = 1$.

3. Plot these parametric equations.

$$x = 5 \cos t$$
$$y = 3 \sin t$$

Describe verbally how the resulting **ellipse** is related to the unit circle in Problem 1.

4. Plot these parametric equations.

$$x = 2 + 5 \cos t$$
$$y = -1 + 3 \sin t$$

How is the graph related to the one in Problem 3?

5. Plot the **unit hyperbola** with these parametric equations and sketch the result.

$$x = \sec t$$
$$y = \tan t$$

6. Square both sides of both equations in Problem 5, and then combine the two equations in a way that shows that $x^2 - y^2 = 1$.

7. Plot this unit hyperbola, and explain how it is related to the one in Problem 5.

$$x = \tan t$$
$$y = \sec t$$

8. Plot this hyperbola. Sketch on the given axes.

$$x = -4 + 3 \sec t$$
$$y = 1 + 2 \tan t$$

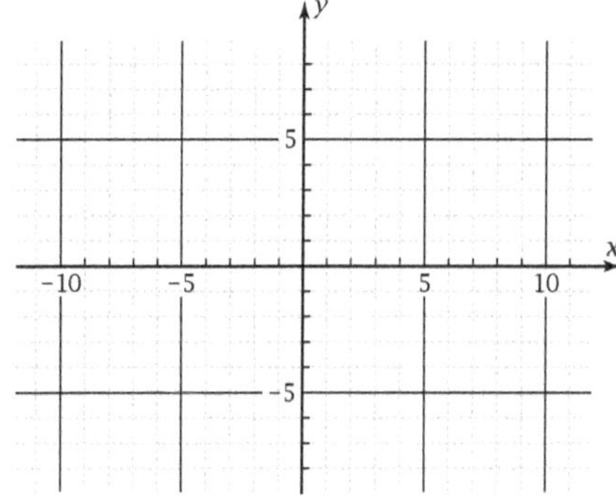

9. What did you learn as a result of doing this Exploration that you did not know before?

Exploration 119: Cartesian Equations of Conic Sections

Date: ___________

Objective: Sketch graphs of dilated and translated conic sections, and confirm by plotting parametrically.

1. For the equation

$$\left(\frac{x-7}{2}\right)^2 + \left(\frac{y+4}{5}\right)^2 = 1$$

which conic section will it be? _______________

2. Sketch the graph of the conic section in Problem 1.

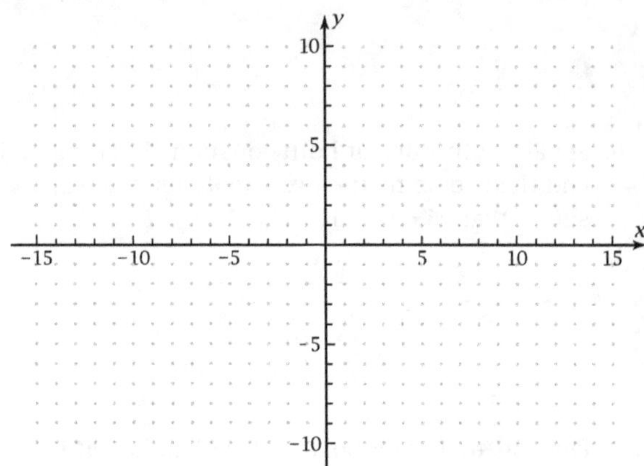

3. Write parametric equations for the conic section in Problem 1.

4. Put your grapher in parametric and radian modes. Set the t-range from 0 to 2π, and use the window shown in Problem 2. Plot the graph. Does it agree with your sketch in Problem 2? _______________

5. Transform the equation in Problem 1 to the form

$$Ax^2 + Cy^2 + Dx + Ey + F = 0$$

6. Return your grapher to function mode. Plot the transformed equation in Problem 5 using the program CONIC. Does the graph agree with those in Problems 2 and 4? _______________

7. For the equation

$$-\left(\frac{x+6}{4}\right)^2 + \left(\frac{y-1}{3}\right)^2 = 1$$

which conic section will it be? _______________

8. Sketch the graph of the conic section in Problem 7.

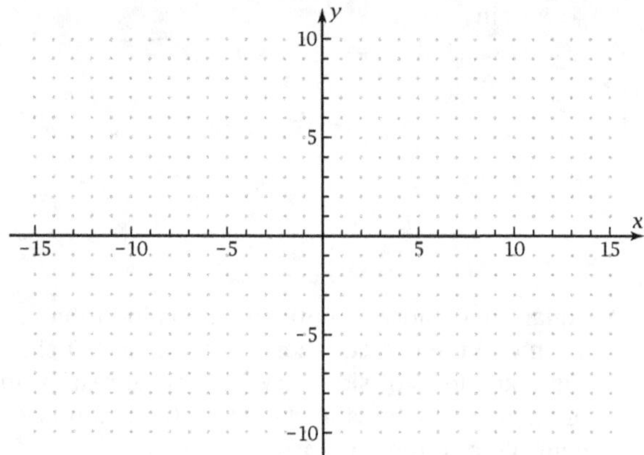

9. Write parametric equations for the conic section in Problem 6. Plot on your grapher. Does the graph agree with your sketch in Problem 8? _________

10. Write parametric equations for the hyperbola graphed here. Do the parametric equations give this graph? _______________

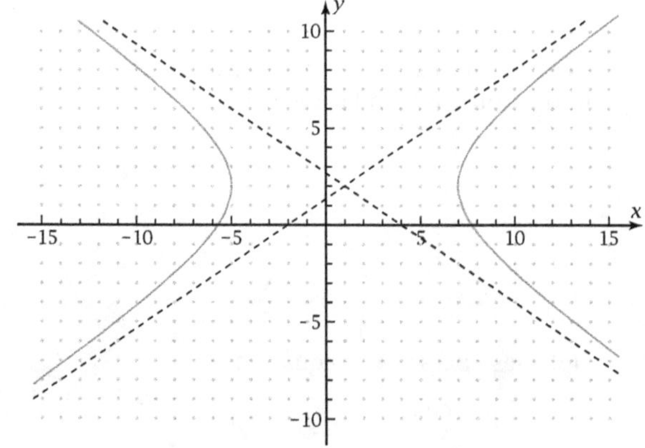

11. Write a Cartesian equation for the hyperbola.

12. What did you learn as a result of doing this Exploration that you did not know before?

Exploration 120: Quadric Surfaces

Objective: Sketch a figure formed by rotating a conic section about one of its axes.

1. The graph shows the parabola $y = x^2$. Sketch the paraboloid formed by rotating this parabola about the y-axis.

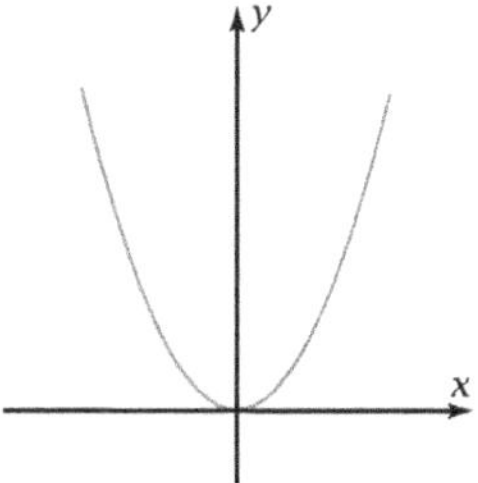

2. The graph shows the hyperbola $4x^2 - y^2 = 4$. Sketch the hyperboloid formed by rotating this hyperbola about the y-axis.

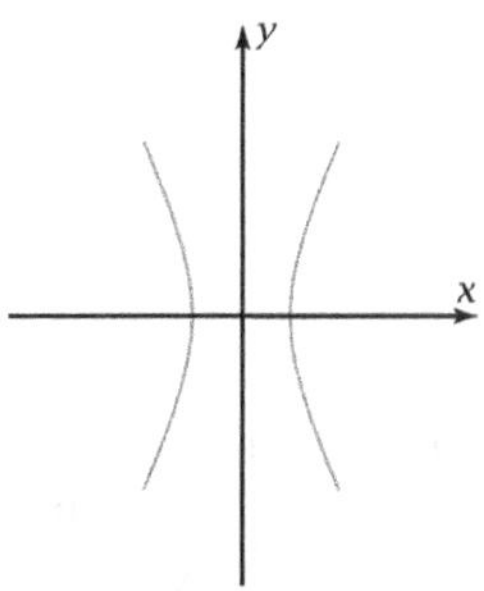

3. The graph shows the hyperbola $-4x^2 + y^2 = 1$. Sketch the hyperboloid formed by rotating this hyperbola about the y-axis.

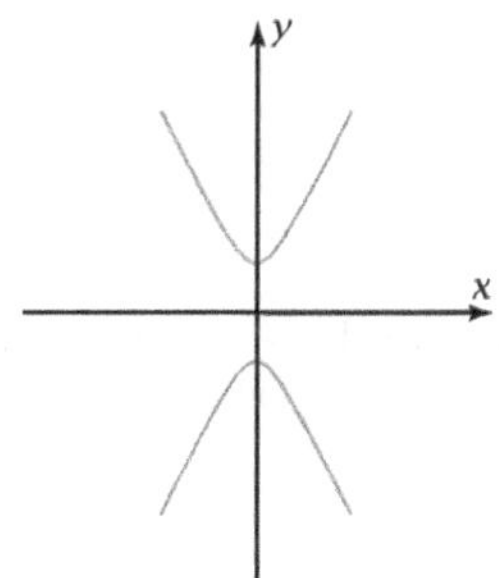

4. Why do you think the surface in Problem 2 is called a **hyperboloid of one sheet** and the surface in Problem 3 a **hyperboloid of two sheets**?

5. The graph shows the ellipse $x^2 + 9y^2 = 9$. Sketch the ellipsoid formed by rotating this ellipse about the x-axis. Do something to make it look three-dimensional.

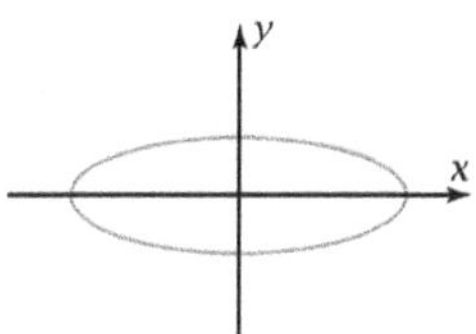

6. On this copy of the ellipse in Problem 5, sketch the ellipsoid formed by rotating the ellipse about the y-axis.

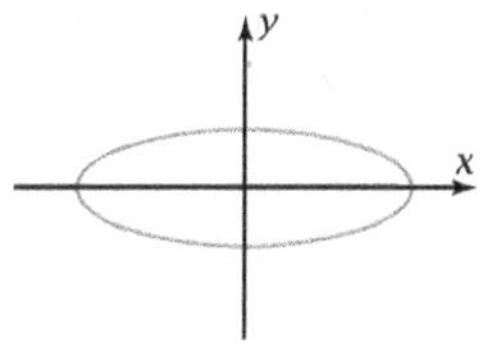

7. What did you learn as a result of doing this Exploration that you did not know before?

Exploration 121: Cylinder in Nose Cone Problem

Objective: Find the maximum volume cylinder that can be inscribed in a given cone.

The figure shows a nose cone being designed for a spaceship. The cone is generated by rotating about the y-axis the part of the line $y = -3x + 6$ that lies in the first quadrant. The dimensions x and y are in feet. A cylindrical container is to be fitted into the nose cone to house radio equipment.

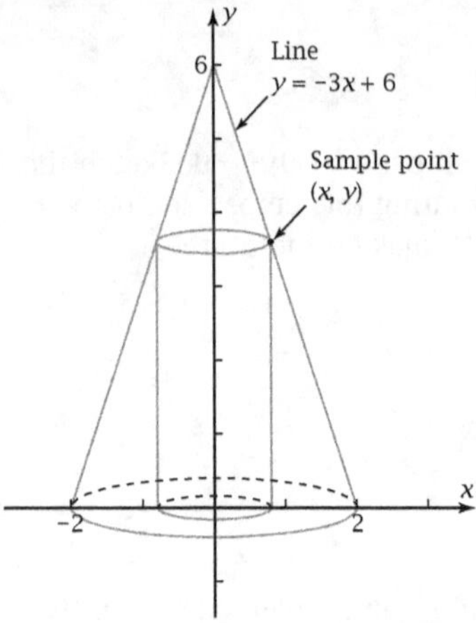

1. The cylinder shown in the figure touches the line generating the cone at the **sample point** (x, y) where $x = 0.8$. What are the radius and altitude of the cylinder? Find the volume of the cylinder shown in the figure.

2. If the sample point is moved to point $(1, 3)$, the radius of the cone will be larger but the altitude will be smaller. Is the volume of the resulting cylinder larger or smaller than the cylinder shown in the figure? By how many cubic feet?

3. Write the volume, V, of the cylinder as a function of the coordinates x and y of the sample point. Transform the equation by substituting for y so that V is a function of x alone.

4. Plot V as a function of x. Sketch the graph.

5. What are the radius and altitude of the largest cylinder that will fit inside this nose cone? What is this maximum volume? Explain how you find your answers.

6. Sketch the maximal cylinder on the given figure. Is the cylinder tall and skinny or short and fat?

7. What did you learn as a result of doing this Exploration that you did not know before?

Exploration 122: Focus and Directrix of an Ellipse

Objective: Find out the meanings of *focus* and *directrix* of an ellipse.

The figure shows an ellipse centered at the point (0, 0).

- One **vertex** is at the point (6, 0).
- One **directrix** is the line $x = 10$.
- One **focus** is at the point (3.6, 0); the other focus is at the point (−3.6, 0).
- The **eccentricity** is $e = 0.6$.

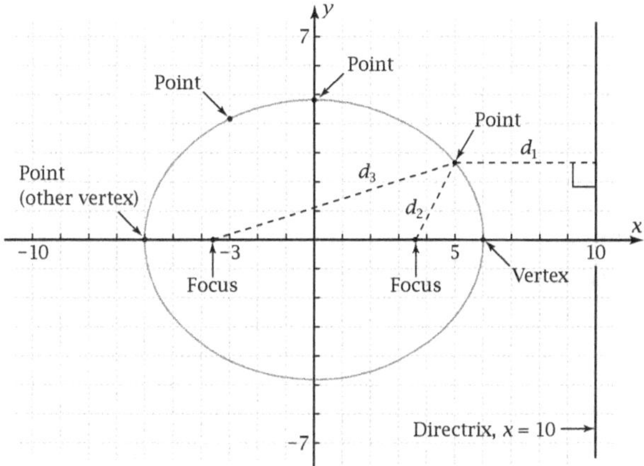

A point is shown on the ellipse at $x = 5$. Its distance from the directrix is d_1, its distance from the focus on the right is d_2, and its distance from the focus on the left is d_3.

1. Measure these distances using the scales on the axes. You can put marks on a piece of paper to measure the slant distance.

   $d_1 =$ __________, $d_2 =$ __________, $d_3 =$ _________

2. Show that $d_2 = 0.6d_1$.

3. Show that $d_2 + d_3 = 12$, the major axis length.

4. Measure d_1, d_2, and d_3 as in Problem 1 for the marked points in this table. Show that in each case, $d_2 = 0.6d_1$ and $d_2 + d_3 = 12$, within the limits of accuracy that you can measure.

x	d_1	d_2	d_3	$d_2 = 0.6d_1$	$d_2 + d_3 = 12$
−3					
0					
6					
−6					

5. The equation for the ellipse shown is

   $$\left(\frac{x}{6}\right)^2 + \left(\frac{y}{4.8}\right)^2 = 1$$

   Use the Pythagorean theorem to show that for the point (0, 4.8), d_2 and d_3 both equal exactly 6, the same as the distance from the center to the vertex, and thus that $d_2 + d_3$ is exactly 12, the length of the major axis.

6. Use the equation in Problem 5 to show that for the point shown, y is exactly $\sqrt{7.04}$ where $x = 5$. Then find d_2 and d_3 exactly, using the Pythagorean theorem. Show that $d_2 + d_3 = 12$.

7. Use the result of Problem 6 to show that $d_2 = 0.6d_1$.

8. Look up the **focus-directrix property** of ellipses. State the property here.

9. Look up the **two-foci** property for ellipses. State the property here.

10. What did you learn as a result of doing this Exploration that you did not know before?

Exploration 123: Focus, Directrix for Parabola and Hyperbola

Objective: Find out the meanings of *focus*, *directrix*, and *eccentricity* for parabolas and hyperbolas.

Problems 1–6 refer to the parabola shown here.

- Its focus is at point (6, 0).
- Its vertex is at point (7, 0).
- Its directrix is the line $x = 8$.

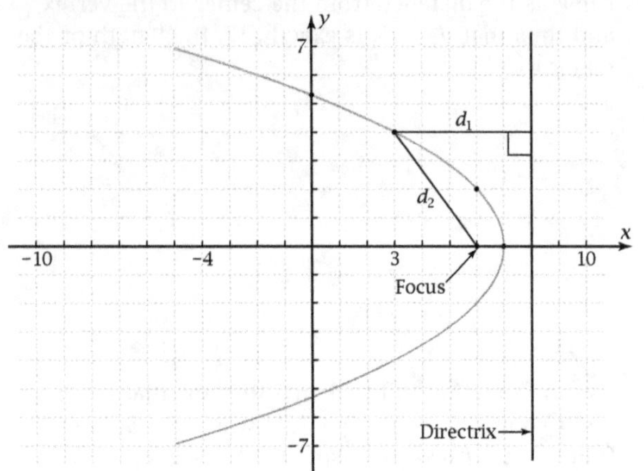

1. A point is shown on the parabola where $x = 3$. Its distance from the directrix is d_1, and its distance from the focus is d_2. Measure these distances using the scales on the axes. Show that $d_2 = d_1$.

2. Measure the distances d_1 and d_2 as in Problem 1 for the marked points in this table. Show that in each case, $d_2 = d_1$.

x	d_1	d_2	Equal?
6			
0			
-4			
7			

3. Pick a point on the ellipse and call it (x, y). Write d_1 from the directrix and d_2 from the focus in terms of x and y. By setting $d_2 = d_1$, find a particular equation of the parabola. Eliminate the radical.

4. How does your equation in Problem 3 reveal algebraically that the graph will be a parabola?

5. Confirm using your grapher that the equation in Problem 3 gives the parabola shown. _________

6. Recall that for an ellipse, $d_2 = ed_1$, where e is the eccentricity and d_1 and d_2 are distances from a point on the ellipse to the directrix and to the focus, respectively. What, then, is the eccentricity of a parabola?

Problems 7–9 refer to the hyperbola shown here.

- Focus: Point (4, 0)
- Vertex: Point (2, 0)
- Directrix: Line $x = 1$

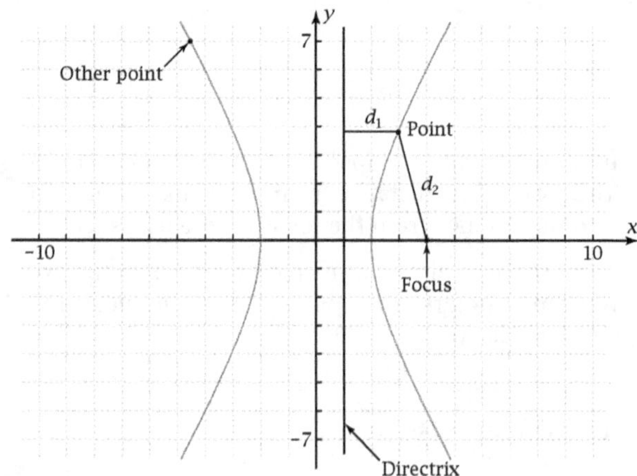

7. Show that $d_2 = 2d_1$ for point P in Quadrant I.

8. Draw d_1 and d_2 for point Q (Quadrant II). Show that $d_2 = 2d_1$ for the other point, too.

9. What is the eccentricity of this hyperbola?

10. What did you learn as a result of doing this Exploration that you did not know before?

Exploration 124: Two-Foci Property for an Ellipse

Date: _______________

Objective: Discover something about the other focus of an ellipse.

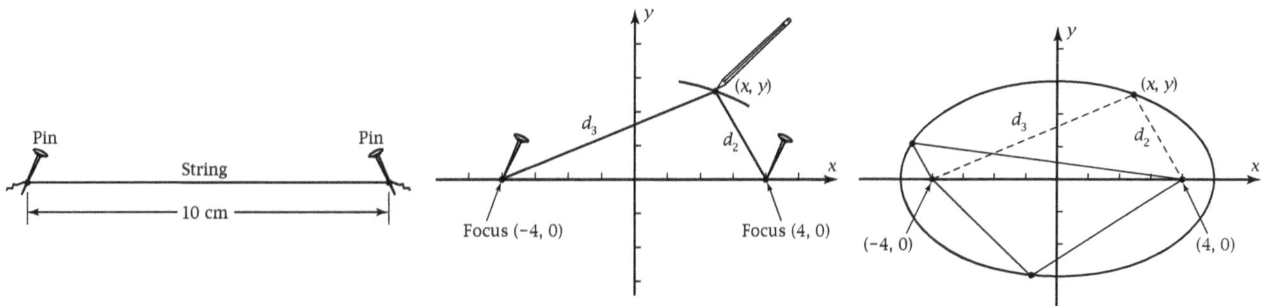

1. The graph on the right shows the ellipse

   $$9x^2 + 25y^2 = 225$$

   Confirm that the equation gives correct values by substituting 4 for x, calculating the two values of y, and showing that the two points are on the graph.

2. Get two pins. Tie a piece of string to the pins so that the pins are exactly 10 spaces apart (the length of the major axis) when the string is stretched tight. Then put the pins at the points (−4, 0) and (4, 0). Trace around as shown in the figure. What do you notice about the path of your pencil?

3. The points in Problem 2 are called **foci** (the plural of *focus*). The distance 4 from the center to each focus is called the **focal radius.** Put your pencil at the point (0, 3) on the ellipse, and draw a right triangle using one focus and the origin. Explain the relationship among the minor radius, the focal radius, and the major radius.

4. Suppose that an ellipse has an x-radius of 30 cm and a y-radius of 70 cm. Sketch the ellipse. Find the focal radius and sketch the two foci.

5. What did you learn as a result of doing this Exploration that you did not know before?

Exploration 125: Pythagorean Property for Hyperbolas

Objective: Show that for a hyperbola, (focal radius)^2 = (transverse radius)^2 + (conjugate radius)^2.

Problems 1–6 refer to the hyperbola graphed here. Its equation is

$$9x^2 - 16y^2 = 144$$

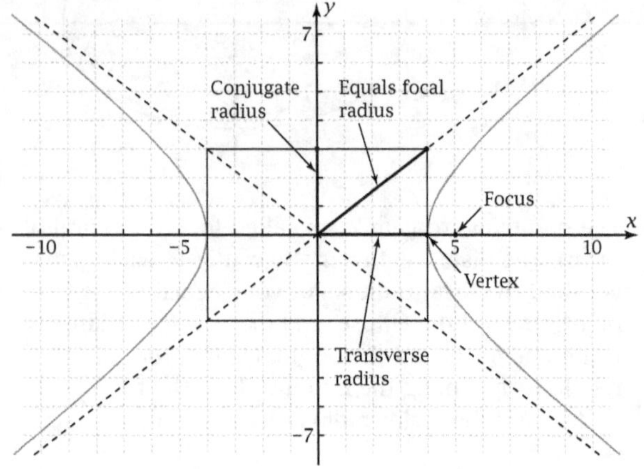

1. Plot the hyperbola on your grapher. Does the result agree with the given figure? _______________

2. Transform the equation to make the right side equal 1. From the result, write the *x*-radius and the *y*-radius.

3. What is the **slope** of the asymptotes? How can you calculate this slope using the *x*- and *y*-radii?

4. The **transverse radius,** *a*, of a hyperbola goes from the center to a vertex. The **conjugate radius,** *b*, is the radius perpendicular to *a*. What are the values of these radii for this hyperbola?

5. One focus of this hyperbola is at the point (5, 0). Thus, the **focal radius** (center to focus) is *c* = 5. By direct measurement, show that the length of the segment from the center along the asymptote to the point (4, 3) also equals the focal radius.

6. The segment you measured in Problem 5 is the hypotenuse of a right triangle. Use the facts that *c* is the length of this hypotenuse and that *a* and *b* are the two legs to write the **Pythagorean property for hyperbolas.** Draw a box around this equation to help you remember it.

For Problems 7–9, the hyperbola graphed here has the *y*-radius equal to the transverse radius.

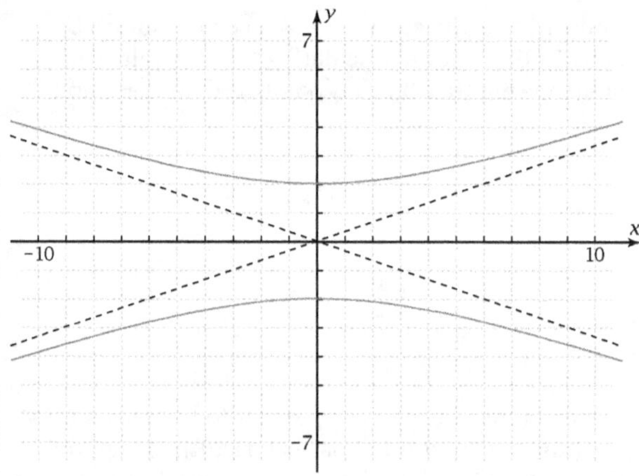

7. Draw the transverse and conjugate radii.

8. Calculate the focal radius. Show both foci.

9. True or false? "The transverse radius of a hyperbola is always longer than the conjugate radius."

10. What did you learn as a result of doing this Exploration that you did not know before?

Exploration 126: More Analytic Properties of Conics   Date: ____________

Objective: Demonstrate that you know the focus-directrix property, the two-foci property, the Pythagorean property, and the eccentricity properties for conic sections.

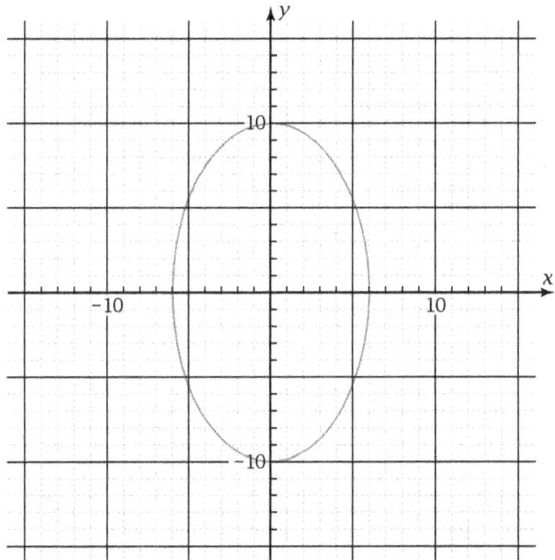

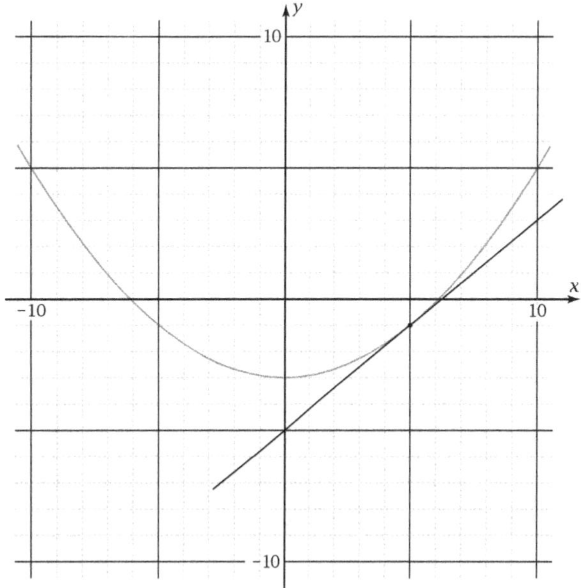

1. The figure shows an ellipse of x-radius 6 and y-radius 10, centered at the origin. Write the Cartesian equation. Transform it so that there are no fractions.

2. Calculate the focal radius, the eccentricity, and the directrix radius.

3. Show the two foci and the two directrices.

4. Plot the point where $x = 5$. Measure the distances from this point to the two foci and to the nearest directrix. Demonstrate that the two-foci property and the focus-directrix property are true.

5. The graph shows a parabola with focus at the origin. What is the eccentricity of a parabola?

6. Draw the directrix of this parabola.

7. Use the focus-directrix property to find the particular equation. Transform it so that y is given in terms of x.

8. Calculate y when x is 5, and show that you get the y-value shown on the graph.

9. The line at the point where $x = 5$ is *tangent* to the graph. Show that a light shining from the focus would be reflected by this line exactly *vertically*.

10. What did you learn as a result of working this Exploration that you did not know before?

Exploration 127: The Discriminant of a Conic Section

Date: _______________

Objective: Figure out how to tell which conic section the graph will be when there is an xy-term.

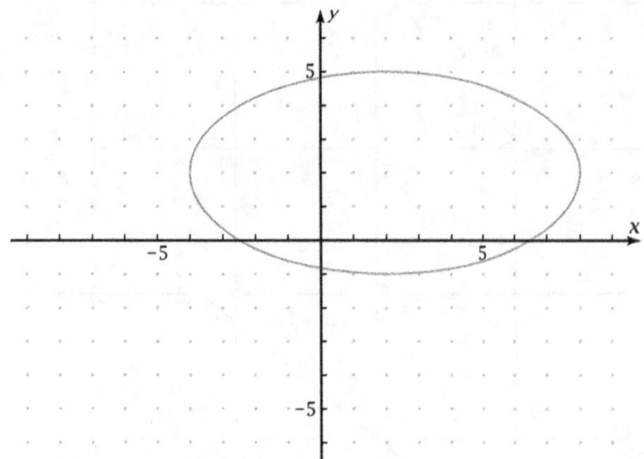

1. The figure shows the graph of

   $$x^2 + 4y^2 - 4x - 16y - 16 = 0$$

   Confirm that this equation produces the given graph by plotting it on your grapher.

2. On the same screen as in Problem 1, plot the equation

   $$x^2 + xy + 4y^2 - 4x - 16y - 16 = 0$$

   Sketch the resulting graph on the given figure.

3. On the same screen, plot the equation in Problem 1 along with the equation

   $$x^2 - xy + 4y^2 - 4x - 16y - 16 = 0$$

   Sketch the result on this next figure.

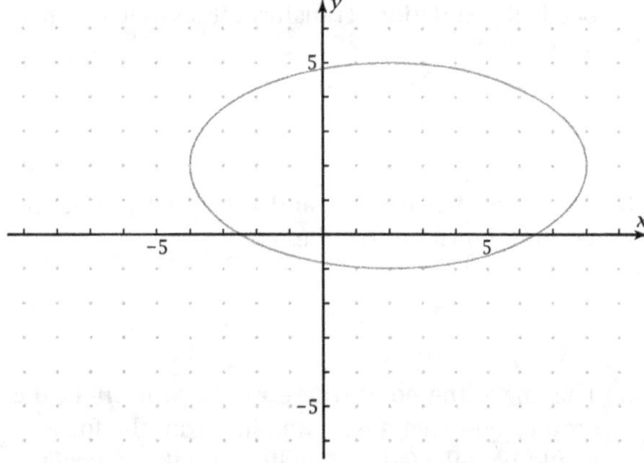

4. What is the major difference in the effect of changing the xy-term to negative as in Problem 3?

5. The **discriminant** for the equation

   $$Ax^2 + Bxy + Cy^2 + Dx + Ey + F = 0$$

   is $B^2 - 4AC$. Calculate the discriminant for each equation in Problems 1, 2, and 3.

6. Find a value of B that will make the discriminant equal zero for

   $$x^2 + Bxy + 4y^2 - 4x - 16y - 16 = 0$$

7. On the same screen, plot the graphs in Problems 1 and 6 (using $B = 4$). Sketch the graphs here.

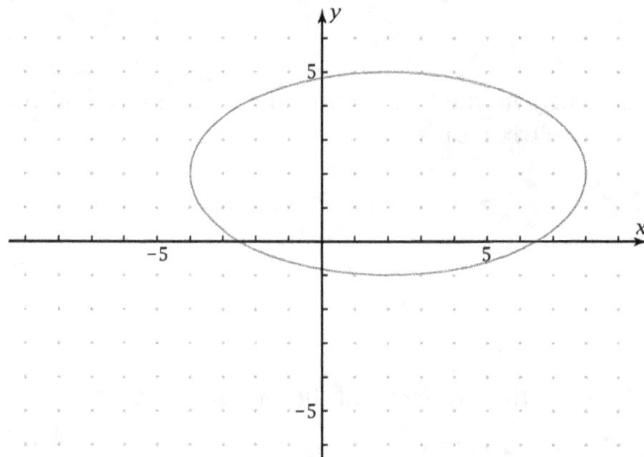

8. Which conic section does the second graph in Problem 7 seem to be?

Exploration 127: The Discriminant of a Conic Section *continued*

9. On the same screen, plot the two equations

$$x^2 + 4y^2 - 4x - 16y - 16 = 0 \quad \text{(from Problem 1)}$$

$$x^2 + 5xy + 4y^2 - 4x - 16y - 16 = 0$$

Sketch the result here.

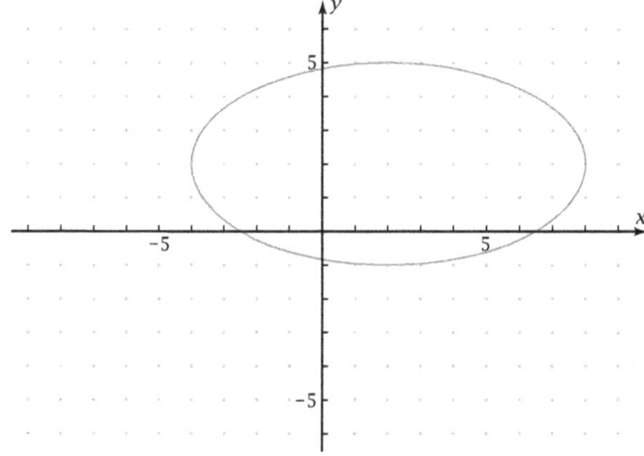

10. Which conic section is the second graph in Problem 9? What does the discriminant equal for this second equation?

11. Write a conclusion about how you can tell from the discriminant just which conic section the graph of a quadratic relation will be.

12. Show that you understand the conclusion in Problem 11 by stating which conic section each of these graphs will be.

$$3x^2 + 7xy + 4y^2 - 8x + 13y + 27 = 0$$

Disc. = _______________ Conic: _______________

$$5x^2 - 2xy + 6y^2 + 22x - 11y - 100 = 0$$

Disc. = _______________ Conic: _______________

$$10x^2 + 50xy + 3y^2 - 91x + 33y + 2001 = 0$$

Disc. = _______________ Conic: _______________

$$x^2 + 2xy + y^2 - 3x - 7y + 29 = 0$$

Disc. = _______________ Conic: _______________

$$x^2 + 3xy + y^2 - 3x - 7y + 29 = 0$$

Disc. = _______ Conic: _______________

13. Show that the old rules for determining whether the graph is a circle, an ellipse, a hyperbola, or a parabola when there is no xy-term are consistent with what you have learned about the discriminant.

14. What did you learn as a result of doing this Exploration that you did not know before?

Exploration 128: A Connection Between Conic Sections and Marketing

Objective: Find the set of values of (x, y) for which it is cheaper to ship from a more expensive warehouse.

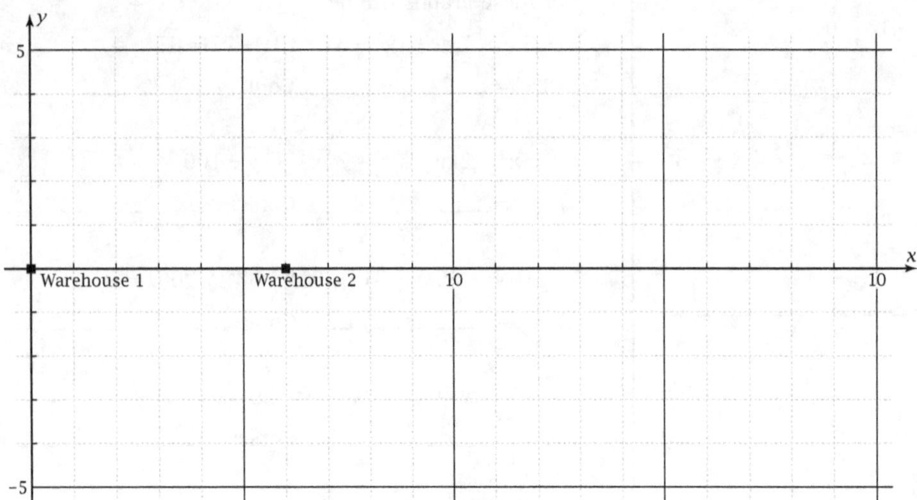

1. Warehouse 1 is located at the origin of a Cartesian coordinate system. Warehouse 2 is located at the point $(6, 0)$, where x and y are in miles. Which warehouse is the point $(12, 5)$ closer to? How much closer?

2. Shipping a truckload of goods from Warehouse 1 costs \$10 per mile. Shipping the same truckload of goods from Warehouse 2 would cost \$20 per mile. If the destination of the truckload is the point $(12, 5)$, from which warehouse is it cheaper to ship, the closer one or the more remote one? How much cheaper?

3. Find a point (not *at* Warehouse 2) for which it would be cheaper to ship from Warehouse 2. Justify your answer.

4. Check with your instructor before proceeding.

   ____

5. Write an inequality that says that the shipping cost from Warehouse 2 is less than or equal to that from Warehouse 1. Transform it to remove any radicals or fractions.

6. Graph the region specified by the inequality in Problem 5. Which conic section is the boundary of the region?

7. What did you learn as a result of doing this Exploration that you did not know before?

Exploration 129: Reflecting Property of Ellipses

Objective: Figure out what happens to a ray of light that starts at a focus and reflects off an elliptical surface.

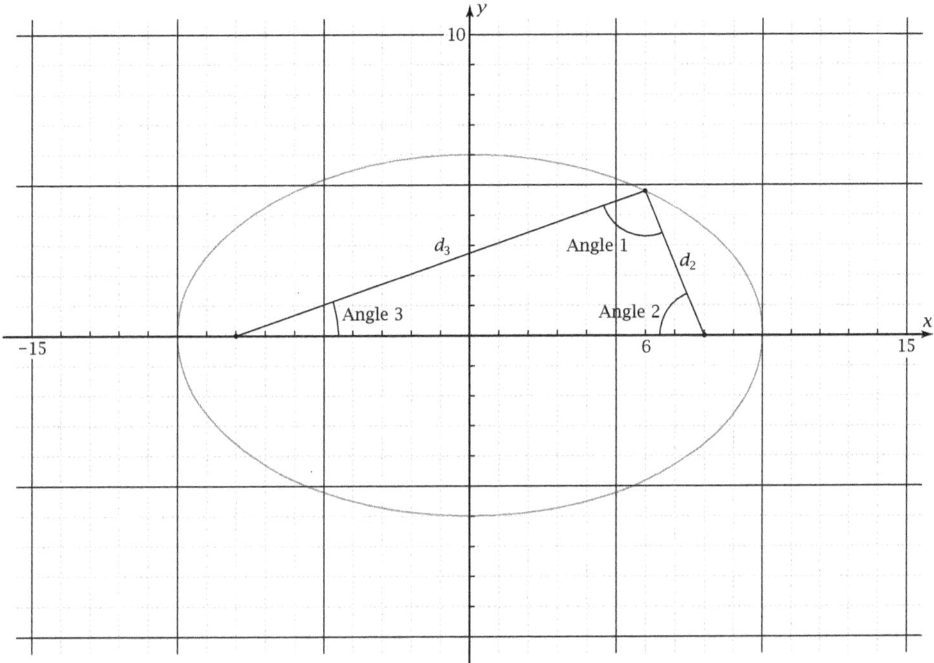

The ellipse $9x^2 + 25y^2 = 900$, shown in the figure, has foci at the points (8, 0) and (−8, 0). Lines of length d_2 and d_3 are drawn from the foci to the point on the ellipse where $x = 6$. The lines make Angle 2 and Angle 3 with the x-axis.

1. Calculate y for the point where $x = 6$.

2. Use the value of y from Problem 1 and appropriate trigonometry to calculate the measures of Angles 2 and 3. Store these without round-off in your grapher's memory.

   Angle 2 = _____________   Angle 3 = _____________

3. Calculate the measure of Angle 1. Store it in your grapher.

4. Calculate the *supplement* of Angle 1. Divide this number by 2.

   Supplement: _______________

   $\frac{1}{2}$ Supplement: _______________

5. Construct angles of measure $\frac{1}{2}$ Supplement with vertex at the point (6, y), one counterclockwise from d_2 and the other clockwise from d_3.

6. Explain why the other sides of the two angles in Problem 5 lie in a line.

7. What is the relationship of the line in Problem 6 to the ellipse?

8. Plot the point where $x = -9$. Show (quickly) that the property in Problem 7 applies at this point.

9. State in words the property you have discovered in this Exploration.

Exploration 130: Computation of a Tangent Line

Objective: Given an ellipse, calculate the equation of a line tangent to it at a given point, and verify the reflecting property.

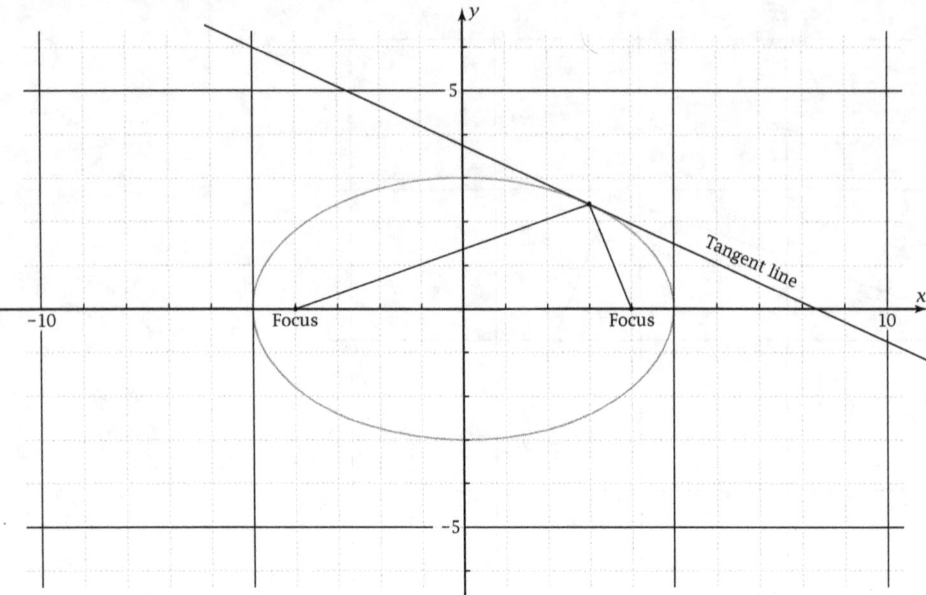

The ellipse shown has equation $9x^2 + 25y^2 = 225$. The line is tangent to the ellipse at the point where $x = 3$.

1. Calculate y for the point shown where $x = 3$.

2. The tangent line has slope of exactly -0.45. Draw the rise and run on the diagram to verify that this is true.

3. Find an equation of the tangent line in Problem 2.

4. Check with your instructor before proceeding.

   ____

5. The foci of the ellipse are at the points (6, 0) and (−6, 0). The segments connect the foci to the point (3, y). Calculate the acute angles these segments make with the tangent line. You will need to be resourceful to do this! What do you notice about the two answers?

6. Describe how the result of Problem 5 explains the reflecting property of ellipses.

Exploration 130: Computation of a
Tangent Line *continued*

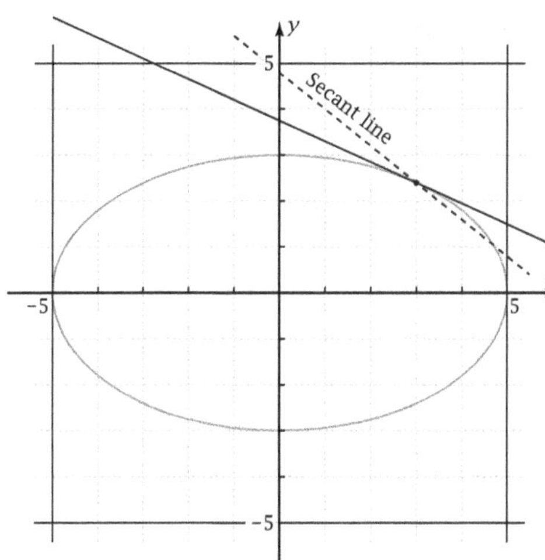

7. The graph shows a secant line (cutting the ellipse in *two* places) as well as the tangent line. Let m be the slope of the secant line. Write the equation of the secant line. (Recall that it contains the point where $x = 3$.) Transform so that y is given in terms of x and m.

8. Solve the system of equations for the line in Problem 7 and the ellipse in Problem 1. Expand and combine the x^2-terms, the x-terms, and the constant term.

9. The quadratic equation in Problem 8 will have exactly *one* solution if m is the slope of the *tangent* line. In this case, the discriminant will equal zero. Find the slope of the tangent line. Be prepared for a lot of calculation!

10. What did you learn as a result of working this Exploration that you did not know before?

Exploration 131: Equation from a Geometric Definition

Objective: Given the geometric definition of a set of points, find the particular equation.

The figure shows the graph of a set of points. For each point (x, y) on the graph, its distance d_1 from the fixed point $(7, 5)$ is *twice* its distance from the fixed point $(1, 2)$.

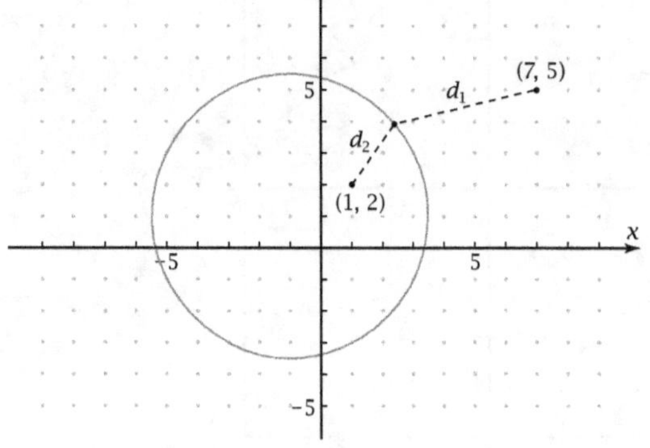

1. Mark a piece of paper with the length d_2 shown in the figure. Does d_1 really equal $2d_2$? _______________

2. Pick another point (x, y) on the graph. Draw lines to it from the points $(7, 5)$ and $(1, 2)$. Does $d_1 = 2d_2$ for this point? _______________

3. Starting with the equation $d_1 = 2d_2$, find each distance in terms of x and y using the **distance formula.** Square both sides of the resulting equation to eliminate the radicals. Then simplify as much as possible.

4. How does your answer to Problem 3 confirm that the graph is really a *circle?*

5. Confirm that your answer to Problem 3 is correct by plotting the graph on your grapher.

____

6. **Complete the square** to find the center and radius of the circle. Confirm on the graph that these quantities are correct.

7. What did you learn as a result of doing this Exploration that you did not know before?

Exploration 132: A Quadratic-Quadratic System

Objective: Apply your knowledge of conic section equations to solving a system of two quadratic equations in two variables.

The figure shows an ellipse and a hyperbola plotted on the same axes. The equations for these graphs form a **system** of equations. A solution of this system is a point (x, y) that satisfies both equations.

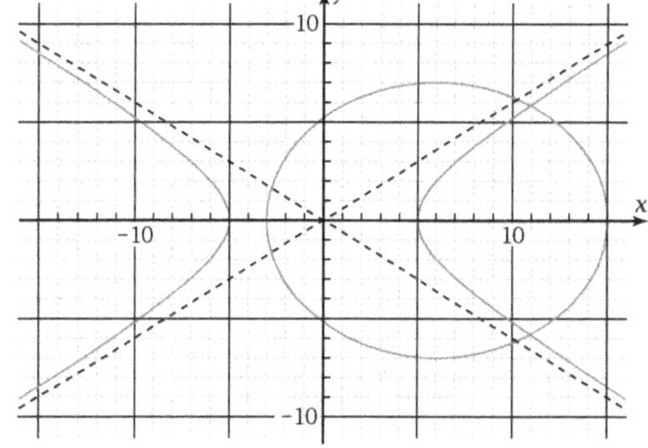

1. Find all solutions of this system **graphically** by finding the points where the graphs cross. Write the coordinates to the nearest decimal place.

2. Write Cartesian equations for the two graphs. Transform each equation to the form $Ax^2 \pm Cy^2 + Dx + Ey + F = 0$.

3. Solve the system **numerically** by plotting both equations on the same screen using the CONIC2 program and using the intersect feature. Do your answers to Problem 1 agree with these more precise answers?

4. Solve the system of equations **algebraically.** To do this, try combining the two equations to eliminate one of the variables. Solve the resulting equation using the quadratic formula. Substitute the two values you get for the one variable into one of the equations to find the values of the other variable. Explain why you get four points, only two of which have real-number coordinates.

5. What did you learn as a result of doing this Exploration that you did not know before?

Exploration 133: Limaçon in Polar Coordinates

Objective: Plot polar curves on paper and on your grapher.

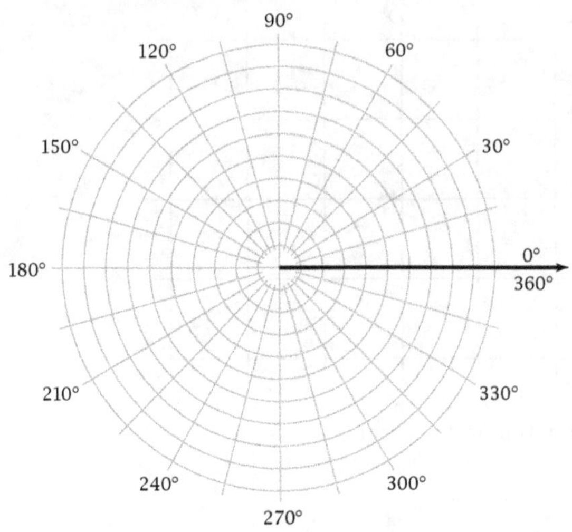

1. Here is a table of values of r and θ for a curve in polar coordinates. On the given polar coordinate paper, plot the points and connect them with a smooth curve.

θ	r	θ	r
0°	3.0	195°	1.2
15°	4.8	210°	−0.5
30°	6.5	225°	−1.9
45°	7.9	240°	−3.1
60°	9.1	255°	−3.8
75°	9.8	270°	−4.0
90°	10.0	285°	−3.8
105°	9.8	300°	−3.1
120°	9.1	315°	−1.9
135°	7.9	330°	−0.5
150°	6.5	345°	1.2
165°	4.8	360°	3.0
180°	3.0		

2. Explain how you plot points for which r is negative.

3. The equation of the curve in Problem 1 is

$$r = 3 + 7 \sin \theta$$

Do you agree that r-values from this equation, rounded to one decimal place, are the same as the values in the table in Problem 1?

4. Plot the graph in Problem 2 on your grapher. Use a θ-range of [0°, 360°] with a θ-step of 5°. Use equal scales on the two axes. Does the graph on your grapher confirm the graph you plotted in Problem 1?

5. What is the name of the geometrical figure in Problems 1 and 4?

6. Set r equal to 0 in Problem 3, and solve the resulting equation for θ. Write the general solution.

7. Give the *two* values of θ in [0°, 360°] at which the graph goes through the pole.

8. Draw a ray on your graph in Problem 1 at each of the two values of θ in Problem 7. How are the rays related to the graph?

9. What did you learn as a result of doing this Exploration that you did not know before?

Exploration 134: Roses and Circles in Polar Coordinates

Objective: Plot polar curves on your grapher, and find places where polar curves intersect.

1. The graph of $r = 10 \cos 2\theta$ is called a **four-leaved rose.** Plot the rose. Sketch the result.

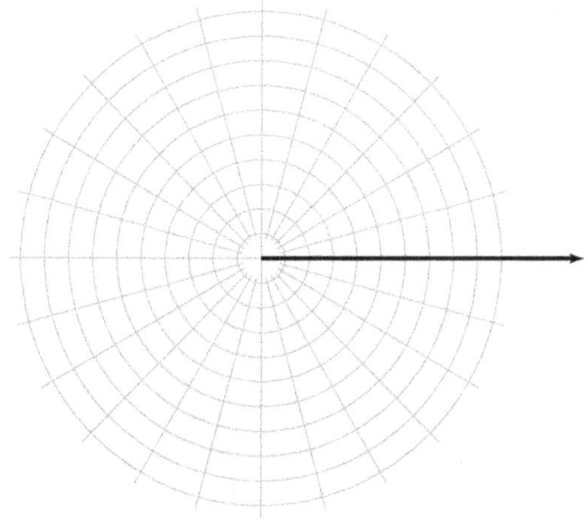

2. The graph of $r = 10 \cos 3\theta$ is a **three-leaved rose.** Plot the rose and sketch it here.

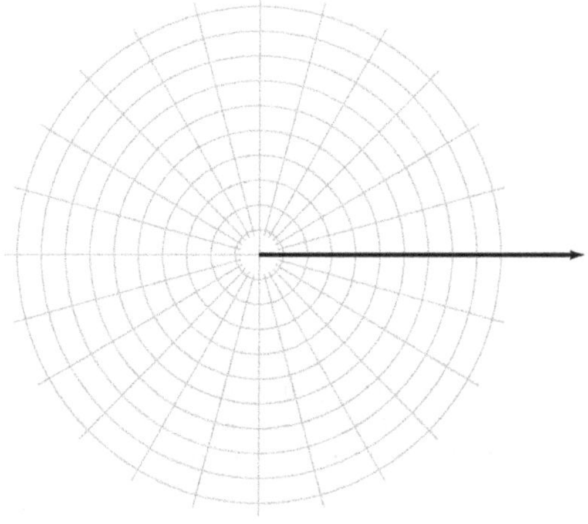

3. Suppose that $r = 10 \cos n\theta$, where n stands for a positive integer. How many leaves would the "rose" have if $n = 4$? $n = 5$? What relationship does n have to the number of leaves? Why is this relationship different for odd and even values of n?

4. Plot the graph of $r = 10 \cos \theta$. Sketch it.

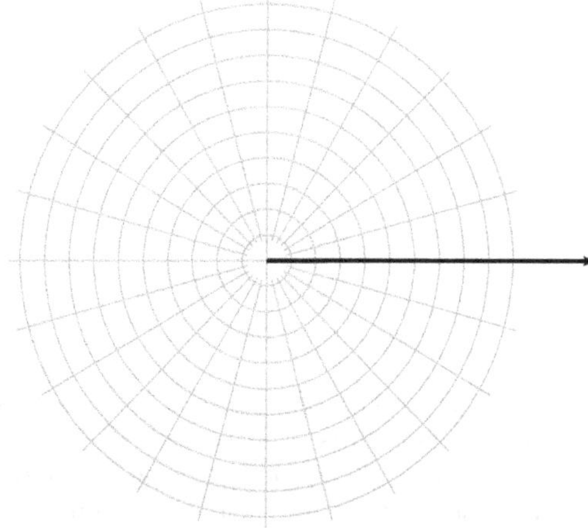

5. Show that the graph in Problem 4 is a circle by transforming the equation to Cartesian form.

6. Explain why the circle in Problem 4 can be called a "one-leaved rose."

7. What did you learn as a result of doing this Exploration that you did not know before?

Exploration 135: Intersections of Polar Curves

Objective: Plot polar curves on your grapher, and find places where polar curves intersect.

The figure shows:

- The limaçon $r_1 = 3 + 2\cos\theta$
- The rose $r_2 = 5\sin 2\theta$

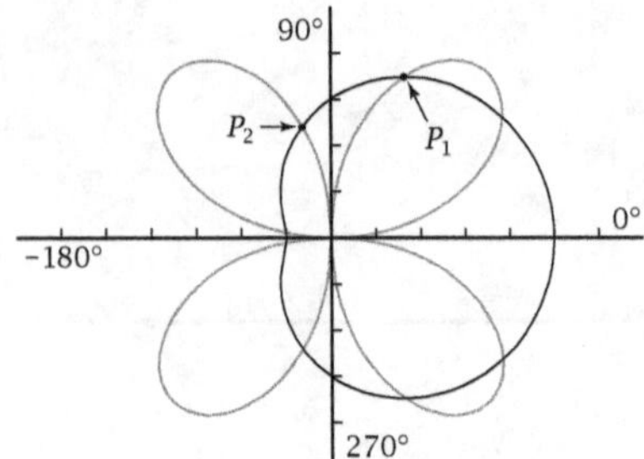

1. Plot the two graphs on your grapher. Use degrees, simultaneous mode, and a fairly small θ-step so that the graphs plot relatively slowly. Pause the plotting when the graphs reach the intersection point P_1. Approximately what does θ equal at this point?

2. Resume the plotting, and then pause it again at the θ-value corresponding to point P_2 on the limaçon. Where is the point on the rose for this value of θ? Explain why P_2 is not an intersection point of the two graphs.

3. Continue the graphing until a complete 360° has been plotted. Which of the apparent intersections in the figure are true intersections, and which are not? What do you notice about the r-values on the rose for the points that are not true intersections?

4. With your grapher in function mode, plot the auxiliary Cartesian graphs

   $$y_1 = 3 + 2\cos\theta$$

   $$y_2 = 5\sin 2\theta$$

   Sketch the result.

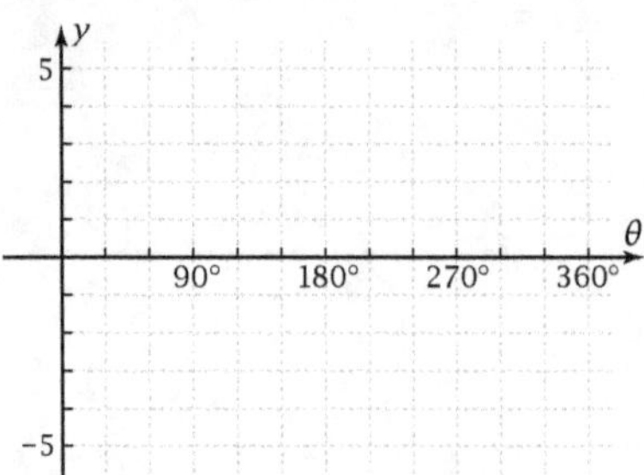

5. Solve numerically to find the first two positive values of θ where the graphs in Problem 4 intersect. Show that these correspond to the two points where the polar graphs intersect.

6. Show on the auxiliary graphs in Problem 4 that the second-quadrant angle θ for point P_2 corresponds to a point on the limaçon but not to a point on the rose.

7. What did you learn as a result of doing this Exploration that you did not know before?

Exploration 136: Products of Complex Numbers

Date: _____________

Objective: Demonstrate the property of the product of two complex numbers.

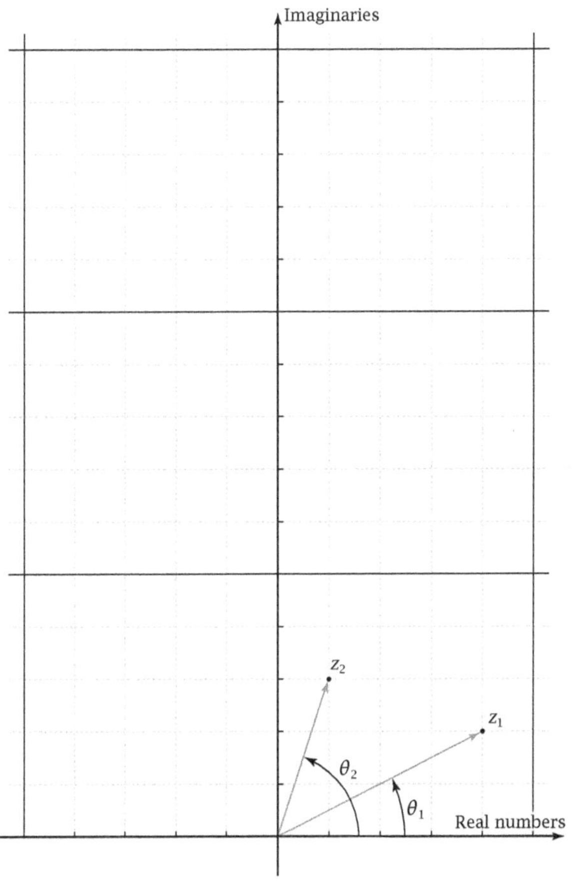

1. The figure shows two complex numbers, z_1 and z_2, in the complex plane. Write each number in the form $a + bi$, where $i = \sqrt{-1}$.

   $z_1 =$ _____________________

   $z_2 =$ _____________________

2. Explain why $i^2 = -1$.

3. Find the product $z_1 z_2$. Simplify. Plot the answer on the figure.

4. Measure with a protractor the arguments θ_1 and θ_2 of z_1 and z_2. Measure the argument θ_3 of $z_1 z_2$. What relationship do you find among the three angles?

5. Calculate θ_1, θ_2, and the argument of $z_1 z_2$. Show numerically that the argument of $z_1 z_2$ is exactly equal to the sum of the arguments of the other two numbers.

6. Calculate the moduli of z_1 and z_2 and the modulus of $z_1 z_2$. Leave all three answers in radical form. Then show that the modulus of $z_1 z_2$ is exactly equal to the product of the other two moduli.

7. What did you learn as a result of doing this Exploration that you did not know before?

Exploration 137: Complex Number Product Proof

Objective: Prove that the product of two complex numbers has the product of the moduli and the sum of the arguments.

1. Let $z_1 = 3 \operatorname{cis} 57°$, and let $z_2 = 5 \operatorname{cis} 41°$. Write $z_1 z_2$ using the definition $\operatorname{cis} \theta = \cos \theta + i \sin \theta$.

2. Complete the multiplication in Problem 1 and simplify. Along the way, do these things.
   - Commute and associate the 3 and the 5.
   - Multiply the two binomials, leaving the 15 factored out.
   - Use the fact that $i^2 = -1$.
   - Commute and associate the real parts (without i) and the imaginary parts (with i).
   - Factor out the i from the imaginary parts.
   - Use the composite argument properties to simplify the result. Leave the arguments as sums.
   - Transform the answer to cis notation.

3. Describe a simple way to multiply complex numbers based on the results of Problems 1 and 2.

4. If you multiply $7 \operatorname{cis} 67°$ by $7 \operatorname{cis}(-67°)$, you get $49 \operatorname{cis} 0°$. What does $\operatorname{cis} 0°$ equal? What kind of number is the answer? What are the two numbers $r \operatorname{cis} \theta$ and $r \operatorname{cis}(-\theta)$ called?

5. What did you learn as a result of doing this Exploration that you did not know before?

Exploration 138: Projectile Motion Problem

Objective: Derive the parametric equations of a curve from its geometrical properties, and explore the graph on your grapher.

The figure shows the path followed by a baseball as it leaves the point $(x, y) = (0, 0)$, 3 feet above home plate. Vector $\vec{v}$, the initial velocity of the ball, makes an angle of α with the horizontal. In this problem, you will analyze the distance the ball goes for various values of α.

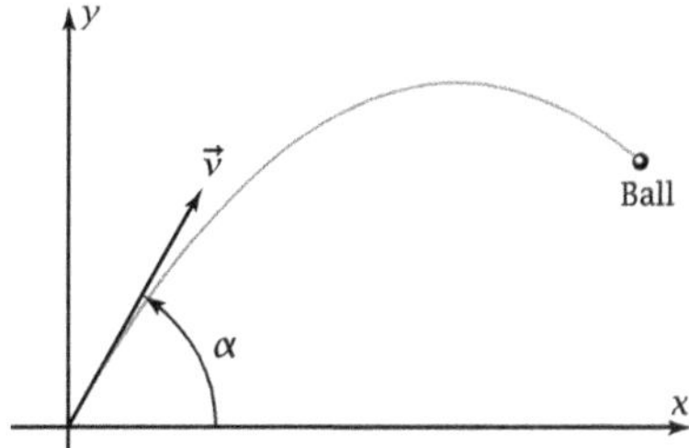

1. Suppose that $|\vec{v}| = 100$ ft/sec and $\alpha = 60°$. Write $\vec{v}$ in terms of its vertical and horizontal components.

2. Neglecting air resistance, the distance the ball travels in the x-direction equals the x-component of $\vec{v}$ multiplied by time, t. Write the parametric equation for x as a function of t.

3. If the ball were not acted on by gravity, the distance it would travel in the y-direction would be the y-component of $\vec{v}$ multiplied by t. However, gravity causes it to fall below this distance by an amount $16t^2$. Write the parametric equation for y as a function of t.

4. Enter the parametric equations for x and y into your grapher. Plot the graph using a window large enough to show the point where the ball hits the ground.

5. Find graphically the approximate time and place at which the ball is back to $y = 0$, the height at which it was hit.

6. Find algebraically the time at which the ball reaches the ground, where $y = -3$.

7. Use the answer to Problem 6 to find the ball's horizontal distance from where it was hit when it reaches the ground.

8. Investigate the time and place at which the ball hits the ground if the angle α is 50°, 40°, 30°, and 20°. What do you observe about the time the ball is in the air and the horizontal distance it travels?

9. What did you learn as a result of doing this Exploration that you did not know before?

Exploration 139: Hyperbola Construction

Objective: Given a geometrical description of a curve in the xy-plane, write parametric equations and verify that they are right.

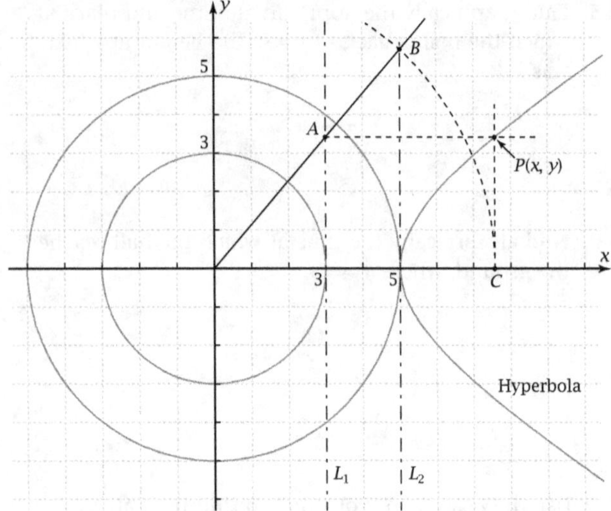

The figure shows a geometrical construction for a hyperbola. Fixed circles centered at the origin have radii 3 and 5 units. Fixed lines L_1 and L_2 are tangent to the circles, parallel to the y-axis. A line at a variable angle t radians to the x-axis intersects L_1 and L_2 at points A and B, respectively. An arc centered at the origin is drawn through B and intersects the x-axis at point C. A vertical line through C and a horizontal line through A intersect at point $P(x, y)$ on the hyperbola.

1. Derive parametric equations for the curve.

2. Verify your equations on your grapher.

3. The point shown in the figure has $t = 0.85$ radians. Use your equations to calculate the point (x, y), and show that the point agrees with the figure.

4. Transform the parametric equations to a Cartesian equation by eliminating the parameter t.

5. Verify your Cartesian equation by calculating y for $x = 7$ and for $x = 2$.

6. Verify that the curve is really a hyperbola.

7. Is the domain of x the same for the parametric equations as it is for the Cartesian equation?

____

8. What did you learn as a result of doing this Exploration that you did not know before?

Exploration 140: The Witch of Maria Agnesi

Objective: Find parametric equations of a curve given by a geometrical construction.

Here you see a figure called the **witch of Maria Agnesi.** A fixed circle of radius 5 units is centered on the *y*-axis and contains the origin. A fixed horizontal line is drawn tangent to the top of the circle. A variable line from the origin at an angle, *t*, intersects the circle at *A* and the tangent line at *B*. A horizontal line from *A* and a vertical line from *B* intersect at point *P(x, y)* on the graph of the witch.

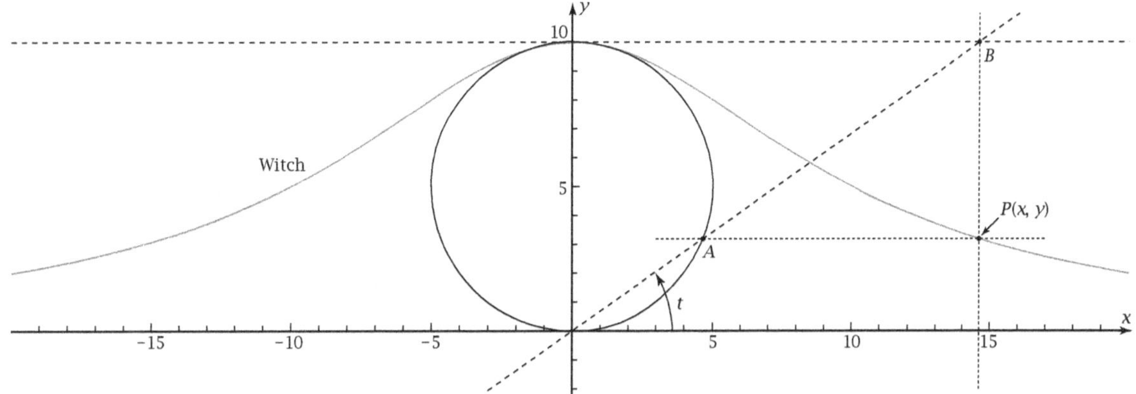

1. Sketch a triangle with vertices at points *A*, (0, 10), and (0, 0). What theorem from geometry explains why the triangle is a *right* triangle?

2. In terms of *t*, how long is the segment from the origin to point *A*?

3. On the given figure, sketch a right triangle including the angle with measure *t* and one leg with length *x*. Then write *x* as a function of *t*.

4. On the given figure, sketch a right triangle including the angle with measure *t* and one leg with length *y*. Then write *y* as a function of *t*.

5. The point shown in the figure has $t = 0.6$ radians. Calculate *x* and *y* for this point, and show that your answers agree with the figure.

6. Plot the witch on your grapher. On the same screen, plot the circle shown. Use a *t*-range of one complete revolution and a window with an *x*-range of $[-20, 20]$ and equal scales on the two axes. Check your graph with your instructor. ________________

7. How do you explain why the witch is retraced as *t* increases from π to 2π?

8. What did you learn as a result of doing this Exploration that you did not know before?

Exploration 141: The Conchoid of Nicomedes

Objective: Given a geometrical description of a curve in the *xy*-plane, write parametric equations and verify that they are right.

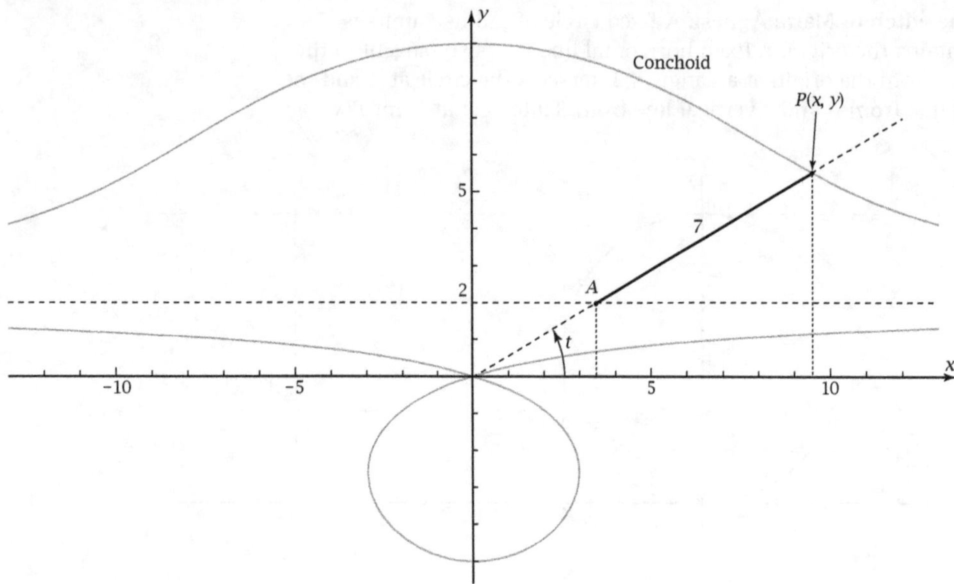

The figure shows a **conchoid of Nicomedes.** It is formed by a ray from the origin that rotates through an angle of *t* degrees with the *x*-axis. The ray intersects the fixed line *y* = 2 at point *A*. From *A* you measure "out" 7 units to point *P* on the graph of the conchoid.

1. Show that you get a point that is *on* the curve when you do the construction just described for:

   a. A value of *t* between the ray shown and $\frac{\pi}{2}$

   b. $t = \frac{\pi}{2}$

   c. A value of *t* between $\frac{\pi}{2}$ and π

2. Show geometrically how *t* = 4 radians gives a point on the lower loop of the conchoid.

3. The parametric equation for *x* has two parts, one for the segment from the origin to the point down from *A* and the other from there to the point down from *P*. Write an equation for *x* as a function of *t*.

4. Write the parametric equation for *y* as a function of *t*. It, too, will have two parts.

5. Plot your parametric equations from Problems 3 and 4 on your grapher. If it does not resemble the figure at page top, find and fix your errors.

6. Plot on the given graph the point for *t* = 4. Show by measurement that this point is also 7 units from the line *y* = 2.

7. The Cartesian equation of this conchoid is:

   $$(x^2 + y^2)(y - 2)^2 = 49y^2$$

   Verify that this equation is correct by calculating the values of *x* for *y* = 8 and showing that the points really are on the conchoid.

8. Just for fun, see if you can derive the Cartesian equation in Problem 7, either from scratch or by transforming the parametric equations.

Precalculus and Trigonometry Explorations
©2004 Key Curriculum Press

Exploration 142: Cardioid Problem

Objective: Find parametric equations of a curve given by a geometrical construction.

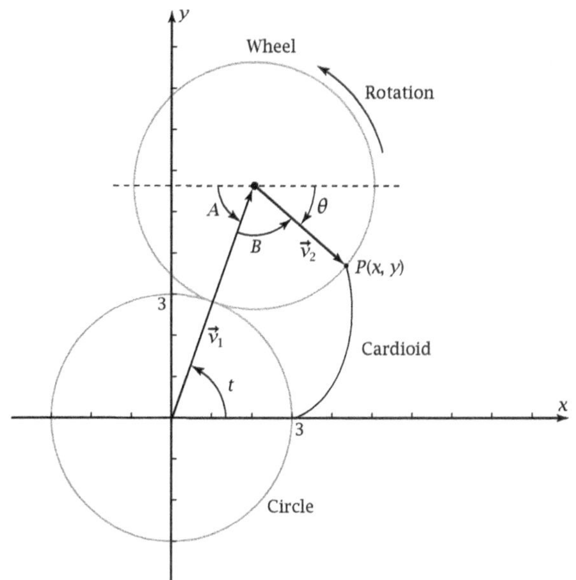

The figure shows a fixed circle of radius 3 centered at the origin. A moving circle of radius 3 rotates counterclockwise around the fixed circle without slipping. Point $P(x, y)$ on the rotating circle traces a geometrical figure called a **cardioid** (which means "heartlike"). Angle t radians in standard position is measured to vector $\vec{v}_1$, which goes from the center of the fixed circle to the center of the rotating circle. Vector $\vec{v}_2$ goes from the center of the rotating circle to point P.

1. Write vector $\vec{v}_1$ in terms of functions of angle t.

2. Write vector $\vec{v}_2$ in terms of functions of angle θ.

3. How is angle A in the figure related to angle t?

4. How is angle B in the figure related to angle t?

5. What does angle θ equal when $t = 0$? What, then, does θ equal in terms of angle t?

6. Let vector $\vec{r}$ be the position vector for point P. Draw $\vec{r}$ on the figure.

7. Write vector $\vec{r}$ in terms of $\vec{v}_1$ and $\vec{v}_2$.

8. Write vector $\vec{r}$ in terms of functions of t.

9. Plot the graph of the cardioid by entering the coefficients of vector $\vec{r}$ as x and y in parametric mode. Use a window with an x-range of $[-15, 15]$ and equal scales on the two axes. Use a t-range large enough to generate the whole graph. Check your graph with your instructor. _____________

10. The given figure is drawn accurately, with $t = 1.22$ radians. Find x and y using your equations, and confirm that the point P shown is correct.

11. What did you learn as a result of doing this Exploration that you did not know before?

Exploration 143: Parametric Equations of Hypocycloids

Objective: Use vectors to get parametric equations of fairly complicated curves.

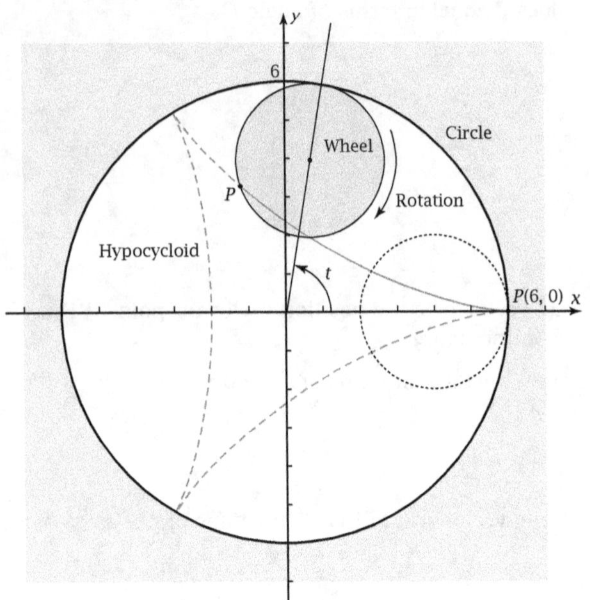

The figure shows a small circle of radius 2 cm rotating clockwise, without slipping, inside a large circle of radius 6 cm. A point $P(x, y)$ on the small circle traces out a **hypocycloid.** P starts at the point (6, 0), as shown by the dotted small circle. In this exercise, your objective is to find an equation for the position vector $\vec{r}$ (not shown) to point P in terms of the parameter t, where t is the angle in standard position to a line through the center of the small circle.

1. This next figure shows vector $\vec{v}_1$ from the origin to the center of the small circle. Write $\vec{v}_1$ in terms of t and the unit vectors $\vec{i}$ and $\vec{j}$.

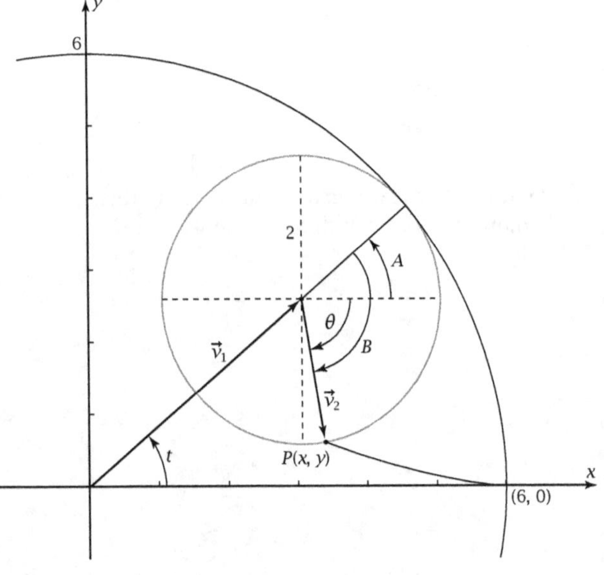

2. Let $\vec{v}_2$ be a vector in a coordinate system with origin at the center of the small circle (dashed axes). Let θ be the angle in standard position for this vector. Write $\vec{v}_2$ in terms of θ, $\vec{i}$, and $\vec{j}$.

3. What does angle A equal in terms of t? What does angle B equal in terms of t? What, then, does θ equal in terms of t?

4. Write $\vec{v}_2$ in terms of t, $\vec{i}$, and $\vec{j}$.

5. Write the position vector $\vec{r}$ to point (x, y) in terms of t, $\vec{i}$, and $\vec{j}$. Simplify by combining the coefficients of $\vec{i}$ and $\vec{j}$. The result is called the **vector equation** of the hypocycloid.

6. Plot the hypocycloid on your grapher. Use equal scales on the two axes. Use a t-range large enough to get one complete cycle. Does the graph agree with the one in the first figure? __________

7. This particular hypocycloid is called a **deltoid** or a **hypocycloid of three cusps.** Explain the meanings of these names.

Date: __________

For Problems 8–11, write the parametric equations, and sketch the graphs on the dot paper.

8. Change the point on the small circle to 1 cm from the center instead of on the circumference.

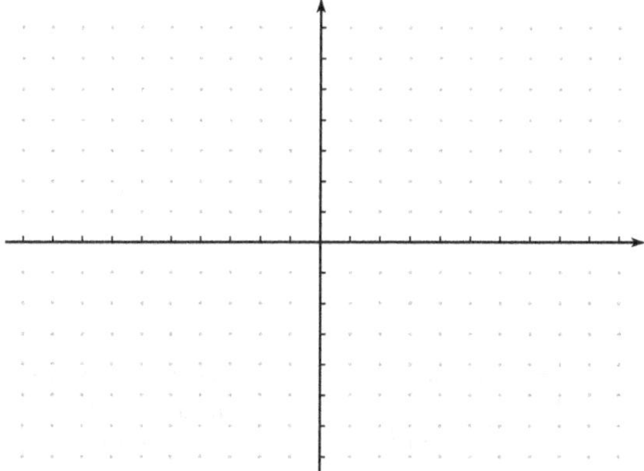

9. Change the point on the small circle to 1 cm beyond the rim of the small circle.

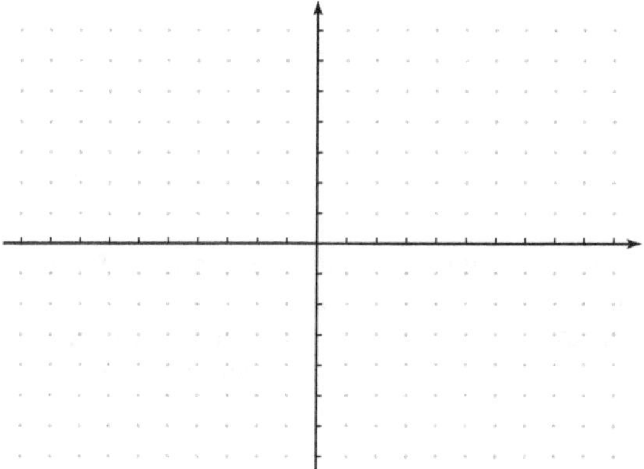

10. Make the hypocycloid have *five* cusps by choosing the appropriate diameter for the small circle.

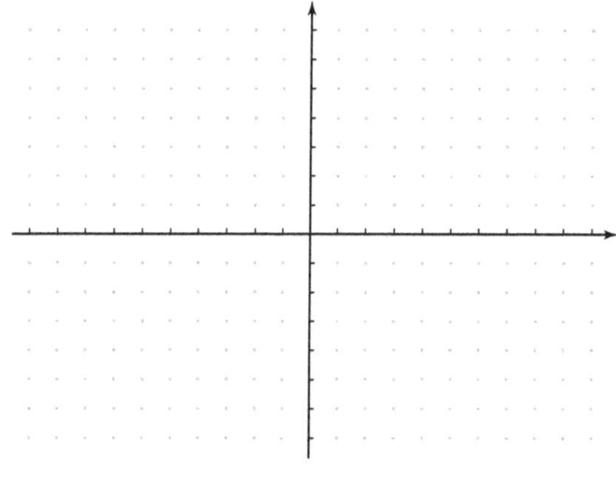

11. Plot the hypocycloid traced by a circle of radius 1.8 cm rotating inside a circle of radius 6 cm. Pick a range for *t* large enough to get the complete graph.

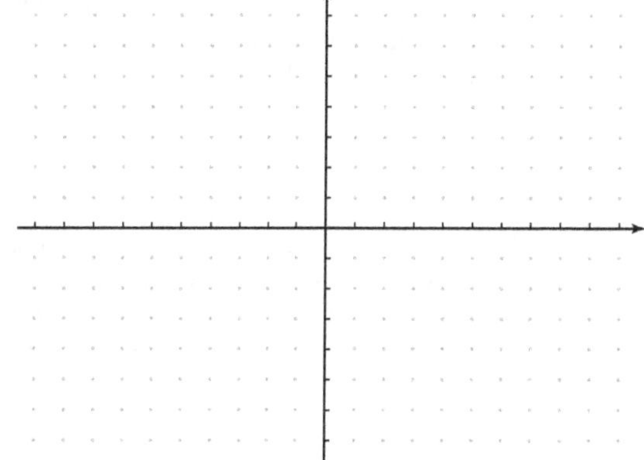

12. What did you learn as a result of doing this Exploration that you did not know before?

Exploration 144: Springs and Moving Ellipses

Objective: Write parametric equations for an ellipse that moves in the *x*-direction as *t* increases, and confirm by grapher.

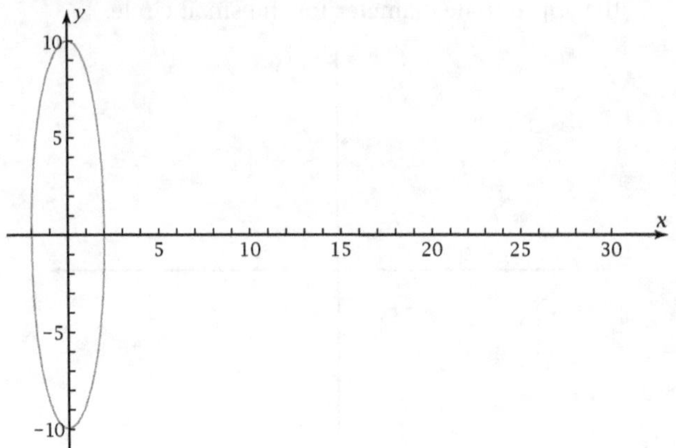

1. Write parametric equations for the given ellipse.

2. Confirm your answer to Problem 1 by plotting on your grapher. Use the window shown. Were your equations right? _____________

3. Write parametric equations for an ellipse congruent to the one in Problem 1 but translated 15 units in the *x*-direction. Confirm by your grapher, and sketch the ellipse on the axes for Problem 1.

4. The "spring" in the next figure was generated using the parametric equations for the ellipse in Problem 1, translated in the *x*-direction by a *variable* amount proportional to *t*. The translation is 4 units for each 2π cycle of *t*. Write parametric equations for this curve.

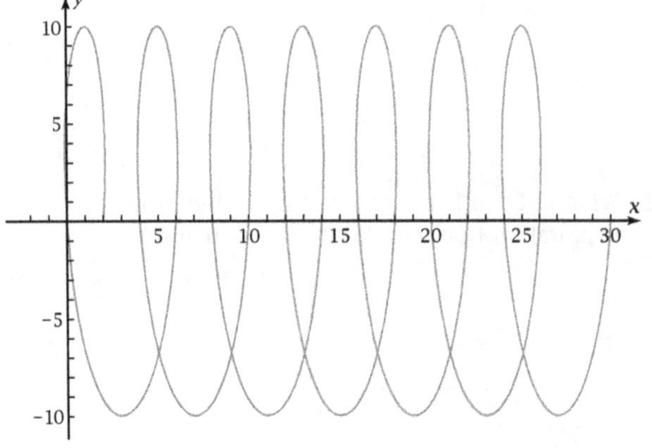

5. Confirm that your answer to Problem 4 is correct by plotting on your grapher. Trace to two consecutive high points, and record the *x*-, *y*-, and *t*-values.

6. How do your answers to Problem 5 show that the ellipse really did move exactly 4 units in the 2π cycle of *t*-values?

7. This next graph shows the spring of Problem 4 translated by just the right amount to form *cusps* (sharp points) rather than loops at the high points. Find the parametric equations of this graph.

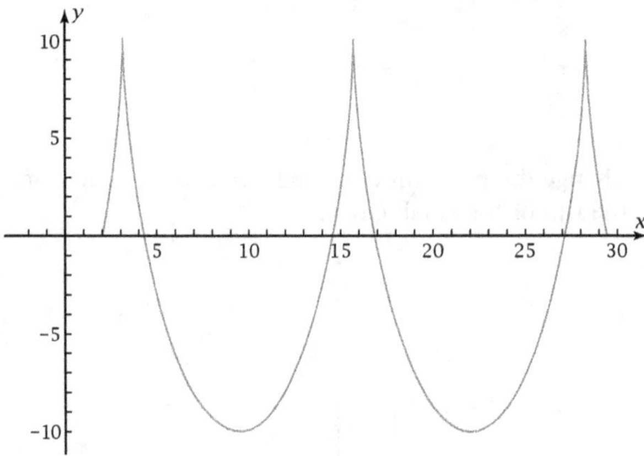

8. Suppose you were hired to do drawings for a commercial art company. Your task is to draw a coil spring for an automobile suspension ad. Explain how you would go about drawing such a spring.

9. What did you learn as a result of doing this Exploration that you did not know before?

Exploration 145: Is a Hanging Chain a Parabola?

Objective: Find the particular equation of a hanging chain, and show that the shape is not a parabola.

A chain hanging under its own weight forms a curve called a **catenary.** The word comes from the Latin *catina,* meaning "chain." The equation of a catenary is

$$y = k \cosh \frac{1}{k}x + C$$

where cosh is the hyperbolic cosine function and k and C stand for constants related to dilation and translation of the figure. You might have to look in your grapher's catalog to find cosh.

1. Plot $y = \cosh x$ on a grapher that will project images onto the wall. Hold up a length of chain to the image on the wall. Does the chain really seem to follow the hyperbolic cosine graph?

2. Hang a chain about 2 m long as shown in the figure so that you can mark points on the board behind it. Construct a y-axis through the vertex (halfway between the points where the chain is attached to the wall). Use the chalk tray as the x-axis. Measure the coordinates of the vertex and the coordinates of the point on the right where the chain is attached to the wall. Use these two points to calculate the values of k and C in the catenary equation. Note that you will have to use a numerical solution to find k because it appears both algebraically as a coefficient and transcendentally in the argument of the hyperbolic cosine.

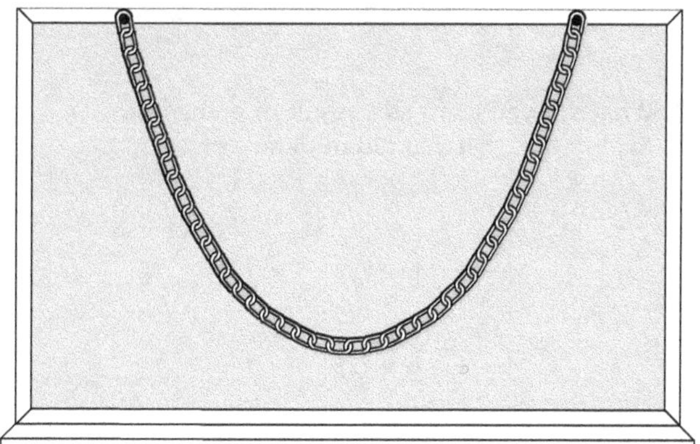

3. Make a table of values of y versus x for each 10 cm or 20 cm. Take down the chain, and plot on the board the points shown in your table. Then rehang the chain. Does it fit the points you plotted?

4. Find the particular equation of the parabola that has vertex at the same point as the catenary and that passes through the two points where the chain is attached to the wall. Plot the catenary and the parabola on the same screen. Sketch the results. How does the parabola differ from the catenary?

5. What did you learn as a result of doing this Exploration that you did not know before?

Exploration 146: Introduction to Sequences

Objective: Find a pattern in a sequence of numbers, and use the pattern to find other terms in the sequence.

1. The following numbers are the first few **terms** of a **sequence.**

   2, 6, 12, 20, 30, 42, . . .

   What are the next two terms? How did you calculate them?

2. The symbol for the first term of the series is t_1, the second term is t_2, and so forth. What does t_5 equal? What does t_{10} equal?

3. Let n be the **term number.** The term number is the same as the subscript in the symbol for the term. It shows the order of the term in the sequence. On this graph paper, plot a graph of t_n as a function of n.

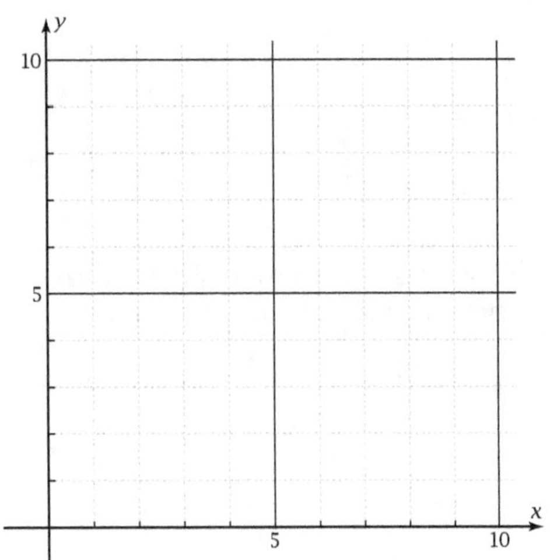

4. What is the major difference between the graph you plotted in Problem 4 and other graphs you have plotted in precalculus?

5. Write the term numbers underneath the corresponding term.

   t_n: 2, 6, 12, 20, 30, 42, . . .

   n:

6. What operation could you do to 6 to get t_6 for the answer? In general, what operation could you do to n to get t_n for the answer? Write a formula for t_n in terms of n.

7. Enter your formula from Problem 6 into your grapher as y_1. Then make a table of values of t_n, starting at t_1. Do all of the values correspond to the given ones? If not, go back and fix your formula.

8. Use your formula to calculate t_{1234}.

9. The number 11,340,056 is a term in this sequence. Write an equation stating that t_n equals this number for some n. Then solve the equation to find out what n equals.

10. What did you learn as a result of doing this Exploration that you did not know before?

Exploration 147: Patterns in Sequences

Objective: Given the first few terms of a sequence, discover a pattern and use it to find more terms or term numbers.

1. An **arithmetic sequence** progresses by adding a constant to the preceding term to get the next term. Could these sequences be arithmetic sequences?

    a. 5, 10, 20, . . . ______________________________

    b. 5, 10, 15, . . . ______________________________

    c. 5, 10, 40, . . . ______________________________

2. The constant added to a term of an arithmetic sequence to get the next term is called the **common difference.** What is the common difference for the arithmetic sequence in Problem 1? Why do you think it is called a "difference"?

3. A **geometric sequence** progresses by multiplying the preceding term by a constant to get the next term. Could these sequences be geometric sequences?

    a. 5, 10, 20, . . . ______________________________

    b. 5, 10, 15, . . . ______________________________

    c. 5, 10, 40, . . . ______________________________

4. The constant that multiplies a term of a geometric sequence to get the next term is called the **common ratio.** What is the common ratio for the geometric sequence in Problem 3? Why do you think it is called a "ratio"?

Problems 5–9 pertain to this sequence.

n: 1 2 3 4 5 6 7 . . .

t_n: 4, 10, 18, 28, 40, 54, 70, . . .

5. By discovering a pattern in the sequence, write the next two terms.

6. Figure out what operation(s) can be performed on the term numbers, n, to get the values of t_n. That is, figure out a formula for t_n as a function of n.

7. On this graph paper, plot the graph of the sequence. Connect the points with a dashed line to show the pattern.

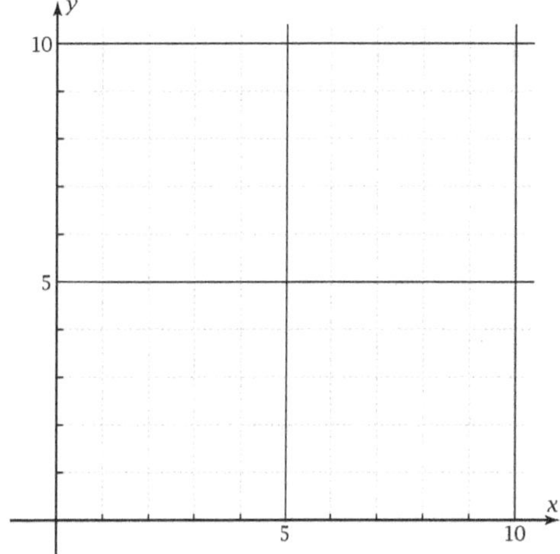

8. Use the formula from Problem 6 to calculate t_{53}.

9. The number 23,868 is a term in the sequence. Calculate its term number.

10. What did you learn as a result of doing this Exploration that you did not know before?

Exploration 148: Arithmetic and Geometric Sequences

Objective: Use patterns to find terms in an arithmetic or geometric sequence or to find the term number of a given term.

Car Manufacturing Problem: Two competing manufacturers come out with new sport-utility vehicles, the Flame which is 200 inches long, and the Seeker, which is only 195 inches long. In the second year, the Seeker is made 2% longer, meaning its length is multiplied by 1.02. In order to maintain its lead in size, the Flame is made 3 inches longer in the second year. The patterns of multiplying the previous length by 1.02 and adding 3 to the previous length are continued for several years.

1. Write the first four terms of each sequence of lengths.

2. What kind of sequence is formed by repeatedly multiplying by 1.02?

3. What kind of sequence is formed by repeatedly adding 3?

4. Will the Seeker's length ever exceed the Flame's length? Give numerical evidence to support your conclusion.

5. Calculate algebraically the length of the Flame in the 40th year if the pattern is maintained each year.

6. Calculate algebraically the length of the Seeker in the 30th year if the pattern is maintained each year.

7. The number 341 is a term in the sequence of Flame lengths. Calculate the term number algebraically.

8. The number 1205.73 is (approximately) a term in the sequence of Seeker lengths. Calculate the term number algebraically.

9. What did you learn as a result of doing this Exploration that you did not know before?

Exploration 149: Introduction to Series

Objective: Find partial sums of a series for which a formula for t_n is known by adding up the terms.

1. Write out the indicated terms of this series.

$$\sum_{k=1}^{6} 3k + 5$$

2. Evaluate the partial sum in Problem 1.

3. Enter a program into your grapher to calculate the partial sum of a series. The formula for t_n should be entered as y_1. When the program runs, your grapher should ask you to input the number of terms. Then it should calculate and display each partial sum up to the one you asked for. Test your program by using it to find the partial sum in Problem 1. You may assume that the program is working correctly if it gives the answer you got in Problem 2.

4. Use your program from Problem 3 to evaluate S_{100} for the series in Problem 1. That is, find

$$\sum_{k=1}^{100} 3k + 5$$

5. Figure out a formula for t_n for this next series. Then use your program in Problem 3 to find the fifth partial sum, S_5, for the series. Confirm that the program gives you the correct answer by actually adding the terms shown here.

$$2 + 5 + 10 + 17 + 26 + \cdots$$

6. Use your program to calculate the 50th partial sum for the series in Problem 5.

7. Write out the first four terms of the geometric series with first term 1000 and common ratio 1.06. Calculate the fourth partial sum by adding up these terms.

8. Write a formula for the nth term of the geometric series in Problem 7. Confirm that your formula is correct by using the program of Problem 3 to find the fourth partial sum you calculated manually in Problem 7.

9. Use your program to find the 30th partial sum of the geometric series in Problem 7. (This number is the amount of money you would have at the end of 30 years if you invested $1000 a year in a savings account that pays an interest of 6% per year, compounded annually.)

10. Calculate the following partial sums of the geometric series with first term 800 and common ratio 0.9.

    S_{10} _______________________________________

    S_{20} _______________________________________

    S_{50} _______________________________________

    S_{100} _______________________________________

    S_{200} _______________________________________

11. The partial sums in Problem 10 **converge to 8000.** What do you think this means?

12. What did you learn as a result of doing this Exploration that you did not know before?

Exploration 150: Partial Sums of Arithmetic Series

Objective: Find partial sums of an arithmetic series *without* actually adding up the terms.

1. This is the tenth partial sum of an arithmetic series.

   $S_{10} = 13 + 20 + 27 + 34 + 41 + 48 + 55 + 62 + 69 + 76$

   What is the common difference? What number does the partial sum S_{10} equal?

2. Add the first and last terms, the second and next-to-last terms, etc. What do you notice? How many pairs of terms can be made this way? How can you calculate the partial sum S_{10} quickly without actually adding the terms one by one?

3. Show that the pattern you described in Problem 2 works for an *odd* number of terms, such as S_{11}, for the series in Problem 1.

4. An arithmetic series has $t_1 = 47$ and $t_{300} = 3927$. Find S_{300}.

5. An arithmetic series has $t_1 = 54$ and common difference $d = 11$. Find t_{47}. Find S_{47} using the pattern in Problem 2. Confirm your answer by actually adding the terms using your SERIES program.

6. Find the sum of the first 100,000 counting numbers quickly. Why would it be impractical to calculate this sum using your SERIES program?

7. The pattern for S_n in an arithmetic series can be generalized to this equation.

   $$S_n = \frac{n}{2}[t_1 + t_n]$$

   Show that you can also write S_n as

   $$S_n = \frac{n}{2}[2t_1 + d(n - 1)]$$

8. 86,697 is a partial sum of the arithmetic series

   $37 + 38.2 + 39.4 + \cdots$

   Use the equation in Problem 7 to calculate the value of n. You may use your quadratic formula program to solve the resulting equation. Confirm the answer using your SERIES program.

9. What did you learn as a result of doing this Exploration that you did not know before?

Exploration 151: Partial Sums of Geometric Series   Date: __________

Objective: Find partial sums of a geometric series *without* actually adding up the terms.

1. This is the sixth partial sum of a geometric series.

   $$S_6 = 7 + 21 + 63 + 189 + 567 + 1701$$

   What is the common ratio? What number does the partial sum equal?

2. Write out the terms of $3S_6$. (Don't add.) What do you notice about the terms in the $3S_6$ series and the terms in the original partial sum, S_6?

3. If you find the difference $S_6 - 3S_6$ without first simplifying, the middle terms will **telescope.** What do you think is meant by "telescoping"?

4. Recall that the nth term, t_n, for a geometric sequence or series is equal to t_1 times $(n - 1)$ common ratios. That is,

   $$t_n = t_1 r^{n-1}$$

   Show that the quantity $S_6 - 3S_6$ you calculated in Problem 3 is equal to $7(1 - 3^6)$.

5. The quantity $S_6 - 3S_6$ is also equal to $-2S_6$. Use this fact and the answer to Problem 4 to calculate S_6. Show that the answer is the same as you got in Problem 1.

6. If you generalize the pattern in Problems 4 and 5, you will find that S_n for a geometric series equals the first term of the series multiplied by a fraction that involves r and n. Write an equation for S_n.

7. Use your equation in Problem 6 to find the 15th partial sum for the series in Problem 1. Confirm your answer by adding up the terms. You may use your SERIES program to do this.

8. Use the equation in Problem 6 to calculate the 30th partial sum of the geometric series with first term 1000 and common ratio 1.06. Confirm your answer using your SERIES program.

9. Use the equation in Problem 6 to calculate the 50th partial sum of the geometric series with first term 800 and common ratio 0.9. Then calculate the 200th partial sum. What happens to the term r^n in the equation when n becomes very large?

10. The series in Problem 9 converges to 8000 as n becomes large. Describe a quick way to calculate this number using the equation in Problem 6.

11. What did you learn as a result of doing this Exploration that you did not know before?

Exploration 152: Binomial Series and the Binomial Formula

Objective: Find patterns that allow you to calculate a single term in a binomial series.

This table shows the terms in the binomial series that comes from $(a + b)^7$.

n	t_n	Pattern in Coefficients
1	$1a^7$	
2	$7a^6b$	
3	$21a^5b^2$	
4	$35a^4b^3$	
5	$35a^3b^4$	
6	$21a^2b^5$	
7	$7ab^6$	
8	$1b^7$	

1. Write out rows 0 through 7 of Pascal's triangle, and thus show that the coefficients in the table are correct.

2. The coefficient 35 for t_4 can be calculated from numbers in the preceding term, t_3. Show that you have found this pattern by showing how the 35 can be calculated.

3. Show that the pattern in Problem 2 works for other terms in the table.

4. The pattern in Problems 2 and 3 is

$$\frac{(\text{coefficient})(\text{exponent of } a)}{(\text{term number})}$$

Using this pattern on $1a^7$ gives

$$t_2 = \frac{7}{1} a^6b$$

Use the pattern to find the coefficient of t_3, but do not simplify.

5. By following the pattern without simplifying, you will find that the fifth term is

$$t_5 = \frac{7}{1} \cdot \frac{6}{2} \cdot \frac{5}{3} \cdot \frac{4}{4} a^3b^4 = \frac{7 \cdot 6 \cdot 5 \cdot 4}{1 \cdot 2 \cdot 3 \cdot 4} a^3b^4$$

The denominator is a factorial, 4!. By clever algebra, you can also write the numerator using factorials. Show how.

6. The factorials in Problem 5 are related to the exponents of a and b and to the exponent 7 of the original binomial. Show that this pattern gives the correct value of the coefficient 21 of t_6.

7. The pattern in Problem 6 is called the **binomial formula.** Use the binomial formula to find the term that has b^9 in the binomial series $(a - b)^{13}$.

8. What did you learn as a result of doing this Exploration that you did not know before?

Exploration 153: Arithmetic and Geometric Series Problems

Objective: Find terms and partial sums of arithmetic or geometric series numerically or algebraically.

Pushups Problem: Emma Strong starts an exercise program. On the first workout, she does five pushups. The next workout, she does eight pushups. She decides to let the number of pushups in each workout be a term in an arithmetic series.

1. Find algebraically the number of pushups she does on the tenth workout.

2. Find algebraically the total number of pushups she has done after ten workouts.

3. Write the numbers of pushups Emma does on workouts 1, 2, 3, . . . , 10. Confirm that your answer to Problem 2 is correct by actually adding these terms.

4. If Emma were able to keep up the arithmetic series of pushups, one of the terms in the series would be 101. Calculate the term number algebraically.

Medication Problem: N. Hale takes 50 mg of allergy medicine each day. By the next day, some of the medicine has decomposed, but the rest is still active in his body. Mr. Hale finds that the amount still in his body after n days is given by this partial sum.

$$S_n = \sum_{k=1}^{n} 50(0.8^{k-1})$$

5. Demonstrate that you know what sigma notation means by writing out the first three terms of this series.

6. Find S_3 for the series in Problem 5.

7. Run your SERIES program to find S_{40}. Write the answer here, and check with your instructor. _________

8. After how many days does the amount in his body first exceed 200 mg?

9. Does the amount in Mr. Hale's body seem to be converging to a certain number, or does it just keep getting bigger without limit? How can you tell?

Loan Problem: Len de Monet borrows $200 from his parents to buy a new calculator. They require him to pay back 10% of his unpaid balance at the end of each month.

10. How much does Len pay at the end of the first month? What, then, is his unpaid balance after that payment?

(Over)

Exploration 153: Arithmetic and Geometric Series Problems *continued*

11. Calculate Len's unpaid balances after 2 months and after 3 months.

12. Is the sequence of unpaid balances in Problems 10 and 11 geometric, arithmetic, or neither? How can you tell?

13. Calculate algebraically the unpaid balance at the end of 1 full year.

14. Len must pay off the rest of the loan when his unpaid balance has dropped below $5. After how many months will this have happened? Describe the method you use. (It is not enough to say, "I used my calculator" or "Guess and check.")

15. From the results of Problems 10 and 11, write the amounts of the payments Len makes at the end of 1, 2, and 3 months. The total Len has paid after n months is the partial sum of the series of payments. Find the third partial sum algebraically. Confirm that the answer is correct by adding the first three terms.

16. If Len were not required to pay off the loan when the balance dropped below $5, his payments would go on infinitely! Show algebraically that the total he would have paid after a long time converges to the original $200 he borrowed.

17. What did you learn as a result of doing this Exploration that you did not know before?

Exploration 154: A Power Series for a Familiar Function

Objective: Learn what a power series is and how it can fit closely a particular function.

Let $P(x)$ be defined by

$$P(x) = 1 + x + \frac{1}{2!}x^2 + \frac{1}{3!}x^3 + \cdots + \frac{1}{n!}x^n + \cdots$$

The letter P is used because the right side of the equation is a **power series.** It is also appropriate because the expression looks like a **polynomial,** except that it has an infinite number of terms. In this Exploration, you will calculate and plot values of $P(x)$ and try to figure out which familiar function P represents.

1. Calculate $P(0.6)$ three times, using 3, 4, and 5 terms of the series (term index $n = 2$, 3, and 4).

2. The values of $P(0.6)$ in Problem 1 are **partial sums** of the series. Use the SUM and SEQUENCE commands on your grapher to enter an equation into y_1 that will calculate $P(0.6)$ for $n = x$ terms. Then make a table of values of $P(0.6)$ using $n = 5$, 6, 7, 8, 9, and 10. Calculate $e^{0.6}$. What do you notice about the partial sums?

3. Change the equation in y_1 so that it calculates the 11th partial sum ($n = 10$) of the series for $P(x)$. Then plot the graph using a window with an x-range of $[-5, 5]$ and a y-range of $[-1, 10]$. Sketch the result.

4. On the same screen as in Problem 3, plot the graph of $y = e^x$. Does the graph support the conjecture that $P(x) = e^x$?

5. Find the 11th partial sum ($n = 10$) for $P(1)$. How close is the answer to e^1?

6. Show that the 11th partial sum for $P(10)$ is *not* close to e^{10}.

7. What did you learn as a result of doing this Exploration that you did not know before?

Exploration 155: Complex Numbers in Exponential Form

Date: _________

Objective: Use the power series for sine, cosine, and e^x to find relationships between trigonometric and exponential functions.

The sine, cosine, and natural exponential functions can be written as **power series,** as follows:

$$\sin x = x - \frac{1}{3!}x^3 + \frac{1}{5!}x^5 - \frac{1}{7!}x^7 + \cdots$$

$$\cos x = 1 - \frac{1}{2!}x^2 + \frac{1}{4!}x^4 - \frac{1}{6!}x^6 + \cdots$$

$$e^x = 1 + x + \frac{1}{2!}x^2 + \frac{1}{3!}x^3 + \frac{1}{4!}x^4 + \cdots$$

1. Use the first five terms of each series to find, approximately, the values of sin 0.6, cos 0.6, and $e^{0.6}$. How close do these approximations come to the actual function values from your calculator?

2. Substitute (ix) for x in the series for e^x. Simplify the result.

3. By appropriate commuting, associating, and factoring, show that

$$e^{ix} = \cos x + i \sin x$$

4. The expression on the right side of the equation in Problem 3 should be familiar to you. What kind of expression is it?

5. Show that $e^{-ix} = \cos x - i \sin x$. Which property of trig functions do you use to show this fact? What special words are used for the relationship between $\cos x + i \sin x$ and $\cos x - i \sin x$?

6. Write these as complex numbers in exponential form.

   a. $\cos 3 + i \sin 3$ ______________

   b. $3(\cos 2 + i \sin 2)$ ______________

   c. $1 + i$ ______________

   d. $\sqrt{3} - i$ ______________

   e. i ______________

7. Show that $e^{i\pi} = -1$. (This remarkable equation combines four of the most mysterious numbers in mathematics!)

8. Show that i^i is a *real* number. Find a decimal approximation for this number.

9. What did you learn as a result of doing this Exploration that you did not know before?

Exploration 156: Cubic Function Graphs

Date: _____________

Objective: Learn the shape and some properties of cubic function graphs.

1. Let $f(x) = x^3 - 4x^2 - 3x + 2$. Plot the graph using a window with x-range of about $[-5, 5]$ and a y-range of $[-30, 30]$. Sketch the graph.

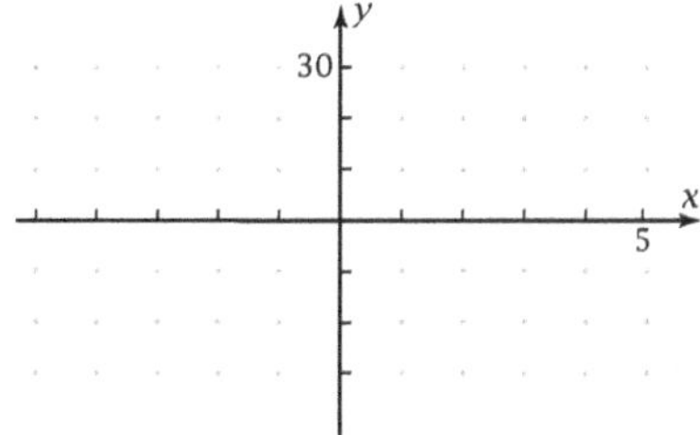

2. The graph in Problem 1 crosses the x-axis at three values of x. One of those values is an integer. Which value? Estimate the other two.

3. The integer value in Problem 2 can be used to write a linear factor of $f(x)$. By dividing this factor into $f(x)$, find the other (quadratic) factor.

4. With the help of the quadratic formula, find the *exact* values of the other two x-intercepts. Confirm that your answers in Problem 2 agree with these exact values.

5. Let $g(x) = x^3 - 4x^2 - 3x + 18$. Plot the graph of g on your grapher. Sketch on the same axes as in Problem 1. What similarities and differences do you notice in the graphs of f and g?

6. Factor $g(x)$ into three linear factors. By setting each factor equal to zero, find the three x-intercepts. What do you notice about two of the x-intercepts?

7. Let $h(x) = x^3 - 4x^2 - 3x + 54$. Plot the graph of h. Sketch on the same axes as in Problem 1.

8. Find one linear factor of $h(x)$. Use it to find two other values of x that make $h(x) = 0$. How do the results agree with the graph? Why do you think values of x that make a polynomial function equal 0 are called **zeros** of the function rather than just x-intercepts?

9. What did you learn as a result of doing this Exploration that you did not know before?

Exploration 157: Synthetic Substitution

Date: _________

Objective: Learn how to evaluate $f(x)$ quickly by pencil and paper and how to use the result to factor $f(x)$.

1. Let $f(x) = 3x^3 + 4x^2 + 11x - 10$. Plot the graph using a window with an x-range of about $[-5, 5]$ and a y-range of $[-100, 100]$. Sketch the graph here.

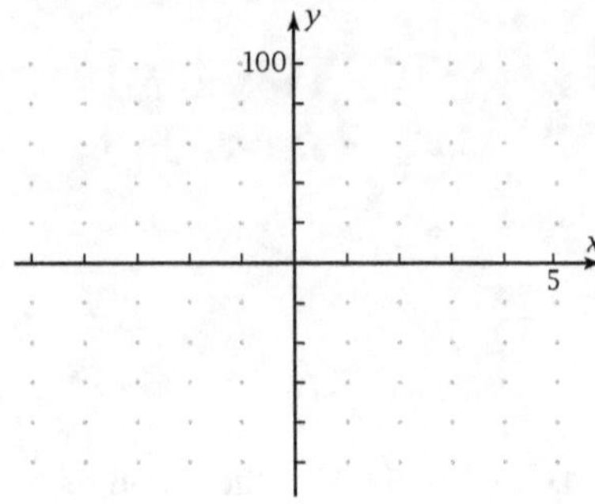

2. Find $f(2)$. Mark the corresponding point on your graph.

3. Use long division to divide $f(x)$ by $x - 2$. What is the quotient? What is the remainder?

4. What do you notice about the remainder and the value of $f(2)$?

5. Find $f(2)$ by synthetic substitution.

6. How could you write the answer to Problem 3 directly from the synthetic substitution results?

7. Find $f\left(\frac{2}{3}\right)$ by synthetic substitution. How does the result agree with your graph?

8. Use the result of Problem 7 to find the other two zeros of $f(x)$. Write the complex zeros in terms of i, and simplify as much as possible.

9. What did you learn as a result of doing this Exploration that you did not know before?

Exploration 158: Sum and Product of the Zeros of a Polynomial

Objective: Discover properties relating the sum and the product of the zeros of a polynomial function.

The figure shows the graph of

$$f(x) = 5x^3 - 33x^2 + 58x - 24$$

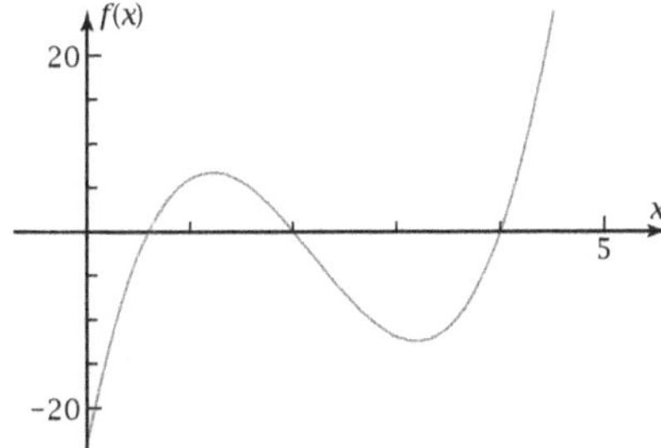

1. Find z_1, z_2, and z_3, the three zeros of this function. Describe your method.

2. Find the product of the zeros, $z_1 z_2 z_3$.

3. If you factor out 5 from each term in the equation of $f(x)$, you get a second factor with leading coefficient equal to 1.

$$f(x) = 5\left(x^3 - \frac{33}{5}x^2 + \frac{58}{5}x - \frac{24}{5}\right)$$

$$= 5(x^3 - 6.6x^2 + 11.6x - 4.8)$$

How does the product of the zeros you found in Problem 2 relate to the coefficients inside the parentheses?

4. Find the sum of the zeros, $z_1 + z_2 + z_3$. How does the answer relate to the coefficients in Problem 3?

5. Find the sum of the pairwise products of the zeros,

$$z_1 z_2 + z_1 z_3 + z_2 z_3$$

How does the answer relate to the coefficients in Problem 3?

6. Use the patterns you observe in Problems 1–5 to find the particular equation of the cubic function $g(x)$ if the leading coefficient is 1 and the zeros are

$$5, 2 + \sqrt{3}, 2 - \sqrt{3}$$

7. Find the particular equation of the cubic function $h(x)$ that is a vertical dilation of $g(x)$ by a factor of 3.

8. Plot the graphs of g and h on the same screen. Do both functions have the same zeros? Sketch the graphs here.

9. What did you learn as a result of doing this Exploration that you did not know before?

Exploration 159: Fitting a Polynomial Function to Points

Objective: Given a set of points, find the particular equation of a polynomial function that fits the points.

The figure shows the graph of a polynomial function P. The table shows plotting points for this function.

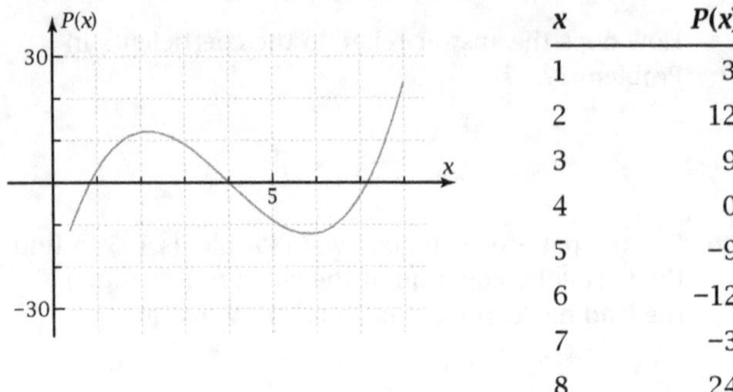

x	$P(x)$
1	3
2	12
3	9
4	0
5	−9
6	−12
7	−3
8	24

1. Show that the **third differences** between the $P(x)$ values are constant. You may do this by entering the x- and y-data in lists on your grapher and using operations on the y-list to make lists of first, second, and third differences.

   Third differences = _________________

2. If the third differences between y-values are constant, then a **cubic function** fits the data.

   $$P(x) = ax^3 + bx^2 + cx + d$$

   Find the particular equation of the cubic function that fits the first four data points. You may do this by substituting 1, 2, 3, and 4 for x and then solving for a, b, c, and d by matrices.

3. Show that the equation of Problem 2 fits the other four data points.

4. Find the particular equation of Problem 2 again, using the cubic regression feature of your grapher. Does the equation come out the same? What statistic tells you that the fit is perfect?

5. Use your equation to predict the value of $P(20)$.

6. Find numerically the largest value of x for which $P(x) = 10$. Does the answer agree with the graph?

7. The number 4 is said to be a **zero** of $P(x)$ because $P(4) = 0$. This fact implies that $(x - 4)$ is a **factor** of $P(x)$,

   $$P(x) = (x - 4)(\text{other factor})$$

   Find the other (quadratic) factor.

8. By setting the quadratic factor in Problem 7 equal to zero, find algebraically the other two zeros of $P(x)$. Do the answers agree with the graph?

9. What did you learn as a result of doing this Exploration that you did not know before?

Precalculus and Trigonometry Explorations
©2004 Key Curriculum Press

Exploration 160: Rational Functions and Discontinuities

Objective: Find and classify discontinuities in the graph of a rational algebraic function.

1. On this graph paper, plot *quickly* the graph of the rational function $f(x) = \frac{1}{x}$ (no grapher).

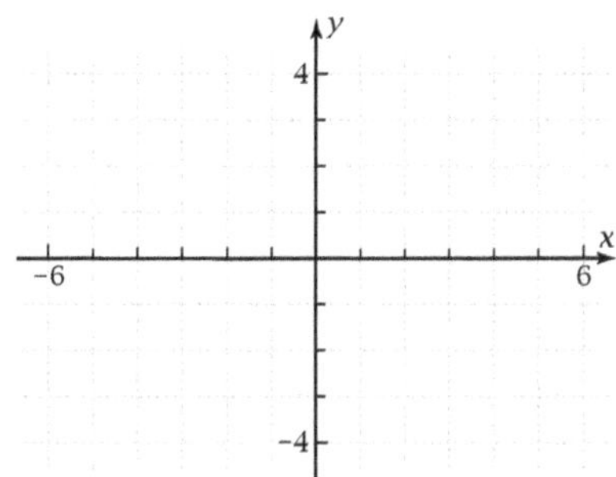

2. The graph in Problem 1 has a **vertical asymptote** at $x = 0$. Give an algebraic reason why there is such an asymptote there.

3. Identify a transformation that maps the graph of f onto the graph of g shown here. (There are at least three ways to answer this!)

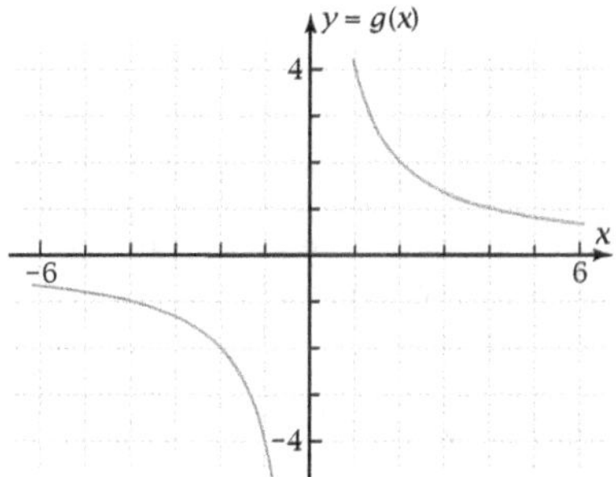

4. Let $h(x) = \frac{1}{x-3}$. What transformation of the graph of f is this? On the graph paper here, plot quickly the graph of h (no grapher).

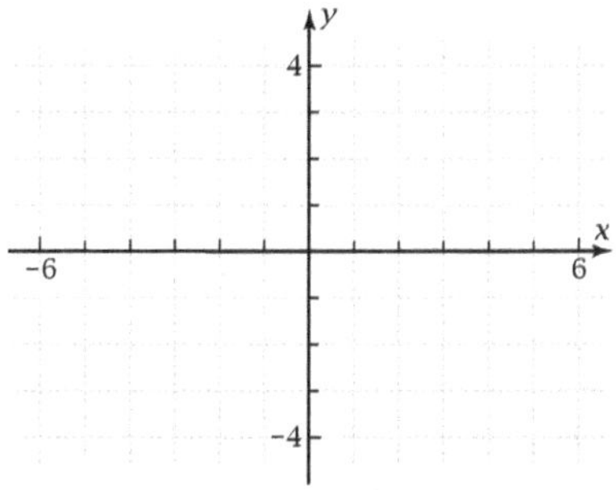

5. Function r is called a **rational function** because $r(x)$ equals a ratio of two polynomials.

$$r(x) = \frac{x-1}{x^2 - 4x + 3}$$

Plot the graph of r on your grapher. Use a friendly window from about $x = -5$ to $x = 5$ that includes each integer as a grid point. Sketch the result on this graph paper.

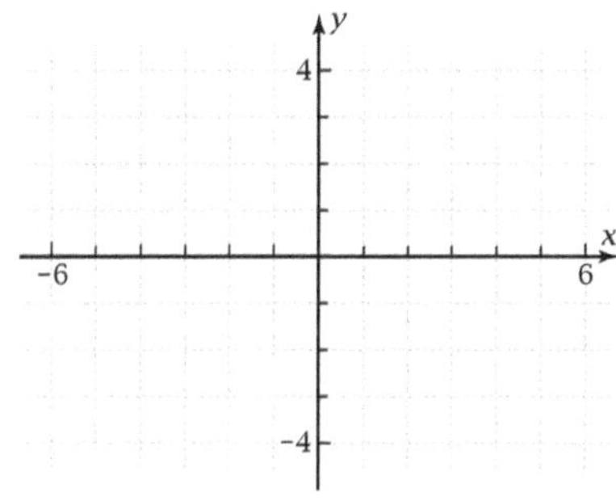

6. In what way is the graph of r similar to the graph of h in Problem 4? In what way is it different?

7. The graph of r has a **removable discontinuity** at $x = 1$. Explain algebraically why there is a discontinuity here.

8. By factoring the denominator in the equation for $r(x)$, show how the discontinuity at $x = 1$ can be "removed" algebraically.

9. Without plotting the graph, how can you tell which kind of discontinuity, removable or asymptote, the graph of a rational function will have at a value of x that makes the denominator equal zero?

10. What did you learn as a result of doing this Exploration that you did not know before?

Exploration 161: Limits and Curved Asymptotes   Date: __________

Objective: Analyze graphs of rational algebraic functions with the help of synthetic substitution and factoring.

1. Let
$$f(x) = \frac{x^3 + 2x^2 - 8x + 5}{x - 1}$$

   Plot the graph of f as y_1. Use a friendly window with an x-range of about $x = -6$ to $x = 4$ that includes $x = 1$ as a grid point. Make sure the grid is off. Sketch the graph.

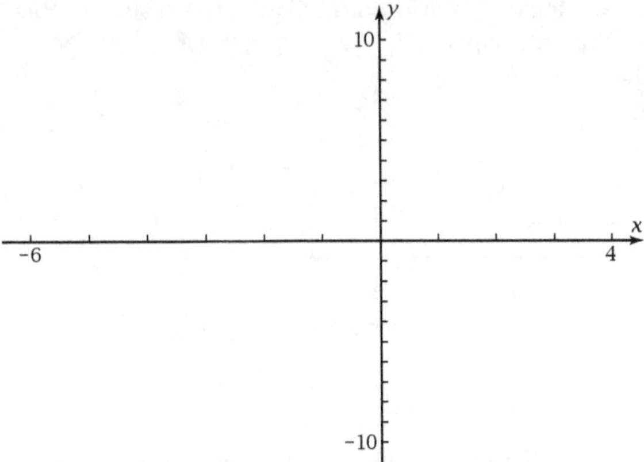

5. Let
$$g(x) = \frac{x^3 + 2x^2 - 8x + 6}{x - 1}$$

   Plot the graph of g as y_2 using "thick" style. What feature does the graph of g have at $x = 1$? Sketch the graphs of $f(x)$ and $g(x)$ here, showing their relationship to each other.

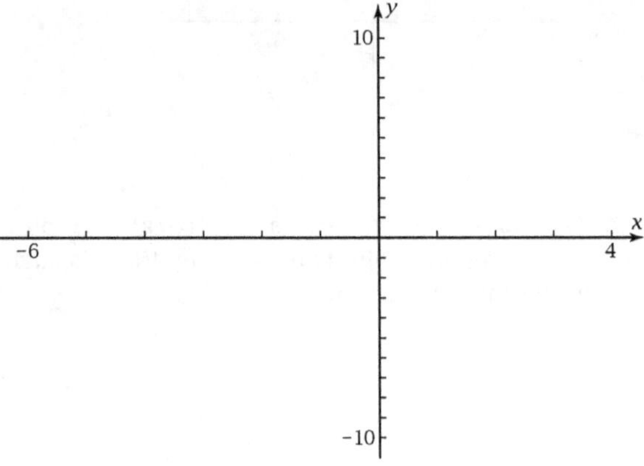

2. $f(1)$ is undefined because of division by zero. Trace to $x = 1$ on your graph. What feature does the graph have at this point? By tracing closer and closer to $x = 1$, find the **limit** $f(x)$ seems to be approaching as x approaches 1. Is the limit as x approaches 1 from the left side the same as when x approaches 1 from the right side?

6. Try to find $\lim_{x \to 1} g(x)$ by tracing to x-values closer and closer to 1. Try x-values on both sides of 1. What happens to the quotient as x approaches 1?

3. Remove the discontinuity at $x = 1$ algebraically by factoring the numerator and reducing the fraction. Evaluate the resulting quotient polynomial at $x = 1$. Is the answer equal to the limit you found in Problem 2?

7. To understand why the graph of $g(x)$ resembles the graph of $f(x)$, simplify the equation for $g(x)$ by synthetic substitution. Write the equation in "mixed-number" form as

$$g(x) = (\text{polynomial}) + \frac{\text{remainder}}{x - 1}$$

   What relationship do you notice between the equations for $f(x)$ and $g(x)$?

4. How do you read $\lim_{x \to 1} f(x) = -1$? What does this equation mean?

Exploration 161: Limits and Curved Asymptotes *continued*

8. Let

$$h(x) = \frac{x^3 + 2x^2 - 8x + 4}{x - 1}$$

Plot the graph of $h(x)$ as y_3.

9. The graphs of f and g from Problem 5 should look like this. On this figure, sketch the graph of h.

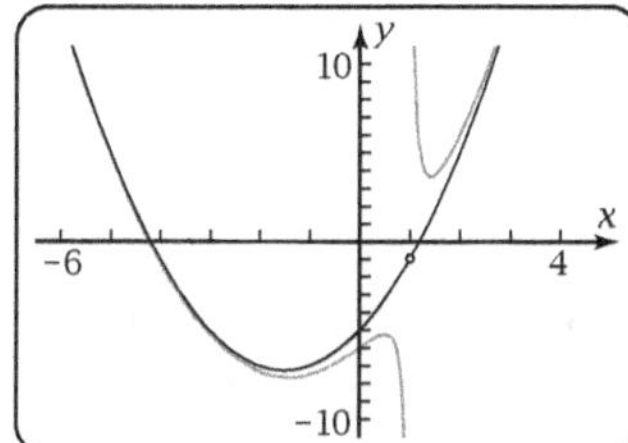

10. Simplify the equation for $h(x)$ by long division or synthetic substitution. Write the result in mixed-number form as

$$h(x) = (\text{polynomial}) + \frac{\text{remainder}}{x - 1}$$

11. How is the polynomial part of $h(x)$ in Problem 10 related to the graph of $h(x)$?

12. Why is it important for your work to be 100% correct in problems like these that are sequential?

13. Write a paragraph summarizing the things you have learned about rational functions, removable discontinuities, asymptotes, etc., as a result of doing this Exploration.

Exploration 162: Rate of Change of a Polynomial Function

Date: _________

Objective: Find the instantaneous rate of change of a polynomial function at a given value of x.

Astronaut Spencer Spacey takes off from the planet Alderaan. He starts his rocket countdown at time $x = 0$ minutes. Shortly thereafter his spaceship takes off. It rises for a while, then drops while his second-stage rocket engine is starting up, and then rises again. Spencer's computer finds that his distance, $f(x)$ miles, above the surface is given by

$$f(x) = x^3 - 10x^2 + 32x - 22$$

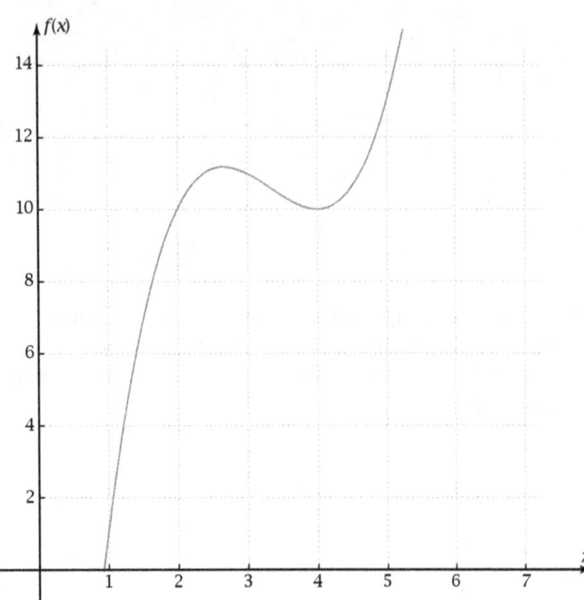

1. How far is Spencer from the surface at $x = 2$ min? at $x = 2.1$ min? How far did he go in this time interval? What was his average velocity for this time interval?

2. What is Spencer's average velocity for the time interval from 2 minutes to $x = 2.001$ minutes?

3. Spencer's **instantaneous velocity** at 2 minutes is the **limit** of his average velocity between 2 and x minutes as x approaches 2. From your answers to Problems 1 and 2, what do you conjecture that his instantaneous velocity equals at 2 minutes?

4. Between time 2 and x minutes, Spencer travels $f(x) - f(2)$ miles. So his average rate $r(x)$ is

$$r(x) = \frac{f(x) - f(2)}{x - 2}$$

Substitute for $f(x)$ and $f(2)$ and simplify the numerator. Then use synthetic substitution to do the division, thus removing the discontinuity at $x = 2$. Write a simplified equation for $r(x)$.

5. Use this simplified equation for $r(x)$ to find the limit as x approaches 2. Does this confirm the instantaneous velocity you found in Problem 3?

$$\lim_{x \to 2} r(x) =$$

6. Plot a line through the point $(2, f(2))$ that has a slope equal to the instantaneous velocity you found in Problem 5. Take into account that the two axes have different scales. How does the line relate to the graph?

Relationship: _________________________________

____

Exploration 162: Rate of Change of a
Polynomial Function *continued*

7. Find Spencer's instantaneous velocity at time 3 minutes by starting with

$$r(x) = \frac{f(x) - f(3)}{x - 3}$$

Explain the fact that the answer is *negative*.

8. Let $g(x) = 3x^2 - 20x + 32$. Find $g(2)$ and $g(3)$. How do your answers compare with the instantaneous velocities at times $x = 2$ and $x = 3$?

9. There is an algebraic way to find the instantaneous rate of change of a polynomial function. Suppose that $P(x) = x^5$. If $r(x)$ is the average rate of change between the fixed time c and the variable time x, then

$$r(x) = \frac{P(x) - P(c)}{x - c} = \frac{x^5 - c^5}{x - c}$$

Divide by means of synthetic substitution. Write a polynomial equation for $r(x)$. (It will involve the constant c.)

$$\underline{c}\begin{array}{cccccc} 1 & 0 & 0 & 0 & 0 & -c^5 \end{array}$$

10. By appropriate calculations, show that

$$\lim_{x \to c} r(x) = 5c^4$$

11. Explain how you can get the 5 and the 4 in $5c^4$ from the original polynomial $P(x) = x^5$.

12. Suppose that $f(x) = 17x^{10}$. What do you think would be an equation for the function $g(x)$ that gives the instantaneous velocity of $f(x)$?

13. Show that you can use the patterns you found in Problems 11 and 12 on the function

$$f(x) = x^3 - 10x^2 + 32x - 22$$

to get the **velocity function** from Problem 8,

$$g(x) = 3x^2 - 20x + 32$$

14. What is the special name given to the velocity function?

15. What did you learn as a result of doing this Exploration that you did not know before?

Solutions to the Explorations

Functions and Mathematical Models

Exploration 1

1. Answers will vary.

2. Answers will vary.

3. Answers will vary. The stack height should increase by the same amount for each additional cup.

4. Answers will vary; no, the 10-cup stack would not be twice as tall as a 5-cup stack.

5. Answers will vary.

6. Linear function

7. Answers will vary. Plug in 3 for x in your equation.

8. Answers will vary.

9. Within: Interpolation
   Outside: Extrapolation

10. Answers will vary.

11. Answers will vary.

Exploration 2

1.

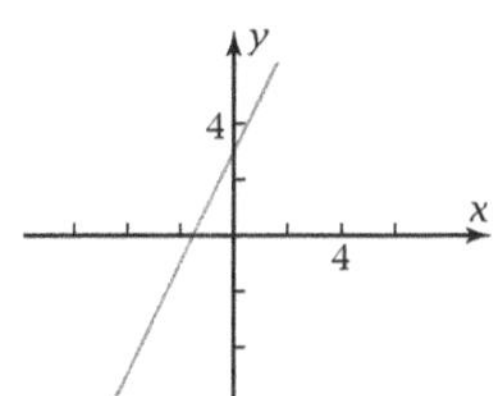

The graph is a straight line.

2.

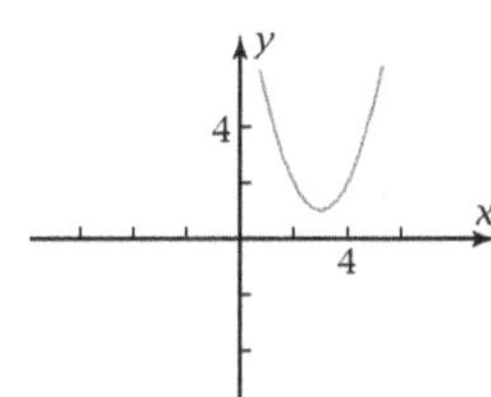

Both come from the Latin word for "square," which in turn comes from the Latin word for "four," referring to the four sides of a square.

3.

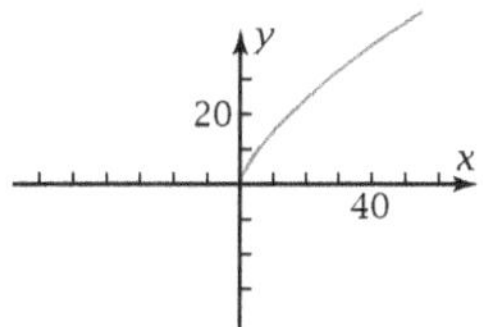

The variable (x in this case) is raised to a power.

4.

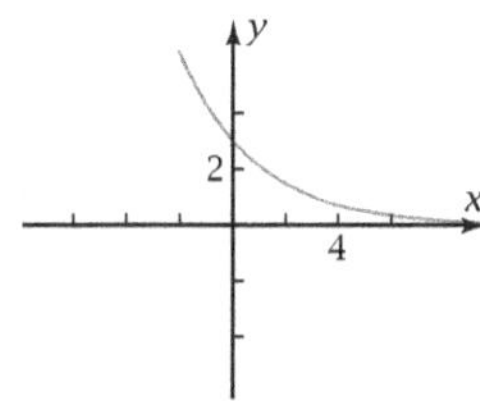

Algebraically, a power function has the variable as the base and a constant as the exponent, whereas an exponential function has a constant as the base and the variable (x in this case) as an exponent. Graphically, the parent power function passes through the point (0, 0) and increases to the right (and increases to the left if the power is an integer), whereas the parent exponential function passes through the point (0, 1), increases without bound to the left, and approaches 0 to the right.

5.

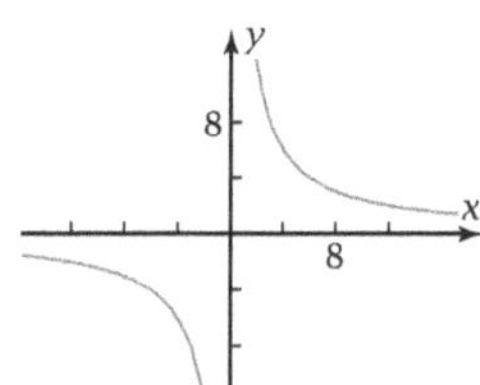

y gets smaller as x gets larger, and vice versa. It can be written as $y = 24x^{-1}$, involving the -1 power of x.

6. It is a fourth-degree polynomial. The largest value of x at which the graph crosses the x-axis is 7.

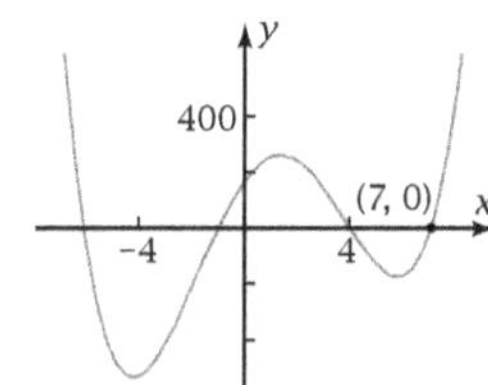

7.

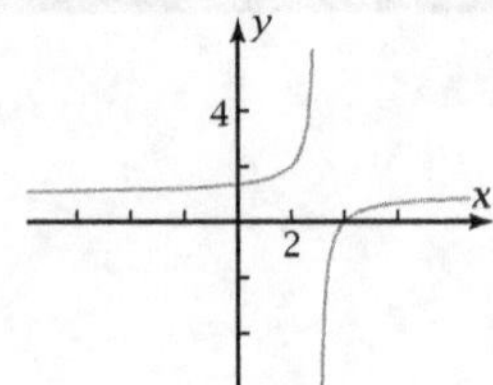

It has a discontinuity and an asymptote at $x = 3$. It is expressed as $\frac{p(x)}{q(x)}$, where $p(x)$ and $q(x)$ are both functions.

8. Answers will vary.

Exploration 3

1. Let y be the weight in pounds. Let x be the distance, in miles, from center.

$$y = \frac{x}{20} \text{ for } 0 \le x \le 4000$$

2. $y = \dfrac{32 \times 10^8}{x^2}$ for $x \ge 4000$

3. The graphs look the same. They cross at the point (4000, 200).

4. $y = \dfrac{32 \times 10^8}{x^2(x \ge 4000)}$

   For $x < 4000$, the calculator is dividing by 0, which gives no answer.

5. $y_1 = 0.05x/(0 \le x \text{ and } x \le 4000)$
   The complete graph now matches the graph in Problem 4.

6. x-values: Domain
   y-values: Range

7. Answers will vary.

Exploration 4

1. The values for $g(x)$ are 5, 6, 4, 1, 3, 4.

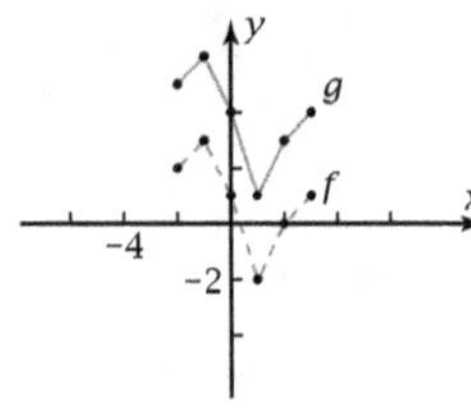

2. The graph is slid vertically without changing its shape or proportions.

3. The values for $g(x)$ are 2, 3, 1, −2, 0, 1.

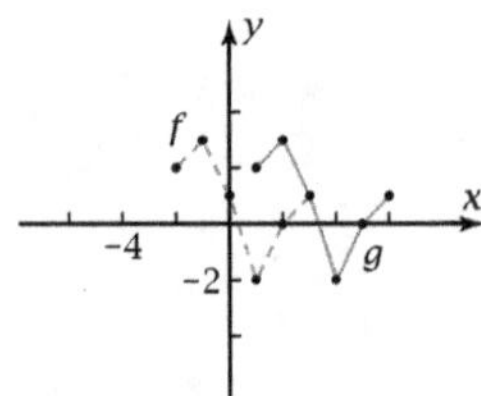

4. Horizontal translation by 3 units (right)

5. The values for $g(x)$ are 4, 6, 2, −4, 0, 2.

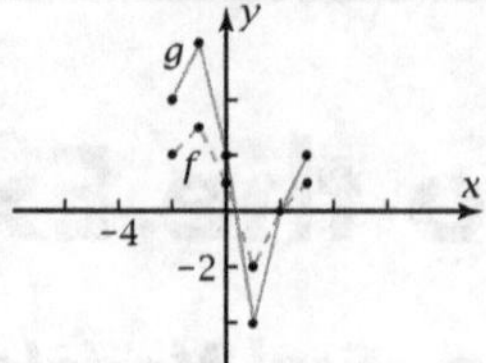

6. The graph is stretched or squashed vertically but not shifted vertically as a whole.

7. The values for $g(x)$ are 2, 3, 1, −2, 0, 1.

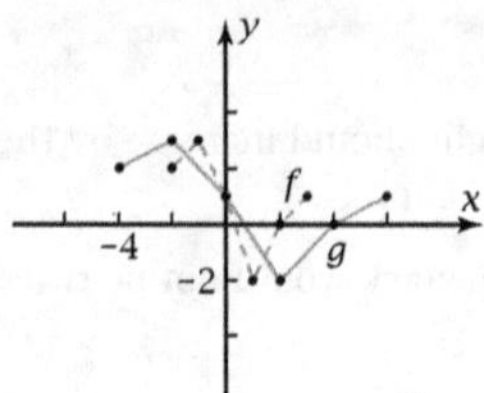

8. By a factor of 2. $2 = \dfrac{1}{\frac{1}{2}}$

9. Answers will vary.

Exploration 5

1.

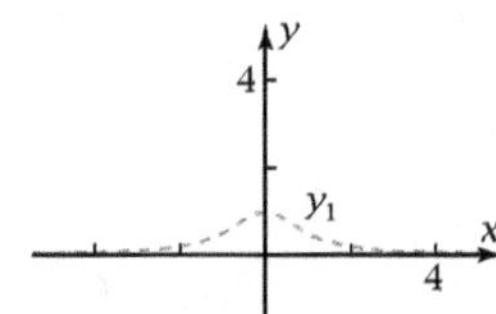

2.

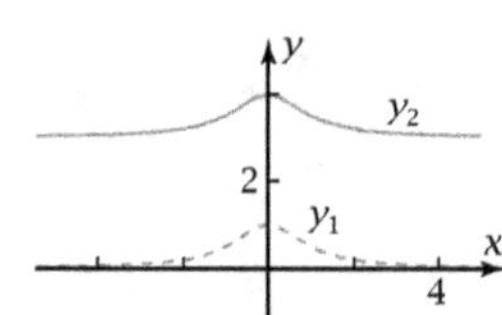

3. The graph is slid vertically without changing its shape or proportions.

4.

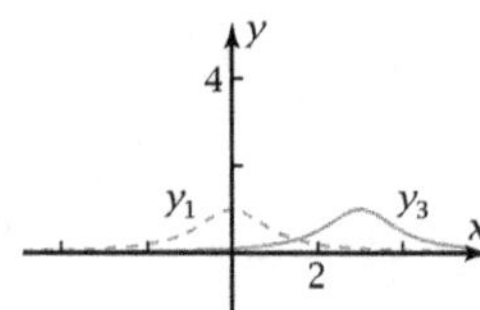

5. Horizontal translation by 3 units (right)

6.

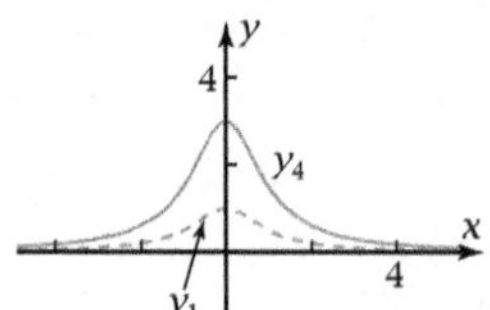

7. The graph is stretched or squashed vertically but not shifted vertically as a whole.

8.

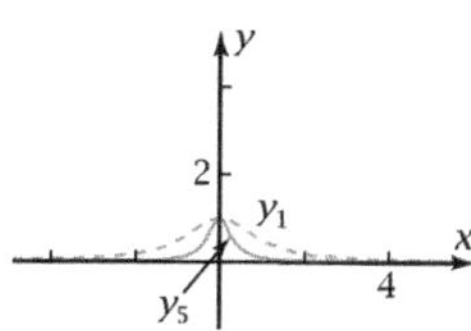

9. By a factor of $\frac{1}{3}$

10. Answers will vary.

Exploration 6

1. Vertical translation by -6; $g(x) = f(x) - 6$

2. Horizontal translation by $+10$; $g(x) = f(x - 10)$

3. Vertical dilation by 3; $g(x) = 3f(x)$

4. Horizontal dilation by 2; $g(x) = f\left(\frac{1}{2}x\right)$

5. Reflection across the x-axis of that part of the graph that is below the x-axis; $g(x) = |f(x)|$

6. Reflection across the y-axis; $g(x) = f(-x)$

7. Answers will vary.

Exploration 7

1.

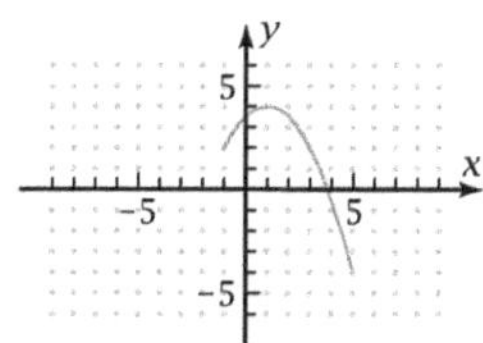

2. Horizontal translation by -6
$g(x) = f(x + 6)$

3. Vertical translation by 3
$g(x) = 3 + f(x)$

4. Vertical dilation by factor of $\frac{1}{2}$

$g(x) = \frac{1}{2}f(x)$

5. Horizontal dilation by factor of 2

$g(x) = f\left(\frac{1}{2}x\right)$

6. Horizontal dilation by factor of 2 and vertical translation by 3

$g(x) = 3 + f\left(\frac{1}{2}x\right)$

7. Answers will vary.

Exploration 8

1. f: Domain: $1 \le x \le 5$ Range: $2 \le y \le 6$
g: Domain: $2 \le x \le 8$ Range: $0 \le y \le 3$

2., 3.

x	$g(x)$	$f(g(x))$
0	None	None
1	None	None
2	0	None
3	0.5	None
4	1	6
5	1.5	5.5
6	2	5
7	2.5	4.5
8	3	4
9	None	None

4. The lines in which $x = 2$ and $x = 3$

5.

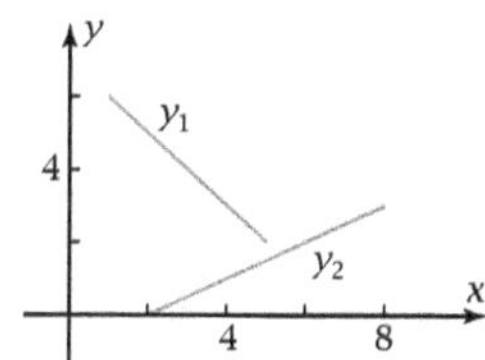

6. Domain: $4 \le x \le 8$
Range: $4 \le y \le 6$

7. $f(x) = -x + 7$

$g(x) = \frac{1}{2}x - 1$

8.

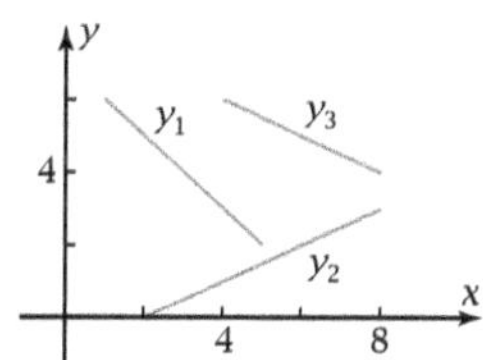

Yes

9.

Yes

10. $f(g(x)) = -g(x) + 7$

$= -\left(\frac{1}{2}x - 1\right) + 7$

$= -\frac{1}{2}x + 8$

$f(g(x)) = -\frac{1}{2}x + 8$

11. Answers will vary.

1. $x = 2y - 5$
   $x + 5 = 2y$
   $y = \dfrac{1}{2}x + 2.5$

2.

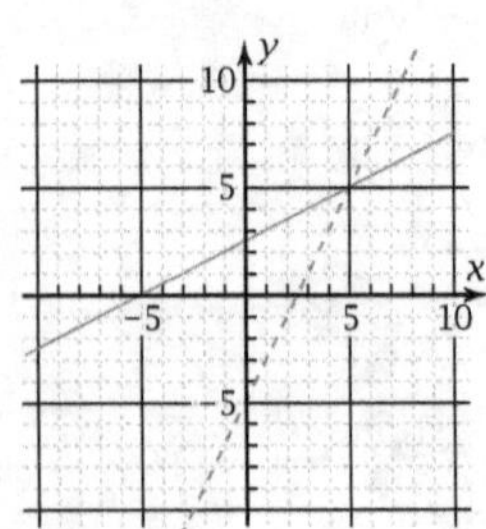

3. There is no place where there are two different ys for the same x.

4. $f^{-1}(x) = \dfrac{1}{2}x + 2.5$

5. $f(3) = 2(3) - 5 = 1$
   $f^{-1}(1) = \dfrac{1}{2} \cdot 1 + 2.5 = 3$
   $f(x) = y \Rightarrow f^{-1}(y) = x$

6.

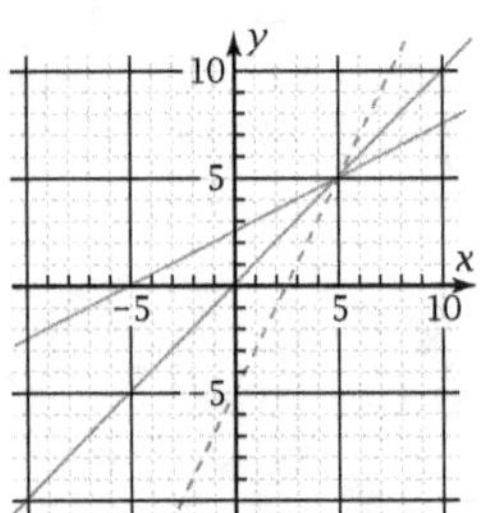

They are reflections in $y = x$.

7.

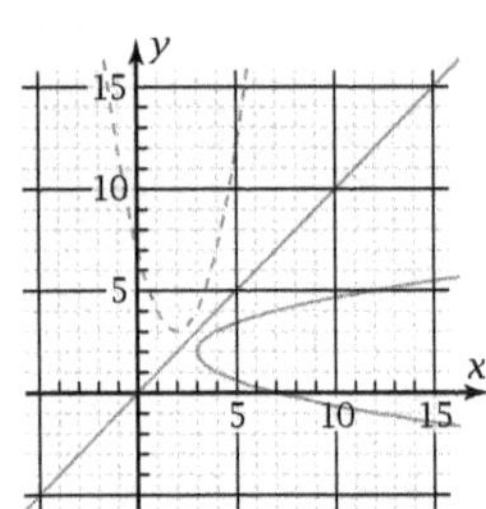

8. There are two values of y for the same value of x.

9. $f(0) = 1;\ f(1) = 2;\ f(2) = 4;\ f(3) = 8$

10. $f^{-1}(1) = 0;\ f^{-1}(2) = 1,\ f^{-1}(4) = 2;\ f^{-1}(8) = 3$

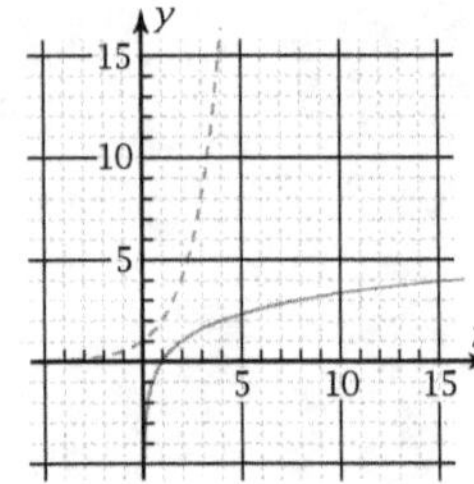

11. Answers will vary.

---

1. Vertical dilation by $\frac{1}{2}$

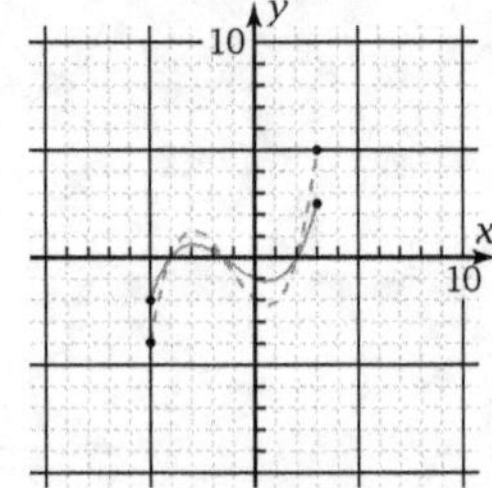

2. Horizontal dilation by 2

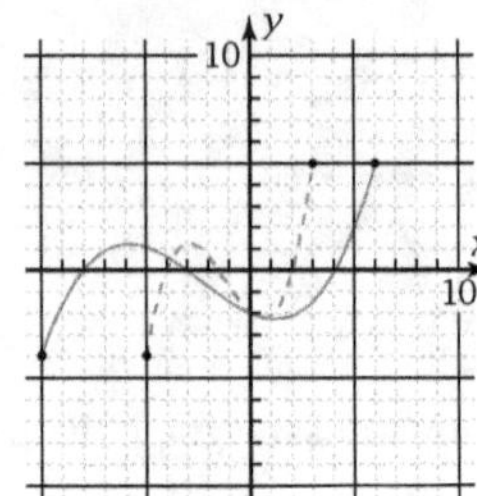

3. Vertical translation by −4

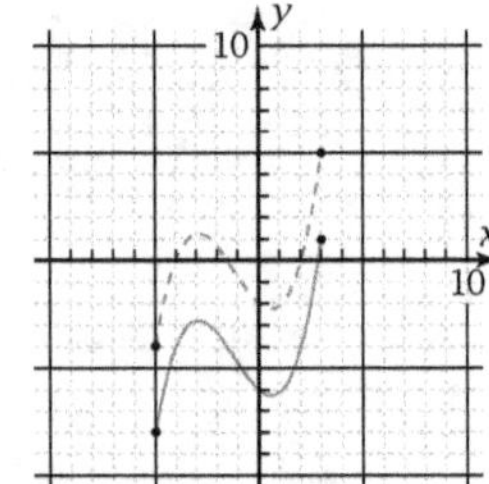

4. Horizontal translation by +7

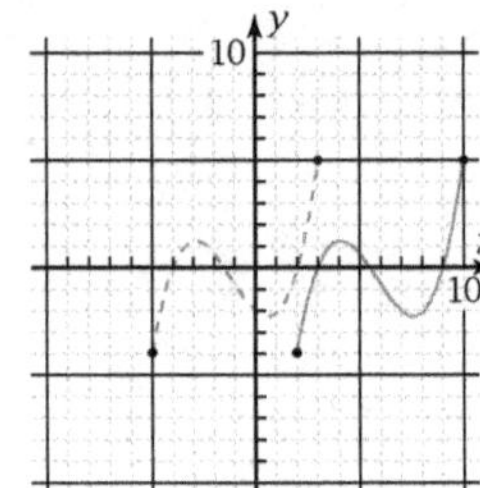

5. Horizontal dilation by −1 (because $\dfrac{1}{-1} = -1$)

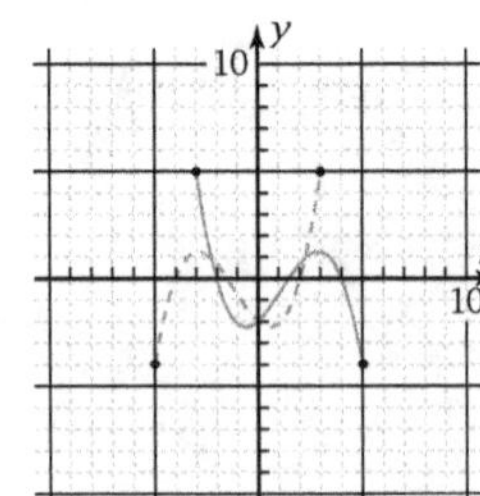

6. Reflection across the y-axis

---

7. Vertical dilation by -1

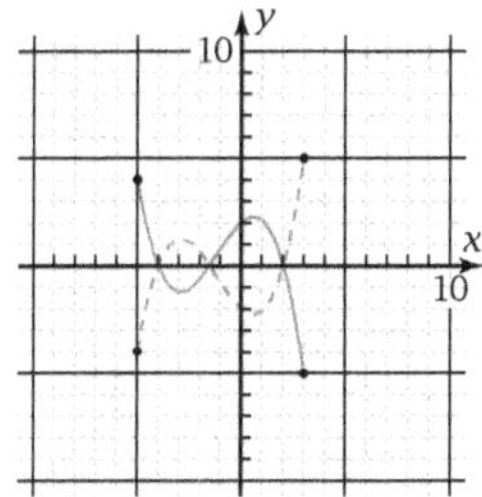

8. Reflection across the x-axis

9. y-direction

10.

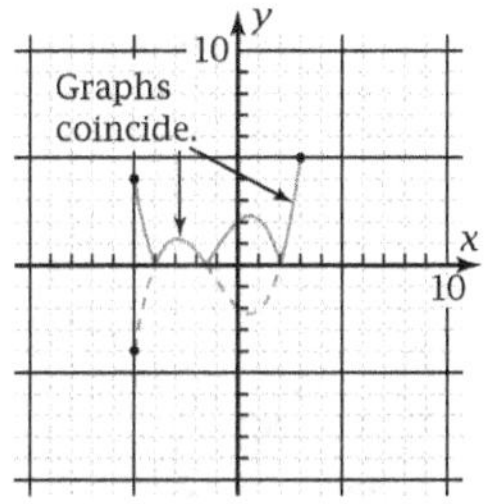

11.

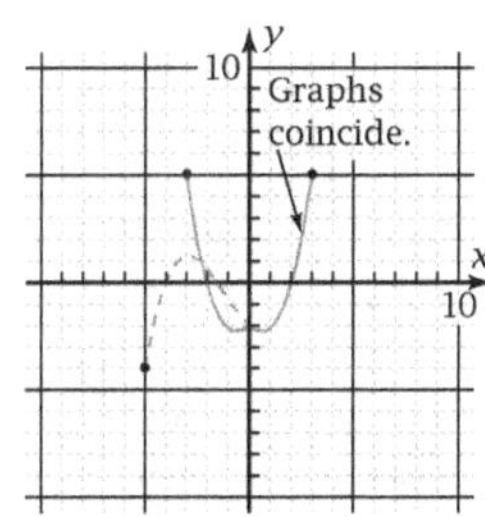

12. Answers will vary.

Periodic Functions and Right Triangle Problems

Exploration 11

1. Horizontal translation by 2
   $y = g(x) = f(x - 2)$

2. Vertical dilation by factor of 3
   $y = g(x) = 3f(x)$

3. Horizontal dilation by factor of $\frac{1}{2}$
   $y = g(x) = f(2x)$

4. Vertical translation by -5
   $y = g(x) = f(x) - 5$

5. Vertical translation by -5; horizontal translation by 2
   $y = g(x) = f(x - 2)$

6. Vertical dilation by factor of 3; horizontal translation by 2
   $y = g(x) = 3f(x - 2)$

7. Answers will vary.

Exploration 12

1. $\theta_{ref} = 180° - 152° = 28°$

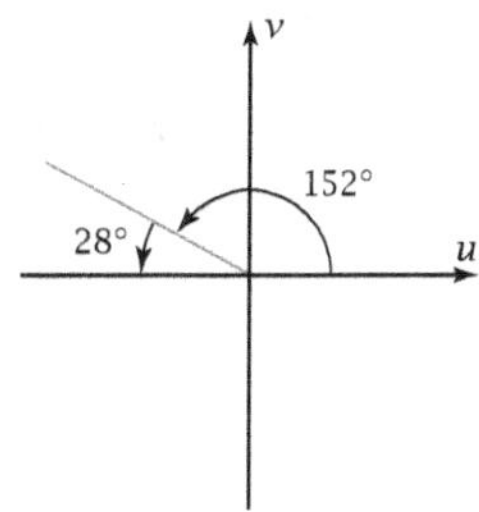

2. $\theta_{ref} = 250° - 180° = 70°$

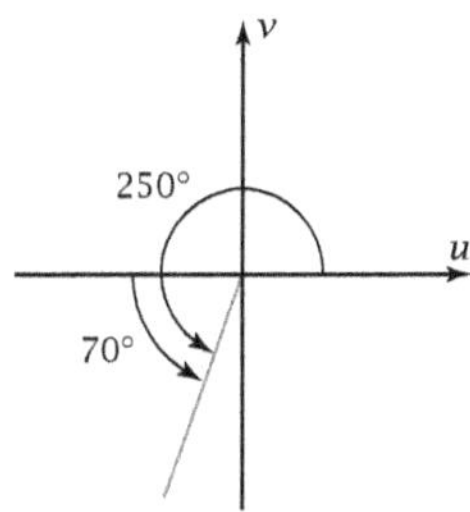

3. Because the angle must be counterclockwise so that its measure will be positive

4. Because it must go to the nearest side of the *horizontal* axis

5. $\theta_{ref} = 360° - 310° = 50°$

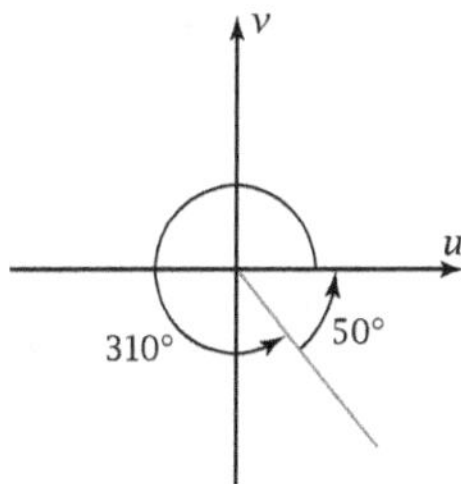

6. $\theta_{ref} = \theta$

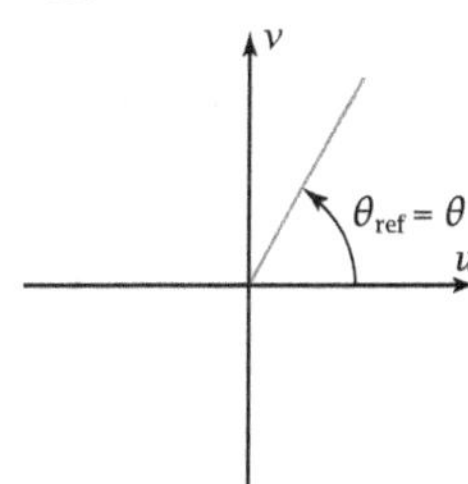

7. $\theta_{ref} = 180° + (-150°) = 30°$

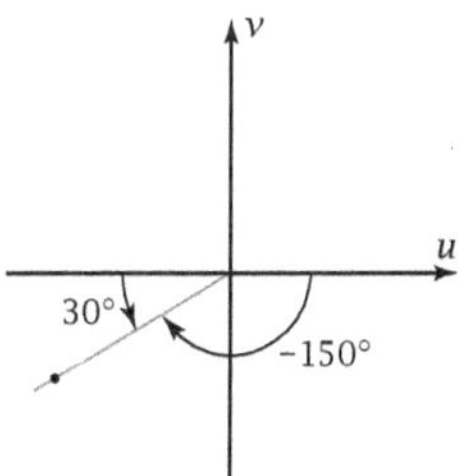

8. Duplicating the triangle above itself makes an angle of 60° at each vertex, so the large triangle is equiangular and therefore equilateral. So all sides are of length 2, and the left (vertical) leg of the original triangle is half of 2, or 1 (-1 because it is below the horizontal axis). So the other (horizontal) leg is $\sqrt{2^2 - 1^2} = \sqrt{3}$ ($-\sqrt{3}$ because it is to the left of the vertical axis).

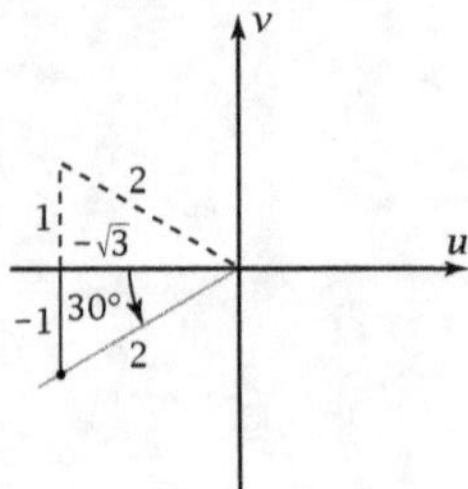

9. Answers will vary.

Exploration 13

1. $\theta = 37°$

2. $u = 4.6$ cm; $v = 3.5$ cm; $r = 5.8$ cm

3. $\sin 37° \approx 0.6034...$; $\cos 37° \approx 0.7931...$

4. $\sin 37° = 0.6018...$; $\cos 37° = 0.7986...$
   Approximate answers are reasonably close.

5. Graph, $\theta_{ref} = 55°$

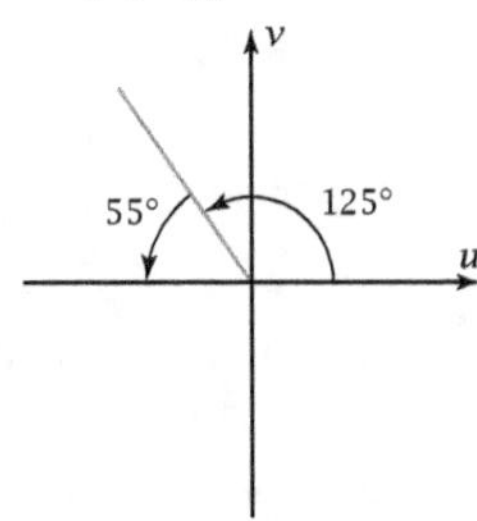

6. $\sin \theta_{ref} = 0.8191...$; $\cos \theta_{ref} = 0.5735...$

7. $\sin 125° = 0.8191...$; $\cos 125° = -0.5735...$; $125°$ terminates in Quadrant II to the left of the y-axis, where the x-coordinates are negative.

8.
Quadrant I	sine + cosine +
Quadrant II	sine + cosine −
Quadrant III	sine − cosine −
Quadrant IV	sine − cosine +

9. Answers will vary.

Exploration 14

1. $(0.50, 0.87)$

2. $\cos 60° = 0.5 =$ the u-coordinate; $\sin 60° = 0.8660... =$ the v-coordinate

3. Graph.

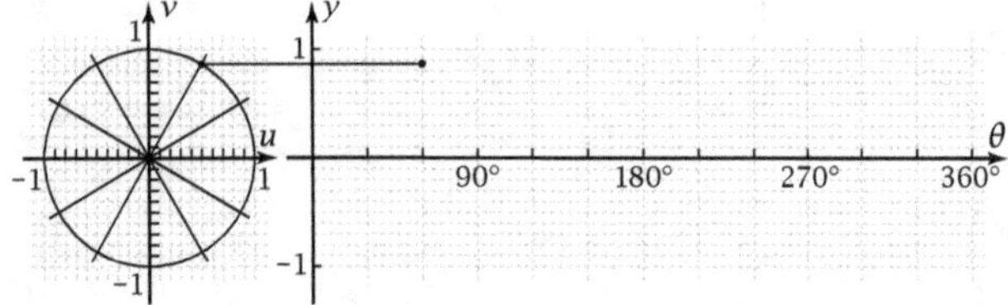

4. Graph.

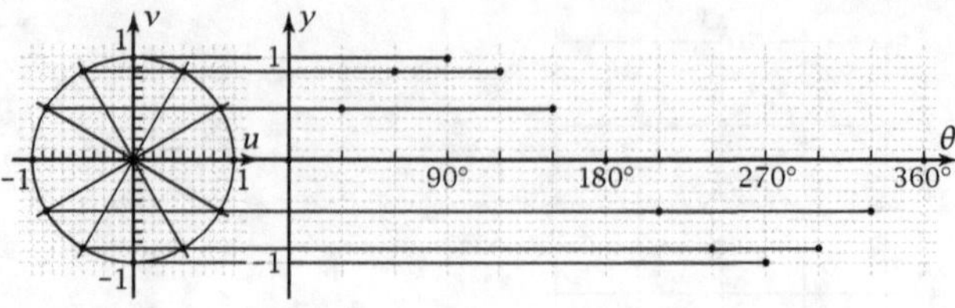

5. Graph.

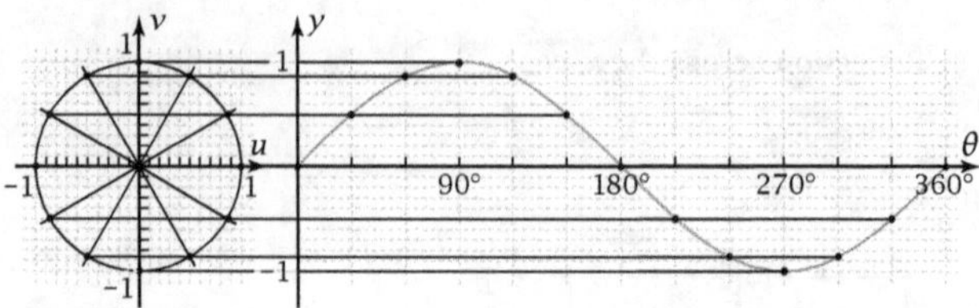

Sinusoid

6. Graph.

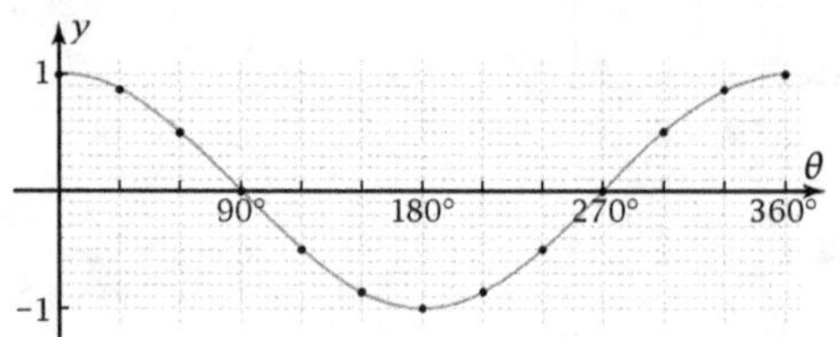

7. Horizontal translation by $-90°$

8. In the uv-diagram, θ appears as an angle in standard position. In the θ y-diagram, it appears as the horizontal coordinate.

9. To emphasize the difference between the two ways of representing an angle and its functions. In the uv-diagram, the vertical v is *not* a function of the horizontal u (both are functions of the central angle θ), while in the θ y-diagram, the vertical y *is* a function of the horizontal θ.

10. Answers will vary.

Exploration 15

1. Yes, the graph agrees.

2. Amplitude = 1; $Y_2 = 5 \sin \theta$

3. Yes

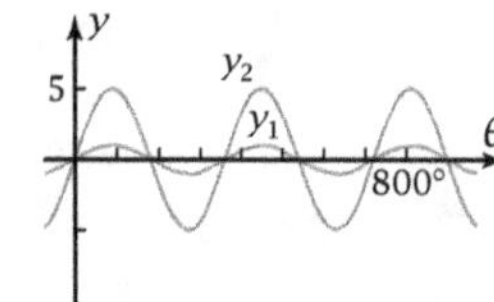

4. x-dilation of $\frac{1}{3}$; $Y_3 = \sin 3\theta$

5. $Y_4 = 8 + \sin \theta$

6. θ-translation of $+60°$; y-dilation of 4; y-translation of -5; $Y_5 = -5 + 4 \sin(\theta - 60°)$

7. $360°$ represents a return to the starting point in a rotation.

8. Answers will vary.

Exploration 16

1. $\sin\theta = \dfrac{v}{r}$ $\csc\theta = \dfrac{r}{v}$

   $\cos\theta = \dfrac{u}{r}$ $\sec\theta = \dfrac{r}{u}$

   $\tan\theta = \dfrac{v}{u}$ $\cot\theta = \dfrac{u}{v}$

2. Sketch.

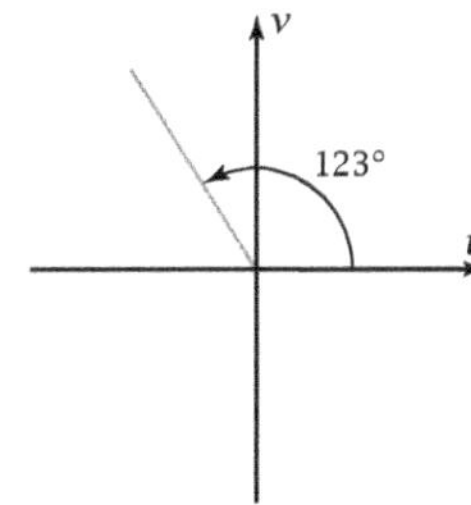

   $\sin 123° = 0.8386...$ $\csc 123° = 1.1923...$
   $\cos 123° = -0.5446...$ $\sec 123° = -1.8360...$
   $\tan 123° = -1.5398...$ $\cot 123° = -0.6494...$

3. In Quadrant II, (u, v) is (negative, positive) and r is always positive, so $\sin\theta = \dfrac{v}{r} = \dfrac{\text{positive}}{\text{positive}} = $ positive, but $\tan\theta = \dfrac{v}{u} = \dfrac{\text{positive}}{\text{negative}} = $ negative.

4. $\sin\theta = -\dfrac{7\sqrt{58}}{58}$ $\csc\theta = -\dfrac{\sqrt{58}}{7}$

   $\cos\theta = -\dfrac{3\sqrt{58}}{58}$ $\sec\theta = -\dfrac{\sqrt{58}}{3}$

   $\tan\theta = \dfrac{7}{3}$ $\cot\theta = \dfrac{3}{7}$

5. $\sin 300° = -\dfrac{\sqrt{3}}{2}$ $\csc 300° = -\dfrac{2\sqrt{3}}{3}$

   $\cos 300° = \dfrac{1}{2}$ $\sec 300° = 2$

   $\tan 300° = -\sqrt{3}$ $\cot 300° = -\dfrac{\sqrt{3}}{3}$

6. Answers will vary.

Exploration 17

1.

θ	r	u	v
15°	10 cm	9.7 cm	2.6 cm
30°	10 cm	8.7 cm	5.0 cm
45°	10 cm	7.1 cm	7.1 cm
60°	10 cm	5.0 cm	8.7 cm
75°	10 cm	2.6 cm	9.7 cm

2.

θ	$\sin\theta$	$\cos\theta$	$\tan\theta$
15°	0.97	0.26	0.27
30°	0.87	0.50	0.57
45°	0.71	0.71	1.00
60°	0.50	0.87	1.74
75°	0.26	0.97	3.73

3.

θ	$\sin\theta$	$\cos\theta$	$\tan\theta$
15°	0.9659...	0.2588...	0.2679...
30°	0.8660...	0.5	0.5773...
45°	0.7071...	0.7071...	1
60°	0.5	0.8660...	1.7320...
75°	0.2588...	0.9659...	3.7320...

4. The answers should be close.

5. Answers will vary.

Exploration 18

1. Measurements are correct.

2. $\tan A = \dfrac{3}{4} = 0.75$

3. $A = \tan^{-1}\dfrac{3}{4} = 36.8698...°$

4. Measure of $A \approx 37°$ agrees with the calculated answer.

5. Hypotenuse = 5 cm

6. $\cos A = \dfrac{4}{5}$

7. $\cos A = 0.8$. Answers agree.

8. Draw as directed by the text.

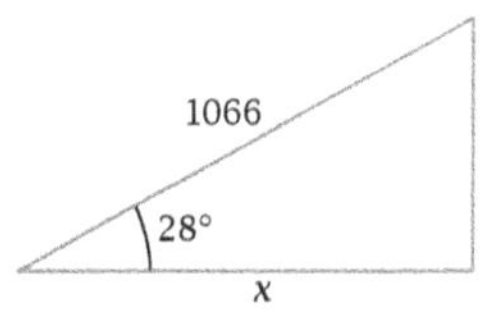

9. $\dfrac{x}{1066} = \cos 28°$

10. $x = 1066 \text{ ft} \cdot \cos 28° = 941.2221... \text{ ft}$

11. $1066 \text{ ft} \cdot \sin 28° = 500.4566... \text{ ft}$
    $\sqrt{(1066 \text{ ft})^2 - (941.2221... \text{ ft})^2} = 500.4566... \text{ ft}$

12. Answers will vary.

Exploration 19

1. Measurements are correct.

2. $10 \text{ cm} \cdot \cos 68° = 3.7460... \text{ cm}$

3. Measurement is correct.

4. Measurements are correct.

5. $\tan^{-1}\dfrac{6}{9} = 33.6900...°$

6. $9 \text{ cm} \cdot \sec 33.6900...° = 10.8166... \text{ cm}$
   or $6 \text{ cm} \cdot \csc 33.6900...° = 10.8166... \text{ cm}$

7. $\sqrt{(6 \text{ cm})^2 + (9 \text{ cm})^2} = \sqrt{117} \text{ cm} = 10.8166... \text{ cm}$

   Answers agree.

8. Answers will vary.

Exploration 20

1. Draw as directed by the text.

2. $x \approx 580$ m, $y \approx 450$ m

3. $\tan 27° = \dfrac{y}{307 + x}$, $\tan 38° = \dfrac{y}{x}$

4. By rewriting the equations as $\cot 27° = \dfrac{307\,m + x}{y} = \dfrac{307\,m}{y} + \dfrac{x}{y}$
   and $\cot 38° = \dfrac{x}{y}$, you get

   $y = \dfrac{307\,m}{\cot 27° - \cot 38°} = 449.7055\ldots$ m ≈ 450 m

   $x = \dfrac{307\,m \cdot \cot 38°}{\cot 27° - \cot 38°} = 575.5968\ldots$ m ≈ 576 m

5. Answers are reasonably close.

6. The actual height is 1454 ft, or 443.2 m.

7. Answers will vary.

Applications of Trigonometric and Circular Functions

Exploration 21

1. Use December's temperatures for month 0.

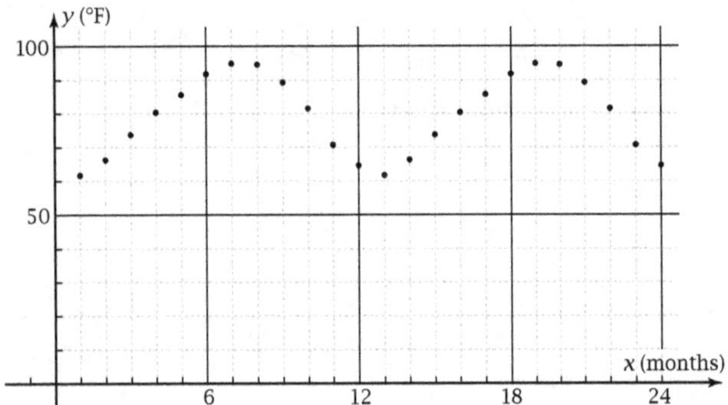

2. θ-dilation of $\dfrac{12°}{360°} = \dfrac{1}{30}$; $y = \cos 30\theta$

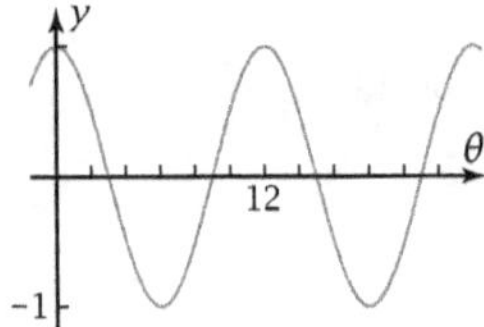

3. In Problem 1, the θ-dilation is $\dfrac{12°}{360°} = \dfrac{1}{30}$. Here the
   t-dilation (if t represents time in months) is
   $\dfrac{12\text{ months}}{360°} = \dfrac{1}{30}$ months/degree, so $y = \cos 30t$.

4. θ-translation of $+7°$; $y = 30 \cos (\theta - 7)$

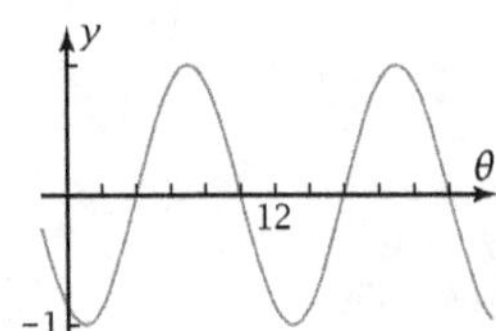

5. $y = 78.3 + \cos 30(\theta - 7°)$

6. $y = 78.3 + 16.6 \cos 30(\theta - 7°)$. Actually, this should be
   $y = 78.3 + 16.6 \cos 30(t - 7)$, where t is time in months.

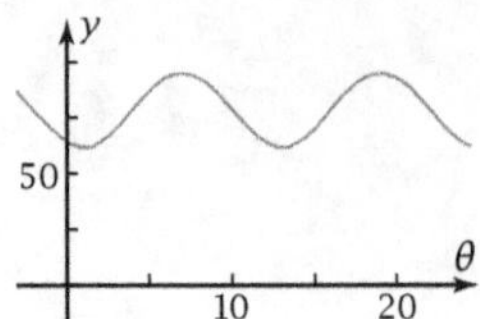

7. The fit is shown for only the first year. The second year is the
   same. The fit is good but not perfect.

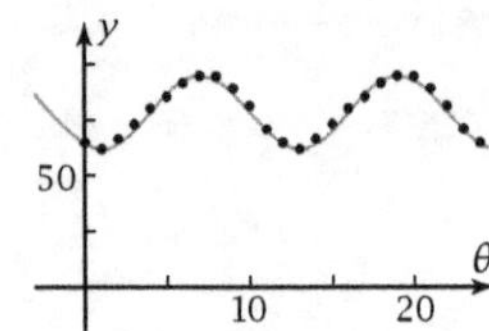

8. Answers will vary.

Exploration 22

1.

X	Y_1		X	Y_1
0	0		180	0
10	.17		270	−1
20	.34		360	0
30	.5		450	1
40	.64		540	0
50	.77		630	−1
60	.87		720	0
70	.94			
80	.98			
90	1			

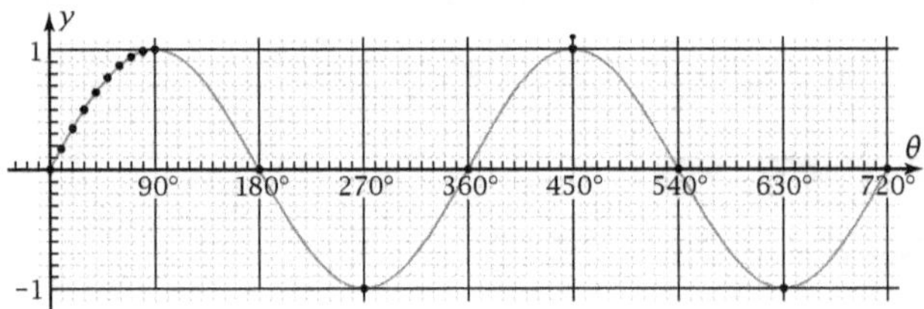

2.

X	Y_1		X	Y_1
0	1		180	-1
10	.98		270	0
20	.94		360	1
30	.87		450	0
40	.77		540	-1
50	.64		630	0
60	.5		720	1
70	.34			
80	.17			
90	0			

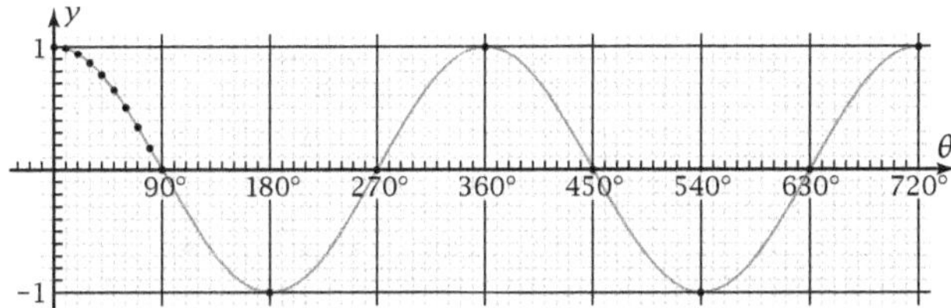

3. $\sin 45° = 0.7071...$; $\cos 65° = .4226...$; the point $(45°, 0.71)$ is on the first graph, and the point $(65°, 0.42)$ is on the second.

4. $\sin^{-1} 0.4 = 23.5781...°$; $\cos^{-1} 0.8 = 36.8698...°$; these correspond to the points $(24°, 0.4)$ on the first graph and $(37°, 0.8)$ on the second.

5. $-1 \le y \le 1$

6. Answers will vary. Many examples are given in the text.

7. Answers will vary.

Exploration 23

1. Horizontal dilation: $\frac{1}{2}$
   Period: 180°
   Amplitude: 3
   Phase displacement: +70°
   Vertical displacement: +4

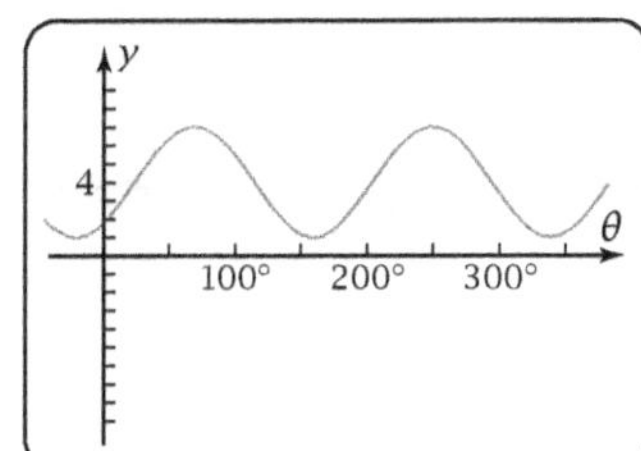

2. Horizontal dilation: $\frac{1}{30}$
   Period: 12°
   Amplitude: 4
   Phase displacement: $-1°$
   Vertical displacement: -2

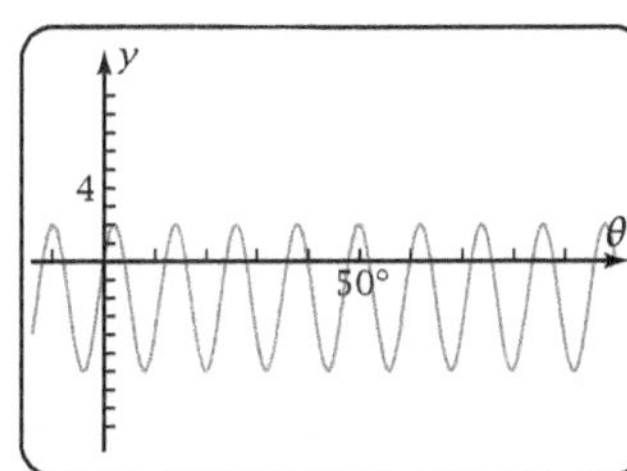

3. Horizontal dilation: $\frac{1}{6}$
   Period: 60°
   Amplitude: 5
   Phase displacement: 10°
   Vertical displacement: -3
   $y = -3 + 5 \cos 6(\theta - 10°)$

4. The graph should agree.

5. $y = -7.3301270189222$

6. Answers will vary.

Exploration 24

1.

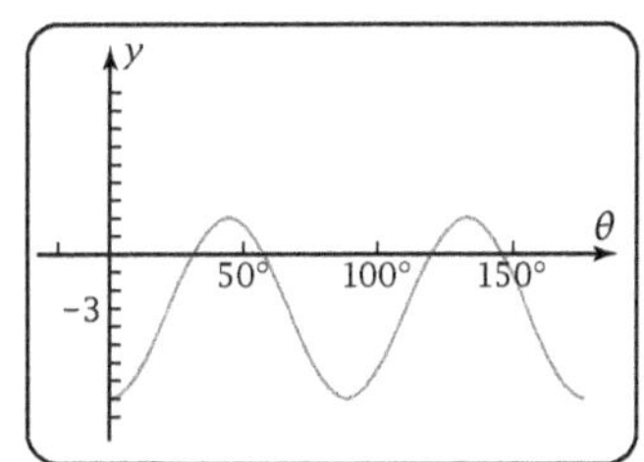

2. $y = 55 + 45 \cos 12(\theta + 3°)$

3. $y = 55 + 45 \sin 12(\theta - 19.5°)$

4. The graphs match.

5. $y = 0.5 + 3.5 \sin 0.9(\theta - 500°)$

6. $y = -60 + 40 \sin 7.5(\theta - 36°)$

7.

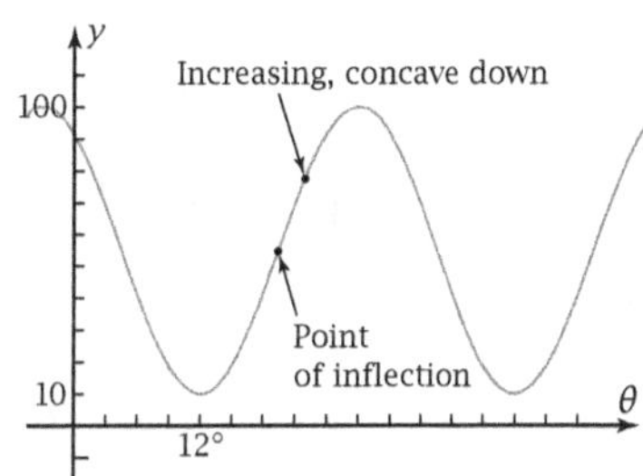

8. Answers will vary.

Exploration 25

1.

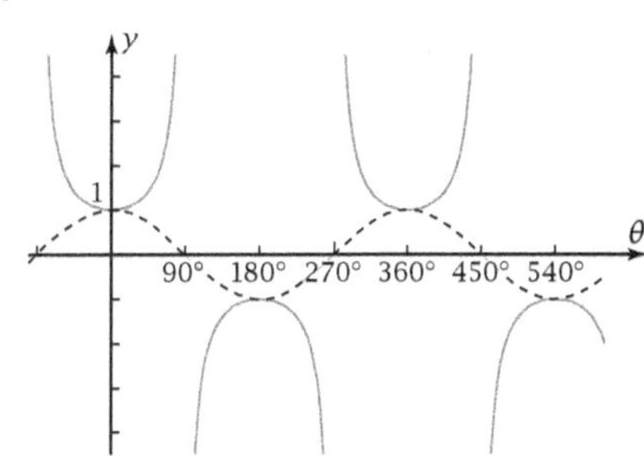

2. $\tan \theta = \dfrac{\sin \theta}{\cos \theta}$

3. Asymptotes are at $\theta = 90° + 180°n$, where $\cos \theta = 0$.

4. Intercepts are at $\theta = 0° + 180°n$, where $\sin \theta = 0$.

5. $\tan 45° = 1$
Graph for 3, 4, and 5

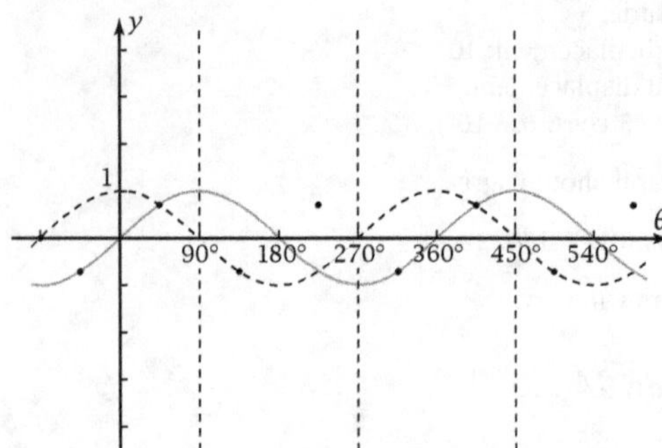

6.

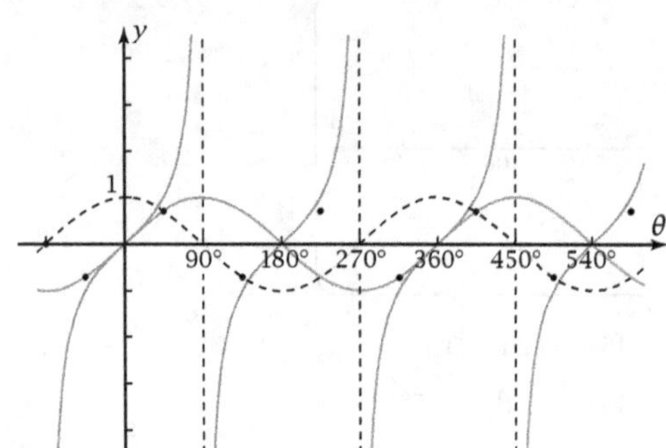

7. The graphs should match.

8.

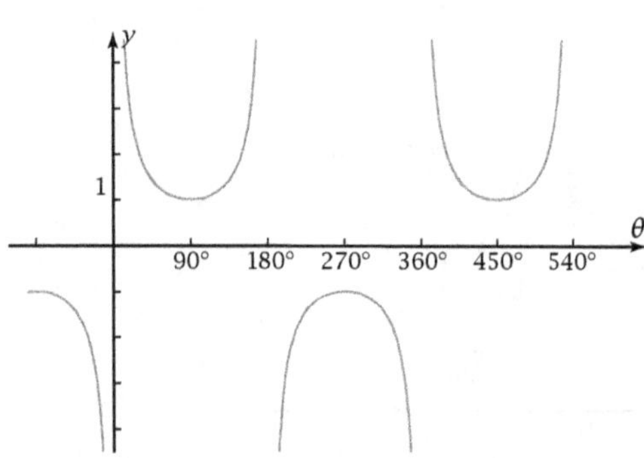

9.

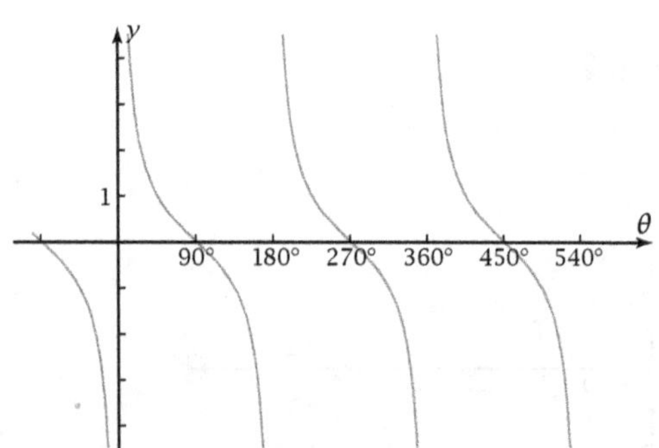

10. Points of inflection are at $\theta = 0° + 180°n$. $\tan \theta$ has no points of inflection because it is constantly decreasing, except where it changes from low values back to high ones discontinuously rather than through points that are on the graph.

11. $\sec \theta$ goes from concave up to concave down (and vice versa) discontinuously rather than through points that are on the graph.

12. Answers will vary.

Exploration 26

1. Horizontal dilation: $\frac{1}{5}$
   Period: 36°
   Horizontal translation: +7°
   Vertical dilation: $\frac{1}{2}$
   Vertical translation: +3

2.

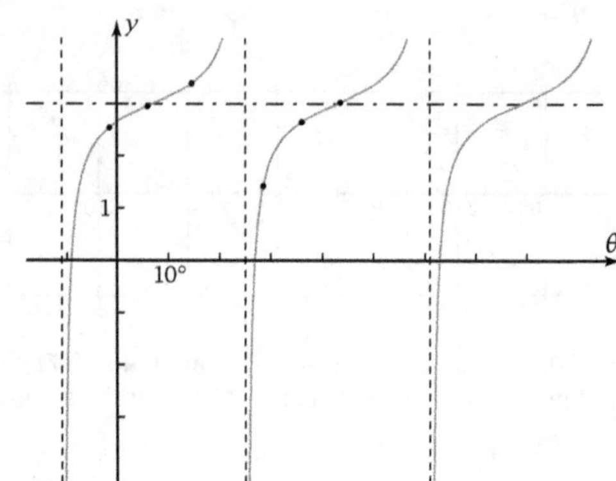

3. Horizontal dilation: $\frac{1}{6}$
   Period: 30°
   Horizontal translation: −9°
   Vertical dilation: 2
   Vertical translation: −1

4. $y = -1 + 2 \cot 6(x + 9°)$

5. Horizontal dilation: $\frac{1}{4}$
   Period: 90°
   Horizontal translation: −10°
   Vertical dilation: 3
   Vertical translation: +1

6.

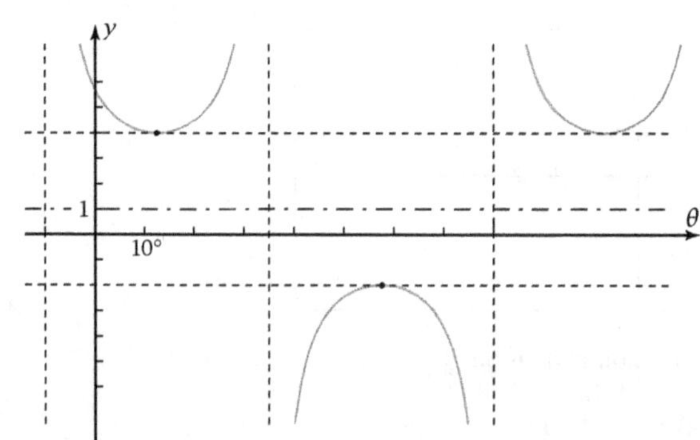

7. Horizontal dilation: $\frac{1}{2}$
   Period: 180°
   Horizontal translation: +25°
   Vertical dilation: 3
   Vertical translation: +4

8. $4 + 3 \sec 2(\theta - 25°)$

9. Answers will vary.

Exploration 27

1. See drawing in Problem 2.

2. Student drawings

3. 1 radian = 57.2957...°

4. $1 \text{ radian} = \dfrac{360°}{2\pi} = \dfrac{180°}{\pi} = 57.2957...°$

5. $3 \text{ radians} = 3 \cdot \dfrac{180°}{\pi} = 171.8873...°$

6. Answers will vary.

Exploration 28

1.–4. Student drawings.

5. 1 radian = 57.2957...°

6. 2π radians

7. $1 \text{ radian} = \dfrac{360°}{2\pi} = \dfrac{180°}{\pi} = 57.2957...°$

8. $3 \text{ radians} = 3 \cdot \dfrac{180°}{\pi} = 171.8873...°$

9. The proportion of arc length to radius is the same for any-size circle.

10. Answers will vary.

Exploration 29

1.

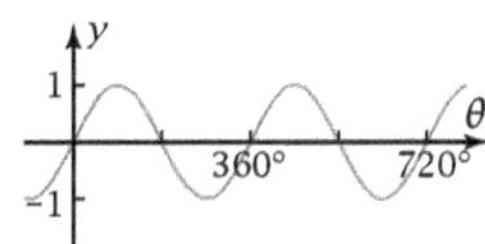

2.

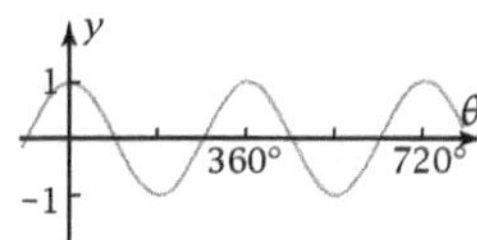

3.

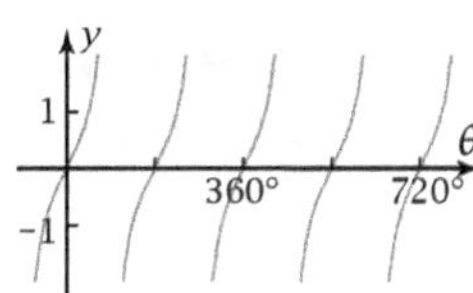

4.

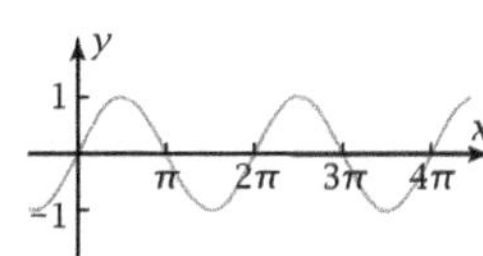

5.

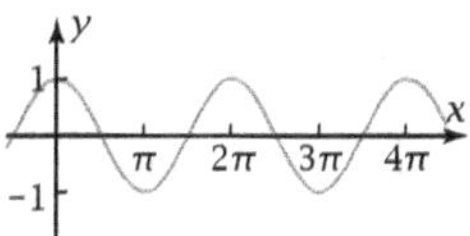

6.

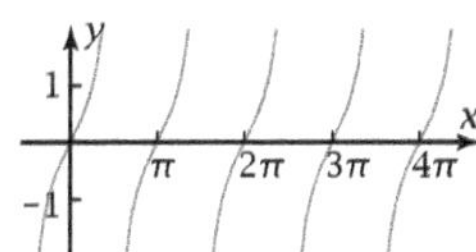

7. Period for sine and cosine = 360° = 2π radians; period for tangent = 180° = π radians.

8. $y = 3 + 2 \cos \dfrac{2\pi}{10}(x - 1)$

9. Answers will vary.

Exploration 30

1. $x = -4.5, -0.5, 8.5, 12.5, 21.5, 25.5$

2. $y = 9 + 7\cos \dfrac{2\pi}{13}(x - 4)$

3.

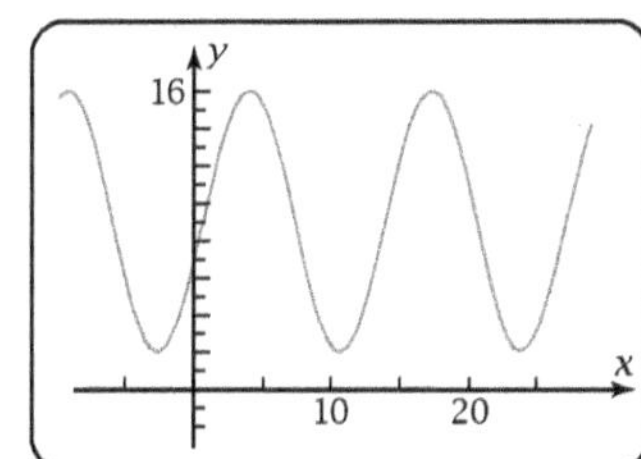

The graphs match.

4. Yes

5. $x = -4.492$

6. $x = -0.508, 8.508$

7. $x = 12.492$

8. $n = 2$: $x = -4.492 + 26 = 21.508$
   $n = 1$: $x = 12.492 + 13 = 25.492$

9. $n = 76$: $x = 1000.4915...$
   Sinusoid is going up because the multiple of the period was added onto 12.4915, where the sinusoid is going up.

10. Answers will vary.

Exploration 31

1. $y = 12.9764\ldots$

2. $9 + 7 \cos \dfrac{2\pi}{13}(x - 4) = 5$

   $7 \cos \dfrac{2\pi}{13}(x - 4) = -4$

   $\cos \dfrac{2\pi}{13}(x - 4) = -\dfrac{4}{7}$

   $\dfrac{2\pi}{13}(x - 4) = \arccos\left(-\dfrac{4}{7}\right)$

   $x - 4 = \dfrac{13\pi}{2}\left(\arccos\left(-\dfrac{4}{7}\right)\right)$

   $x = 4 + \dfrac{13\pi}{2}\arccos\left(-\dfrac{4}{7}\right)$

   $x = 4 + \dfrac{13}{2\pi}\left(\pm\cos^{-1}\left(-\dfrac{4}{7}\right) + 2\pi n\right)$

   $x = 4 + \dfrac{13}{2\pi}\cos^{-1}\left(-\dfrac{4}{7}\right) + 13n$ $\quad$ or

   $4 - \dfrac{13}{2\pi}\cos^{-1}\left(-\dfrac{4}{7}\right) + 13n$

3.

n	x_1	x_2
-1	$-4.491\ldots$	$-13.508\ldots$
0	$8.508\ldots$	$-0.508\ldots$
1	$21.508\ldots$	$12.491\ldots$
2	$34.508\ldots$	$25.491\ldots$

4. $x = -4.491\ldots, -0.508\ldots, 8.508\ldots, 12.491\ldots, 21.508\ldots, 25.491\ldots$
   See table in Problem 3 for n-values.

5. $x = 1308.508\ldots, 1299.491\ldots$

6. $x = 1000.491\ldots; n = 77$

7. Answers will vary.

Exploration 32

1.

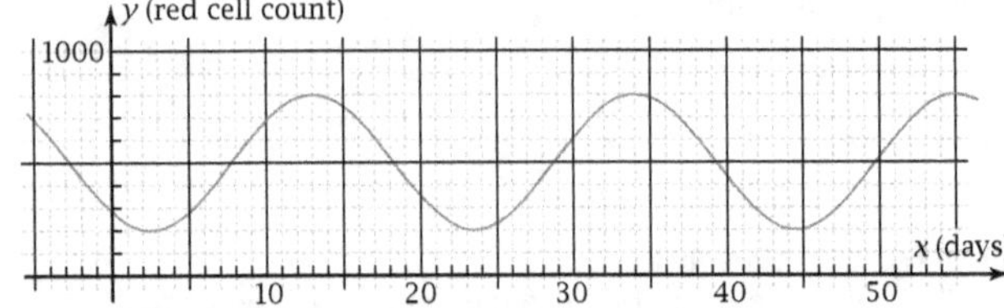

2. $R = 500 + 300 \cos \dfrac{2\pi}{21}(x - 13)$

3. Graph matches sketch from part a.

4. March 19 = day 78 (or 79 in a leap year)
   $R(78) = 747.8716\ldots$
   $R(79) = 687.0469\ldots$
   The patient will feel good on her birthday if this is not a leap year, so-so otherwise.

5. Graph is above 700 on March 19 on a non-leap year, below otherwise.

6. $x = 13 \pm \dfrac{21}{2\pi}\left(\cos^{-1}\dfrac{R - 500}{300} + 2\pi n\right)$
   $31.1889\ldots$ days $\leq x \leq 36.8110\ldots$ days
   Jan. 31 $\leq x \leq$ Feb. 5

7. Answers will vary.

Exploration 33

1. $y = 2000 + 500 \cos \dfrac{\pi}{70}(x + 30)$

2.

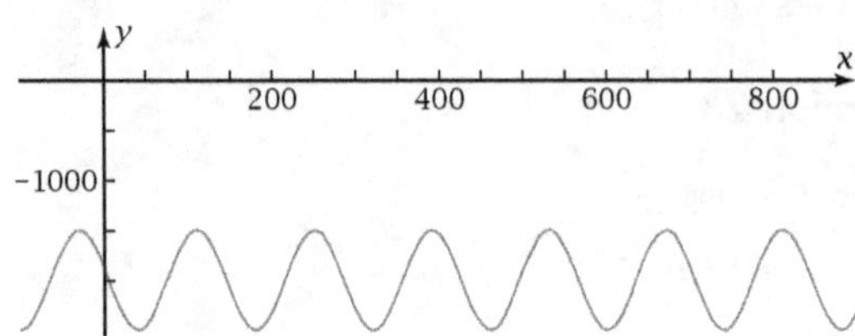

3. $795.6617\ldots \leq x \leq 824.3382\ldots$

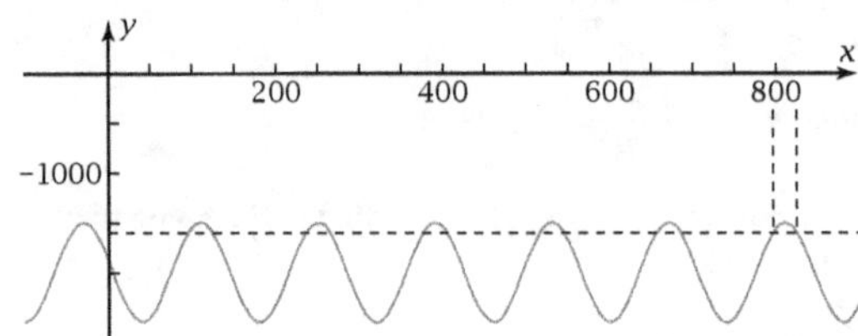

4. $-2000 + 500 \cos \dfrac{\pi}{70}(x + 30) = -1600$

   $500 \cos \dfrac{\pi}{70}(x + 30) = 400$

   $\cos \dfrac{\pi}{70}(x - 30) = 0.8$

   $\dfrac{\pi}{70}(x + 30) = \pm 0.6435\ldots + 2\pi n$

   $x + 30 = \pm 14.3382\ldots + 140n$
   $x = -15.6617\ldots + 140n$ or $-44.3382\ldots + 140n$
   $n = 6$: $x = 840 - 15.6617\ldots$ or $840 - 44.3382\ldots$
   $x = 795.6617\ldots$ or 824.3382

5. It would make a difference because the period would be
   $2 \cdot 68$, or 136, instead of 140.

6. Answers will vary.

Trigonometric Function Properties, Identities, and Parametric Functions

Exploration 34

1. $\tan x = \dfrac{\sin x}{\cos x}$

   $\cot x = \dfrac{\cos x}{\sin x}$

   $\sec x = \dfrac{1}{\cos x}$

   $\csc x = \dfrac{1}{\sin x}$

2. One function is the reciprocal of the other.

3. Quotient property

4. $\cos^2 0.6 + \sin^2 0.6 = 1$

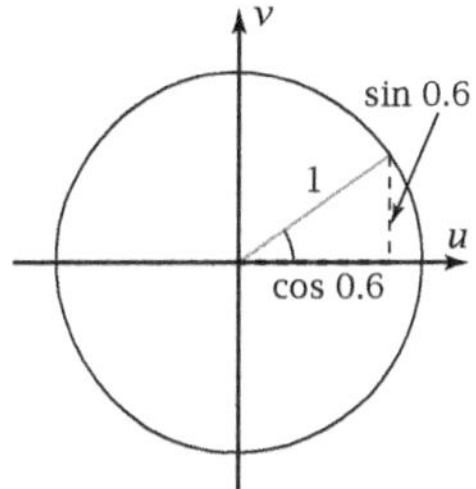

5. Pythagorean property

6. $\cos^2 x + \sin^2 x = 1$

$$\frac{1}{\cos^2 x} \cdot (\cos^2 x + \sin^2 x) = \frac{1}{\cos^2 x}$$

$$\frac{\cos^2 x}{\cos^2 x} + \frac{\sin^2 x}{\cos^2 x} = \frac{1}{\cos^2 x}$$

$$1 + \left(\frac{\sin x}{\cos x}\right)^2 = \left(\frac{1}{\cos x}\right)^2$$

$$1 + \tan^2 x = \sec^2 x$$

7. $\cos^2 x + \sin^2 x = 1$

$$\frac{1}{\sin^2 x} \cdot (\cos^2 x + \sin^2 x) = \frac{1}{\sin^2 x}$$

$$\frac{\cos^2 x}{\sin^2 x} + \frac{\sin^2 x}{\sin^2 x} = \frac{1}{\sin^2 x}$$

$$b\left(\frac{\cos x}{\sin x}\right)^2 + 1 = \left(\frac{1}{\sin x}\right)^2$$

$$\cot^2 x + 1 = \csc^2 x$$

8. $\tan x = \dfrac{\sin x}{\cos x} = \dfrac{1/\cos x}{1/\sin x} = \dfrac{\sec x}{\csc x}$

9. $\csc x \cdot \tan x = \dfrac{1}{\sin x} \cdot \dfrac{\sin x}{\cos x} = \dfrac{1}{\cos x} = \sec x$

10. Answers will vary.

11. $\csc x \cdot \tan x \cdot \cos x = \dfrac{1}{\sin x} \cdot \dfrac{\sin x}{\cos x} \cdot \cos x$

$$= \dfrac{1}{\cos x} \cdot \cos x = 1$$

12. Answers will vary.

Exploration 35

1. $\cot x = \dfrac{1}{\tan x}$

$\sec x = \dfrac{1}{\cos x}$

$\csc x = \dfrac{1}{\sin x}$

2. $\tan x = \dfrac{\sin x}{\cos x} = \dfrac{\sec x}{\csc x}$

$\cot x = \dfrac{\cos x}{\sin x} = \dfrac{\csc x}{\sec x}$

3. $\cos^2 x + \sin^2 x = 1$

$1 + \tan^2 x = \sec^2 x$

$\cot^2 x + 1 = \csc^2 x$

4. $\sec x - \cos x \Rightarrow \sin x \tan x$

$\sec x - \cos x$

$$= \dfrac{1}{\cos x} - \cos x$$

Make fractions to add for 1 term.

$$= \frac{1}{\cos x} - \frac{\cos^2 x}{\cos x}$$

Get common denominator.

$$= \frac{1 - \cos^2 x}{\cos x}$$

Add to get 1 term.

$$= \frac{\sin^2 x}{\cos x}$$

When you see squares of functions, think Pythagorean.

$$= \sin x \cdot \frac{\sin x}{\cos x}$$

If you see something you want in the answer, guard it.

$\sin x \cdot \tan x$

Familiar property

5. Answers will vary.

Exploration 36

1. $\sec x = \dfrac{1}{\cos x},\ \csc x = \dfrac{1}{\sin x},\ \cot x = \dfrac{1}{\tan x}$

$\tan x = \dfrac{\sin x}{\cos x},\ \cot x = \dfrac{\cos x}{\sin x}$

$\cos^2 x + \sin^2 x = 1,\ \tan^2 x + 1 = \sec^2 x,$

$\cot^2 x + 1 = \csc^2 x$

2. $(1 + \cos A)(1 - \cos A) = 1 - \cos^2 A = \sin^2 A$

3. $\cot B + \tan B = \dfrac{\cos B}{\sin B} + \dfrac{\sin B}{\cos B}$

$$= \frac{\cos^2 B}{\sin B \cos B} + \frac{\sin^2 B}{\sin B \cos B}$$

$$= \frac{\cos^2 B + \sin^2 B}{\sin B \cos B}$$

$$= \frac{1}{\sin B \cos B}$$

$$= \csc B \sec B$$

4. $\csc P \cos^2 P + \sin P = \csc P \left(\cos^2 p + \dfrac{1}{\csc P} \sin P\right)$

$$= \csc P (\cos^2 P + \sin^2 P)$$

$$= \csc P$$

5. y_2 appears to be $1 + y_1$. This is consistent with the property $\sec^2 x = \tan^2 x + 1$.

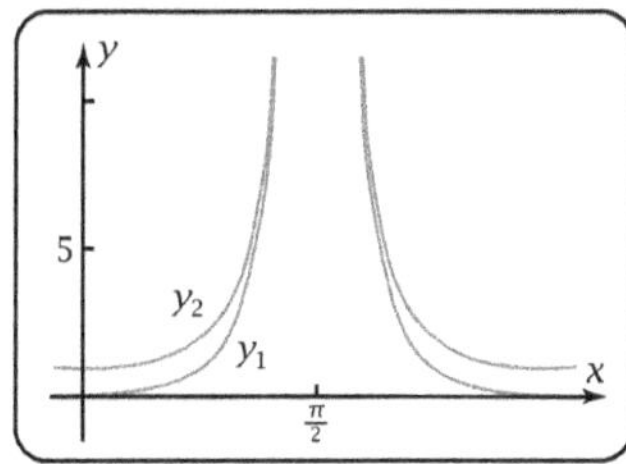

6. Answers will vary.

Exploration 37

1. $\sec x = \dfrac{1}{\cos x},\ \csc x = \dfrac{1}{\sin x},\ \cot x = \dfrac{1}{\tan x}$

2. $\tan x = \dfrac{\sin x}{\cos x},\ \cot x = \dfrac{\cos x}{\sin x}$

3. $\cos^2 x + \sin^2 x = 1$, $\tan^2 x + 1 = \sec^2 x$, $\cot^2 x + 1 = \csc^2 x$

4. $\tan x = \dfrac{\sin x}{\cos x}$

    $= \dfrac{1/\csc x}{1/\sec x}$

    $= \dfrac{\sec x}{\csc x}$

5. $\cot x = \dfrac{1}{\tan x} = \dfrac{1}{\sec x/\csc x} = \dfrac{\csc x}{\sec x}$

6. $\dfrac{\sec A}{\sin A} - \dfrac{\sin A}{\cos A} = \dfrac{1/\cos A}{\sin A} - \dfrac{\sin^2 A}{\cos A \sin A}$

    $= \dfrac{1}{\cos A \sin A} - \dfrac{\sin^2 A}{\cos A \sin A}$

    $= \dfrac{1 - \sin^2 A}{\cos A \sin A}$

    $= \dfrac{\cos^2 A}{\cos A \sin A}$

    $= \dfrac{\cos A}{\sin A}$

    $= \cot A$

7. $\dfrac{1}{1 - \cos B} + \dfrac{1}{1 + \cos B}$

    $= \dfrac{1 + \cos B}{(1 + \cos B)(1 - \cos B)} + \dfrac{1 - \cos B}{(1 - \cos B)(1 + \cos B)}$

    $= \dfrac{1 + \cos B}{1 - \cos^2 B} + \dfrac{1 - \cos B}{1 - \cos^2 B}$

    $= \dfrac{1 + \cos B + 1 - \cos B}{1 - \cos^2 B}$

    $= \dfrac{2}{\sin^2 B}$

    $= 2\csc^2 B$

8. Answers will vary.

Exploration 38

1. arccos $0.4 = 66.4218\ldots°$

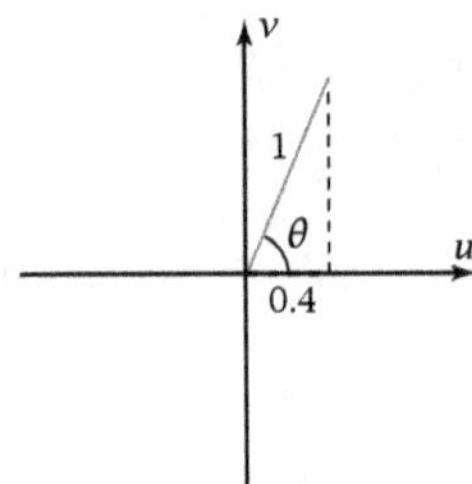

2. arccos $0.4 = -66.4218\ldots°$

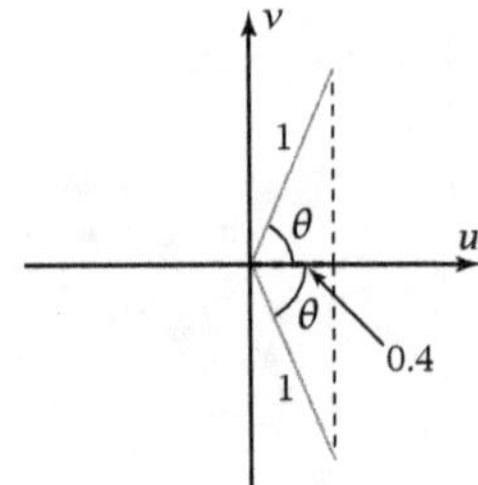

3. arccos $0.4 = \pm 66.4218\ldots° + 360n°$

   Original and new θ have the same reference angle.

4. arcsin $0.3 = 17.4576\ldots°$

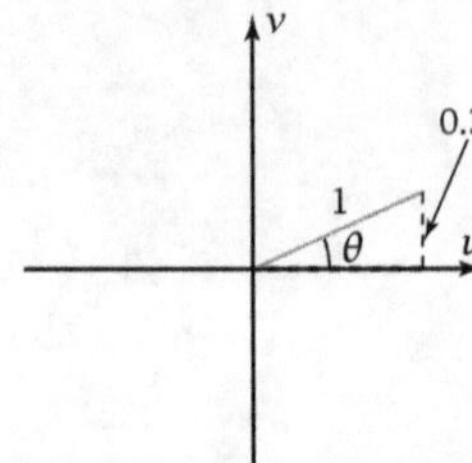

5. arcsin $0.3 = 162.5423\ldots°$

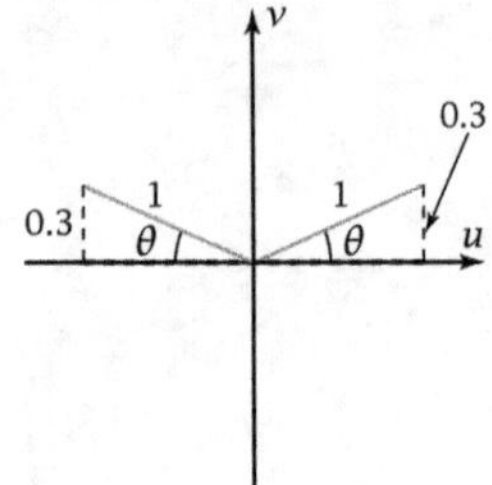

6. arcsin $0.3 = 17.4576\ldots° + n360°$
   or $162.5423\ldots° + n360°$

7. $\sin^{-1}(-0.8) = -53.1301\ldots°$
   arcsin $(-0.8) = 180° - (-53.1301\ldots°)$
               $= 233.1301\ldots°$

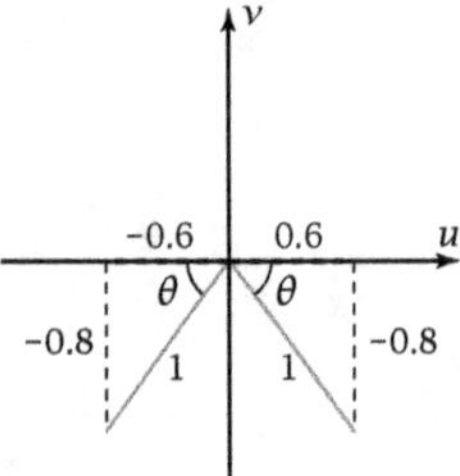

8. $\tan^{-1} 5 = 78.6900\ldots°$
   arctan $5 = 78.69000\ldots° + 180°$
          $= 158.6900\ldots°$

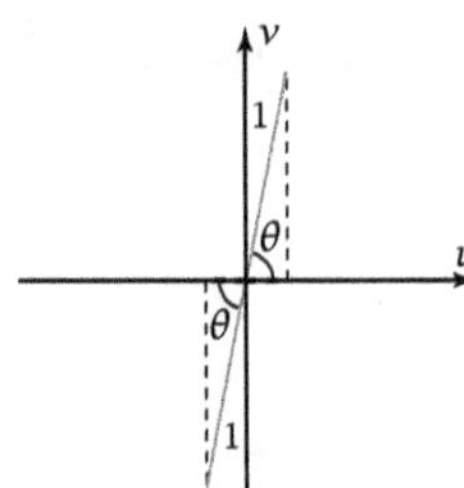

9. arctan $5 = 78.6900\ldots° + n180°$
   or $158.6900\ldots° + n360°$

10. arctan $(-0.6) = -30.9637\ldots°$ or $149.6362\ldots°$

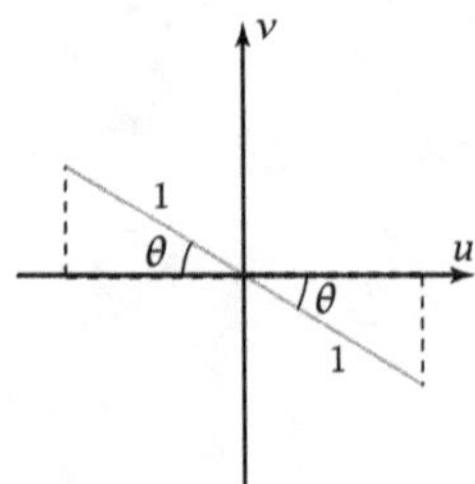

11. Answers will vary.

Exploration 39

1. $2 \cos(\theta - 17°) = 1 \Rightarrow \cos(\theta - 17°) = \dfrac{1}{2}$

   $\Rightarrow \theta - 17° = \pm 60 + n360°$
   $\Rightarrow \theta = 53° + n360°$ or $-77° + n360°$
   $\quad = 53°, 283°, 413°, 643°$

2. $\tan^2 \theta - 2 \tan \theta - 3$
   $= (\tan \theta - 3)(\tan \theta + 1) = 0$
   $\Rightarrow \tan \theta = 3$ or $\tan \theta = -1$
   $\Rightarrow \theta = 71.5650...° + n180°$ or
   $\quad \theta = -45° + n180°$
   $\theta = -288.4349...°, -225°, -108.4349...°,$
   $\quad -45°, 71.5650...°, 135°, 251.5650...°, 315°$

3. $-1 - 5 \sin \theta = 2 \cos^2 \theta = 2 - 2 \sin^2 \theta$
   $\Rightarrow 2 \sin^2 \theta - 5 \sin \theta - 3 = 0$
   $\Rightarrow (2 \sin \theta + 1)(\sin \theta - 3) = 0$
   $\Rightarrow \sin \theta = -\dfrac{1}{2}$ (Note: $\sin \theta < 3$ for all θ)
   $\Rightarrow \theta = -30° + n360°$ or
   $\quad \theta = -150° + n360°$
   $\theta = -150°, -30°, 210°, 330°, 570°, 690°$

4.

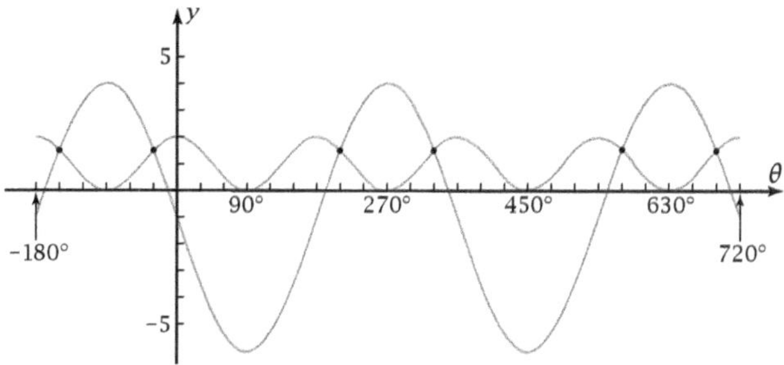

5. Answers will vary.

Exploration 40

1. Example

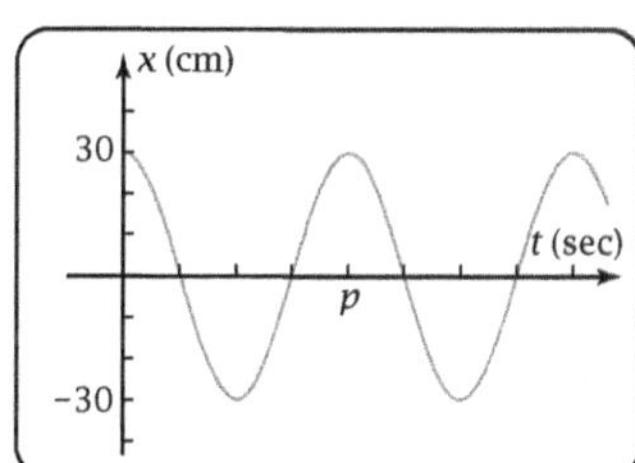

2. Example

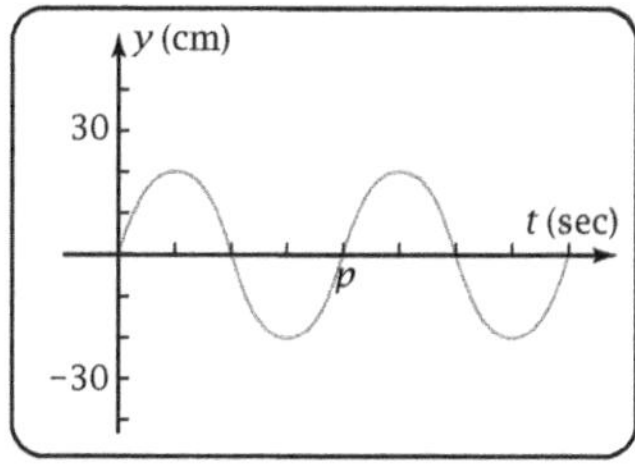

   Period should be the same.

3. Path is an ellipse.

4. If the period is P seconds, then
   $$x = 30 \cos \frac{2\pi}{P} t$$
   $$y = 20 \sin \frac{2\pi}{P} t$$

5.

t	x	y
0	30	0
$\dfrac{P}{8}$	$21.2132...$	$14.1421...$
$\dfrac{P}{4}$	0	20
$\dfrac{3P}{8}$	$-21.2132...$	$14.1421...$
$\dfrac{P}{2}$	-30	0
$\dfrac{5P}{8}$	$-21.2132...$	$-14.1421...$
$\dfrac{3P}{4}$	0	-20
$\dfrac{7P}{8}$	$21.2132...$	$-14.1421...$
P	30	0

6.

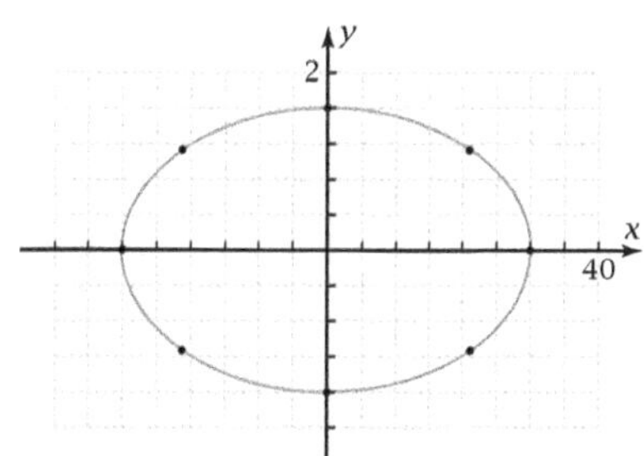

7. The graph is an ellipse.

8. Parameter

9. Parametric function

10.

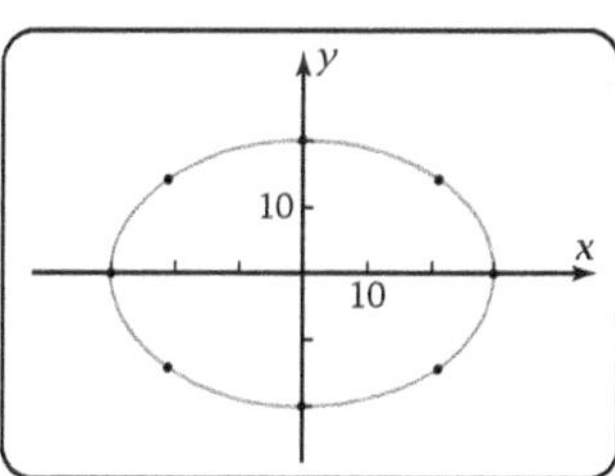

11. $\cos \dfrac{2\pi}{P} t = \dfrac{x}{30}$, $\sin \dfrac{2\pi}{P} t = \dfrac{y}{20}$

   $\cos^2 \dfrac{2\pi}{P} t + \sin^2 \dfrac{2\pi}{P} t = 1$

   $\Rightarrow \left(\dfrac{x}{30}\right)^2 + \left(\dfrac{y}{20}\right)^2 = 1$

Exploration 41

1. $x = 5 + 4 \cos t$
   $y = 3 + 2 \sin t$

2. Equations are correct.

3. $x = 4 + 2 \cos t$, $y = 5 + 0.4 \sin t$

4. Top:
   $x = 4 + 2 \cos t$, $y = 5 + 0.4 \sin t$

   Bottom: (solid)
   $x = 4 + 3 \cos t/(t \geq 0 \text{ and } t \leq \pi)$,
   $y = 1 + 0.6 \sin t/(t \geq 0 \text{ and } t \leq \pi)$

   Bottom: (dashed)
   $x = 4 + 3 \cos t/(t \geq \pi \text{ and } t \leq 2\pi)$,
   $y = 1 + 0.6 \sin t/(t \geq \pi \text{ and } t \leq 2\pi)$

5. (solid)
   $x = 1 + 0.4 \cos t/(t \geq \frac{\pi}{2} \text{ and } t \leq \frac{3\pi}{2})$,
   $y = 3 + 2 \sin t/(t \geq \frac{\pi}{2} \text{ and } t \leq \frac{3\pi}{2})$

   (dashed)
   $x = 1 + 0.4 \cos t/(t \geq -\frac{\pi}{2} \text{ and } t \leq \frac{\pi}{2})$,
   $y = 3 + 2 \sin t/(t \geq -\frac{\pi}{2} \text{ and } t \leq \frac{\pi}{2})$

6. Answers will vary.

Exploration 42

1. $x = \sin 4 = -0.7568\ldots$

2. $\arcsin x = \sin^{-1} x + 2\pi n$ or $(\pi - \sin^{-1} x) + 2\pi n$
   $y = \arcsin 0.4 = 0.4115\ldots + 2\pi n$ or $2.7300\ldots + 2\pi n$
   $y = -5.8716\ldots, -3.5531\ldots, 0.4115\ldots, 2.7300\ldots$
   $y \approx -5.9, -3.6, 0.4, 2.7$

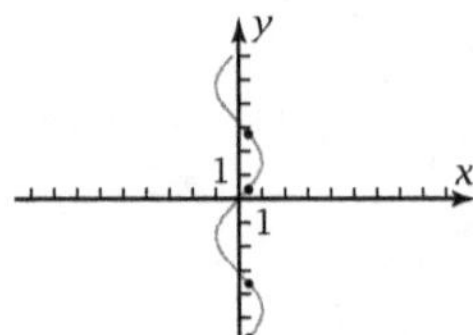

3.

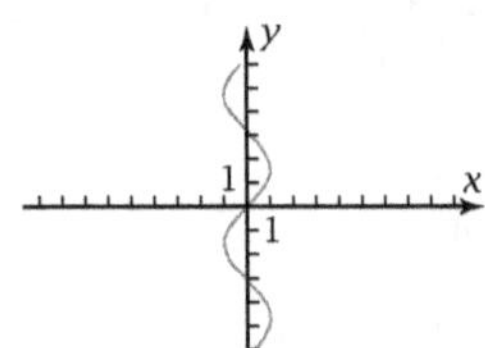

4.

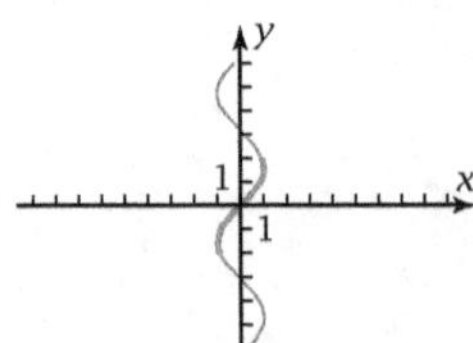

5.

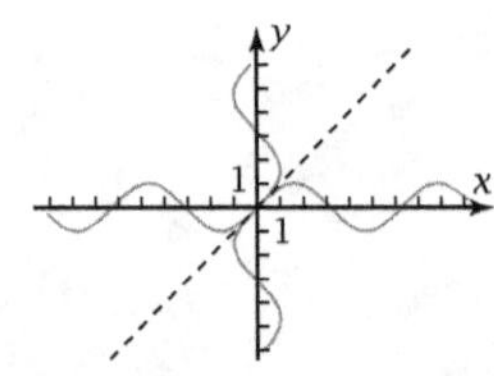

6. They are inverses of each other. Their graphs are reflections across the line $y = x$.

7.

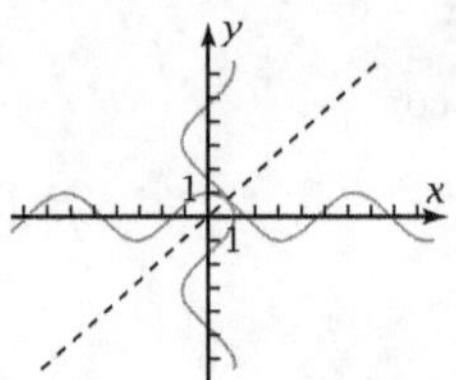

8. Answers will vary.

Exploration 43

1. Range: $-\frac{\pi}{2} \leq y \leq \frac{\pi}{2}$

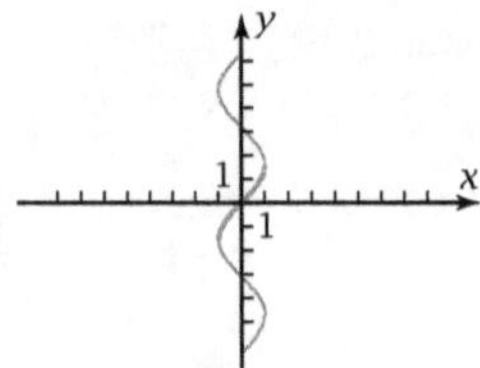

2. Range: $0 \leq y \leq \pi$

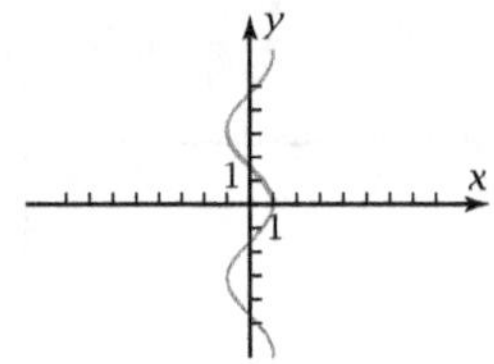

3. If the range of $\cos^{-1}$ were the same as that of $\sin^{-1}$, then $\cos^{-1}$ would not be a function.

4. Range: $-\frac{\pi}{2} \leq y \leq \frac{\pi}{2}$

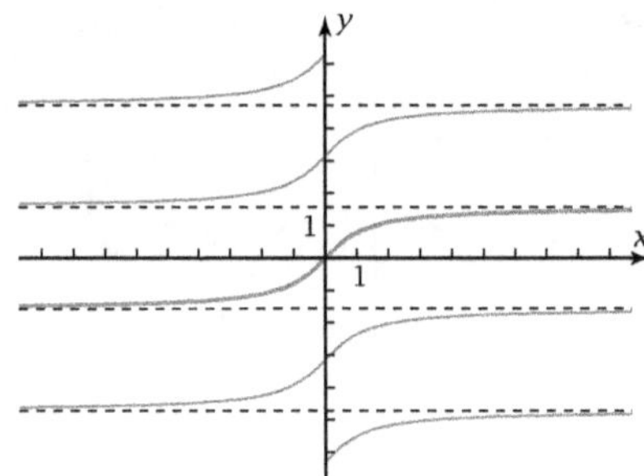

5. $\tan \frac{\pi}{2}$ and $\tan\left(-\frac{\pi}{2}\right)$ are undefined. So the range of $y = \tan^{-1} x$ cannot include these numbers.

6. The graph would not be continuous.

7. $x = \tan 5$; $y = t$

8. Range: $0 < y < \pi$

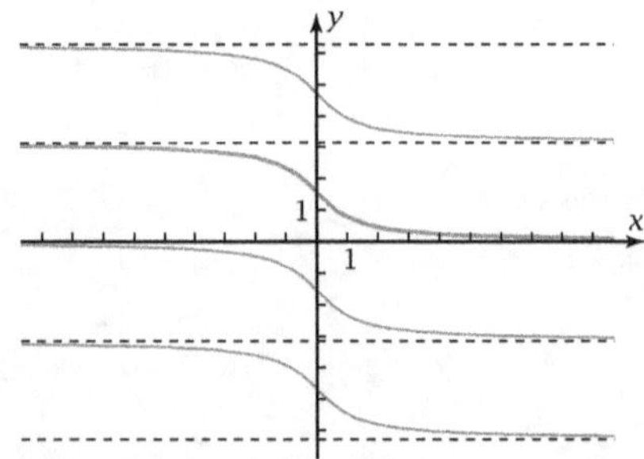

9. Range: $0 \leq y \leq \pi$, and $y \neq \frac{\pi}{2}$

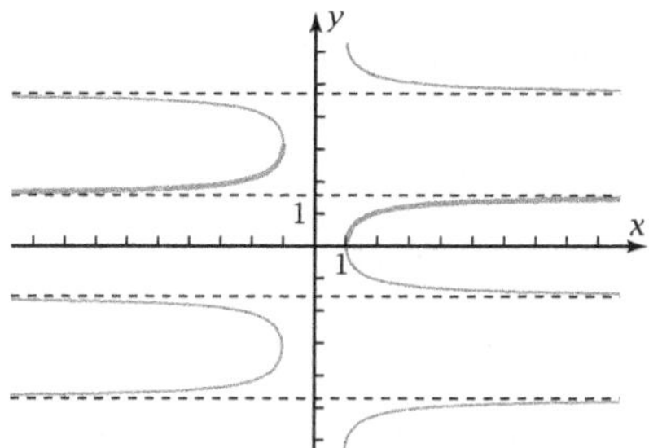

10. It is the commonly accepted branch.

11. Range: $-\frac{\pi}{2} \leq y \leq \frac{\pi}{2}$, and $y \neq 0$

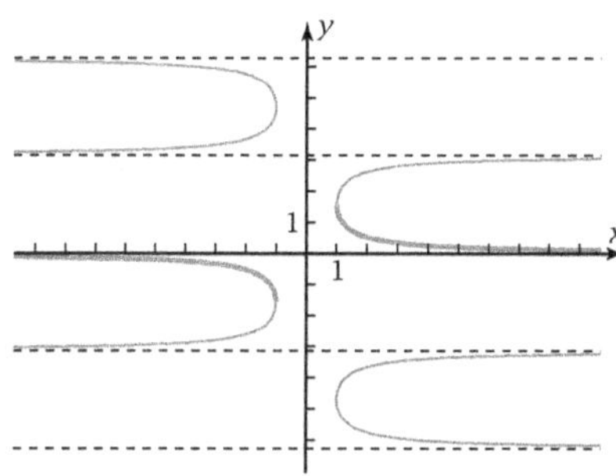

12. Agrees

13. 1. Must be a function

    2. Must use entire domain

    3. Should be continuous

    4. Should be centrally located

    5. If there is a choice, choose the positive branch.

14. Answers will vary.

Exploration 44

1. arcsin $0.8 = 0.9272... + 2\pi n$ or $2.2142... + 2\pi n$
   $\approx -5.4, -4.1, 0.9, 2.2, 7.2, 8.5$

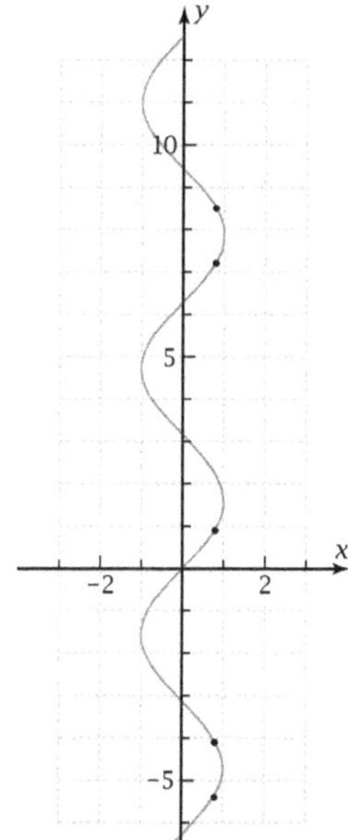

2. arccos $(-0.3) = \pm 1.8754... + 2\pi n$
   $\approx -44, -1.9, 1.9, 4.4, 8.2, 10.7$

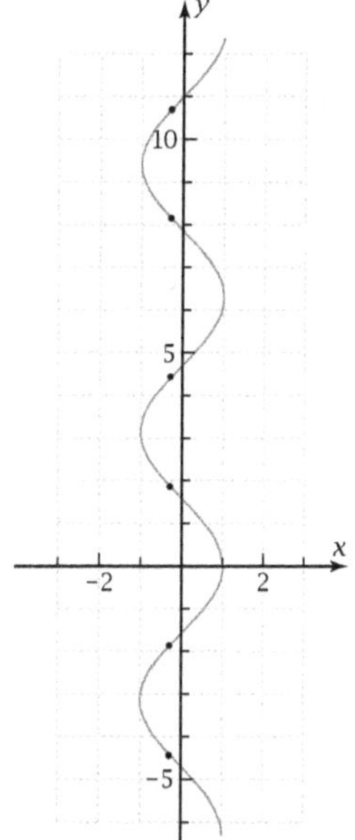

3. arctan $4 = 1.3258... + \pi n$
   $\approx -5.0, -1.8, 1.3, 4.5, 7.6, 10.7$

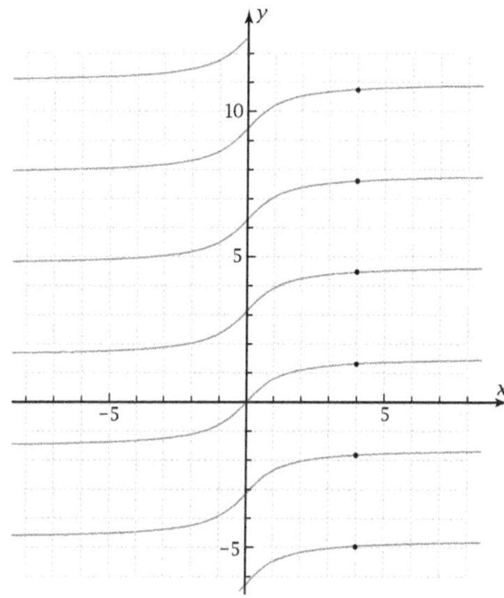

4. The values are on the principal branches.

5. Answers will vary.

Properties of Combined Sinusoids

Exploration 45

1. y_1 is the solid graph, y_2 the dashed graph.

2.

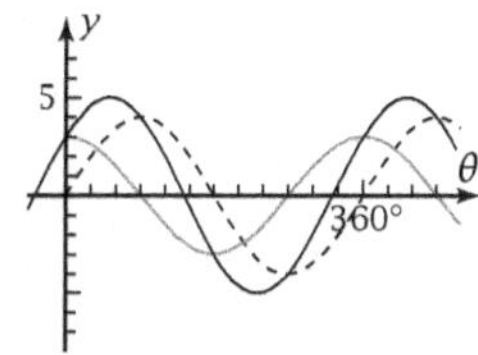

3. $A = 5$
   $D = 53.1301...°$

4. $y_4 = 5 \cos (\theta - 53.1301...°)$
   The graph coincides.

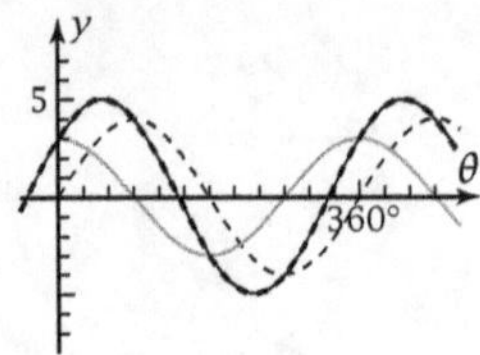

5. $A \cos D = 3$
   $A \sin D = 4$
   $A = \sqrt{A^2} = \sqrt{A^2(\cos^2 D + \sin^2 D)}$
   $\quad = \sqrt{(A \cos D)^2 + (A \sin D)^2}$
   $\quad = \sqrt{3^2 + 4^2}$
   $\quad = 5$

6. $3 = A \cos D = 5 \cos D \Rightarrow \cos D =$
   $4 = A \sin D = 5 \sin D \Rightarrow \sin D =$
   $D = \cos^{-1} \frac{3}{5} = \sin^{-1} \frac{4}{5} = 53.1301...°$

7. $y = -12 \cos \theta + 5 \sin \theta$
   $\quad = \sqrt{12^2 + 5^2} \cos \left(\theta - \cos^{-1} \frac{12}{13} \right)$
   $\quad = 13 \cos (\theta - 22.6198...°)$

8. $y = \sqrt{157} \cos (x - 61.3895...°)$

9. $y = 12 \cos (x + 37.8749...°)$

10. Answers will vary.

Exploration 46

1. $\cos (58° - 20°) = \cos 38° = 0.7880...$
   $\cos 58° - \cos 20° = 0.5299... - 0.9396... = -0.4097...$

2. $\cos (58° - 20°) = \cos 58° \cos 20° + \sin 58° \sin 20°$

3. $\cos (A - B) = \cos A \cos B + \sin A \sin B$

4. The conjecture should work.

5. cosine (first angle − second angle)
   $\quad$ = cosine (first angle) · cosine (second angle)
   $\quad$ + sine (first angle) · sine (second angle)

6. The argument of the cosine is a composite of two numbers.

7. $y_1 = 6 \cos (\theta - 70°)$
   $\quad = 6 (\cos \theta \cos 70° + \sin \theta \sin 70°)$
   $\quad = 6 \cos 70° \cos \theta + 6 \sin 70° \sin \theta$
   $y_2 = (20.5212... \cos \theta) + (5.6381... \sin \theta)$

8. Yes

9. Answers will vary.

Exploration 47

1. Graph with $(\cos A, \sin A)$ on the top label and $(\cos B, \sin B)$ on the lower label.

2. $d^2 = (\cos A - \cos B)^2 + (\sin A - \sin B)^2$
   $\quad = \cos^2 A - 2 \cos A \cos B + \cos^2 B$
   $\quad\quad + \sin^2 A + 2 \sin A \cos A + \sin^2 A$
   $\quad = (\cos^2 A + \sin^2 A) + (\cos^2 B + \sin^2 B)$
   $\quad\quad - 2 \cos A \cos B - 2 \sin A \sin B$
   $\quad = 1 + 1 - 2 \cos A \cos B - 2 \sin A \sin B$
   $d^2 = 2 - 2 \cos A \cos B - 2 \sin A \sin B$

3. $d^2 = (\cos (A - B) - 1)^2 + (\sin (A - B) - 0)^2$
   $\quad = (\cos^2 (A - B) - 2 \cos (A - B) + 1 + \sin^2 (A - B)$
   $\quad = (\cos^2 (A - B) + \sin^2 (A - B) - 2 \cos (A - B) + 1$
   $\quad = 1 - 2 \cos (A - B) + 1$
   $d^2 = 2 - 2 \cos (A - B)$

4. $2 - 2 \cos (A - B) = 2 - 2 \cos A \cos B - 2 \sin A \sin B$
   $\cos (A - B) = \cos A \cos B + \sin A \sin B$

5. Answers will vary.

Exploration 48

1. y_1 is the tall single arch, y_2 the short wiggly graph.

2. See answer to Problem 3.

3.

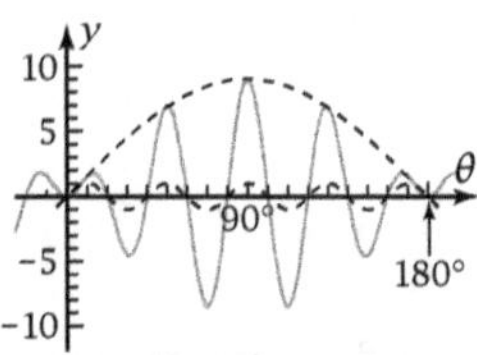

4. The vertical dilation at any θ is the value of $9 \sin \theta$.

5. See answer to Problem 6.

6.

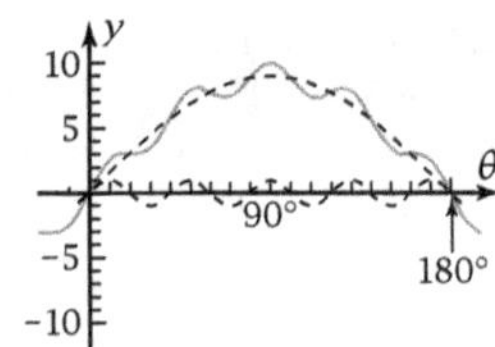

7. The vertical translation at any θ is the value of $9 \sin \theta$.

8. a. . . . a sinusoid with variable sinusoidal axis.

   b. . . . a sinusoid with variable amplitude.

9. Answers will vary.

Exploration 49

1. Added

2. $y = 3 \sin \theta + 2 \cos 13\theta$

3. The graphs match.

4. Multiplied

5. $y = 5 \cos \theta \sin 15\theta$

6. The graphs match.

7. Answers will vary.

Exploration 50

1. $y = 4 \cos 4\theta + 2 \cos 28\theta$

2. $y = 5 \cos 3\theta \cdot \sin 36\theta$

3. $y = 2 \sin \dfrac{\pi}{4}x + 3 \cos 5\pi x$

4. $y = 6 \sin x \cdot \sin 13x$

5. Answers will vary.

Exploration 51

1. $y = 4 \cos 4\theta + 2 \sin 32\theta$

2. $y = 5 \cos 3\theta \cdot \cos 21\theta$

3. $y = 3 \cos x \cdot \sin 15x$

4. $y = 2 \sin \dfrac{\pi}{6}x + 4 \sin \dfrac{11\pi}{6}x$

5. Answers will vary.

Exploration 52

1.

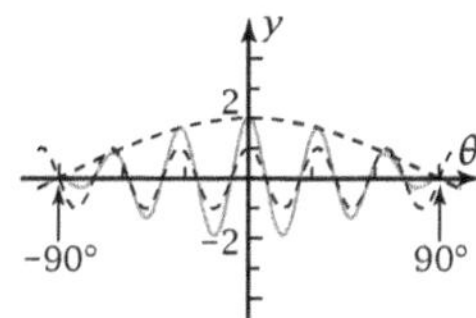

2.

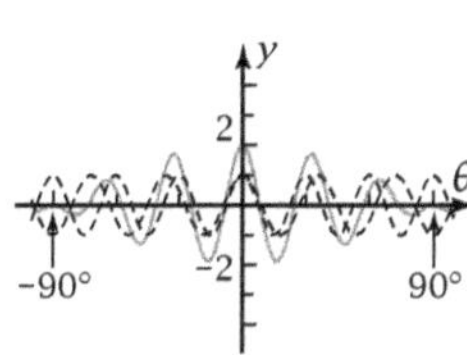

3. $\cos 12\theta = \cos(11\theta + \theta)$
$\qquad = \cos 11\theta \cos \theta - \sin 11\theta \sin \theta$
$\cos 10\theta = \cos(11\theta - \theta)$
$\qquad = \cos 11\theta \cos \theta + \sin 11\theta \sin \theta$
$\cos 12\theta + \cos 10\theta$
$\quad = \cos 11\theta \cos \theta - \sin 11\theta \sin \theta$
$\qquad + \cos 11\theta \cos \theta + \sin 11\theta \sin \theta$
$\quad = 2 \cos 11\theta \cos \theta$

4. $y = 2 \cos 20\theta \cos 2\theta = \cos 22\theta + \cos 18\theta$

5.

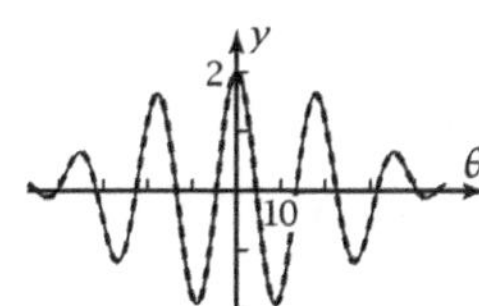

6. Answers will vary.

Exploration 53

1. $\cos(A + B) = \cos A \cos B - \sin A \sin B$
$\sin(A + B) = \sin A \cos B + \cos A \sin B$

2. $\sin 2\theta = \sin(\theta + \theta)$
$\qquad = \sin \theta \cos \theta + \cos \theta \sin \theta$
$\qquad = 2 \sin \theta \cos \theta$

3. $\cos 2\theta = \cos(\theta + \theta)$
$\qquad = \cos \theta \cos \theta - \sin \theta \sin \theta$
$\qquad = \cos^2 \theta - \sin^2 \theta$

4. $\cos 2\theta = \cos^2 \theta - \sin^2 \theta - 1 + \cos^2 \theta + \sin^2 \theta$
$\qquad = 2 \cos^2 \theta - 1$

5. $\cos 2\theta = \cos^2 \theta - \sin^2 \theta + 1 - \cos^2 \theta - \sin^2 \theta$
$\qquad = 1 - 2 \sin^2 \theta$

6.

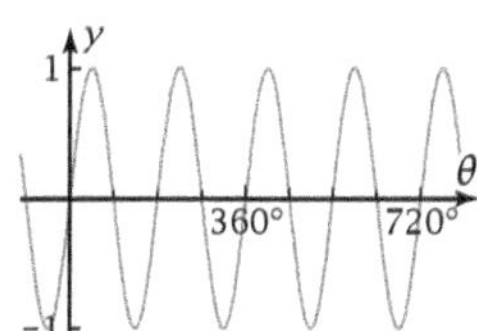

7. The graph is the parent sine graph with a horizontal dilation of $\frac{1}{2}$.

8. $\cos 2\theta + \cos \theta = 1$
$(2 \cos^2 \theta - 1) + \cos \theta = 1$
$2 \cos^2 \theta + \cos \theta - 2 = 0$
$$2\left(\cos \theta + \frac{1 + \sqrt{17}}{4}\right)\left(\cos \theta + \frac{1 - \sqrt{17}}{4}\right) = 0$$
$$\cos \theta = \frac{1 + \sqrt{17}}{4} = -1.2807$$
$$\text{or } \cos \theta = \frac{1 - \sqrt{17}}{4} = 0.7807\ldots$$
$$-1 \le \cos \theta \le 1, \text{ so } \cos \theta = \frac{1 - \sqrt{17}}{4} = 0.7807\ldots$$
$$\theta = \pm\cos^{-1}\frac{1 - \sqrt{17}}{4} + 360°n$$
$\qquad = -38.6682\ldots°, 38.6682\ldots°, 321.3317\ldots°,$
$\qquad 398.6682\ldots°, 681.3317\ldots°, 758.6682\ldots°$

9.

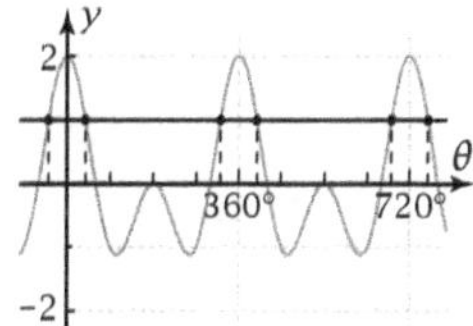

10. The curve intersects the line at the points found in Problem 8.

11. Answers will vary.

Exploration 54

1. $\sin 2x = 2 \sin x \cos x$

2. $\sin^2 x = \dfrac{1}{2}(1 - \cos 2x)$

3. $\sin^2 x = 1 - \cos^2 x$

4. $\cos^2 x = 1 - \sin^2 x$

5. $\cos^2 x = \dfrac{1}{2}(1 + \cos 2x)$

6. $\cos 2x = 2 \cos^2 x - 1$

7. $\cos 2x = 1 - 2 \sin^2 x$

8. $\cos 2x = \cos^2 x - \sin^2 x$

9. $\tan 2x = \dfrac{2 \tan x}{1 - \tan^2 x}$

10. $\sin \dfrac{1}{2}x = \pm\sqrt{\dfrac{1}{2}(1 - \cos x)}$

11. $\cos \dfrac{1}{2}x = \pm\sqrt{\dfrac{1}{2}(1 + \cos x)}$

12. $\tan \dfrac{1}{2}x = \pm\sqrt{\dfrac{1 - \cos x}{1 + \cos x}}$

13. $\tan \dfrac{1}{2}x = \dfrac{\sin x}{1 + \cos x} = \dfrac{1 - \cos x}{\sin x}$

14. $\sin 3x = \sin(2x + x)$
$\quad = \sin 2x \cos x + \cos 2x \sin x$
$\quad = 2 \sin x \cos x \cos x + (1 - 2 \sin^2 x) \sin x$
$\quad = 2 \sin x(1 - \sin^2 x) + \sin x - 2 \sin^3 x$
$\quad = 2 \sin x - 2 \sin^3 x + \sin x - 2 \sin^3 x$
$\quad = 3 \sin x - 4 \sin^3 x$

15. $\cos 4x = \cos 2(2x)$
$\quad = 2 \cos^2 2x - 1$

16. $\cos 6x = \cos 2(3x)$
$\quad = 2 \cos^2 3x - 1$

17. $\sin x \cos x = \dfrac{1}{2}(2 \sin x \cos x) = \dfrac{1}{2} \sin 2x$

18. $\sin x \cos y = \dfrac{1}{2}(2 \sin x \cos y)$
$\quad = \dfrac{1}{2}\big(\sin(x + y) + \sin(x - y)\big)$

19. $\cos x \cos y = \dfrac{1}{2}(2 \cos x \cos y)$
$\quad = \dfrac{1}{2}\big(\cos(x + y) + \cos(x - y)\big)$

20. $\cos x + \cos y = 2 \cos \dfrac{1}{2}(x + y) \cos \dfrac{1}{2}(x - y)$

21. $\sin x + \sin y = 2 \sin \dfrac{1}{2}(x + y) \cos \dfrac{1}{2}(x - y)$

22. $\sin x + \cos x = \sqrt{2} \cos\left(x - \dfrac{\pi}{4}\right)$

23. $\sin 3x \sin 7x = \dfrac{1}{2}(2 \sin 3x \sin 7x)$
$\quad = \dfrac{1}{2}\big(-\cos(3x + 7x) + \cos(3x - 7x)\big)$
$\quad = -\dfrac{1}{2}\cos 10x + \dfrac{1}{2}\cos(-4x)$
$\quad = \dfrac{1}{2}\cos 4x - \dfrac{1}{2}\cos 10x$

24. $\sin 3x + \sin 7x = 2 \sin \dfrac{1}{2}(3x + 7x) \cos \dfrac{1}{2}(3x - 7x)$
$\quad = 2 \sin 5x \cos(-2x)$
$\quad = 2 \sin 5x \cos 2x$

25. $\sqrt{3} \cos x - \sin x = 2 \cos\left(x + \dfrac{\pi}{6}\right)$

26. $4 \cos x - 4 \sin x = 4 \cos\left(x + \dfrac{\pi}{4}\right)$

27. Answers will vary.

Triangle Trigonometry

Exploration 55

1. Measurements are correct.

2. Measurements are correct.

3. Answers will vary slightly but should be approximately

$\angle A$	a (cm)
30°	2.1
60°	3.6
90°	5.0
120°	6.1
150°	6.8

4. 180°: $a = 4 + 3 = 7$ cm
   0°: $a = 4 - 3 = 1$ cm

5.

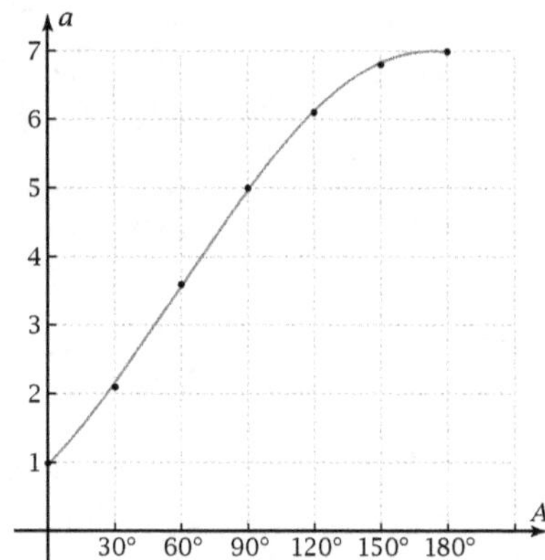

6. Answers may vary. The actual answer is $24 \cos \theta$.

7.

$\angle A$	a (cm)
30°	2.0531...
60°	3.6055...
90°	5.0000...
120°	6.0828...
150°	6.7664...

8. Answers will vary.

Exploration 56

1. $(c, 0)$; $(b \cos A, b \sin A)$

2. $a^2 = (b \cos A - c)^2 + (b \sin A - 0)^2$
   $\quad = b^2 \cos^2 A - 2bc \cos A + c^2 + b^2 \sin^2 A$

3. $a^2 = b^2 \cos^2 A - 2bc \cos A + c^2 + b^2 \sin^2 A$
   $\quad = b^2 + c^2 - 2bc \cos A$

4. $\sqrt{4.8^2 + 2.3^2 - 2 \cdot 4.8 \cdot 2.7 \cos 115°} = 6.4252...$ cm

5. Measurements are correct.

6. The unknown side is opposite the given angle. The two given sides include the given angle.

7. Answers will vary.

Exploration 57

1. $b^2 = a^2 + c^2 - 2ac \cos B$

2. $\cos B = \dfrac{a^2 + c^2 - b^2}{2ac}$

3. $a = 5.3$ cm; $b = 7.5$ cm; $c = 4.3$ cm

4. $B = \cos^{-1} \dfrac{5.3^2 + 4.3^2 - 7.5^2}{2(5.3)(4.3)} \approx 102°$

5. Answers should be close to $102°$.

6. Answers may vary.

7. $A = \cos^{-1} \dfrac{7.5^2 + 4.3^2 - 5.3^2}{2(7.5)(4.3)} \approx 44°$

8. $C = \cos^{-1} \dfrac{5.3^2 + 7.5^2 - 4.3^2}{2(5.3)(7.5)} \approx 34°$

9. $102° + 44° + 34° = 180°$

10. Answers should be close to $A = 44°$ and $C = 34°$.

11. Answers may vary.

12. Answers may vary. By Hero's formula,
$$s = \frac{5.3 + 7.5 + 4.3}{2} = 8.55.$$
$$A_{\triangle ABC} = \sqrt{s(s - 5.3)(s - 7.5)(s - 4.3)}$$
$$\approx 11.1 \text{ cm}^2$$

13. Answers will vary.

Exploration 58

1.

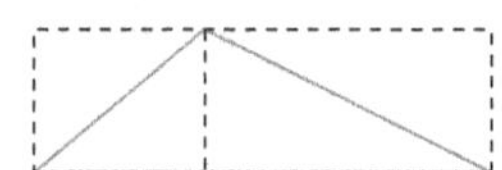

The two small upper triangles added by the rectangle are equal in area to the two small triangles that make up the original triangle.

2. The base $XZ = 8.1$ cm. The altitude h from Y to XZ is 4.4 cm.
$$A_{\triangle XYZ} = \frac{1}{2} bh = \frac{1}{2}(8.1)(4.4) \approx 17.8 \text{ cm}^2$$

3. $h = z \sin X$

4. $A_{\triangle XYZ} = \frac{1}{2} bh = \frac{1}{2} y(z \sin X) = \frac{1}{2} yz \sin X$

5. $X = 38°$; $z = 7.0$ cm;
$$A_{\triangle XYZ} = \frac{1}{2}(8.1)(7.0) \sin 38° \approx 17.5 \text{ cm}^2$$
This agrees to within round-off and measurement error.

6. $Z = 59°$; $x = 5.0$ cm;
$$\frac{1}{2} xy \sin Z = \frac{1}{2}(5.0)(8.1) \sin 59° \approx 17.4 \text{ cm}^2$$

7. $Y = 83°$;
$$\frac{1}{2} xz \sin Y = \frac{1}{2}(5.0)(7.0) \sin 83° \approx 17.4 \text{ cm}^2$$

8. $\frac{1}{2} \cdot 500 \cdot 700 \sin 70° \approx 164{,}446 \text{ ft}^2$

9. The area would actually be smaller by approximately $51{,}958 \text{ ft}^2$. $\frac{1}{2} \cdot 500 \cdot 700 \sin 140° \approx 112{,}448 \text{ ft}^2$. Although two vertices of the triangle are now much farther apart, the base is still the same and the altitude is considerably shorter.

10. Answers will vary.

Exploration 59

1. $8^2 = 7^2 + 11^2 - 2 \cdot 7 \cdot 11 \cos A$
$$\cos A = \frac{7^2 + 11^2 - 8^2}{2 \cdot 7 \cdot 11} = \frac{106}{154} = 0.6883\ldots$$

2. $A = \cos^{-1} \dfrac{7^2 + 11^2 - 8^2}{2 \cdot 7 \cdot 11} = 46.5033\ldots°$
$$\text{Area} = \frac{1}{2} \cdot 7 \cdot 11 \sin A = 27.9284\ldots$$

3. $s = \dfrac{1}{2}(7 + 11 + 8) = 13$
$$\text{Area} = \sqrt{13(13 - 7)(13 - 11)(13 - 8)} = 27.9284\ldots$$

4. $\text{Area} = \dfrac{1}{2} bc \sin A = \sqrt{\left(\dfrac{1}{2} bc \sin A\right)^2}$
$$= \sqrt{\frac{1}{4} b^2 c^2 \sin^2 A} = \sqrt{\frac{1}{4} b^2 c^2 (1 - \cos^2 A)}$$

5. $(1 - \cos^2 A) = (1 + \cos A)(1 - \cos A)$
$$\text{Area} = \sqrt{\frac{1}{4} b^2 c^2 (1 + \cos A)(1 - \cos A)}$$
$$= \sqrt{\frac{1}{2} bc (1 + \cos A) \cdot \frac{1}{2} bc (1 - \cos A)}$$

6. $\cos A = \dfrac{b^2 + c^2 - a^2}{2bc}$

7. $\dfrac{1}{2} bc(1 + \cos A) = \dfrac{1}{2} bc \left(1 + \dfrac{b^2 + c^2 - a^2}{2bc}\right)$
$$= \frac{bc}{2} + \frac{b^2 + c^2 - a^2}{4} = \frac{2bc}{4} + \frac{b^2 + c^2 - a^2}{4}$$
$$= \frac{b^2 + 2bc + c^2 - a^2}{4} = \frac{(b + c)^2 - a^2}{4}$$
$$= \frac{((b + c) + a)((b + c) - a)}{4}$$
$$= \frac{b + c + a}{2} \cdot \frac{b + c - a}{2}$$

8. $\dfrac{b + c - a}{2} = \dfrac{b + c + a - 2a}{2} = \dfrac{a + b + c}{2} - a = (s - a)$
$$\frac{a - b + c}{2} = \frac{a + b + c - 2b}{2} = \frac{a + b + c}{2} - b = (s - b)$$
$$\frac{a + b - c}{2} = \frac{a + b + c - 2c}{2} = \frac{a + b + c}{2} - c = (s - c)$$
So $\sqrt{\dfrac{b + c + a}{2} \cdot \dfrac{b + c - a}{2} \cdot \dfrac{a - b + c}{2} \cdot \dfrac{a + b - c}{2}}$
$$= \sqrt{s(s - a)(s - b)(s - c)}$$

9. $s = \dfrac{1}{2}(50 + 60 + 80) = 95$ cm
$$\text{Area} = \sqrt{95(95 - 50)(95 - 60)(95 - 80)}$$
$$= 1498.1238\ldots \text{ cm}^2$$

10. $s = \dfrac{1}{2}(10 + 12 + 26) = 24$ cm
$$\text{Area} = \sqrt{24(24 - 10)(24 - 12)(24 - 26)} = \sqrt{-8064}$$
Hero's formula results in a negative number under the square root, so there is no possible area.

11. Answers will vary.

Exploration 60

1. Measurements are correct.

2. $\dfrac{6.0}{\sin 57°} = 7.1541...; \dfrac{7.0}{\sin 78°} = 7.1563...$

3. Yes, to within measurement error

4. $45°$

5. $x = \dfrac{6.0 \sin 45°}{\sin 57°} \approx 5.1$ cm

6. Yes

7. $\dfrac{1}{2} ab \sin C; \dfrac{1}{2} bc \sin A; \dfrac{1}{2} ac \sin B$

8. $\dfrac{1}{2} bc \sin A = \dfrac{1}{2} ac \sin B = \dfrac{1}{2} ab \sin C$

9. $\dfrac{1}{abc/2}\left(\dfrac{1}{2} bc \sin A = \dfrac{1}{2} ac \sin B = \dfrac{1}{2} ab \sin C\right)$

   $\dfrac{1}{abc} bc \sin A = \dfrac{1}{abc} ac \sin B = \dfrac{1}{abc} ab \sin C$

   $\dfrac{\sin A}{a} = \dfrac{\sin B}{b} = \dfrac{\sin C}{c}$

10. The statements are equivalent because if the parts of an equation are the same and nonzero, then the reciprocals of the parts of the equation are equal and nonzero.

11. Answers will vary.

Exploration 61

1. $A = \cos^{-1} \dfrac{4^2 + 10^2 - 7^2}{2 \cdot 4 \cdot 10} = 33.12294020...°$

2. $C = \sin^{-1} \dfrac{10 \sin A}{7} = 51.31781254...°$

   But this is not correct—see Problem 4.

3. $C = \cos^{-1} \dfrac{4^2 + 7^2 - 10^2}{2 \cdot 4 \cdot 7} = 128.68218745...°$

4. No. We should have used
   $180° - 51.31781254...° = 128.68218745...°$, the complement of the answer in Problem 3. ($51.31781254...°$ is what angle C would be if B were to the left of AC, with AB still horizontal.)

5. $X = \sin^{-1} \dfrac{11 \sin 40°}{8} \approx 62.1°$ or $117.9°$

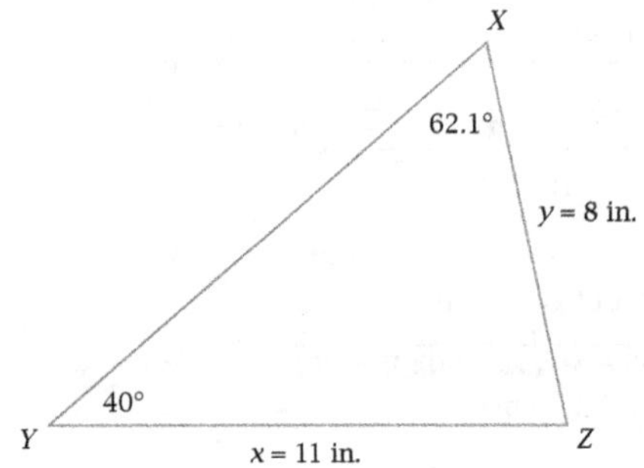

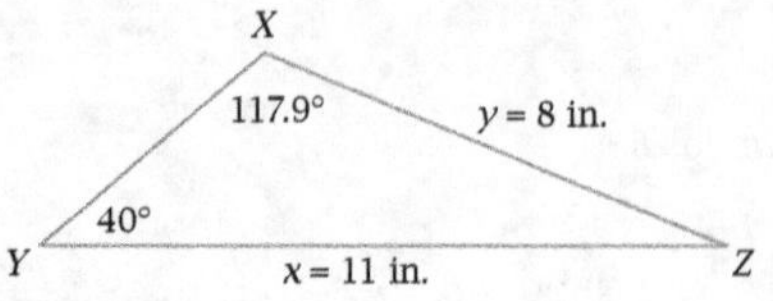

6. Answers will vary.

Exploration 62

1. Measurements are correct.

2. Answers should be close to 10.8 cm and 3.6 cm.

3. "Having two or more possible meanings"; there are two possible answers.

4. Now there is only one answer, 15.5 cm. (Student answers should be close to this.)

5. Now there is no answer. Side x is too short.

6. $\dfrac{x}{8} = \sin 26°$, so $x = 8 \sin 26° \approx 3.5$ cm.

7. $5^2 = y^2 + 8^2 - 2 \cdot y \cdot 8 \cos 26°$
   $y^2 - 16y \cos 26° + 64 - 25 = 0$
   $y^2 + (-16 \cos 26°)y + 39 = 0$
   $y = \dfrac{16 \cos 26° \pm \sqrt{(-16 \cos 26°)^2 - 4 \cdot 1 \cdot 39}}{2 \cdot 1}$
   ≈ 10.8 cm or 3.6 cm

8. $9^2 = y^2 + 8^2 - 2 \cdot y \cdot 8 \cos 26°$
   $y^2 - 16y \cos 26° + 64 - 81 = 0$
   $y^2 + (-16 \cos 26°)y - 17 = 0$
   $y = \dfrac{16 \cos 26° \pm \sqrt{(-16 \cos 26°)^2 - 4 \cdot 1 \cdot (-17)}}{2 \cdot 1}$
   ≈ 15.5 cm or -1.1 cm (The negative answer represents the triangle that would result if Z were to the left of X.)

9. $3^2 = y^2 + 8^2 - 2 \cdot y \cdot 8 \cos 26°$
   $y^2 - 16y \cos 26° + 64 - 9 = 0$
   $y^2 + (-16 \cos 26°)y + 55 = 0$
   The discriminant $(-16 \cos 26°)^2 - 4 \cdot 1 \cdot 55 \approx -13.2$, so there is no solution.

10. Answers will vary.

Exploration 63

1.

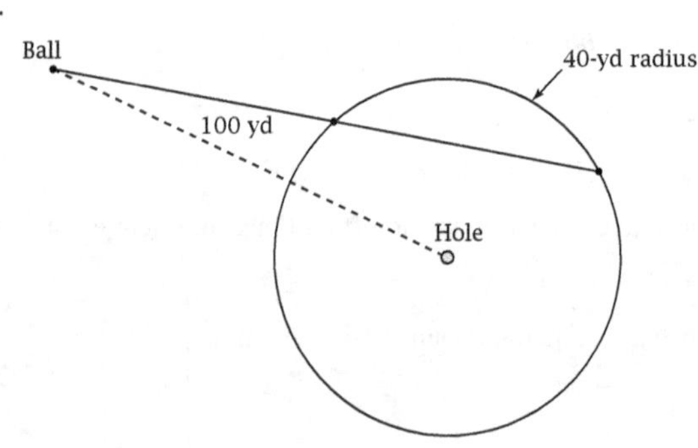

2. Let x be the distance, in feet, the ball traveled.
   $40^2 = 100^2 + x^2 - 2 \cdot 100 \cdot x \cdot \cos 15°$
   $x^2 - (200 \cos 15°)x + 8400 = 0$
   $x = 127.0905...$ or $66.0945...$
   Distances are about 127.09 ft and 66.09 ft.

3.

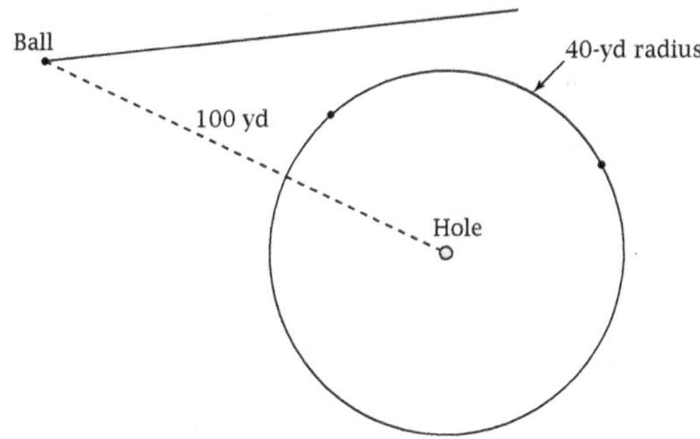

$$40^2 = 100^2 + x^2 - 2 \cdot 100 \cdot x \cos 30°$$
Discriminant $b^2 - 4ac = -3600$
$\therefore$ no possible values of x.

4.

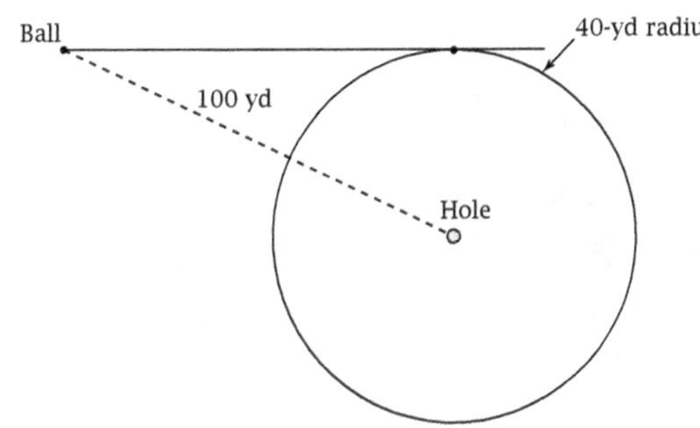

$$\frac{40}{100} = \sin \theta$$
$$\sin \theta = \frac{2}{5}$$
$$\theta = 23.578...°$$

5. Answers will vary.

Exploration 64

1.

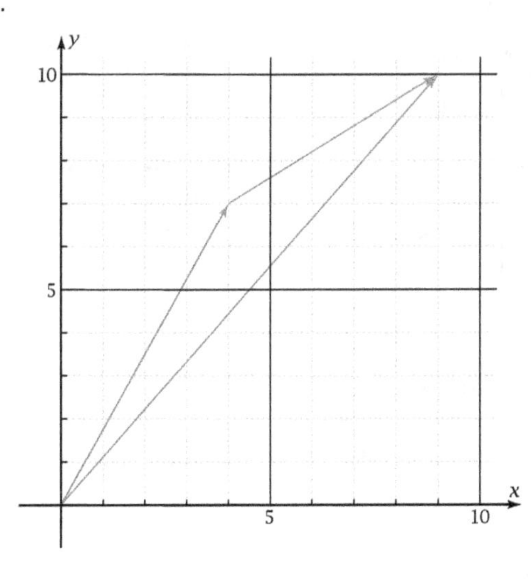

or

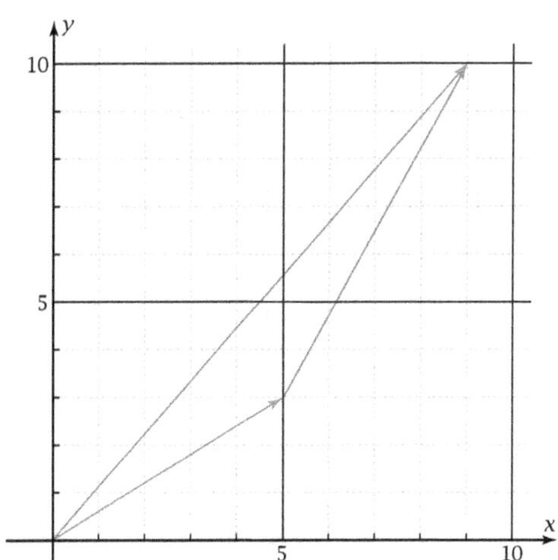

2. $|\vec{r}| = \sqrt{9^2 + 10^2} = \sqrt{181} \approx 13.5$

$\theta = \tan^{-1}\frac{10}{9} \approx 48.0°$

3.

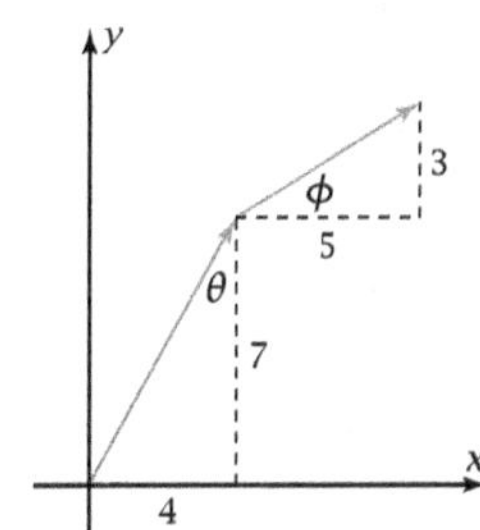

$\theta + 90° + \phi = \tan^{-1}\frac{4}{7} + 90° + \tan^{-1}\frac{3}{5} \approx 150.7°$

4. $\theta - \phi = \tan^{-1}\frac{7}{4} - \tan^{-1}\frac{3}{5} \approx 29.3°$

5. See *both* graphs in Problem 1.

6. $4\vec{i} + 7\vec{j}$

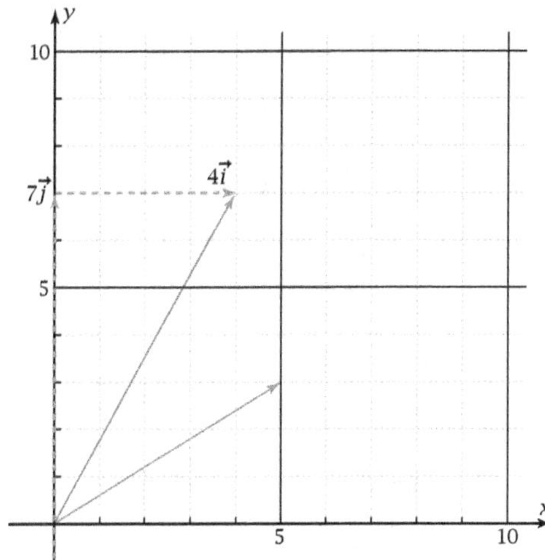

7. The $\vec{i}$-component is the sum of the $\vec{i}$-components, and the $\vec{j}$-component is the sum of the $\vec{j}$-components.

8. Answers will vary.

1. $0° = 360°$

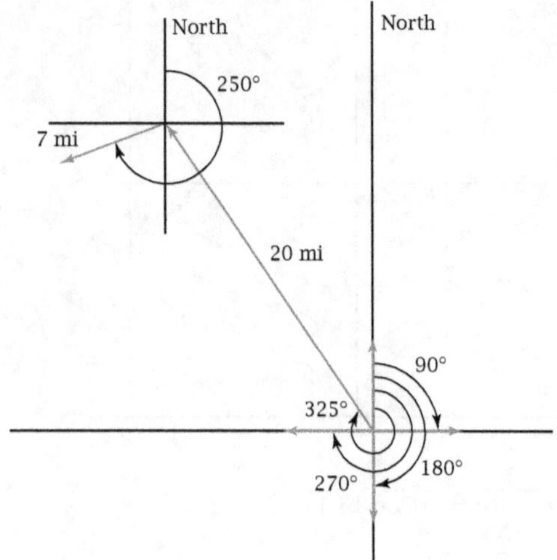

2. $360° - 325° + 90° = 125°$
$360° - 250° + 90° = 200°$

First vector:
$(20 \cos 125°)\vec{i} + (20 \sin 125°)\vec{j} \approx -11.5\vec{i} + 16.4\vec{j}$

Second vector:
$(7 \cos 200°)\vec{i} + (7 \sin 200°)\vec{j} \approx -6.6\vec{i} - 2.4\vec{j}$

3. $(20 \cos 125° + 7 \cos 200°)\vec{i}$
$+ (20 \sin 125° + 7 \sin 200°)\vec{j} \approx -18.0\vec{i} + 14.0\vec{j}$

4. $\tan^{-1} \dfrac{20 \sin 125° + 7 \sin 200°}{20 \cos 125° + 7 \cos 200°} \approx 142.2°$ because $(-18.0, 14.0)$
is in the second quadrant.
$\beta = 360° - 142.2° + 90° = 307.8°$

5. $\sqrt{(20 \cos 125° + 7 \cos 200°)^2 + (20 \sin 125° + 7 \sin 200°)^2}$
≈ 22.8 mi
22.8 mi at 307.8°

6.

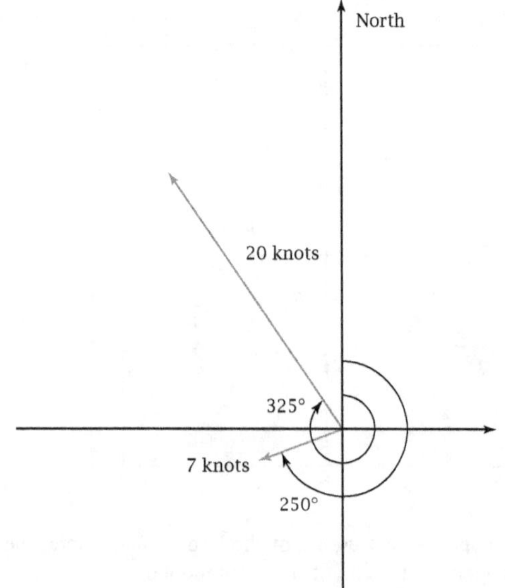

7. This answer is the same as in Problem 5, with different units
and the vector of length 7 units translated, which makes no
difference to the answer. So the answer is the same (just with
different units), 22.8 knots at 307.8°.

8. Answers will vary.

1. Sketch not to scale.

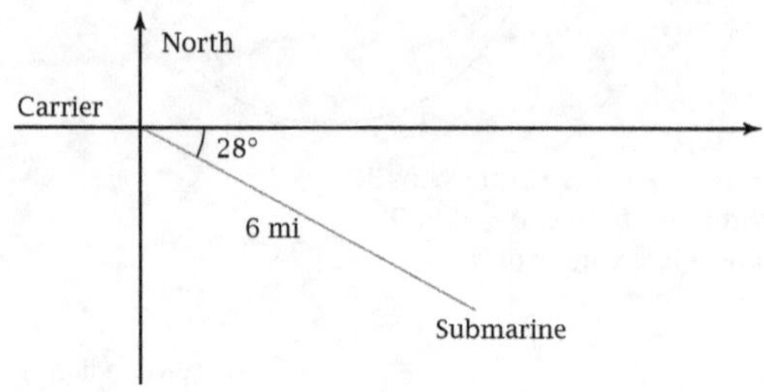

2. (See sketch in Problem 3.)
$4^2 = x^2 + 6^2 - 2 \cdot x \cdot 6 \cos 28°$
$x^2 - 12x \cos 28° + 36 - 16 = 0$
$x^2 + (-12 \cos 28°)x + 20 = 0$
$x = \dfrac{12 \cos 28° \pm \sqrt{(-12 \cos 28°)^2 - 4 \cdot 1 \cdot 20}}{2 \cdot 1} \approx 2.5$ mi
(the answer for this problem) or 8.1 mi (the answer for
Problem 4)

3. Sketch not to scale.

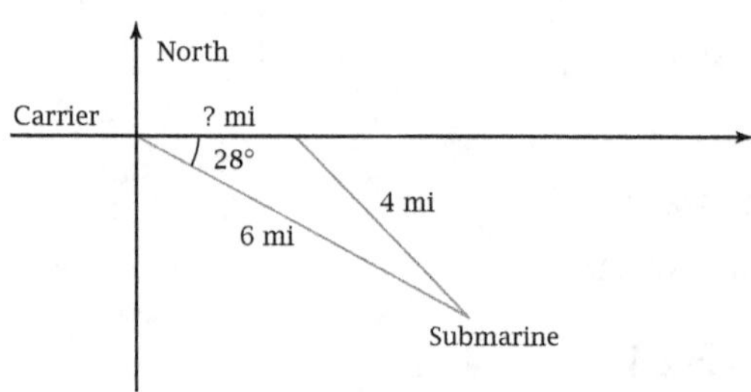

When drawn to scale, the distance is 2.5 cm.

4. 8.1 mi (See Problem 2.)

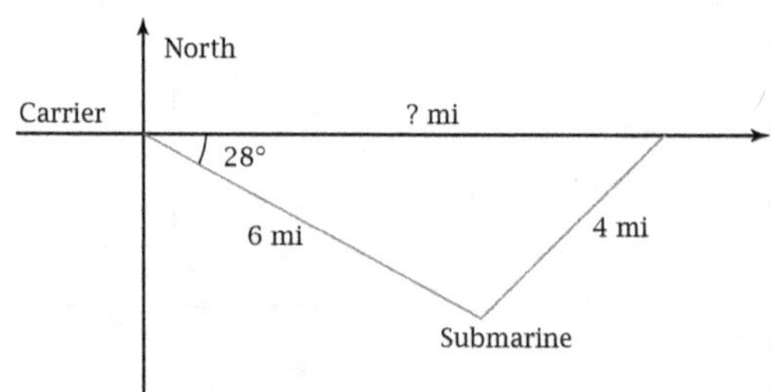

5. $\dfrac{\sin \theta}{8.1376...} = \dfrac{\sin 28°}{4}$
$\theta = \sin^{-1} \dfrac{(8.1376...) \sin 28°}{4} \approx 107.2°$ because the angle is
obtuse.

6. $\dfrac{1}{2} \cdot 6 \cdot 4 \sin \theta \approx 11.5$ mi^2

7. $2^2 = x^2 + 6^2 - 2 \cdot x \cdot 6 \cos 28°$
   $x^2 - 12x \cos 28° + 36 - 4 = 0$
   $x^2 + (-12 \cos 28°)x + 32 = 0$
   The discriminant is $(-12 \cos 28°)^2 - 4 \cdot 1 \cdot 32 \approx -15.7$. So there is no real solution.

8. Answers will vary.

Exploration 67

1. $\theta = \dfrac{360°}{8} = 45°$

   $A_{\text{triangle}} = \dfrac{1}{2} \cdot 10 \cdot 10 \sin 45° = 25\sqrt{2}$

   $\approx 35.4 \text{ units}^2$

   $A_{\text{octagon}} = 8 \cdot A_{\text{triangle}} = 8 \cdot 25\sqrt{2}$

   $\approx 282.8 \text{ units}^2$

2. $n \cdot \dfrac{1}{2} \cdot 10 \cdot 10 \sin \dfrac{360°}{n} = 50n \sin \dfrac{360°}{n}$

3.

n	Area
3	129.9038...
4	200
5	237.7641...
6	259.8076...
7	273.6410...
8	282.8427...
9	289.2544...
10	293.8926...
11	297.3524...
12	300

4. Square: $200 \sin 90° = 200 \cdot 1 = 200 \text{ units}^2$

   Dodecagon: $600 \sin 30° = 600 \cdot \dfrac{1}{2} = 300 \text{ units}^2$

5. Answers will vary.

6. More and more digits from the left remain the same.

7. 314.1592..., that is, 100π. The n-gons are approaching a circle with radius 10, whose area would be $\pi \cdot 10^2 = 100\pi$.

8. Answers will vary.

Properties of Elementary Functions

Exploration 68

1. Linear
   $y = mx + b$
   $3m + b = 25$
   $8m + b = 13$
   $5m = -12$
   $m = -2.4$
   $3(-2.4) + b = 25$
   $b = 25 + 7.2$
   $b = 32.2$
   $y_2 = -2.4x + 32.2$

2. The answer checks.

3. Neither

4. The graph is not a power function because the y-intercept is not zero.
   $y = a \cdot b^x$
   $a \cdot b^3 = 4$
   $a \cdot b^8 = 9$
   $\dfrac{a \cdot b^8}{a \cdot b^3} = \dfrac{9}{4}$
   $b^5 = 2.25$
   $b = 2.25^{1/5} = 1.1760...$
   $a(1.1760...)^3 = 4$
   $a = \dfrac{4}{(1.1760...)^3} = 2.4589...$
   $y = 2.4589... (1.1760...)^x$

5. The answer checks.

6. Concave up

7. The y-axis appears to be a vertical asymptote, which indicates a power function with a negative exponent.
   $y = ax^b$
   $a \cdot 3^b = 20$
   $a \cdot 8^b = 2$
   $\dfrac{a \cdot 8^b}{a \cdot 3^b} = \dfrac{2}{20} = 0.1$
   $\left(\dfrac{8}{3}\right)^b = 0.1$
   $\log \left(\dfrac{8}{3}\right)^b = \log 0.1$
   $b \cdot \log \dfrac{8}{3} = \log 0.1$
   $b = \dfrac{\log 0.1}{\log \frac{8}{3}} = -2.3475...$
   $a \cdot 3^{-2.3475...} = 20$
   $a = \dfrac{20}{3^{-2.3475...}} = 263.7030...$
   $\therefore y = 263.7030...x^{-2.3475...}$

8. The answer checks.

9. Concave up

10. Quadratic
    $y = ax^2 + bx + c$
    $9a + 3b + c = 26.6$
    $64a + 8b + c = 38.6$
    $196a + 14b + c = 13.4$
    $\begin{bmatrix} 9 & 3 & 1 \\ 64 & 8 & 1 \\ 196 & 14 & 1 \end{bmatrix}^{-1} \begin{bmatrix} 26.6 \\ 38.6 \\ 13.4 \end{bmatrix} = \begin{bmatrix} -0.6 \\ 9 \\ 5 \end{bmatrix}$
    $\therefore y = 0.6x^2 + 9x + 5$

11. The answer checks.

12. Concave down

13. Answers will vary.

Exploration 69

1.

x	y
2	1.8
4	16.2
6	145.8
8	1312.2
10	11809.8
12	106288.2

2. $\dfrac{y2}{y1} = \dfrac{y(6)}{y(4)} = \dfrac{145.8}{16.2} = 9$

3. $\dfrac{y2}{y1} = \dfrac{y(10)}{y(8)} = \dfrac{11809.8}{1312.2} = 9$

4. $\dfrac{106288.2}{11809.8} = \dfrac{11809.8}{1312.2} = \dfrac{1312.2}{145.8} = \dfrac{145.8}{16.2} = \dfrac{16.2}{1.8} = 9$

5. Add–multiply

6.

x	y
2	40
4	320
6	1080
8	2560
10	5000
12	8640

7. $\dfrac{y(4)}{y(2)} = \dfrac{320}{40} = 8$

$\dfrac{y(8)}{y(4)} = \dfrac{2560}{320} = 8$

$\dfrac{y(12)}{y(6)} = \dfrac{8640}{1080} = 8$

8. ... multiplies y by 8.

9. If f is a power function, multiplying x by a constant multiplies $f(x)$ by a constant.

10.

x	y
2	31
4	45
6	59
8	73
10	87
12	101

11. $y(6) = 59 = 31 + 28 = y(2) + 28$
$y(8) = 73 = 45 + 28 = y(4) + 28$
$y(10) = 87 = 59 + 28 = y(6) + 28$
$y(12) = 101 = 73 + 28 = y(8) + 28$

12. ... adds 28 to y.

13. The add–add property of linear functions states that if f is a linear function, adding a constant to x adds a constant to $f(x)$.

14. Given $y = a \cdot x^n$ and $x_2 = cx_1$:
$y_2 = a \cdot x_2^n = a \cdot (cx_1)^n = c^n \cdot a \cdot x_1^n = c^n \cdot y_1$
$= k \cdot y_1$, where $k = c^n$.

Exploration 70

1. The values are the same as in the table shown.

2. First differences:
$12.0 - 12.2 = -0.2$
$13.4 - 12.0 = 1.4$
$16.4 - 13.4 = 3.0$
$21.0 - 16.4 = 4.6$

Second differences:
$1.4 - (-0.2) = 1.6$
$3.0 - 1.4 = 1.6$
$4.6 - 3.0 = 1.6$

All the second differences are the same!

3. $12.2 = 4a + 2b + c$
$12.0 = 16a + 4b + c$
$13.4 = 36a + 6b + c$
$a = 0.2,\ b = -1.3,\ c = 14$
$q(x) = 0.2x^2 - 1.3x + 14$

4.

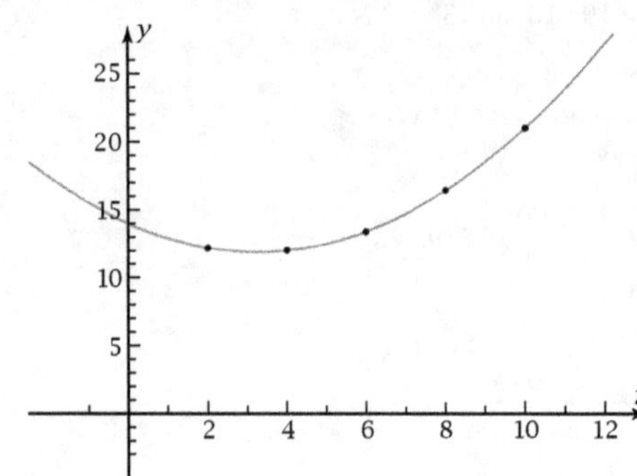

5. Yes, the five points lie on the graph.

6. First differences:
$34.8 - 12.3 = 22.5$
$44.7 - 34.8 = 9.9$
$42.0 - 44.7 = -2.7$
$26.7 - 42.0 = -15.3$

Second differences:
$9.9 - 22.5 = -12.6$
$-2.7 - 9.9 = -12.6$
$-15.3 - (-2.7) = -12.6$
$f(x) = -0.7x^2 + 11x + 2$
$f(1) = -0.7 \cdot 1 + 11 \cdot 1 + 2 = 12.3$
$f(4) = -0.7 \cdot 16 + 11 \cdot 4 + 2 = 34.8$
$f(7) = -0.7 \cdot 49 + 11 \cdot 7 + 2 = 44.7$
$f(10) = -0.7 \cdot 100 + 11 \cdot 10 + 2 = 42.0$
$f(13) = -0.7 \cdot 169 + 11 \cdot 13 + 2 = 26.7$

7. Answers will vary.

Exploration 71

1. Function has add–multiply property of exponential functions: $f(x + 2) = 0.7 \cdot f(x)$ for all x.
$f(x) = 6000 \cdot 0.7^x$
$f(13.7) = 6000 \cdot 0.7^{13.7} = 45.2891\ldots$

2. Function has multiply–multiply property of power functions: $g(2x) = 8 \cdot g(x)$ for all x.
$g(x) = 0.7x^3$
$g(22.3) = 0.7 \cdot 22.3^3 = 7762.6969$

3. Function has add–add property of linear functions: $h(x + 2) = h(x) + 34.6$ for all x.
$h(x) = 17.3x + 52.9$
$h(100) = 17.3 \cdot 100 + 52.9 = 1782.9$

4. Function has constant-second-differences property of quadratic functions.
$q(x) = 0.4x^2 - 0.7x + 9$
$q(-7.9) = 0.4 \cdot 7.9^2 - 0.7 \cdot 7.9 + 9 = 34.494$

5. Answers will vary.

Exploration 72

1.

x	$y_1 = \log x$	x as Power of 10
0.1	-1	10^{-1}
1	0	10^0
10	1	10^1
100	2	10^2
1000	3	10^3

2. See table in Problem 1.

3. Multiplying the value of x by a constant adds a constant to the corresponding value of y. In this case, multiplying x by 10 adds 1 to y.

4. Multiplying x by 5 adds 0.6989... to y.

x	$y_1 = \log x$
1	0
5	0.6989...
25	1.3979...
125	2.0969...
625	2.7958...

5.

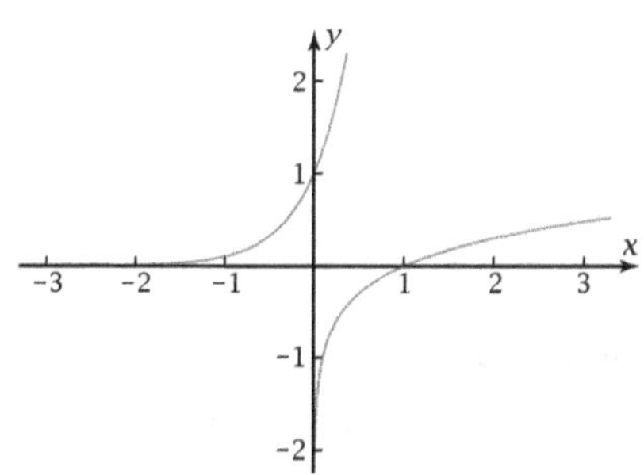

6. $\log 1776 = 3.2494...$
$10^{\log 1776} = 1776$
$10^{\log x} = x$

7. $10^{1.8} = 63.0957...$
$\log 10^{1.8} = 1.8$
$\log 10^x = x$

8. $\log_b x = y \Leftrightarrow b^y = x;\ b > 0,\ b \neq 1,\ x > 0$
Base = 10

9.

x	y
9	2
81	4
129	6
6561	8
59049	10

Multiplying the value of x by a constant adds a constant to the corresponding value of y. In this case, multiplying x by 9 adds 2 to y.

10. $\log_b 9 = 2 \Rightarrow b^2 = 9 \Rightarrow b = 3$

11. The logarithm (to the base B) of x is the power (exponent) to which B must be raised to give x.

12. Answers will vary.

Exploration 73

1.

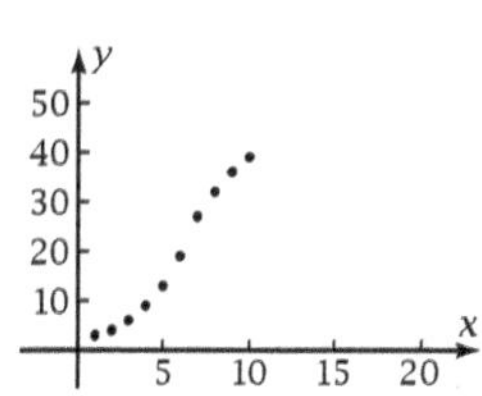

2. Sample equation: $y = 1.9741... \cdot 1.4568...^x$

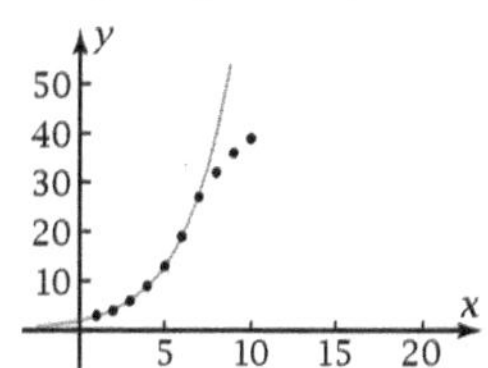

3. $2 = \dfrac{43}{1 + ae^{-6}}$ $\qquad$ $39 = \dfrac{43}{1 + ae^{-10b}}$
$2ae^{-b} = 41$ $\qquad\qquad$ $39ae^{-10b} = 4$
$ae^{-b} = \dfrac{41}{2}$ $\qquad\qquad$ $ae^{-10b} = \dfrac{4}{39}$
$\ln a - b = \ln \dfrac{41}{2}$ $\qquad$ $\ln a - 10b = \ln \dfrac{4}{39}$

Solving the system of equations for b gives
$b = 0.5886...$
Plugging b into the first equation gives
$a = 36.9312...$
So $y = \dfrac{43}{1 + 36.9312... \, e^{-0.5886...x}}$

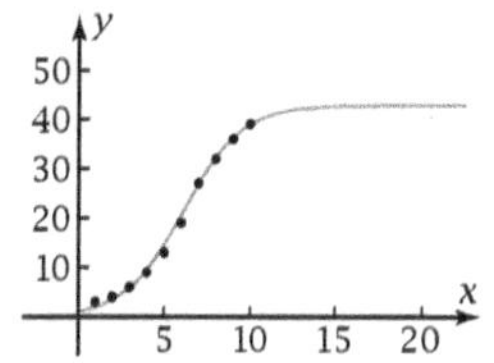

4. $y(0) = \dfrac{43}{1 + 36.9312...} = 1.1336$
There were approximately 1134 people.

5. The graph seems to be leveling off at about 43,000.

6. From Greek *logos*, meaning "word" or "calculation"

7. Answers will vary.

Fitting Functions to Data

Exploration 74

1. $\hat{y} = 1.4x + 3.8$

2. See table in Problem 5.

3. See table in Problem 5.

4. $0.4 - 1.8 + 3.0 - 2.2 + 0.6 = 0$

5.

$\hat{y}$	$y - \hat{y}$	$(y - \hat{y})^2$
6.6	0.4	0.16
10.8	−1.8	3.24
15.0	3.0	9.00
19.2	−2.2	4.84
23.4	0.6	0.36

$SS_{res} = 17.65$

6.

$\hat{y}$	$y - \hat{y}$	$(y - \hat{y})^2$
6.7	0.3	0.09
10.9	−1.9	3.61
15.1	2.9	8.41
19.3	−2.3	5.29
23.5	0.5	0.25

$SS_{res} = 17.65$

The new equation doesn't fit the data as well.

7.

$\hat{y}$	$y - \hat{y}$	$(y - \hat{y})^2$
6.8	0.2	0.04
11.3	−2.3	5.29
15.8	2.2	4.84
20.3	−3.3	10.89
24.8	−0.8	0.25

$SS_{res} = 21.70$

The two modified equations do not fit as well as the actual regression equation.

8. Answers will vary.

9. Answers will vary.

Exploration 75

1. $\hat{y} = -2x + 46.4$
   y decreases as x increases.

2.

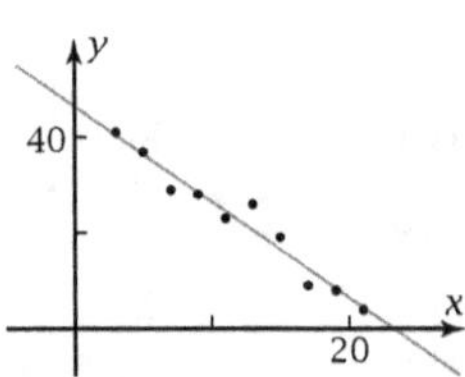

It seems to fit fairly well.

3.

$\hat{y}$	$y - \hat{y}$	$(y - \hat{y})^2$
40.4	0.6	0.36
36.4	0.6	0.36
32.4	−3.4	11.56
28.4	−0.4	0.16
24.4	−1.4	1.96
20.4	5.6	31.36
16.4	2.6	6.76
12.4	−3.4	11.56
8.4	−0.4	0.16
4.4	−0.4	0.16

4. $SS_{res} = 64.40$

5.

$\hat{y}$	$y - \hat{y}$	$(y - \hat{y})^2$
40	1	1
36	1	1
32	−3	9
28	0	0
24	−1	1
20	6	36
16	3	9
12	−3	9
8	0	0
4	0	0

$SS_{res} = 66.00$

The modified equation doesn't fit the data as well.

6. $(\bar{x}, \bar{y}) = (12, 22.4); \hat{y}(\bar{x}) = -2(12) + 46.4 = 22.4 = \bar{y}$

7. No, it's impossible to tell which is better just by looking.

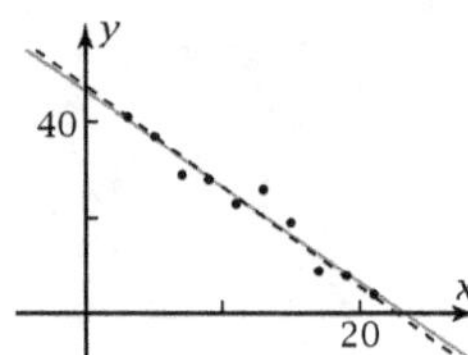

8.

$\hat{y}$	$y - \hat{y}$	$(y - \hat{y})^2$
41.3	−0.3	0.09
37.1	−0.1	0.01
32.9	−3.9	15.21
28.7	−0.7	0.49
24.5	−1.5	2.25
20.3	5.7	32.49
16.1	2.9	8.41
11.9	−2.9	8.41
7.7	0.3	0.09
3.5	0.5	0.25

$SS_{res} = 67.70$

Its SS_{res} is larger than that of the regression equation.

9. Answers will vary.

Exploration 76

1. $\bar{y} = \dfrac{1}{5}(7 + 9 + 18 + 17 + 24) = \dfrac{75}{5} = 15$

   See graph in Problem 2.

2.

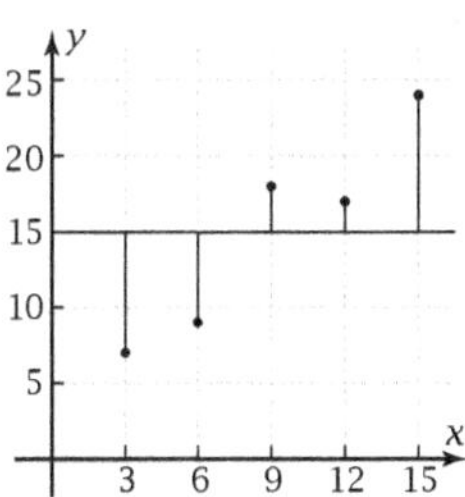

3.

x	y	$y - \bar{y}$	$(y - \bar{y})^2$
3	7	-8	64
6	9	-6	36
9	18	3	9
12	17	2	4
15	24	9	81

   $SS_{\text{dev}} = 194$

4. $\hat{y}(x) = 1.4x + 2.4$
   $r^2 = 0.9092\ldots,\ r = 0.9535\ldots$

5.

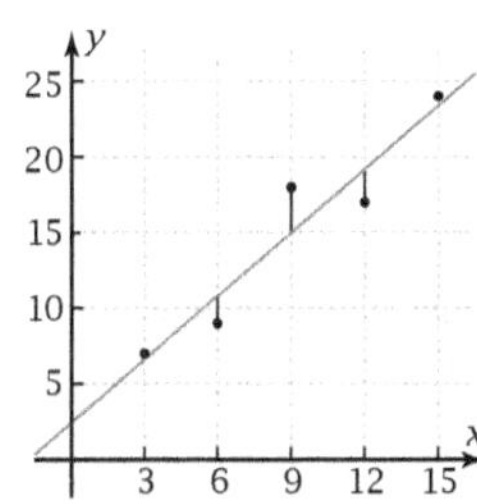

6. See Problem 5.

7.

x	y	$\hat{y}$	$(y - \hat{y})^2$
3	7	6.6	0.16
6	9	10.8	3.24
9	18	15	9
12	17	19.2	4.84
15	24	23.4	0.36

   $SS_{\text{res}} = 17.6$

8. $\dfrac{SS_{\text{dev}} - SS_{\text{res}}}{SS_{\text{dev}}} = \dfrac{194 - 17.6}{194} = \dfrac{176.4}{194} = 0.9092\ldots$

9. The coefficient of determination, $\dfrac{SS_{\text{dev}} - SS_{\text{res}}}{SS_{\text{dev}}}$, shows up as r^2;

   r is calculated by taking $\pm\sqrt{\dfrac{SS_{\text{dev}} - SS_{\text{res}}}{SS_{\text{dev}}}}$, using the positive branch if the regression function is increasing and using the negative branch if the regression function is decreasing.

Exploration 77

1. $\bar{x} = \dfrac{1}{3}(2 + 3 + 7) = \dfrac{12}{3} = 4$ hr

   $\bar{y} = \dfrac{1}{3}(20 + 17 + 8) = \dfrac{45}{3} = 15$

x (hr)	y (gal)	$x - \bar{x}$	$y - \bar{y}$
2	20	-2	5
3	17	-1	2
7	8	3	-7

2.

x (hr)	y (gal)	$(x - \bar{x})^2$	$(y - \bar{y})^2$	$(x - \bar{x})(y - \bar{y})$
2	20	4	25	-10
3	17	1	4	-2
7	8	9	49	-21

   $\sum(x - \bar{x})^2 = 14\ \text{hr}^2$
   $\sum(y - \bar{y})^2 = 78\ \text{gal}^2$
   $\sum(x - \bar{x})(y - \bar{y}) = -33\ \text{hr} \cdot \text{gal}$

3. $\dfrac{[-33\ \text{hr} \cdot \text{gal}]^2}{(14\ \text{hr}^2)(78\ \text{gal}^2)} = 0.9972\ldots$

4. $\hat{y} = -2.3571\ldots x + 24.4285\ldots$
   $r = -0.9986\ldots,\ r^2 = 0.9972\ldots$

5. $\bar{y} = m\bar{x} + b \Rightarrow b = \bar{y} - m\bar{x}$
   For any point on the regression line,
   $\hat{y} = mx + b = mx + \bar{y} - m\bar{x} = \bar{y} + m(x - \bar{x})$.

   Substitute this into the expression for SS_{res}:
   $SS_{\text{res}} = \sum(y - \hat{y})^2 = \sum[y - (\bar{y} + m(x - \bar{x}))]^2$
   $SS_{\text{res}} = \sum[(y - \bar{y}) - m(x - \bar{x})]^2$

   Square:
   $\sum[(y - \bar{y}) - m(x - \bar{x})]^2$
   $\quad = \sum[(y - \bar{y})^2 - 2m(x - \bar{x})(y - \bar{y}) + m^2(x - \bar{x})^2]$

   Rewrite the expression:
   $\sum[(y - \bar{y}) - m(x - \bar{x})]^2$
   $\quad = \sum(y - \bar{y})^2 - 2m\sum(x - \bar{x})(y - \bar{y}) + m^2\sum(x - \bar{x})^2$

6. $\sum(x - \bar{x})(y - \bar{y}) = \sum[xy - x\bar{y} - \bar{x}y + \bar{x}\bar{y}]$
   $\quad = \sum xy - \sum \bar{x}y - \sum x\bar{y} + \sum \bar{x}\bar{y}$
   $\quad = \sum xy - \bar{x}\sum y - \bar{y}\sum x + n\bar{x}\bar{y}$
   $\quad = \sum xy - \left(\dfrac{1}{n}\sum x\right)\sum y - \left(\dfrac{1}{n}\sum y\right)\sum x + n\bar{x}\bar{y}$
   $\quad = \sum xy - 2 \cdot \dfrac{1}{n}\sum x\sum y + n\bar{x}\bar{y}$
   $\quad = \sum xy - 2 \cdot \dfrac{1}{n}\sum x\sum y + n\left(\dfrac{1}{n}\sum x\right)\left(\dfrac{1}{n}\sum y\right)$
   $\quad = \sum xy - 2 \cdot \dfrac{1}{n}\sum x\sum y + \dfrac{1}{n}\sum x\sum y$
   $\quad = \sum xy - \dfrac{1}{n}\sum x\sum y$

7. First note that $\bar{x} = \dfrac{1}{n}\Sigma x \Rightarrow \Sigma x = n\bar{x} = \Sigma\bar{x}$.

Simplify the denominator:

$$n\Sigma x^2 - (\Sigma x)^2 = n\Sigma x^2 - \Sigma x \Sigma x$$
$$= n\Sigma x^2 - n\bar{x}\Sigma x$$
$$= n\Sigma x^2 - 2n\bar{x}\Sigma x + n\bar{x}\Sigma x$$
$$= n\Sigma x^2 - n\Sigma 2\bar{x}x + n\Sigma\bar{x}^2$$
$$= n\Sigma(x^2 - 2\bar{x}x + \bar{x}^2)$$
$$= n\Sigma(x - \bar{x})^2$$
$$\therefore m = \frac{n\Sigma xy - \Sigma x\Sigma y}{n\Sigma x^2 - (\Sigma x)^2} = \frac{n\Sigma xy - \Sigma x\Sigma y}{n\Sigma(x - \bar{x})^2}$$
$$= \frac{n\left(\Sigma xy - \dfrac{1}{n}\Sigma x\Sigma y\right)}{n\Sigma(x - \bar{x})^2}$$

Then using the result of Problem 6,

$$m = \frac{n\Sigma(x - \bar{x})(y - \bar{y})}{n\Sigma(x - \bar{x})^2} = \frac{\Sigma(x - \bar{x})(x - \bar{y})}{\Sigma(x - \bar{x})^2}.$$

8. Using the expression for m from Problem 7, the middle term in the expression for SS_{res} from Problem 5 becomes:

$$2m\Sigma(x - \bar{x})(y - \bar{y}) = 2\frac{\Sigma(x - \bar{x})(y - \bar{y})}{\Sigma(x - \bar{x})^2}\Sigma(x - \bar{x})(y - \bar{y})$$
$$= 2\frac{[\Sigma(x - \bar{x})(y - \bar{y})]^2}{\Sigma(x - \bar{x})^2}$$

Next, the third term in the expression for SS_{res} from Problem 5 becomes:

$$m^2\Sigma(x - \bar{x})^2 = \left[\frac{\Sigma(x - \bar{x})(y - \bar{y})}{\Sigma(x - \bar{x})^2}\right]^2\Sigma(x - \bar{x})^2$$
$$= \frac{[\Sigma(x - \bar{x})(y - \bar{y})]^2}{[\Sigma(x - \bar{x})^2]^2}\Sigma(x - \bar{x})^2$$
$$= \frac{[\Sigma(x - \bar{x})(y - \bar{y})]^2}{\Sigma(x - \bar{x})^2}$$
$$\therefore -2m\Sigma(x - \bar{x})(y - \bar{y}) + m^2\Sigma(x - \bar{x})^2$$
$$= -2\frac{[\Sigma(x - \bar{x})(y - \bar{y})]^2}{\Sigma(x - \bar{x})^2} + \frac{[\Sigma(x - \bar{x})(y - \bar{y})]^2}{\Sigma(x - \bar{x})^2}$$
$$= -\frac{[\Sigma(x - \bar{x})(y - \bar{y})]^2}{\Sigma(x - \bar{x})^2}$$
$$\therefore SS_{\text{res}} = \Sigma(y - \bar{y})^2 - \frac{[\Sigma(x - \bar{x})(y - \bar{y})]^2}{\Sigma(x - \bar{x})^2}$$

9. Simplify the numerator of the expression for r^2.

$$SS_{\text{dev}} - SS_{\text{res}} = \Sigma(y - \bar{y})^2 - \Sigma(y - \bar{y})^2 + \frac{[\Sigma(x - \bar{x})(y - \bar{y})]^2}{\Sigma(x - \bar{x})^2}$$
$$= \frac{[\Sigma(x - \bar{x})(y - \bar{y})]^2}{\Sigma(x - \bar{x})^2}$$

Then:

$$r^2 = \frac{SS_{\text{dev}} - SS_{\text{res}}}{SS_{\text{dev}}} = \frac{\dfrac{[\Sigma(x - \bar{x})(y - \bar{y})]^2}{\Sigma(x - \bar{x})^2}}{\Sigma(y - \bar{y})^2} = \frac{[\Sigma(x - \bar{x})(y - \bar{y})]^2}{\Sigma(x - \bar{x})^2\Sigma(y - \bar{y})^2}$$

10. Recall from Problem 2 that:

$$\Sigma(x - \bar{x})^2 = 14\ \text{hr}^2$$
$$\Sigma(y - \bar{y})^2 = 78\ \text{gal}^2$$
$$\Sigma(x - \bar{x})(y - \bar{y}) = -33\ \text{hr} \cdot \text{gal}$$

So $r = \dfrac{\Sigma(x - \bar{x})(y - \bar{y})}{\sqrt{\Sigma(x - \bar{x})^2}\sqrt{\Sigma(y - \bar{y})^2}}$

$$= \frac{-33\ \text{hr} \cdot \text{gal}}{\sqrt{14\ \text{hr}^2}\sqrt{78\ \text{gal}^2}} = \frac{-33}{\sqrt{14}\sqrt{78}} = -0.9986...$$

Note that $r < 0$, indicating that the regression equation is a decreasing function.

11. Answers will vary.

Exploration 78

1. $\hat{y} = 17{,}364.89...x + 358{,}350.81...$
   $r = 0.999623...$ (very close to 1)
   $\hat{y}(5) = 445{,}175.28...$, which is close to 451,310

2. 1985: $x = 15$
   $\hat{y}(15) = 618{,}824.24...$ (not close to 497,560)
   1997: $x = 27$
   $\hat{y}(27) = 827{,}202.99...$ (not close to 717,750)

3. Yes

4. $y_2 = 17{,}356.82...x + 314{,}672.82...$
   $r = 0.960182...$

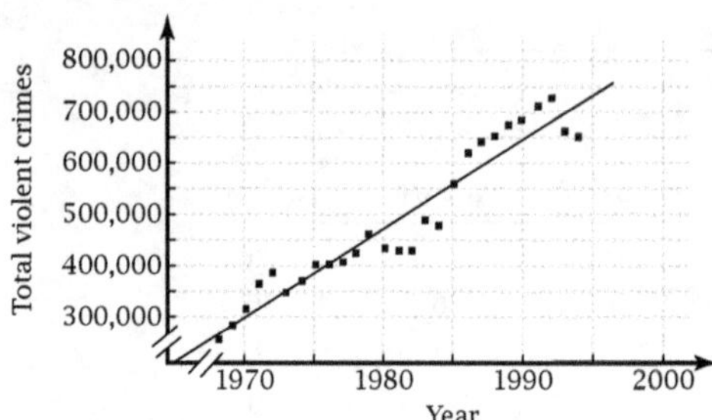

5. $y_3 = -1038.8041...x^2 + 28{,}552.4376...x + 304{,}266.307...$
   ($R^2 = 0.9190...$)
   $y_3(27) = 317{,}893.9...$, which is much different from 717,750.

6. Answers will vary.

Exploration 79

1. Because both kinds of function can be decreasing and concave up

2. Both approach the x-axis as an asymptote as x becomes large. The exponential function has a y-intercept for the temperature at $x = 0$. The power has a vertical asymptote at $x = 0$, which is the wrong behavior.

3. Exponential: $r = -0.9977...$
   Power: $r = -0.9576...$
   $\therefore$ Exponential fits better because r is closer to -1.

4.

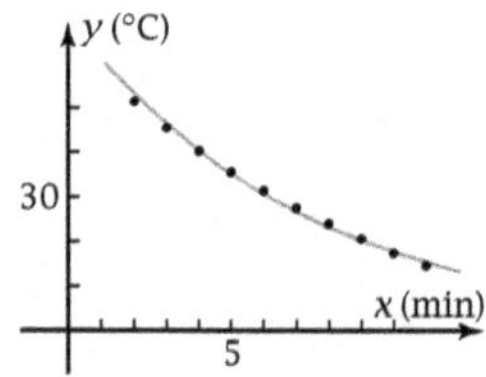

5.

x (min)	Residuals
2	$-1.5864\ldots$
3	$-0.7467\ldots$
4	$-0.0883\ldots$
5	$0.4024\ldots$
6	$0.7244\ldots$
7	$0.7637\ldots$
8	$0.6955\ldots$
9	$0.3852\ldots$
10	$-0.1102\ldots$
11	$-0.8413\ldots$

6.

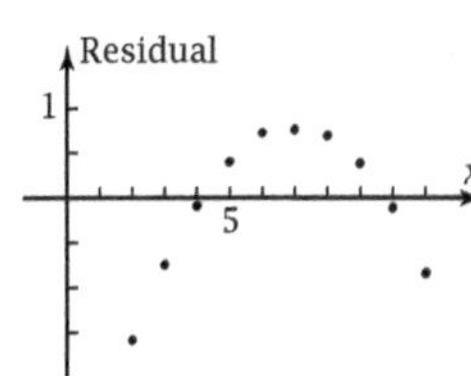

7. There is something in the data that is not accounted for by exponential function. If there is no pattern, the residuals are accounted for by random measurement errors.

8. Answers will vary.

Exploration 80

1. Both are increasing and concave up.

2. Exponential: $y_1 = 21.1772\ldots(1.009142\ldots)^x$
   $r = 0.9091\ldots$
   Power: $0.00000002117\ldots x^{4.0340\ldots}$

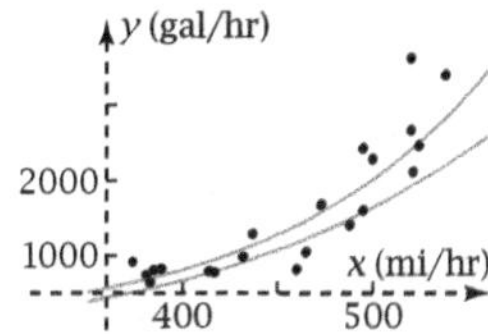

3.

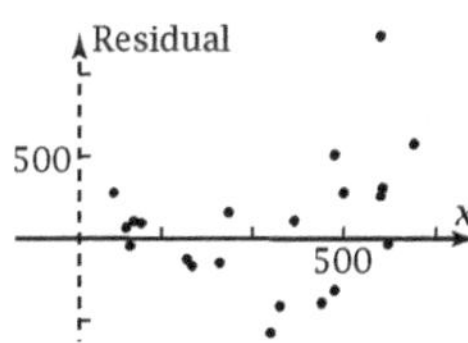

Somewhat of a pattern.

4.

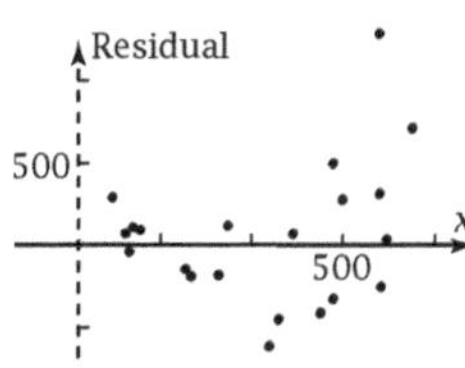

About the same.

5. Answers will vary.

Exploration 81

1. $y_1 = 320.9749\ldots(1.00501\ldots)^x$
   $r = 0.9977\ldots$, indicating a good fit.

2.

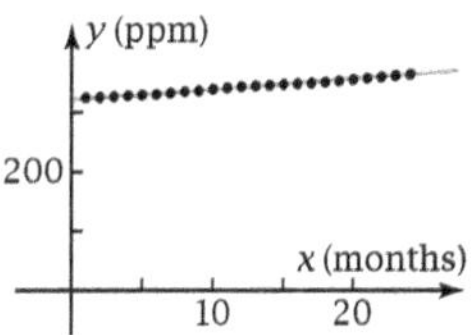

The function fits well.

3. There is a sinusoidal pattern.

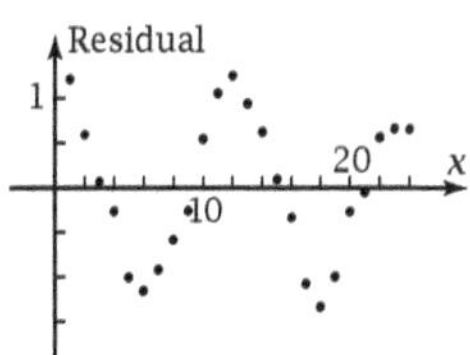

4. $y_2 = 1.0906\ldots \sin(0.5321\ldots x + 1.4039\ldots) - 0.01625\ldots$

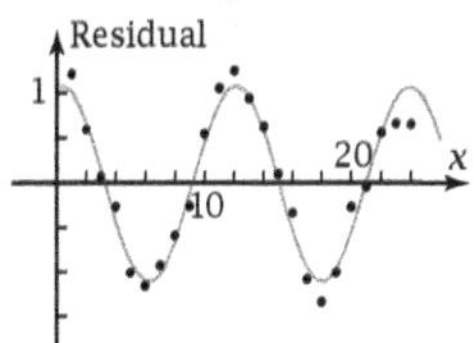

5. The graph goes through more points.

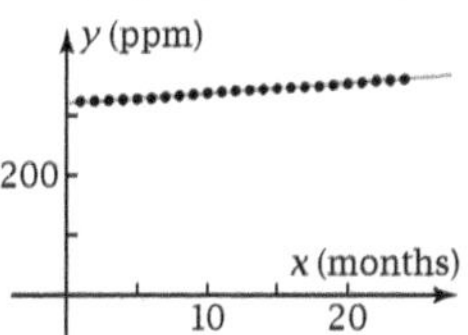

6.

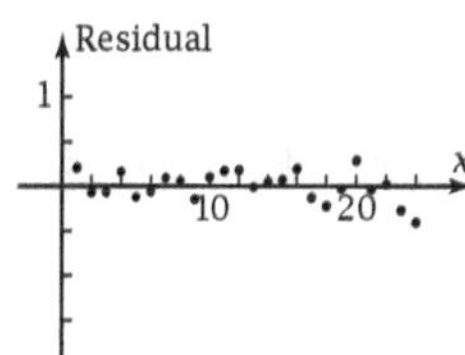

The randomness of this residual plot indicates that y_3 accounts for all but random fluctuations.

7. Answers will vary.

Probability, and Functions of a Random Variable

Exploration 82

1. $n(A) = 12$, $n(B) = 10$

2. $n(A \text{ or } B) = 12 + 10 = 22$
   Add $n(A)$ and $n(B)$.

3. The occurrence of one of them excludes the possibility that the other will occur.

4. $n(A \text{ and } B) = 12 \cdot 10 = 120$
   Multiply $n(A)$ and $n(B)$.

5. The way one occurs does not affect the way the other could occur.

6. $n(C) = 7$

7. $n(A \text{ and } C) = 3$

8. $n(A \text{ or } C) = n(A) + n(C) - n(A \text{ and } C)$
   $= 12 + 10 - 3 = 19$

9. $n(X) + n(Y) - n(X \text{ and } Y)$

10. $n(X) \cdot n(Y)$

11. Answers will vary.

Exploration 83

1. A rearrangement of their order

2. $7! = 5040$

3. $5 \cdot 5! \cdot 2 = 1200$

4. $\dfrac{1200}{5040} = \dfrac{5}{21} \approx 23.8\%$

5. $\dfrac{5 \cdot 5! \cdot 4}{5040} = \dfrac{2400}{5040} = \dfrac{10}{21} \approx 47.6\%$

6. ${}_7P_3 = 210$

7. $1 \cdot {}_6P_2 = 30$

8. $\dfrac{6!}{3! \cdot 2!} = 60$

9. $1 \cdot 6! = 720$

10. $7 \cdot 1 \cdot 5! = 840$

11. Answers will vary.

Exploration 84

1. $P(E \text{ and } H) = P(E) \cdot P(H) = 0.8 \cdot 0.7 = 0.56$

2. $P(E) + P(H) = 0.8 + 0.7 = 1.5 > 1$
   Because $P(E \text{ or } H) \leq 1$, the preceding calculation shows $P(E \text{ or } H) \neq P(E) + P(H)$.

3. $P(E \text{ or } H) = P(E) + P(H) - P(E \text{ and } H) = 0.94$

4. $P(\text{not } E) = 1 - 0.8 = 0.2$
   $P(\text{not } H) = 1 - 0.7 = 0.3$

5. $P(\text{not } E \text{ and not } H) = P(\text{not } E) \cdot P(\text{not } H) = 0.06$

6. $P(E \text{ and not } H) = P(E) \cdot P(\text{not } H) = 0.8 \cdot 0.3 = 0.24$

7. $P(H \text{ and not } E) = P(H) \cdot P(\text{not } E) = 0.7 \cdot 0.2 = 0.14$

8. $0.56 + 0.24 + 0.14 + 0.06 = 1$

9. $P(H|E)$ is the probability that H occurs after event E has already occurred, that is, the probability that you got an "A" in History if you got an "A" in English. If $P(H|E) = P(H)$, then H is independent of E. So your chance of getting an "A" in history is not affected by whether you get an "A" in English.

10. $P(E \cup H) = P(E \text{ or } H)$ is the probability that you got an "A" in either English or History or both. $P(E \cap H) = P(E \text{ and } H)$ is the probability that you got "A"s in both History and English.
    $P(E \cup H) = P(E) + P(H) - P(E \cap H)$

11. Answers will vary.

Exploration 85

1. Answers will vary.

In the remaining questions, let $p =$ the probability found in Problem 1.

2. Numerical answers will vary but should be p^{10}.

3. Numerical answers will vary but should be $(1 - p)^{10}$.

4. Numerical answers will vary but should be
   ${}_{10}C_1 p^1 (1 - p)^9 = 10p(1 - p)^9$.

5. Numerical answers will vary but should be
   ${}_{10}C_0 p^0 (1 - p)^{10} + {}_{10}C_1 p^1 (1 - p)^9$
   $= (1 - p)^{10} + 10p(1 - p)^9 = (9p + 1)(1 - p)^9$.

6. Numerical answers will vary but should be
   $P(x) = {}_{10}C_x p^x (1 - p)^{10-x}$.

7. Answers will vary but should have a peak near $10p$.

8. Numerical answers will vary but should be $\displaystyle\sum_{x=4}^{7} P(10 - x)$.

9. Answers will vary.

Exploration 86

1. $(0.95) \cdot 4 + (0.05) \cdot 3 + (0.00) \cdot 2 = 3.95$

2.

English	$(0.90) \cdot 4 + (0.10) \cdot 3 + (0.00) \cdot 2 = 3.90$
Spanish	$(0.80) \cdot 4 + (0.15) \cdot 3 + (0.05) \cdot 2 = 3.75$
Chemistry	$(0.70) \cdot 4 + (0.20) \cdot 3 + (0.10) \cdot 2 = 3.60$
History	$(0.55) \cdot 4 + (0.30) \cdot 3 + (0.15) \cdot 2 = 3.40$

3. $\dfrac{3.95 + 3.90 + 3.75 + 3.60 + 3.40}{5} = 3.72$

4.

$x =$ Number of Heads	$P(x)$
0	0.07776
1	0.25920
2	0.34560
3	0.23040
4	0.07680
5	0.01024

5.

Payoff (x)	$P(x) \cdot$ Payoff (x)
$0 - 10 = -10$	-0.7776
$0 - 10 = -10$	-2.5012
$0 - 10 = -10$	-3.4560
$20 - 10 = 10$	2.3040
$50 - 10 = 40$	3.0710
$100 - 10 = 90$	0.9216

$\displaystyle\sum_{x=0}^{5} P(x) \cdot \text{Payoff}(x) = -0.5280$

6. Answers will vary.

Three-Dimensional Vectors

Exploration 87

1.

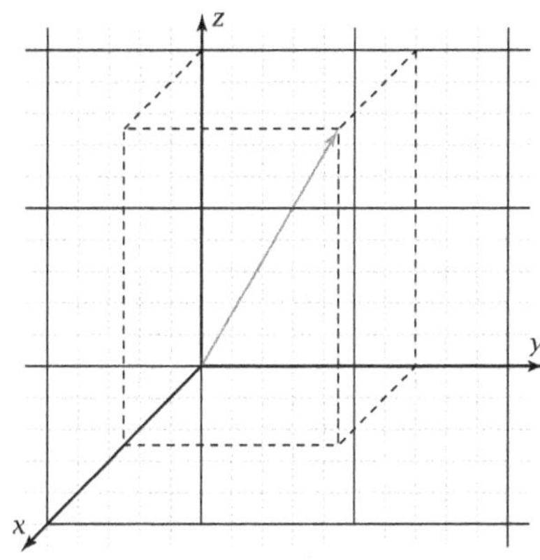

2.

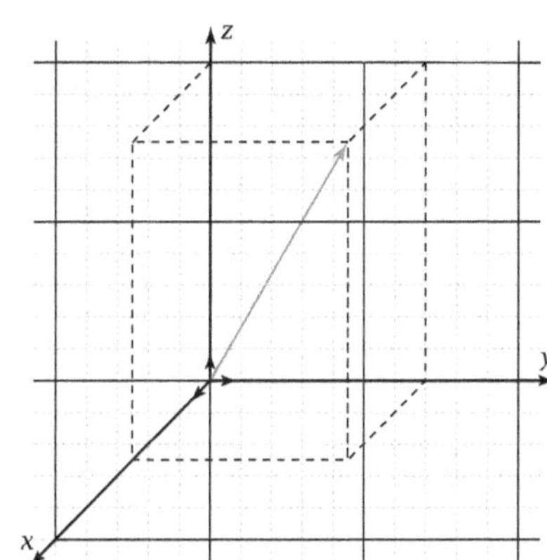

3. $|\vec{v}| = \sqrt{5^2 + 7^2 + 10^2} = \sqrt{174} = 13.1909\ldots$

4. $3\vec{v} = (3 \cdot 5)\vec{i} + (3 \cdot 7)\vec{j} + (3 \cdot 10)\vec{k}$
   $= 15\vec{i} + 21\vec{j} + 30\vec{k}$

5. $|3\vec{v}| = \sqrt{15^2 + 21^2 + 30^2}$
   $= \sqrt{3^2(5^2 + 7^2 + 10^2)} = 3\sqrt{5^2 + 7^2 + 10^2}$
   $= 3\sqrt{174} = 3|\vec{v}|$

6. $\dfrac{5}{\sqrt{174}}\vec{i} + \dfrac{7}{\sqrt{174}}\vec{j} + \dfrac{10}{\sqrt{174}}\vec{k}$

7. $\vec{v} + \vec{a} = (5 + 4)\vec{i} + [7 + (-3)]\vec{j} + (10 + 8)\vec{k}$
   $= 9\vec{i} + 4\vec{j} + 8\vec{k}$

8. $\vec{b} = (7 - 1)\vec{i} + (3 - 5)\vec{j} + (11 - 2)\vec{k}$
   $= 6\vec{i} - 2\vec{j} + 9\vec{k}$

9. $(\vec{i} + 5\vec{j} + 2\vec{k}) + 0.3(6\vec{i} - 2\vec{j} + 9\vec{k})$
   $= [1 + 0.3(6)]\vec{i} + [5 + 0.3(-2)]\vec{j} + [2 + 0.3(9)]\vec{k}$
   $= 2.8\vec{i} - 4.4\vec{j} + 4.7\vec{k}$

10. Answers will vary.

Exploration 88

1. $6 \cdot 8 \cdot \cos 0° = 6 \cdot 8 \cdot 1 = 48$

2. $6 \cdot 8 \cdot \cos 180° = 6 \cdot 8 \cdot (-1) = -48$

3. $6 \cdot 8 \cdot \cos 90° = 6 \cdot 8 \cdot 0 = 0$

4. $6 \cdot 8 \cdot \cos 50° = 30.8538\ldots$

5. $6 \cdot 8 \cdot \cos 110° = -16.4169\ldots$

6. $\vec{a} \cdot \vec{b} = |\vec{a}||\vec{b}|\cos \theta$, where θ is the angle between $\vec{a}$ and $\vec{b}$ placed tail-to-tail.

7. $\vec{i} \cdot \vec{i} = \vec{j} \cdot \vec{j} = \vec{k} \cdot \vec{k} = 1 \cdot 1 \cdot \cos 0° = 1 \cdot 1 \cdot 1 = 1$;
   $\vec{i} \cdot \vec{j} = \vec{j} \cdot \vec{k} = \vec{i} \cdot \vec{k} = 1 \cdot 1 \cdot \cos 90° = 1 \cdot 1 \cdot 0 = 0$

8. $\vec{a} \cdot \vec{b} = (2\vec{i} + 5\vec{j} + 7\vec{k}) \cdot (9\vec{i} + 3\vec{j} + 4\vec{k})$
   $= 2 \cdot 9 \cdot \vec{i} \cdot \vec{i} + 2 \cdot 3 \cdot \vec{i} \cdot \vec{j} + 2 \cdot 4 \cdot \vec{i} \cdot \vec{k}$
   $+ 5 \cdot 9 \cdot \vec{j} \cdot \vec{i} + 5 \cdot 3 \cdot \vec{j} \cdot \vec{j} + 5 \cdot 4 \cdot \vec{j} \cdot \vec{k}$
   $+ 7 \cdot 9 \cdot \vec{k} \cdot \vec{i} + 7 \cdot 3 \cdot \vec{k} \cdot \vec{j} + 7 \cdot 4 \cdot \vec{k} \cdot \vec{k}$
   $= 18 \cdot 1 + 6 \cdot 0 + 8 \cdot 0 + 45 \cdot 0 + 15 \cdot 1 + 20 \cdot 0$
   $+ 63 \cdot 0 + 21 \cdot 0 + 28 \cdot 1$
   $= 18 + 15 + 28 = 61$

9. Scalar product, inner product

10. $\vec{c} \cdot \vec{d} = 4 \cdot 2 - 6 \cdot 5 + 9 \cdot (-3) = -49$

11. $|\vec{c}| = \sqrt{4^2 + 6^2 + 9^2} = \sqrt{133}$;
    $|\vec{d}| = \sqrt{2^2 + 5^2 + 3^2} = \sqrt{38}$; $\vec{c} \cdot \vec{d} = |\vec{c}||\vec{d}| \cos \theta$
    $\Rightarrow \theta = \cos^{-1} \dfrac{\vec{c} \cdot \vec{d}}{|\vec{c}||\vec{d}|} = \cos^{-1} \dfrac{-49}{\sqrt{133} \cdot \sqrt{38}}$
    $= 133.5709\ldots°$

Exploration 89

1. $p = |\vec{a}| \cos \theta = 5 \cos 33° = 4.1933\ldots$

2. $|\vec{b}| = \sqrt{9^2 + 3^2 + 4^2} = \sqrt{106} = 10.2956\ldots$
   $\vec{u} = \dfrac{9}{\sqrt{106}}\vec{i} + \dfrac{3}{\sqrt{106}}\vec{j} + \dfrac{4}{\sqrt{106}}\vec{k}$

3. $\vec{p} = p\vec{u} = 4.1933\ldots\vec{u}$
   $= 3.6656\ldots\vec{i} + 1.2218\ldots\vec{j} + 1.6291\ldots\vec{k}$

4. $\vec{b} \cdot \vec{c} = 18 + 15 + 28 = 61$
   $|\vec{b}| = \sqrt{106}$
   $|\vec{c}| = \sqrt{2^2 + 5^2 + 7^2} = \sqrt{78}$
   $\theta = \cos^{-1} \dfrac{61}{\sqrt{106} \cdot \sqrt{78}} = 47.8667\ldots°$

5. $p = |\vec{c}| \cos \theta = \sqrt{78} \cos 47.8667\ldots°$
   $= 5.9248\ldots$

6. $\vec{p} = \dfrac{p}{|\vec{b}|}\vec{b} = \dfrac{5.9248\ldots}{\sqrt{106}}\vec{b} = 0.5754\ldots\vec{b}$
   $= 5.1792\ldots\vec{i} + 1.7264\ldots\vec{j} + 2.3018\ldots\vec{k}$

7.

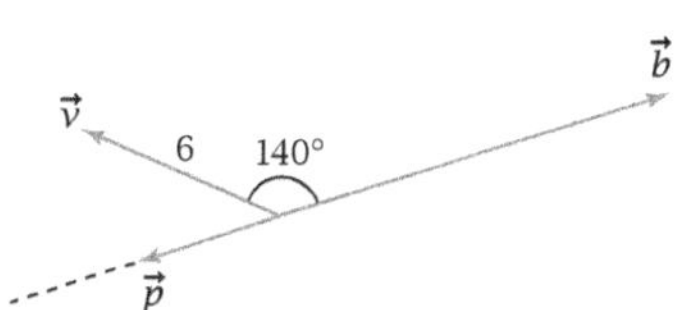

8. $p = |\vec{v}| \cos \theta = 6 \cos 140° = -4.5962\ldots$
   p is negative because $\vec{p}$ and $\vec{b}$ point in opposite directions.

9. $\vec{p} = \dfrac{p}{|\vec{b}|}\vec{b} = \dfrac{-4.5962\ldots}{\sqrt{106}}\vec{b} = -0.4464\ldots\vec{b}$
   $= -4.0178\ldots\vec{i} - 1.3392\ldots\vec{j} - 1.7857\ldots\vec{k}$

10. Scalar projections are negative if the angle is obtuse. Lengths of vectors are never negative.

11. $\vec{r} \cdot \vec{s} = 3 \cdot 4 - 5 \cdot 8 - 2 \cdot 7 = -42$

$|\vec{r}| = \sqrt{3^2 + 5^2 + 2^2} = \sqrt{38}$,

$|\vec{s}| = \sqrt{4^2 + 8^2 + 7^2} = \sqrt{129}$

$\theta = \cos^{-1} \dfrac{\vec{r} \cdot \vec{s}}{|\vec{r}||\vec{s}|} = \cos^{-1} \dfrac{-42}{\sqrt{38}\sqrt{129}} = 126.8611...°$

$p = |\vec{r}| \cos \theta = \sqrt{38} \cos 126.8611...°$

$\qquad = -3.6978...$

$\vec{p} = \dfrac{p}{|\vec{s}|}\vec{s} = \dfrac{-3.6978...}{\sqrt{129}}\vec{s} = -0.3255...\vec{s}$

$\qquad = -1.3023...\vec{i} - 2.6046...\vec{j} + 2.2790...\vec{k}$

12. $\vec{p} = \dfrac{\vec{r} \cdot \vec{s}}{|\vec{s}|^2}\vec{s} = \dfrac{-42}{129}\vec{s} = -0.3255...\vec{s}$

13. Answers will vary.

Exploration 90

1. $\vec{a} + \vec{b}$

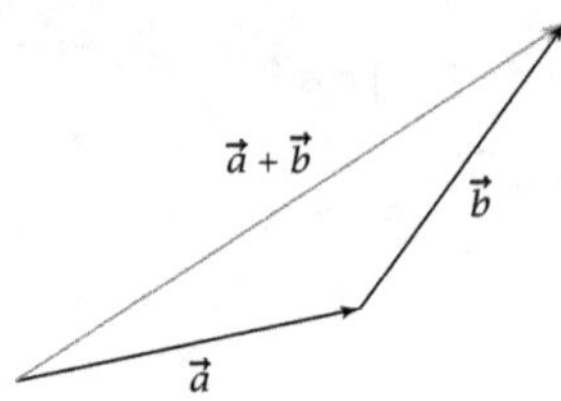

2. $\vec{a} - \vec{b}$

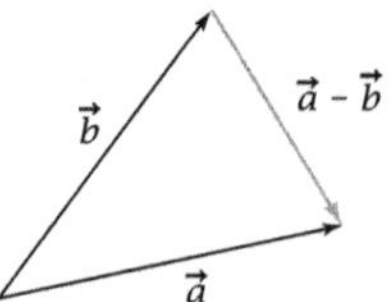

3. $\vec{a}$

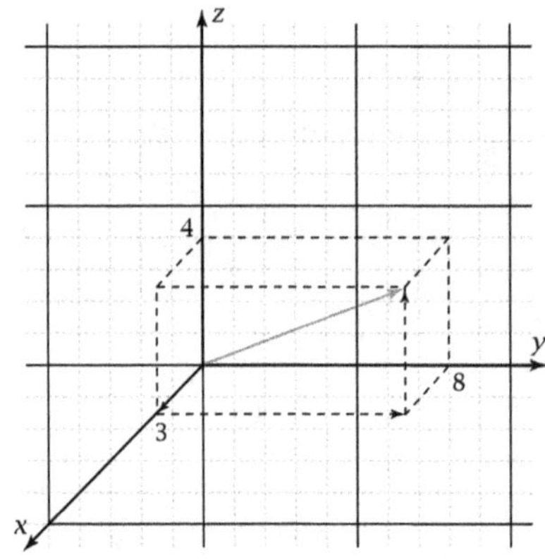

4. $|\vec{a}| = \sqrt{89}$, $|\vec{b}| = \sqrt{134}$

5. $\vec{a} + \vec{b} = 5\vec{i} + \vec{j} + 13\vec{k}$

$\vec{a} - \vec{b} = \vec{i} + 15\vec{j} - 5\vec{k}$

6. Distance $= |\vec{a} + \vec{b}| = \sqrt{195}$

7. Vector is $\vec{b} - \vec{a} = -\vec{i} - 15\vec{j} + 5\vec{k}$.

8. $\vec{a} \cdot \vec{b} = -14$

9. Scalar product, inner product

10. Obtuse. The dot product is negative.

11. By definition, $\vec{a} \cdot \vec{b} = |\vec{a}||\vec{b}| \cos \theta$

$\therefore \cos \theta = \dfrac{-14}{\sqrt{89}\sqrt{134}}$, so $\theta = 97.3654...°$.

12. $\vec{p}$

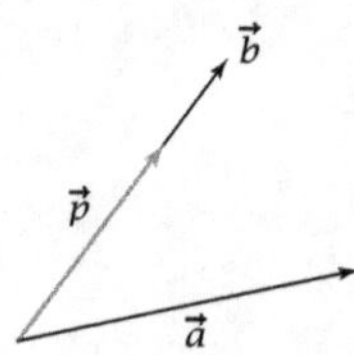

13. $p = |\vec{a}| \cos \theta = \sqrt{89} \cos 97.36...° = -1.2094...$

p is negative because $\vec{p}$ points in the opposite direction from $\vec{a}$ in this case.

14. $\vec{u} = \dfrac{\vec{b}}{|\vec{b}|} = \dfrac{1}{\sqrt{134}} (2\vec{i} - 7\vec{j} + 9\vec{k})$

$\qquad = \dfrac{2}{\sqrt{134}}\vec{i} - \dfrac{7}{\sqrt{134}}\vec{j} + \dfrac{9}{\sqrt{134}}\vec{k}$

$\qquad = 0.1727...\vec{i} - 0.6047...\vec{j} - 0.7774...\vec{k}$

15. $\vec{p} = p\vec{u}$

$\qquad = -1.2094...(0.172...\vec{i} - 0.604...\vec{j} + 0.777...\vec{k})$

$\qquad = -0.2089...\vec{i} - 0.7313...\vec{j} - 0.9402...\vec{k}$

16.

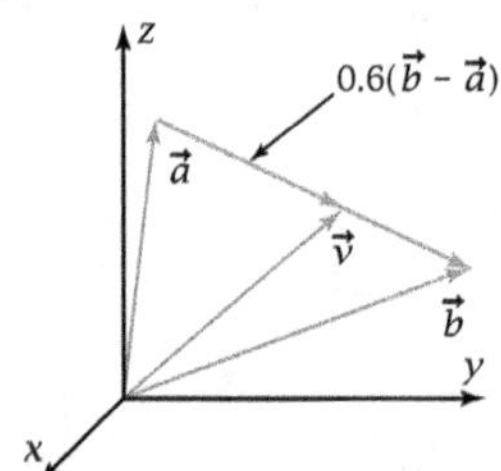

Vector from $\vec{a}$ to $\vec{b}$ is

$\vec{b} - \vec{a} = -\vec{i} - 15\vec{j} + 5\vec{k}$

0.6 of this is $0.6(-\vec{i} - 15\vec{j} + 5\vec{k}) = -0.6\vec{i} - 9\vec{j} + 3\vec{k}$

$\therefore \ \vec{v} = \vec{a} + (-0.6\vec{i} - 9\vec{j} + 3\vec{k})$

$\qquad = (3\vec{i} + 8\vec{j} + 4\vec{k}) + (-0.6\vec{i} - 9\vec{j} + 3\vec{k})$

$\qquad = 2.4\vec{i} - \vec{j} + 7\vec{k}$

17. Answers will vary.

Exploration 91

1.

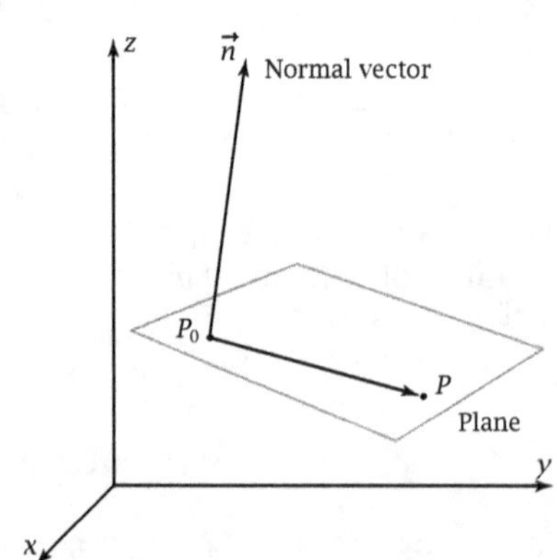

2. $\vec{n} \cdot \vec{P_0P} = |\vec{n}||\vec{P_0P}| \cos \theta$, where $\theta = 90°$.

So $\theta = 0$. Therefore, $\vec{n} \cdot \vec{P_0P} = 0$.

3. $P_0\vec{P} = \vec{P} - \vec{P}_0 = \left(x\vec{i} + y\vec{j} + z\vec{k}\right) - \left(8\vec{i} + 9\vec{j} + 4\vec{k}\right)$
   $= (x - 8)\vec{i} + (y - 9)\vec{j} + (z - 4)\vec{k}$

4. $\left(3\vec{i} + 7\vec{j} + 10\vec{k}\right) \cdot \left[(x - 8)\vec{i} + (y - 9)\vec{j} + (z - 4)\vec{k}\right] = 0$
   $\Rightarrow 3(x - 8) + 7(y - 9) + 10(z - 4) = 0$
   $\Rightarrow 3x + 7y + 10z = 127$

5. $3x + 7y + 10z = D$
   $\Rightarrow 3(8) + 7(9) + 10(4) = D$
   $\Rightarrow 24 + 63 + 40 = D$
   $\Rightarrow D = 127$
   $\therefore 3x + 7y + 10z = 127$

6. $3(5) + 7(2) + 10z = 127 \Rightarrow z = 9.8$

7. $3x + 7(0) + 10(0) = 127 \Rightarrow x = 42.3333\ldots$
   x-intercept is $(42.3333\ldots, 0, 0)$

8. A variable point in the plane means an arbitrary point $p(x, y, z)$ whose coordinates are not fixed except that they must satisfy the equation of the plane.

9. Answers will vary.

Exploration 92

1. $|\vec{a}| = \sqrt{3^2 + 5^2 + 7^2} = \sqrt{83}$;
   $|\vec{b}| = \sqrt{11^2 + 2^2 + 13^2} = \sqrt{294}$;
   $|\vec{a}||\vec{b}| = \sqrt{83} \cdot \sqrt{294} = \sqrt{24{,}402}$
   $|\vec{a} \times \vec{b}| = \sqrt{51^2 + 38^2 + 49^2} = \sqrt{6446} \neq |\vec{a}||\vec{b}|$

2. $\theta = \cos^{-1}\dfrac{\vec{a} \cdot \vec{b}}{|\vec{a}||\vec{b}|} = \cos^{-1}\dfrac{3 \cdot 11 + 5 \cdot 2 + 7 \cdot 13}{\sqrt{83} \cdot \sqrt{294}}$
   $= \cos^{-1}\dfrac{134}{\sqrt{24{,}402}} = 30.9282\ldots°$;
   $|\vec{a}||\vec{b}| \sin\theta = \sqrt{83} \cdot \sqrt{294} \cdot \sin 30.9282\ldots°$
   $= 80.2869\ldots = \sqrt{6446} = |\vec{a} \times \vec{b}|$

3. $\left(\vec{a} \times \vec{b}\right) \cdot \vec{a} = 51 \cdot 3 + 38 \cdot 5 - 49 \cdot 7 = 0$;
   $\left(\vec{a} \times \vec{b}\right) \cdot \vec{b} = 51 \cdot 11 + 38 \cdot 2 - 49 \cdot 13 = 0$; $\vec{a} \times \vec{b}$
   is perpendicular to both $\vec{a}$ and $\vec{b}$.

4. Student demonstration

5. $\vec{a} \times \vec{b}$ is a vector perpendicular to both $\vec{a}$ and $\vec{b}$, with direction given by the right-hand rule and with magnitude $|\vec{a} \times \vec{b}| = |\vec{a}||\vec{b}| \sin\theta$, where θ is the angle between $\vec{a}$ and $\vec{b}$, placed tail-to-tail.

6. $|\vec{i} \times \vec{i}| = |\vec{i}||\vec{i}| \sin 0° = 1 \cdot 1 \cdot 0 = 0$, and similarly for $\vec{j}$ and $\vec{k}$.

7. $|\vec{i} \times \vec{j}| = |\vec{i}||\vec{j}| \sin 90° = 1 \cdot 1 \cdot 1 = 1$, and when the fingers of your right hand curl from $\vec{i}$ to $\vec{j}$, your thumb points in the direction of $\vec{k}$; similarly for $\vec{j} \times \vec{k}$ and $\vec{k} \times \vec{i}$.

8. If you curl the fingers of your right hand from $\vec{j}$ to $\vec{i}$, your thumb points in the opposite direction than if you curl from $\vec{i}$ to $\vec{j}$.

9. $\vec{a} \times \vec{b} = \left(3\vec{i} + 5\vec{j} + 7\vec{k}\right) \times \left(11\vec{i} + 2\vec{j} + 13\vec{k}\right)$
   $= 3\vec{i} \times 11\vec{i} + 3\vec{i} \times 2\vec{j} + 3\vec{i} \times 13\vec{k}$
   $\quad + 5\vec{j} \times 11\vec{i} + 5\vec{j} \times 2\vec{j} + 5\vec{j} \times 13\vec{k}$
   $\quad + 7\vec{k} \times 11\vec{i} + 7\vec{k} \times 2\vec{j} + 7\vec{k} \times 13\vec{k}$
   $= 33\left(\vec{i} \times \vec{i}\right) + 6\left(\vec{i} \times \vec{j}\right) + 39\left(\vec{i} \times \vec{k}\right)$
   $\quad + 55\left(\vec{j} \times \vec{i}\right) + 10\left(\vec{j} \times \vec{j}\right) + 65\left(\vec{j} \times \vec{k}\right)$
   $\quad + 77\left(\vec{k} \times \vec{i}\right) + 14\left(\vec{k} \times \vec{j}\right) + 91\left(\vec{k} \times \vec{k}\right)$
   $= 33.0 + 6\vec{k} + 39\left(-\vec{j}\right)$
   $\quad + 55\left(-\vec{k}\right) + 10.0 + 65\vec{i}$
   $\quad + 77\vec{j} + 14\left(-\vec{i}\right) + 91.0$

10. $\vec{a} \times \vec{b} = 65\vec{i} - 14\vec{i} + 77\vec{j} - 39\vec{j} + 6\vec{k} - 55\vec{k}$
    $= 51\vec{i} + 38\vec{j} - 49\vec{k}$

11. The terms on the "inner" diagonal (top left to bottom right) vanish, leaving only the "outer" terms.

12. The result is a vector.

13. $\begin{vmatrix} \vec{i} & \vec{j} & \vec{k} \\ 3 & 5 & 7 \\ 11 & 2 & 13 \end{vmatrix} = \vec{i}\begin{vmatrix} 5 & 7 \\ 2 & 13 \end{vmatrix} - \vec{j}\begin{vmatrix} 3 & 7 \\ 11 & 13 \end{vmatrix} + \vec{k}\begin{vmatrix} 3 & 5 \\ 11 & 2 \end{vmatrix}$
    $= \vec{i}(5 \cdot 13 - 7 \cdot 2) - \vec{j}(3 \cdot 13 - 7 \cdot 11) + \vec{k}(3 \cdot 2 - 5 \cdot 11)$
    $= 51\vec{i} + 38\vec{j} - 49\vec{k}$

14. $\begin{vmatrix} \vec{i} & \vec{j} & \vec{k} \\ 4 & -5 & 8 \\ 3 & 6 & -7 \end{vmatrix} = \vec{i}(35 - 48) - \vec{j}(-28 - 24) + \vec{k}(24 + 15)$
    $= -13\vec{i} + 52\vec{j} + 39\vec{k}$

15. Answers will vary.

Exploration 93

1. $\overrightarrow{AB} = \vec{B} - \vec{A} = (4 - 5)\vec{i} + (-2 - 7)\vec{j} + (6 - 3)\vec{k} = -\vec{i} - 9\vec{j} + 3\vec{k}$;
   $\overrightarrow{AC} = \vec{C} - \vec{A} = (2 - 5)\vec{i} + (-6 - 7)\vec{j} + (1 - 3)\vec{k} = -3\vec{i} - 13\vec{j} - 2\vec{k}$

2. $\overrightarrow{AB} \cdot \overrightarrow{AC} = 1 \cdot (-3) - 9 \cdot (-13) + 3 \cdot (-2) = 114$

3. $\overrightarrow{AB} \cdot \overrightarrow{AC} = |\overrightarrow{AB}||\overrightarrow{AC}| \cos\theta$, so
   $\theta = \cos^{-1}\dfrac{\overrightarrow{AB} \cdot \overrightarrow{AC}}{|\overrightarrow{AB}||\overrightarrow{AC}|}$
   $= \cos^{-1}\dfrac{114}{\sqrt{1^2 + 9^2 + 3^2} \cdot \sqrt{3^2 + 13^2 + 2^2}}$
   $= \cos^{-1}\dfrac{114}{\sqrt{91} \cdot \sqrt{182}} = 27.6466\ldots°$

4. $\overrightarrow{AB} \times \overrightarrow{AC} = \begin{vmatrix} \vec{i} & \vec{j} & \vec{k} \\ -1 & -9 & 3 \\ -3 & -13 & -2 \end{vmatrix}$
   $= \vec{i}\begin{vmatrix} -9 & 3 \\ -13 & -2 \end{vmatrix} - \vec{j}\begin{vmatrix} -1 & 3 \\ -3 & -2 \end{vmatrix} + \vec{k}\begin{vmatrix} -1 & -9 \\ -3 & -13 \end{vmatrix}$
   $= \vec{i}[-9 \cdot (-2) - 3 \cdot (-13)] - \vec{j}[-1 \cdot (-2) - 3 \cdot (-3)]$
   $\quad + \vec{k}[-1 \cdot (-13) - (-9) \cdot (-3)]$
   $= \vec{i}(18 + 39) - \vec{j}(2 + 9) + \vec{k}(13 - 27)$
   $= 57\vec{i} - 11\vec{j} - 14\vec{k}$

5. $|\overrightarrow{AB} \times \overrightarrow{AC}| = \sqrt{57^2 + 11^2 + 14^2} = \sqrt{3566} = 59.7159\ldots$
   $|\overrightarrow{AB}||\overrightarrow{AC}| \sin\theta = \sqrt{91}\sqrt{182} \sin 27.6466\ldots° = 59.7159\ldots$

6. $\left(\overrightarrow{AB} \times \overrightarrow{AC}\right) \cdot \overrightarrow{AB} = 57(-1) - 11(-9) - 14 \cdot 3 = 0$;
   $\left(\overrightarrow{AB} \times \overrightarrow{AC}\right) \cdot \overrightarrow{AC} = 57(-3) - 11(-13) - 14(-2) = 0$

7. $\vec{n} = \overrightarrow{AB} \times \overrightarrow{AC} = 57\vec{i} - 11\vec{j} - 14\vec{k}$
   $57x - 11y - 14z = D$;
   $D = 57(5) - 11(7) - 14(3) = 166$;
   $57x - 11y - 14z = 166$

8. $57(2) - 11(-6) - 14(1) = 166$

9. $y = z = 0 \Rightarrow 57x - 11 \cdot 0 - 14 \cdot 0$
   $\Rightarrow 57x = 166 \Rightarrow x = \dfrac{166}{57} = 2.9122\ldots$

10. Let the base $b = |\overrightarrow{AB}|$ and h = altitude. Then

$$\frac{h}{|\overrightarrow{AC}|} = \sin\theta \Rightarrow h = |\overrightarrow{AC}|\sin\theta,$$

so $A = \frac{1}{2}bh = \frac{1}{2}|\overrightarrow{AB}| \cdot |\overrightarrow{AC}|\sin\theta$

$$= \frac{1}{2}|\overrightarrow{AB} \times \overrightarrow{AC}| = \frac{1}{2}\sqrt{3566} = 29.8579...$$

11. Answers will vary.

Exploration 94

1. $\vec{i} \cdot \vec{v} = |\vec{i}||\vec{v}|\cos\alpha = |\vec{v}|\cos\alpha$

$$\Rightarrow \alpha = \cos^{-1}\frac{\vec{i} \cdot \vec{v}}{|\vec{v}|} = \cos^{-1}\frac{1\cdot 4 + 0\cdot 5 + 0\cdot 12}{\sqrt{4^2 + 5^2 + 12^2}}$$

$$= \cos^{-1}\frac{4}{\sqrt{185}} = 72.8972...°;$$

similarly,

$$\beta = \cos^{-1}\frac{\vec{j} \cdot \vec{v}}{|\vec{v}|} = \cos^{-1}\frac{0\cdot 4 + 1\cdot 5 + 0\cdot 12}{\sqrt{185}}$$

$$= \cos^{-1}\frac{5}{\sqrt{185}} = 68.4318...°$$

$$\gamma = \cos^{-1}\frac{\vec{k} \cdot \vec{v}}{|\vec{v}|} = \cos^{-1}\frac{0\cdot 4 + 0\cdot 5 + 1\cdot 12}{\sqrt{185}}$$

$$= \cos^{-1}\frac{12}{\sqrt{185}} = 28.0840...°$$

2. $\cos^2\alpha + \cos^2\beta + \cos^2\gamma = 1$

3. $\cos\alpha = \frac{4}{\sqrt{185}}$; $\cos\beta = \frac{5}{\sqrt{185}}$; $\cos\gamma = \frac{12}{\sqrt{185}}$;

$$\left(\frac{4}{\sqrt{185}}\right)^2 + \left(\frac{5}{\sqrt{185}}\right)^2 + \left(\frac{12}{\sqrt{185}}\right)^2 = \frac{185}{185} = 1$$

4. $\vec{u} = \frac{\vec{v}}{|\vec{v}|} = \frac{\vec{v}}{\sqrt{185}} = \frac{4}{\sqrt{185}}\vec{i} + \frac{5}{\sqrt{185}}\vec{j} + \frac{12}{\sqrt{185}}\vec{k}$

$$= \cos\alpha\vec{i} + \cos\beta\vec{j} + \cos\gamma\vec{k} = c_1\vec{i} + c_2\vec{j} + c_3\vec{k}$$

5. $\left(\frac{9}{17}\right)^2 + \left(-\frac{8}{17}\right)^2 + \left(\frac{12}{17}\right)^2 = \frac{81}{289} + \frac{64}{289} + \frac{144}{289} = \frac{289}{289} = 1$

$$\cos\alpha = \frac{9}{17} \Rightarrow \alpha = \cos^{-1}\frac{9}{17} = 58.0342...°,$$

$$\cos\beta = \frac{8}{17} \Rightarrow \beta = \cos^{-1}\frac{-8}{17} = 118.0724...°,$$

$$\cos\gamma = \frac{12}{17} \Rightarrow \gamma = \cos^{-1}\frac{12}{17} = 45.0991...°,$$

6. $c_3 = \pm\sqrt{1 - c_1^2 - c_2^2}$

$$= \pm\sqrt{1 - (0.3)^2 - (-0.4)^2} = \pm\frac{\sqrt{3}}{2};$$

$\gamma = \cos^{-1}c_3 = 30°$ and $150°$

7. Answers will vary.

Exploration 95

1. $|\vec{u}| = \sqrt{\left(\frac{3}{7}\right)^2 + \left(\frac{6}{7}\right)^2 + \left(\frac{2}{7}\right)^2} = \sqrt{\frac{9}{49} + \frac{36}{49} + \frac{4}{49}} = \sqrt{\frac{49}{49}} = 1$

2. $c_1 = \frac{3}{7}$; $c_2 = \frac{6}{7}$; $c_3 = \frac{2}{7}$

3.

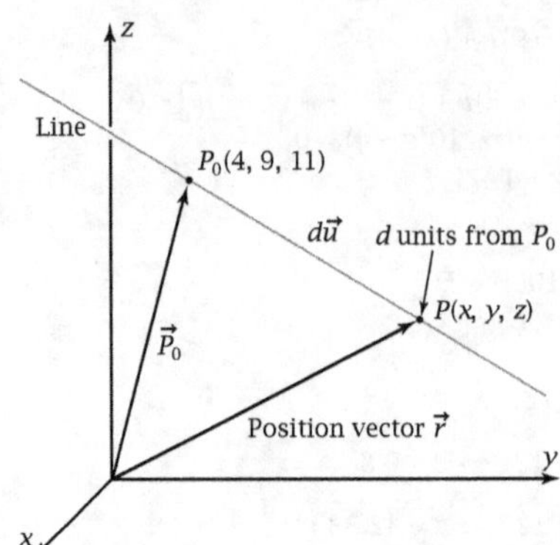

4. $\vec{r}$ is the resultant of $\vec{P}_0$, the position vector to P_0, followed by d repetitions of the unit vector in the direction of the line.

5. $\vec{r} = (4\vec{i} + 9\vec{j} + 11\vec{k}) + d\left(\frac{3}{7}\vec{i} + \frac{6}{7}\vec{j} + \frac{2}{7}\vec{k}\right)$

$$= \left(4 + \frac{3}{7}d\right)\vec{i} + \left(9 + \frac{6}{7}d\right)\vec{j} + \left(11 + \frac{2}{7}d\right)\vec{k}$$

6. $\vec{r}(21) = \left(4 + \frac{3}{7}\cdot 21\right)\vec{i} + \left(9 + \frac{6}{7}\cdot 21\right)\vec{j} + \left(11 + \frac{2}{7}\cdot 21\right)\vec{k}$

$$= 13\vec{i} + 27\vec{j} + 17\vec{k};\text{ the point } (13, 27, 17)$$

7. $x = 4 + \frac{3}{7}d = 50 \Rightarrow d = \frac{322}{3}$ units; it is in the direction $\vec{u}$ from P_0.

8. $\vec{r}\left(\frac{322}{3}\right) = \left(4 + \frac{3}{7}\cdot\frac{322}{3}\right)\vec{i} + \left(9 + \frac{6}{7}\cdot\frac{322}{3}\right)\vec{j} + \left(11 + \frac{2}{7}\cdot\frac{322}{3}\right)\vec{k}$

$$= 50\vec{i} + 101\vec{j} + \frac{125}{3}\vec{k};\text{ the point }\left(50, 101, \frac{125}{3}\right)$$

9. $z = 11 + \frac{2}{7}d = 0 \Rightarrow d = -\frac{77}{2};$

$$\vec{r}\left(-\frac{77}{2}\right) = \left(4 + \frac{3}{7}\cdot\frac{-77}{2}\right)\vec{i} + \left(9 + \frac{6}{7}\cdot\frac{-77}{2}\right)\vec{j} + \left(11 + \frac{2}{7}\cdot\frac{-77}{2}\right)\vec{k}$$

$$= -12.5\vec{i} - 24\vec{j} + 0\vec{k};\text{ the point } (-12.5, -24, 0)$$

10. Answers will vary.

Exploration 96

1. $|\vec{u}| = \sqrt{\left(\frac{3}{7}\right)^2 + \left(\frac{6}{7}\right)^2 + \left(\frac{2}{7}\right)^2} = \sqrt{\frac{9}{49} + \frac{36}{49} + \frac{4}{49}} = \sqrt{\frac{49}{49}} = 1$

2. $\vec{r} = \vec{P}_0 + d\vec{u}$

$$= (7\vec{i} + 11\vec{j} + 3\vec{k}) + d\left(\frac{3}{7}\vec{i} + \frac{6}{7}\vec{j} + \frac{2}{7}\vec{k}\right)$$

$$= \left(7 + \frac{3}{7}d\right)\vec{i} + \left(11 + \frac{6}{7}d\right)\vec{j} + \left(3 + \frac{2}{7}d\right)\vec{k}$$

3. $\overrightarrow{AB} = \vec{B} - \vec{A} = (6 - 5)\vec{i} + (3 - 10)\vec{j} + (6 - 1)\vec{k} = \vec{i} - 7\vec{j} + 5\vec{k};$

$\overrightarrow{AC} = \vec{C} - \vec{A}$

$$= (12 - 5)\vec{i} + (1 - 10)\vec{j} + (4 - 1)\vec{k}$$

$$= 7\vec{i} - 9\vec{j} + 3\vec{k}$$

4. $\vec{n} = \overrightarrow{AB} \times \overrightarrow{AC} = \begin{vmatrix} \vec{i} & \vec{j} & \vec{k} \\ 1 & -7 & 5 \\ 7 & -9 & 3 \end{vmatrix} = 24\vec{i} + 32\vec{j} + 40\vec{k}$

Divide by the LCD, 8, to get the easier-to-work-with $3\vec{i} + 4\vec{j} + 5\vec{k}$, which is still a normal because it points in the same direction.

5. $3x + 4y + 5z = D$;
$D = 3(5) + 4(10) + 5(1) = 60$;
$3x + 4y + 5z = 60$

6. $3\left(7 + \dfrac{3}{7}d\right) + 4\left(11 + \dfrac{6}{7}d\right) + 5\left(3 + \dfrac{2}{7}d\right) = 60$

$\Rightarrow 80 + \dfrac{43}{7}d = 60 \Rightarrow d = -\dfrac{140}{43}$

7. $\vec{r}\left(\dfrac{-140}{43}\right) = \left(7 + \dfrac{3}{7} \cdot \dfrac{-140}{43}\right)\vec{i} + \left(11 + \dfrac{6}{7} \cdot \dfrac{-140}{43}\right)\vec{j}$

$+ \left(3 + \dfrac{2}{7} \cdot \dfrac{-1405}{43}\right)\vec{k}$

$= \dfrac{241}{43}\vec{i} + \dfrac{353}{43}\vec{j} + \dfrac{89}{43}\vec{k}$;

$\left(\dfrac{241}{43}, \dfrac{353}{43}, \dfrac{89}{43}\right) = (5.6046\ldots, 8.2093\ldots, 2.0697\ldots)$

8. Answers will vary.

Exploration 97

1. $\vec{r}(2) = \vec{P}_0 + 2\vec{v}$
$= (4 + 2 \cdot 2)\vec{i} + (9 + 2 \cdot 5)\vec{j} + (11 - 2 \cdot 3)\vec{k}$
$= 8\vec{i} + 19\vec{j} + 5\vec{k}$
$\vec{r}(-1.7) = \vec{P}_0 - 1.7\vec{v}$
$= (4 - 1.7 \cdot 2)\vec{i} + (9 - 1.7 \cdot 5)\vec{j} + (11 + 1.7 \cdot 3)\vec{k}$
$= 0.6\vec{i} + 0.5\vec{j} + 16.1\vec{k}$

2. $\left|\vec{P_0P}\right| = \left|\vec{r}(2) - \vec{P}_0\right| = |2\vec{v}| = 2\sqrt{2^2 + 5^2 + 3^2} = 2\sqrt{38}$

3. Direction numbers are direction cosines multiplied by the length of the direction vector.

4. $\vec{r} = \vec{P}_0 + t\vec{v} = \left(4\vec{i} + 9\vec{j} + 11\vec{k}\right) + t\left(2\vec{i} + 5\vec{j} - 3\vec{k}\right)$
$= (4 + 2t)\vec{i} + (9 + 5t)\vec{j} + (11 - 3t)\vec{k}$

5. The equation in Problem 4 is easier to work with because it has only integer coefficients. But for this equation, the distance along the line equals the parameter d.

6. $x = 4 + 2 \cdot 2 = 8$
$y = 9 + 5 \cdot 2 = 19$
$z = 11 - 3 \cdot 2 = 5$
$\vec{r} = 8\vec{i} + 19\vec{j} + 6\vec{k}$, the same result as in Problem 1.

7. $x = 4 + 2t \Rightarrow t = \dfrac{x - 4}{2}$

$y = 9 + 5t \Rightarrow t = \dfrac{y - 9}{5}$

$z = 11 - 3t \Rightarrow t = \dfrac{z - 11}{-3}$

$t = \dfrac{x - 4}{2} = \dfrac{y - 9}{5} = \dfrac{z - 11}{-3}$

8. $\dfrac{7 - 4}{2} = \dfrac{z - 11}{-3} \Rightarrow z = -3 \cdot \dfrac{3}{2} + 11 = 6.5$

9. Parametric equations allow you to work more easily with individual coordinates.

10. For symmetric equations, you need only one coordinate to find both others (unless the line is parallel to a coordinate axis.)

11. Answers will vary.

Exploration 98

1. $x = 1$ and $y = 2 \Rightarrow 6(1) + 5(2) - 4z = -12$
$\Rightarrow 16 - 4z = -12 \Rightarrow z = 7; (1, 2, 7)$

2. $\vec{n} = 6\vec{i} + 5\vec{j} - 4\vec{k}; \vec{P} = \vec{i} + 2\vec{j} + 7\vec{k}$;
$\vec{n} \cdot \vec{P} = 6 \cdot 1 + 5 \cdot 2 - 4 \cdot 7 = -12$

3. Because $\vec{n} \cdot \vec{P}$ is negative, so is $\cos\theta$, where θ is the angle between $\vec{n}$ and $\vec{P}$, so θ is obtuse and is therefore θ_2 in the drawing, so $\vec{n}$ is $\vec{n}_2$ rather than $\vec{n}_1$. A positive normal vector is $\vec{n}_1 = -6\vec{i} - 5\vec{j} + 4\vec{k}$.

4. $-(6x + 5y - 4z) = -(-12)$
$\Rightarrow 6x - 5y + 4z = 12; D$ becomes positive.

5.

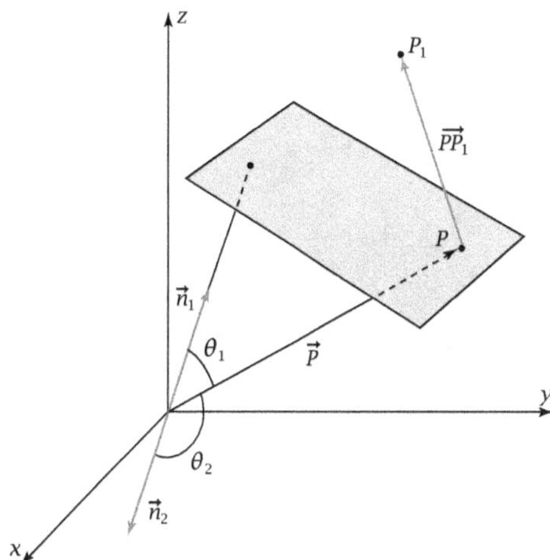

6. $\vec{PP_1} = \vec{P}_1 - \vec{P}$
$= (10 - 1)\vec{i} + (20 - 2)\vec{j} + (30 - 7)\vec{k}$
$= 9\vec{i} + 18\vec{j} + 23\vec{k}$
$\vec{PP_1} \cdot \vec{n} = 9 \cdot (-6) + 18 \cdot (-5) + 23 \cdot 4 = -52$

7. The angle is obtuse; $\vec{P}_1$ lies on the same side of the plane as the origin.

8. $\vec{n} = 3\vec{i} + 6\vec{j} + 2\vec{k}$ because $D = 36 > 0$.
Let $x = y = 0$. Then $3(0) + 6(0) + 2z = 36 \Rightarrow z = 18$, so
$P = (0, 0, 18)$ is a point on the plane.
$\vec{PP_0} = \vec{P}_0 - \vec{P} = (1 - 0)\vec{i} + (2 - 0)\vec{j} + (3 - 18)\vec{k}$
$= \vec{i} + 2\vec{j} - 15\vec{k}. \vec{n} \cdot \vec{PP_0} = 3 \cdot 1 + 6 \cdot 2 + 2 \cdot (-15)$
$= 3 \cdot 1 + 6 \cdot 2 + 2 \cdot (-15) = -15$
$\vec{P}_1$ lies on the same side of the plane as the origin.

9. Answers will vary.

Exploration 99

1. $x = 1$ and $y = 2 \Rightarrow 6(1) + 5(2) + 4z = 28$
$\Rightarrow 16 + 4z = 28 \Rightarrow z = 3; (1, 2, 3)$

2. $\vec{PP_1} = \vec{P}_1 - \vec{P}$
$= (11 - 1)\vec{i} + (22 - 2)\vec{j} + (14 - 3)\vec{k}$
$= 10\vec{i} + 20\vec{j} + 11\vec{k}$

3. $\vec{n} = 6\vec{i} + 5\vec{j} + 4\vec{k}$ because $D = 28 > 0$.

4. $\vec{n} \cdot \overrightarrow{PP_1} = |\vec{n}||\overrightarrow{PP_1}| \cos\theta \Rightarrow \theta = \cos^{-1}\dfrac{\vec{n} \cdot \overrightarrow{PP_1}}{|\vec{n}||\overrightarrow{PP_1}|}$

$\qquad = \cos^{-1}\dfrac{6 \cdot 10 + 5 \cdot 20 + 4 \cdot 11}{\sqrt{6^2 + 5^2 + 4^2} \cdot \sqrt{10^2 + 20^2 + 11^2}}$

$\qquad = \cos^{-1}\dfrac{204}{\sqrt{77} \cdot \sqrt{621}} = 21.1071\ldots°$

5. They are alternate interior angles (because $\vec{d}$ is parallel to $\vec{n}$).

6. $\dfrac{d}{|\overrightarrow{PP_1}|} = \cos\theta \Rightarrow d = |\overrightarrow{PP_1}| \cos\theta$

$\qquad\qquad\qquad = \sqrt{621} \cos 21.1071\ldots° = 23.2479\ldots$

7. $d = |\overrightarrow{PP_1}| \cos\theta = |\overrightarrow{PP_1}| \cdot \dfrac{\vec{n} \cdot \overrightarrow{PP_1}}{|\vec{n}||\overrightarrow{PP_1}|} = \dfrac{\vec{n} \cdot \overrightarrow{PP_1}}{|\vec{n}|}$

$\qquad = \dfrac{204}{\sqrt{77}} = 23.2479\ldots$

8. $\vec{n} = 7\vec{i} + 3\vec{j} + 10\vec{k}$ because $D = 30 > 0$;
$7(1) + 3(1) + 10z = 30 \Rightarrow 10 + 10z = 30 \Rightarrow z = 2$;
$P = (1, 1, 2)$, or $\vec{P} = \vec{i} + \vec{j} + 2\vec{k}$;
$\overrightarrow{PP_1} = \vec{P_1} - \vec{P}$

$\qquad = (-9 - 1)\vec{i} + (-2 - 1)\vec{j} + (17 - 2)\vec{k}$

$\qquad = -10\vec{i} - 3\vec{j} + 15\vec{k}; \ d = \dfrac{\vec{n} \cdot \overrightarrow{PP_1}}{|\vec{n}|}$

$\qquad = \dfrac{7 \cdot (-10) + 3 \cdot (-3) + 10 \cdot 15}{\sqrt{7^2 + 3^2 + 10^2}} = \dfrac{71}{\sqrt{158}} = 5.6484\ldots$

9. Answers will vary.

Exploration 100

1. $\dfrac{d}{|\overrightarrow{PP_1}|} = \sin\theta \Rightarrow d = |\overrightarrow{PP_1}| \sin\theta$

2. $d = |\overrightarrow{PP_1}| \sin\theta = \dfrac{|\vec{v}||\overrightarrow{PP_1}| \sin\theta}{|\vec{v}|} = \dfrac{|\vec{v} \times \overrightarrow{PP_1}|}{|\vec{v}|}$

3. So it won't be confused with d for "distance" (In applications, t is often used because the line describes the motion of a particle and t represents *time*.)

4. $P = (5, 3, -1)$; $\vec{v} = \dfrac{6}{11}\vec{i} - \dfrac{2}{11}\vec{j} + \dfrac{9}{11}\vec{k}$

5. $\overrightarrow{PP_1} = \vec{P_1} - \vec{P}$

$\qquad = (-3 - 5)\vec{i} + (-2 - 3)\vec{j} + [5 - (-1)]\vec{k}$

$\qquad = -8\vec{i} - 1\vec{j} + 6\vec{k}$

6. $\vec{v} \times \overrightarrow{PP_1} = \left(\dfrac{6}{11}\vec{i} - \dfrac{2}{11}\vec{j} + \dfrac{9}{11}\vec{k}\right) \times \left(-8\vec{i} - \vec{j} + 6\vec{k}\right)$

$\qquad = \dfrac{1}{11}\left(6\vec{i} - 2\vec{j} + 9\vec{k}\right) \times \left(-8\vec{i} - \vec{j} + 6\vec{k}\right)$

$\qquad = \dfrac{1}{11}\begin{vmatrix} \vec{i} & \vec{j} & \vec{k} \\ 6 & -2 & 9 \\ -8 & -1 & 6 \end{vmatrix} = \dfrac{1}{11}\left(-3\vec{i} - 108\vec{j} - 22\vec{k}\right)$

$\qquad = \dfrac{-3}{11}\vec{i} - \dfrac{108}{11}\vec{j} - 2\vec{k};$

$|\vec{v} \times \overrightarrow{PP_1}| = \left|\dfrac{1}{11}\left(12\vec{i} - 27\vec{j} - 14\vec{k}\right)\right|$

$\qquad = \dfrac{1}{11}\sqrt{(-3)^2 + 108^2 + 22^2} = \dfrac{1}{11}\sqrt{12,157}$

$\qquad = 10.0235\ldots$

7. $|\vec{v}| = \left|\dfrac{1}{11}\left(6\vec{i} - 2\vec{j} + 9\vec{k}\right)\right| = \dfrac{1}{11}\sqrt{6^2 + 2^2 + 9^2}$

$\qquad = \dfrac{1}{11}\sqrt{121} = \dfrac{11}{11} = 1$; $\vec{v}$ is a unit vector.

$d = \dfrac{|\vec{v} \times \overrightarrow{PP_1}|}{|\vec{v}|} = |\vec{v} \times \overrightarrow{PP_1}| = 5.9446\ldots$), from Problem 6.

8. Answers will vary.

Exploration 101

1. The tail is at point $(0, 20, 0)$; the head is at point $(6, 15, 4)$.
$\vec{v}_1 = (6 - 0)\vec{i} + (15 - 20)\vec{j} + (4 - 0)\vec{k}$
$\qquad = 6\vec{i} - 5\vec{j} + 4\vec{k}$

2. $|\vec{v}_1| = \sqrt{6^2 + 5^2 + 4^2} = \sqrt{77} = 8.7749\ldots$ ft

3. $\vec{v}_2 = 12\vec{i} + 0\vec{j} + 0\vec{k} = 12\vec{i}$

4. $\vec{v}_1 \cdot \vec{v}_2 = 6 \cdot 12 - 5 \cdot 0 + 4 \cdot 0 = 72$
$\theta = \cos^{-1}\dfrac{\vec{v}_1 \cdot \vec{v}_2}{|\vec{v}_1||\vec{v}_2|} = \cos^{-1}\dfrac{72}{\sqrt{77} \cdot 12} = \cos^{-1}\dfrac{6}{\sqrt{77}} = 46.8615\ldots°$

5. $\vec{v}_1 \times \vec{v}_2 = \begin{vmatrix} \vec{i} & \vec{j} & \vec{k} \\ 6 & -5 & 4 \\ -12 & 0 & 0 \end{vmatrix} = 0\vec{i} + 48\vec{j} + 60\vec{k} = 48\vec{j} + 60\vec{k}$

6. $A = \dfrac{1}{2}|\vec{v}_1 \times \vec{v}_2| = \dfrac{1}{2}\sqrt{48^2 + 60^2} = 38.4187\ldots$ ft^2

7. $h = \sqrt{5^2 + 4^2} = \sqrt{41} = 6.4031\ldots$ ft

8. No. The trapezoidal sides slant at a different angle (slope = 4/6) from the triangular ends (slope = 4/5), so the slant heights will be different.

9. $\dfrac{\vec{v}_1}{|\vec{v}_1|} = \dfrac{6\vec{i} - 5\vec{j} + 4\vec{k}}{\sqrt{77}} = \dfrac{6}{\sqrt{77}}\vec{i} - \dfrac{5}{\sqrt{77}}\vec{j} + \dfrac{4}{\sqrt{77}}\vec{k}$

10. $\vec{r} = \dfrac{6d}{\sqrt{77}}\vec{i} + \left(20 - \dfrac{5d}{\sqrt{77}}\right)\vec{j} + \dfrac{4d}{\sqrt{77}}\vec{k}$

11. $|\vec{r}(3)| = \sqrt{20^2 - \dfrac{600}{\sqrt{77}} + 3^2} = 18.4559\ldots$ ft

12. Student project

13. Yes

14. Yes

15. Yes

16. Answers will vary.

Exploration 102

1.

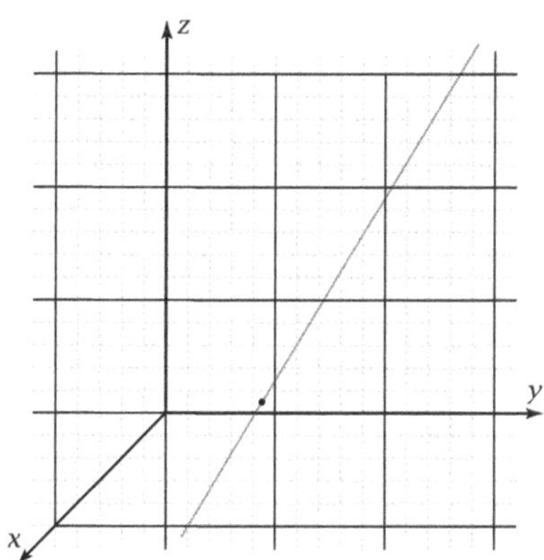

For ease in visualization, the point P_0 from Problem 2 is shown, another point on the line is shown, the height of each point above the xy-plane is shown, and the part of the line below the xy-plane is dotted. Student sketch need not be so detailed.

2. $(5, 7, 3)$

3. $\vec{r}(34) = \left(5 + \dfrac{12}{17} \cdot 34\right)\vec{i} + \left(7 + \dfrac{8}{17} \cdot 34\right)\vec{j} + \left(3 + \dfrac{9}{17} \cdot 34\right)\vec{k}$
   $= 29\vec{i} + 23\vec{j} + 21\vec{k}; \ (29, 23, 21)$

4. $\dfrac{12}{17}\vec{i} + \dfrac{8}{17}\vec{j} + \dfrac{9}{17}\vec{k}$

5. $c_1 = \cos\alpha = \dfrac{12}{17}; \ c_2 = \cos\beta = \dfrac{8}{17}; \ c_3 = \cos\gamma = \dfrac{9}{17};$
   $\cos^2\alpha + \cos^2\beta + \cos^2\gamma = \left(\dfrac{12}{17}\right)^2 + \left(\dfrac{8}{17}\right)^2 + \left(\dfrac{9}{17}\right)^2$
   $= \dfrac{144}{289} + \dfrac{64}{289} + \dfrac{81}{289} = \dfrac{289}{289} = 1$

6. $\alpha = \cos^{-1} c_1 = 45.0991...°;$
   $\beta = \cos^{-1} c_2 = 61.9275...°;$
   $\gamma = \cos^{-1} c_3 = 58.0342...°$

7. $y = 7 + \dfrac{8}{17}d = 23 \Rightarrow d = 34$, so the point is $(29, 23, 21)$, from Problem 3.

8. The point $(29, 23, 21)$ is on the line, but the equation for the line does not have a constant y-coordinate ($\vec{j}$-component).

9.

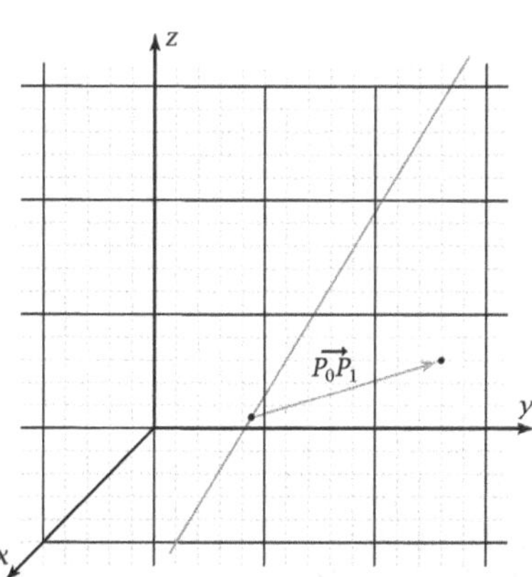

For ease in visualization, an additional point on the line is shown, distances of points above the xy-plane are shown,

and the part of the line that is below the xy-plane is dotted. Student sketches need not be so detailed.

$\overrightarrow{P_0P_1} = \vec{P_1} - \vec{P_0}$
$= (20 - 5)\vec{i} + (23 - 7)\vec{j} + (13 - 3)\vec{k}$
$= 15\vec{i} + 16\vec{j} + 10\vec{k}$

10. $\theta = \cos^{-1} \dfrac{\overrightarrow{P_0P_1} \cdot \vec{u}}{|\overrightarrow{P_0P_1}||\vec{u}|} = \cos^{-1} \dfrac{\overrightarrow{P_0P_1} \cdot \vec{u}}{|\overrightarrow{P_0P_1}|}$

$= \cos^{-1} \dfrac{\left(15\vec{i} + 16\vec{j} + 10\vec{k}\right) \cdot \dfrac{1}{17}\left(12\vec{i} + 8\vec{j} + 9\vec{k}\right)}{|15\vec{i} + 16\vec{j} + 10\vec{k}|}$

$= \cos^{-1} \dfrac{\dfrac{1}{17}(15 \cdot 12 + 16 \cdot 8 + 10 \cdot 9)}{\sqrt{15^2 + 16^2 + 10^2}}$

$= \cos^{-1} \dfrac{\dfrac{398}{17}}{\sqrt{581}} = 13.7640...°$

11. $\dfrac{d}{|\overrightarrow{P_0P_1}|} = \sin\theta \Rightarrow d = |\overrightarrow{P_0P_1}| \sin\theta$
    $= \sqrt{581} \sin 13.7640...° = 5.7349...$

12. $\overrightarrow{P_0P_1} \times \vec{u} = \left(15\vec{i} + 16\vec{j} + 10\vec{k}\right) \times \dfrac{1}{17}\left(12\vec{i} + 8\vec{j} + 9\vec{k}\right)$

$= \dfrac{1}{17}\begin{vmatrix} \vec{i} & \vec{j} & \vec{k} \\ 15 & 16 & 10 \\ 12 & 8 & 9 \end{vmatrix} = \dfrac{1}{17}\left(64\vec{i} - 15\vec{j} - 72\vec{k}\right)$

$= \dfrac{64}{17}\vec{i} - \dfrac{15}{17}\vec{j} - \dfrac{72}{17}\vec{k}$

13. $|\overrightarrow{P_0P_1} \times \vec{u}| = \left|\dfrac{1}{17}\left(64\vec{i} - 15\vec{j} - 72\vec{k}\right)\right|$

$= \dfrac{1}{17}\sqrt{64^2 + 15^2 + 72^2}$

$= \dfrac{1}{17}\sqrt{9505} = 5.7349...$

14. $|\vec{a} \times \vec{b}| = |\vec{a}||\vec{b}| \sin\theta; \ d = |\overrightarrow{P_0P_1}| \sin\theta$
$= \dfrac{|\overrightarrow{P_0P_1}||\vec{u}| \sin\theta}{|\vec{u}|} = \dfrac{|\overrightarrow{P_0P_1} \times \vec{u}|}{1} = |\overrightarrow{P_0P_1} \times \vec{u}|$

15. $2(4) + 4(1) + 3(8) = 36$

16. $2(9) + 4(7) + 3(17) = 97 \neq 36$

17. $2\vec{i} + 4\vec{j} + 7\vec{k}$

18.

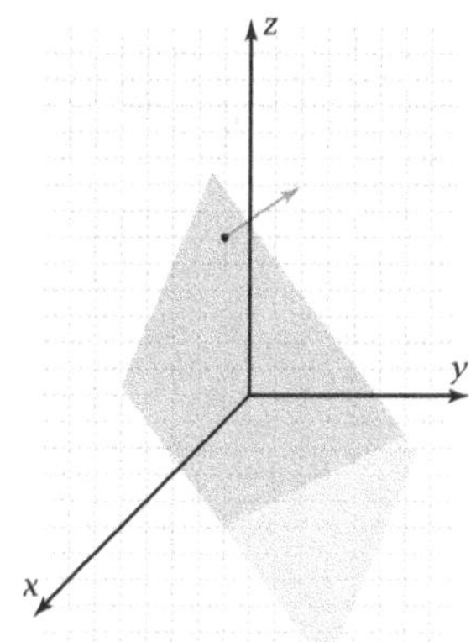

The portion of the plane below the xy-plane is drawn lighter.

19. Positive; $D = 36 > 0$

20.

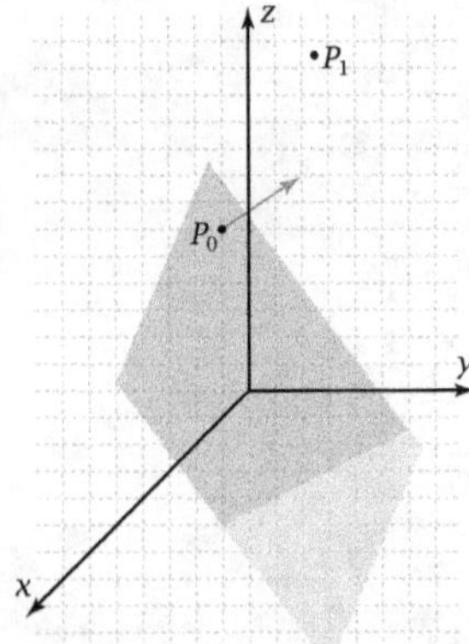

For ease in visualization, the heights of all points above the *xy*-plane are shown, the projection of the given plane onto the *xy*-plane is shown, and the portion of the given plane that is below the *xy*-plane is lighter. Student sketches need not be so detailed.

21. $\overrightarrow{P_0P_1} = \vec{P}_1 - \vec{P}_0 = (9-4)\vec{i} + (7-1)\vec{j} + (17-8)\vec{k}$
    $= 5\vec{i} + 6\vec{j} + 9\vec{k}$

22. $\theta = \cos^{-1}\dfrac{\vec{n}\cdot\overrightarrow{P_0P_1}}{|\vec{n}||\overrightarrow{P_0P_1}|}$

    $= \cos^{-1}\dfrac{2\cdot 5 + 4\cdot 6 + 3\cdot 9}{\sqrt{2^2+4^2+3^2}\cdot\sqrt{5^2+6^2+9^2}}$

    $= \cos^{-1}\dfrac{61}{\sqrt{29}\cdot\sqrt{142}} = 18.0889...°$

23. $\dfrac{d}{|\overrightarrow{P_0P_1}|} = \cos\theta \Rightarrow d = |\overrightarrow{P_0P_1}|\cos\theta$

    $= \sqrt{142}\cdot\cos 18.0889...° = 11.3274...$

24. $\dfrac{|Ax_1 + By_1 + Cz_1 - D|}{\sqrt{A^2+B^2+C^2}} = \dfrac{2(9)+4(7)+3(17)-36}{\sqrt{2^2+4^2+3^2}}$

    $= \dfrac{61}{\sqrt{29}} = 11.3274...$

25. Opposite side; d is positive.

26. $2\left(5 + \dfrac{12}{17}d\right) + 4\left(7 + \dfrac{8}{17}d\right) + 3\left(3 + \dfrac{9}{17}d\right) = 36$

    $\Rightarrow 47 + \dfrac{83}{17}d = 36 \Rightarrow d = -\dfrac{187}{83}$

27. $\vec{r}\left(-\dfrac{187}{83}\right)$

    $= \left(5 + \dfrac{12}{17}\cdot\dfrac{-187}{83}\right)\vec{i} + \left(7 + \dfrac{8}{17}\cdot\dfrac{-187}{83}\right)\vec{j} + \left(3 + \dfrac{9}{17}\cdot\dfrac{-187}{83}\right)\vec{k}$

    $= \dfrac{283}{83}\vec{i} + \dfrac{493}{83}\vec{j} + \dfrac{150}{83}\vec{k}$

    $= (3.4096...)\vec{i} + (5.9397...)\vec{j} + (1.8072...)\vec{k}; \left(\dfrac{283}{83}, \dfrac{493}{83}, \dfrac{150}{83}\right)$

    Check: $2\left(\dfrac{283}{83}\right) + 4\left(\dfrac{493}{83}\right) + 3\left(\dfrac{150}{83}\right) = \dfrac{2988}{83} = 36$

28. Answers will vary.

Matrix Transformations and Fractal Figures

Exploration 103

1. This matrix is called a $\underline{3} \times \underline{4}$ matrix.

2. $\begin{bmatrix} 3 & 2 \\ 4 & 5 \\ 1 & 8 \end{bmatrix}\begin{bmatrix} 1 & 3 & 7 & 2 \\ 4 & 5 & 4 & 1 \end{bmatrix} = \begin{bmatrix} 11 & 19 & 29 & 8 \\ 24 & 37 & 48 & 13 \\ 33 & 43 & 39 & 10 \end{bmatrix}$

3. The number of columns in the first matrix is different from the number of rows in the second matrix.

4. $\begin{bmatrix} 3 & 7 \\ 4 & 2 \end{bmatrix}\begin{bmatrix} 5 & 6 \\ 1 & 8 \end{bmatrix} = \begin{bmatrix} 22 & 74 \\ 22 & 40 \end{bmatrix}$

5. $\begin{bmatrix} 5 & 6 \\ 1 & 8 \end{bmatrix}\begin{bmatrix} 3 & 7 \\ 4 & 2 \end{bmatrix} = \begin{bmatrix} 39 & 47 \\ 35 & 23 \end{bmatrix}$

6. The dimensions of the matrices may not allow commuting the matrices. Commuting the matrices generally results in different linear combinations of elements that become the elements of the product matrix.

7. $\begin{bmatrix} 9 & 4 & 4 \\ 5 & 2 & 1 \\ 4 & 3 & 6 \end{bmatrix}\begin{bmatrix} 1 & 0 & 0 \\ 0 & 1 & 0 \\ 0 & 0 & 1 \end{bmatrix} = \begin{bmatrix} 9 & 4 & 4 \\ 5 & 2 & 1 \\ 4 & 3 & 6 \end{bmatrix}$

8. $\begin{bmatrix} 1 & 0 & 0 \\ 0 & 1 & 0 \\ 0 & 0 & 1 \end{bmatrix}\begin{bmatrix} 9 & 4 & 4 \\ 5 & 2 & 1 \\ 4 & 3 & 6 \end{bmatrix} = \begin{bmatrix} 9 & 4 & 4 \\ 5 & 2 & 1 \\ 4 & 3 & 6 \end{bmatrix}$

9. Multiplying it by any 3×3 matrix, A, gives A as the answer.

10. $\begin{bmatrix} 9 & 4 & 4 \\ 5 & 2 & 1 \\ 4 & 3 & 6 \end{bmatrix}\begin{bmatrix} 1.8 & -2.4 & -0.8 \\ -5.2 & 7.6 & 2.2 \\ 1.4 & -2.2 & -0.4 \end{bmatrix} = \begin{bmatrix} 1 & 0 & 0 \\ 0 & 1 & 0 \\ 0 & 0 & 1 \end{bmatrix}$

11. $\begin{bmatrix} 1.8 & -2.4 & -0.8 \\ -5.2 & 7.6 & 2.2 \\ 1.4 & -2.2 & -0.4 \end{bmatrix}\begin{bmatrix} 9 & 4 & 4 \\ 5 & 2 & 1 \\ 4 & 3 & 6 \end{bmatrix} = \begin{bmatrix} 1 & 0 & 0 \\ 0 & 1 & 0 \\ 0 & 0 & 1 \end{bmatrix}$

12. The product of the two is the identity.

13. $[A]^{-1} = \begin{bmatrix} 1.8 & -2.4 & -0.8 \\ -5.2 & 7.6 & 2.2 \\ 1.4 & -2.2 & -0.4 \end{bmatrix}$

14. Answers will vary.

Exploration 104

1. $\begin{vmatrix} 7 & 9 \\ -6 & 8 \end{vmatrix} = 7(8) - 9(-6) = 110$

2. $2\begin{vmatrix} 5 & 7 \\ -8 & -6 \end{vmatrix} - 3\begin{vmatrix} -1 & 7 \\ 9 & -6 \end{vmatrix} + (-4)\begin{vmatrix} -1 & 5 \\ 9 & -8 \end{vmatrix}$
   $= 2[5(-6) - 7(-8)] - 3[-1(-6) - 7(9)] + (-4)[-1(-8) - 5(9)]$
   $= 2(26) - 3(-57) - 4(-37) = 371$

3. $\begin{vmatrix} 7 & -3 & -4 \\ 3 & -8 & -6 \\ -2 & 1 & -5 \end{vmatrix}$

   $= 7\begin{vmatrix} -8 & -6 \\ 1 & -5 \end{vmatrix} - (-3)\begin{vmatrix} 3 & -6 \\ -2 & -5 \end{vmatrix} + (-4)\begin{vmatrix} 3 & -8 \\ -2 & 1 \end{vmatrix}$
   $= 7[-8(-5) - (-6)(1)] + 3[3(-5) - (-6)(-2)] - 4[3(1) - (-8)(-2)]$
   $= 7(46) + 3(-27) - 4(-13) = 293$

4. Entering the matrix as [A], det([A]) = 293.

5. Answers will vary.

Exploration 105

1. Switch the rows with the columns; that is, the rows of $[M]^T$ are the same as the columns of $[M]$, and vice versa.

2. $\begin{bmatrix} -2 & -5 & -6 \\ 3 & -1 & 8 \\ 4 & -2 & 7 \end{bmatrix} \Rightarrow \begin{bmatrix} 3 & 8 \\ 4 & 7 \end{bmatrix} = 3(7) - 8(4) = -11$

3. $\begin{vmatrix} 1 & -8 \\ -2 & 7 \end{vmatrix} = 7 - 16 = -9$

   $\begin{vmatrix} -3 & -8 \\ 4 & 7 \end{vmatrix} = -21 - (-32) = 11$

4. $\begin{bmatrix} +\begin{vmatrix} 1 & 8 \\ 2 & 7 \end{vmatrix} & -\begin{vmatrix} 3 & 8 \\ 4 & 7 \end{vmatrix} & +\begin{vmatrix} 3 & 1 \\ 4 & 2 \end{vmatrix} \\ -\begin{vmatrix} 5 & 6 \\ 2 & 7 \end{vmatrix} & +\begin{vmatrix} 2 & 6 \\ 4 & 7 \end{vmatrix} & -\begin{vmatrix} 2 & 5 \\ 4 & 2 \end{vmatrix} \\ +\begin{vmatrix} 5 & 6 \\ 1 & 8 \end{vmatrix} & -\begin{vmatrix} 2 & 6 \\ 3 & 8 \end{vmatrix} & +\begin{vmatrix} 2 & 5 \\ 3 & 1 \end{vmatrix} \end{bmatrix}$

   $= \begin{bmatrix} +(-9) & -(-11) & +(2) \\ -(23) & -(-10) & -(-16) \\ +(34) & -(-2) & +(-13) \end{bmatrix}$

   $= \begin{bmatrix} -9 & 11 & 2 \\ -23 & -10 & 16 \\ 34 & 2 & -13 \end{bmatrix}$

5. $\begin{vmatrix} 2 & 3 & 4 \\ 5 & 1 & 2 \\ 6 & 8 & 7 \end{vmatrix} = 2\begin{vmatrix} 1 & 2 \\ 8 & 7 \end{vmatrix} - 3\begin{vmatrix} 5 & 2 \\ 6 & 7 \end{vmatrix} + 4\begin{vmatrix} 5 & 1 \\ 6 & 8 \end{vmatrix}$

   $= 2(-9) - 3(23) + 4(34) = 49$

6. The result should equal adj[M].

7. Answers will vary.

Exploration 106

1. Each column represents the coordinates of a vertex of the triangle: (1, 1), (4, 2), and (1, 3).

2. $[A][M] = \begin{bmatrix} 2 & 0 \\ 0 & 2 \end{bmatrix}\begin{bmatrix} 1 & 4 & 1 \\ 1 & 2 & 3 \end{bmatrix} = \begin{bmatrix} 2 & 8 & 2 \\ 2 & 4 & 6 \end{bmatrix}$

3. The number of columns of [M] is different from the number of rows of [A].

4.

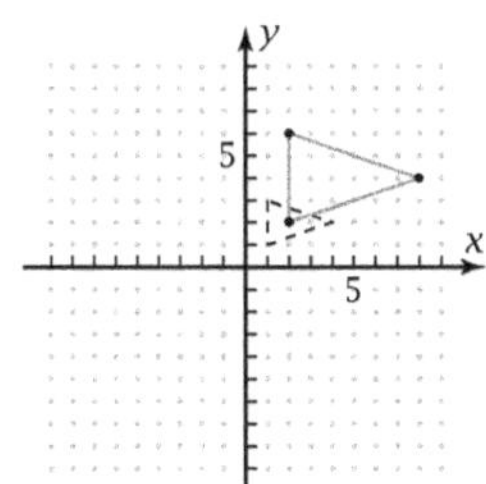

5. Dilation by 2

6. The image is the result of the transformation; [M] is what you have before the transformation.

7. $[B] = \begin{bmatrix} \cos 90° & \cos 180° \\ \sin 90° & \sin 180° \end{bmatrix} = \begin{bmatrix} 0 & -1 \\ 1 & 0 \end{bmatrix}$

   $[B][M] = \begin{bmatrix} 0 & -1 \\ 1 & 0 \end{bmatrix}\begin{bmatrix} 1 & 4 & 1 \\ 1 & 2 & 3 \end{bmatrix} = \begin{bmatrix} -1 & -2 & -3 \\ 1 & 4 & 1 \end{bmatrix}$

   Rotation counterclockwise about the origin

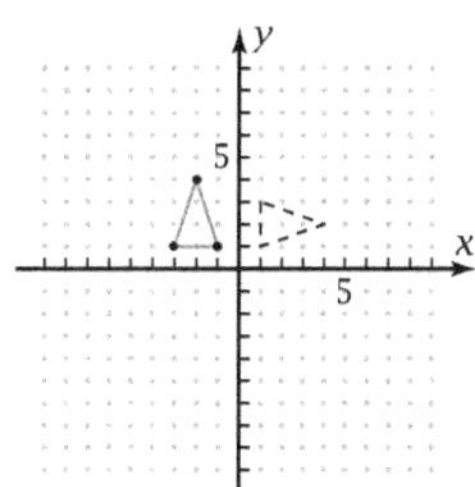

8. $[C] = \begin{bmatrix} \cos 40° & \cos 130° \\ \sin 40° & \sin 130° \end{bmatrix} \approx \begin{bmatrix} 0.8 & -0.6 \\ 0.6 & 0.8 \end{bmatrix}$;

   $[C][M] \approx \begin{bmatrix} 0.1 & 1.8 & -1.2 \\ 1.4 & 4.1 & 2.9 \end{bmatrix}$

9. The pre-image was rotated 40°.

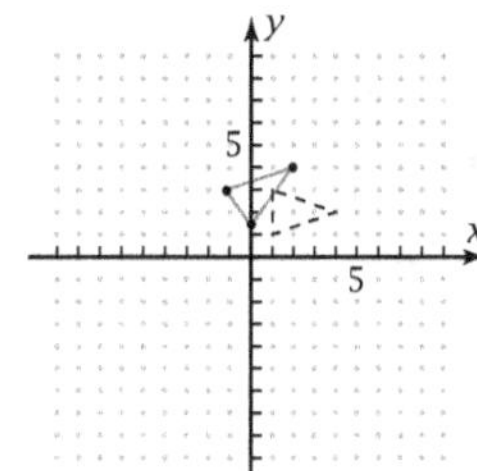

10. $[D] = \begin{bmatrix} 2 & 0 \\ 0 & 2 \end{bmatrix}\begin{bmatrix} \cos 240° & \cos 330° \\ \sin 240° & \sin 330° \end{bmatrix}$

    $= \begin{bmatrix} 2\cos 240° & 2\cos 330° \\ 2\sin 240° & 2\sin 330° \end{bmatrix} \approx \begin{bmatrix} -1 & 1.7 \\ -1.7 & -1 \end{bmatrix}$;

    $[D][M] \approx \begin{bmatrix} 0.7 & -0.5 & 4.2 \\ -2.7 & -8.9 & -4.7 \end{bmatrix}$

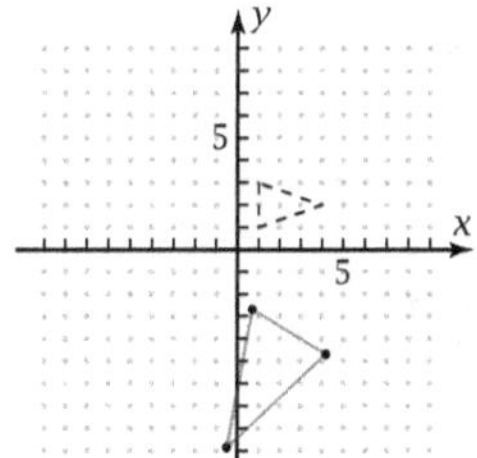

11. Answers will vary.

Exploration 107

1. $[A] = \begin{bmatrix} 0.9\cos 30° & 0.9\cos 120° \\ 0.9\sin 30° & 0.9\sin 120° \end{bmatrix}$

   $= \begin{bmatrix} 0.7794... & -0.45 \\ 0.45 & 0.7794... \end{bmatrix}$

2. $[A][D] = \begin{bmatrix} 0.7794... & -0.45 \\ 0.45 & 0.7794... \end{bmatrix} \begin{bmatrix} 12 & 12 & 6 & 6 \\ 5 & 2 & 2 & 5 \end{bmatrix}$

$= \begin{bmatrix} 7.1030... & 8.4530... & 3.7765... & 2.4265... \\ 9.2971... & 6.9588... & 4.2588... & 6.5971... \end{bmatrix}$

$\approx \begin{bmatrix} 7.1 & 8.5 & 3.8 & 2.4 \\ 9.3 & 7.0 & 4.3 & 6.6 \end{bmatrix}$

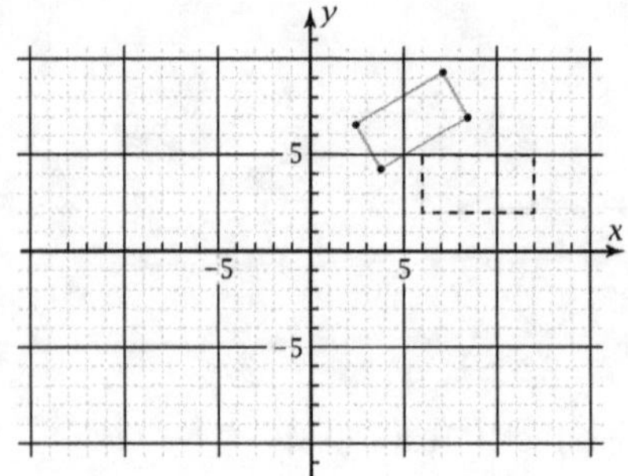

3. $[A]^2[D] = [A][E]$

$= \begin{bmatrix} 1.3525... & 3.4570... & 1.0270... & -1.0774... \\ 10.4427... & 9.2277... & 5.0188... & 6.2338... \end{bmatrix}$

$\approx \begin{bmatrix} 1.4 & 3.5 & 1.0 & -1.1 \\ 10.4 & 9.2 & 5.0 & 6.2 \end{bmatrix}$

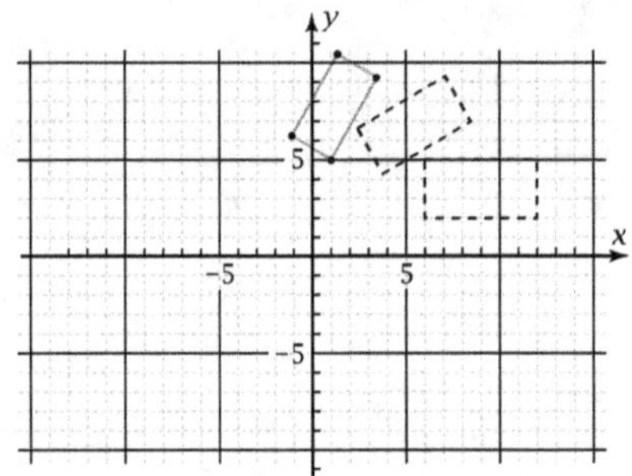

4. $[A]^3[D] = [A][E]$

$= \begin{bmatrix} -3.645 & -1.458 & -1.458 & -3.645 \\ 8.748 & 8.748 & 4.374 & 4.374 \end{bmatrix}$

$\approx \begin{bmatrix} -3.6 & -1.5 & -1.5 & -3.6 \\ 8.7 & 8.7 & 4.4 & 4.4 \end{bmatrix}$

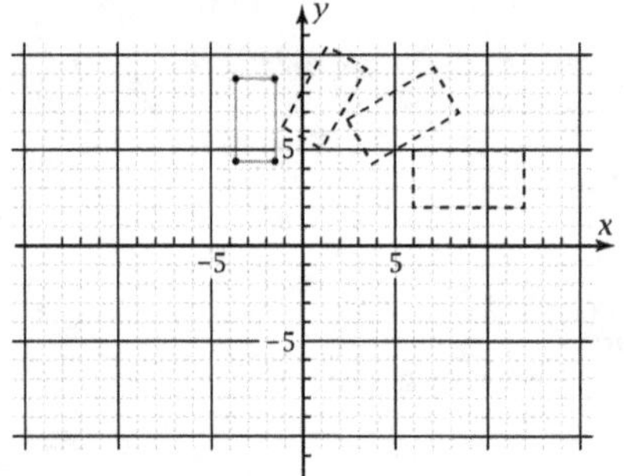

5. The images are spiraling in toward the origin.

6. Student programs will vary. See the Programs for Graphing Calculators appendix for the Itrans program.

7. Yes, the images agree.

8. Images are converging toward the origin.

$[E] \approx \begin{bmatrix} 0.0 & 0.0 & 0.0 & 0.0 \\ 0.0 & 0.0 & 0.0 & 0.0 \end{bmatrix}$

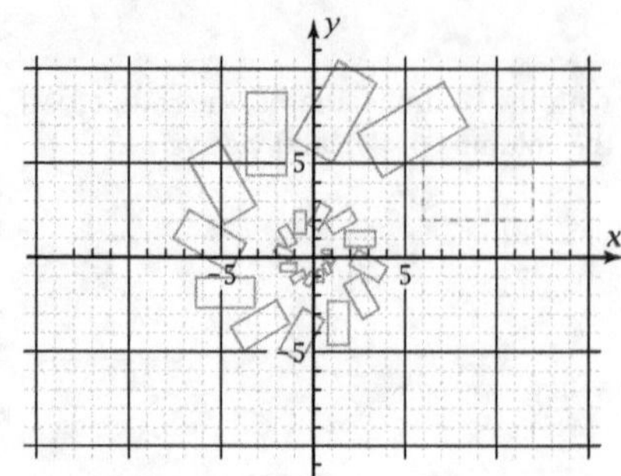

9. $[A] = \begin{bmatrix} 0.95 & 0 \\ 0 & 0.95 \end{bmatrix} \begin{bmatrix} \cos 20° & \cos 110° \\ \sin 20° & \sin 110° \end{bmatrix}$

$= \begin{bmatrix} 0.95 & 0 \\ 0 & 0.95 \end{bmatrix} \begin{bmatrix} 0.9396... & -0.3420... \\ 0.3420... & 0.9396... \end{bmatrix}$

$= \begin{bmatrix} 0.8927... & -0.3249... \\ 0.3249... & 0.8927... \end{bmatrix}$

$[A][D] = \begin{bmatrix} 0.8927... & -0.3249... \\ 0.3249... & 0.8927... \end{bmatrix} \begin{bmatrix} 6 & 8 & 10 & 8 \\ 1 & 2 & 1 & 6 \end{bmatrix}$

$= \begin{bmatrix} 5.0313... & 6.4918... & 8.6021... & 5.1921... \\ 2.8422... & 4.3847... & 4.1418... & 7.9556... \end{bmatrix}$

$\approx \begin{bmatrix} 5.0 & 6.5 & 8.6 & 5.2 \\ 2.8 & 4.4 & 4.1 & 8.0 \end{bmatrix}$

$[A]^2[D] = \begin{bmatrix} 3.5680... & 4.3706... & 6.3334... & 2.0501... \\ 4.1720... & 6.0236... & 6.4925... & 8.7890... \end{bmatrix}$

$\approx \begin{bmatrix} 3.6 & 4.4 & 6.3 & 2.1 \\ 4.2 & 6.0 & 6.5 & 8.8 \end{bmatrix}$

$[A]^3[D] = \begin{bmatrix} 1.8296... & 1.9444... & 3.5443... & -1.0255... \\ 4.8837... & 6.7974... & 7.8537... & 8.5121... \end{bmatrix}$

$\approx \begin{bmatrix} 1.8 & 1.9 & 3.5 & -1.0 \\ 4.9 & 6.8 & 7.9 & 8.5 \end{bmatrix}$

$[A]^4[D] \approx \begin{bmatrix} 0.0 & -0.5 & 0.6 & -3.7 \\ 5.0 & 6.7 & 8.2 & 7.3 \end{bmatrix}$

$[A]^5[D] \approx \begin{bmatrix} -1.6 & -2.6 & -2.1 & -5.6 \\ 4.4 & 5.8 & 7.5 & 5.3 \end{bmatrix}$

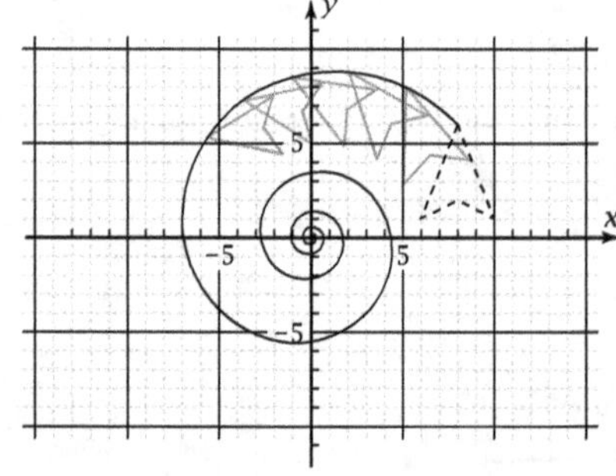

10. Answers will vary.

Exploration 108

1. 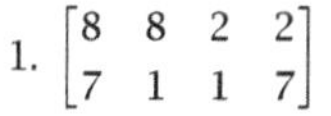
$$\begin{bmatrix} 8 & 8 & 2 & 2 \\ 7 & 1 & 1 & 7 \end{bmatrix}$$

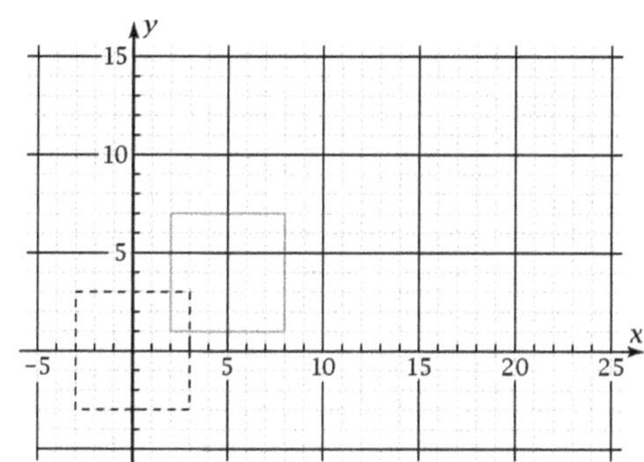

2. $[A][D] = \begin{bmatrix} 1 & 0 & 5 \\ 0 & 1 & 4 \\ 0 & 0 & 1 \end{bmatrix}\begin{bmatrix} 3 & 3 & -3 & -3 \\ 3 & -3 & -3 & 3 \\ 1 & 1 & 1 & 1 \end{bmatrix}$

$= \begin{bmatrix} 8 & 8 & 2 & 2 \\ 7 & 1 & 1 & 7 \\ 1 & 1 & 1 & 1 \end{bmatrix}$

3. The first two rows of $[A][D]$ in Problem 2 are the same as the image matrix of Problem 1!

4. $[A][D] = \begin{bmatrix} 1 & 0 & 5 \\ 0 & 1 & 4 \\ 0 & 0 & 1 \end{bmatrix}\begin{bmatrix} 3 & 3 & -3 & -3 \\ 3 & -3 & -3 & 3 \\ 1 & 1 & 1 & 1 \end{bmatrix}$

$= \begin{bmatrix} 1\cdot3+5\cdot1 & 1\cdot3+5\cdot1 & 1\cdot(-3)+5\cdot1 & 1\cdot(-3)+5\cdot1 \\ 1\cdot3+4\cdot1 & 1\cdot(-3)+4\cdot1 & 1\cdot(-3)+4\cdot1 & 1\cdot3+4\cdot1 \\ 1\cdot1 & 1\cdot1 & 1\cdot1 & 1\cdot1 \end{bmatrix}$

$= \begin{bmatrix} 8 & 8 & 2 & 2 \\ 7 & 1 & 1 & 7 \\ 1 & 1 & 1 & 1 \end{bmatrix}$

They add 5 to each element of the first row of the image, and they add 4 to each element of the second row of the image.

5. The 0 0 1 in the bottom row of $[A]$ ensures that the third row of the image will be the same as the third row of the pre-image so that the transformation can be repeated many times.

6. The upper left corner of $[A]$ is the 2×2 identity matrix. To add a reduction to 70% and a 30° counterclockwise rotation, replace the 2×2 identity matrix with
$\begin{bmatrix} 0.7\cos30° & 0.7\cos120° \\ 0.7\sin30° & 0.7\sin120° \end{bmatrix}.$

$[A] = \begin{bmatrix} 0.7\cos30° & 0.7\cos120° & 5 \\ 0.7\sin30° & 0.7\sin120° & 4 \\ 0 & 0 & 1 \end{bmatrix}$

7.

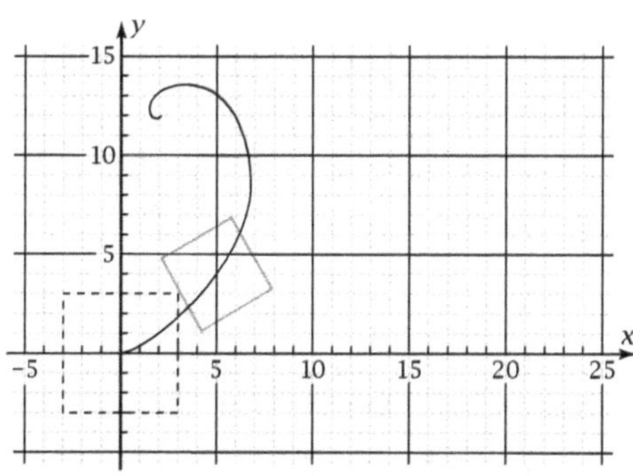

8. Fixed point attractor is approximately (2, 12).

9. Fixed point attractor $\approx$ (2.0496..., 11.9796...).

10.

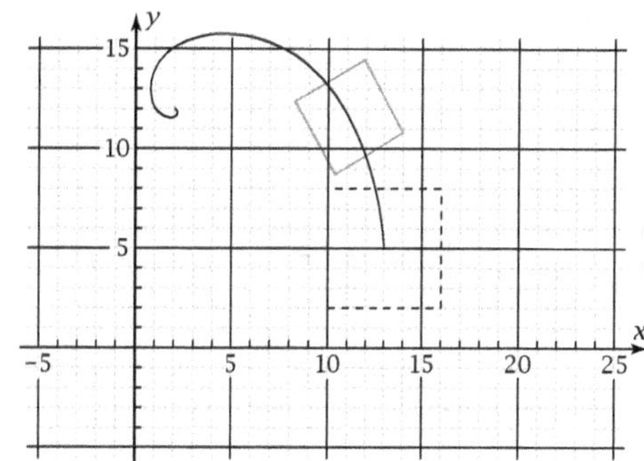

The images converge to the same fixed point as in Problem 8.

11. Answers will vary.

Exploration 109

1. $[A] = \begin{bmatrix} 0.8\cos30° & 0.8\cos120° & 4 \\ 0.8\sin30° & 0.8\sin120° & 2 \\ 0 & 0 & 1 \end{bmatrix}$

$= \begin{bmatrix} 0.6928... & -0.4 & 4 \\ 0.4 & 0.6928... & 2 \\ 0 & 0 & 1 \end{bmatrix}$

2.

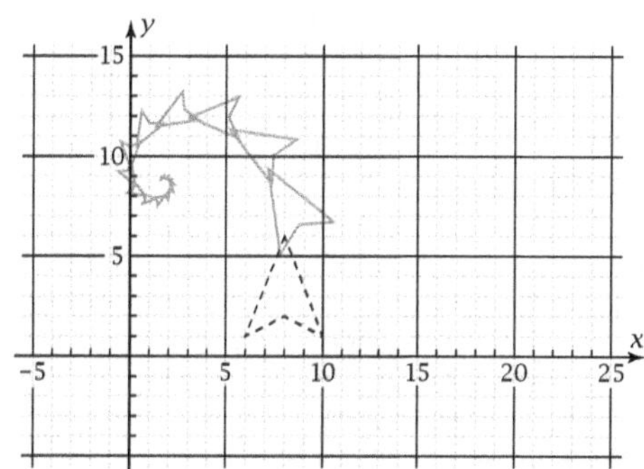

3. The images appear to be attracted to the point (2, 9). The actual fixed point is (1.6854..., 8.7056...), found numerically by raising $[A]$ to a high power, multiplying by $[D]$, and reading the coordinates from the last column.

4. $[A] = \begin{bmatrix} 0.8\cos60° & 0.8\cos150° & 4 \\ 0.8\sin60° & 0.8\sin150° & 2 \\ 0 & 0 & 1 \end{bmatrix}$

$= \begin{bmatrix} 0.4 & -0.6928... & 4 \\ 0.6928... & 0.4 & 2 \\ 0 & 0 & 1 \end{bmatrix}$

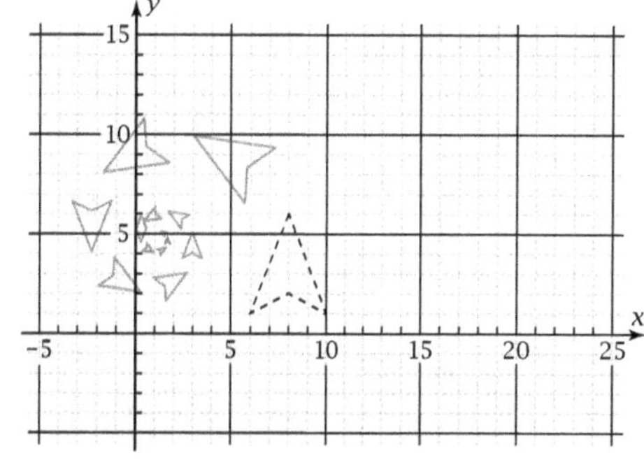

5. The images in Problem 4 appear to be attracted to the point
(1, 5)—the actual fixed point is (1.2075..., 4.7277...)—which is
different from the fixed point in Problem 3, even though the
pre-images were the same.

6.

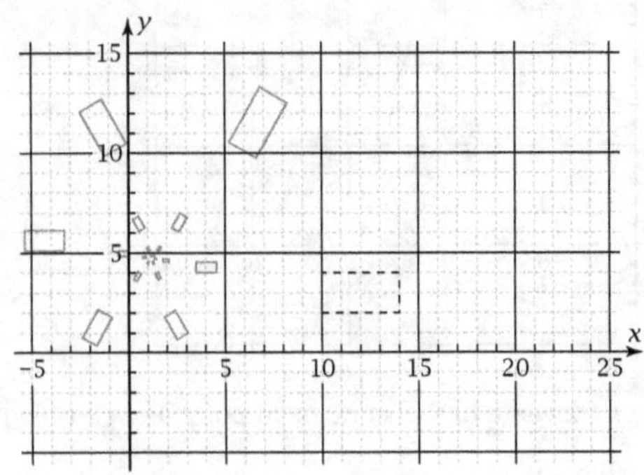

7. The actual fixed point is (1.2075..., 4.7277...), which is the
same as the fixed point in Problem 4, even though the
starting pre-images are different.

8. The transformation matrix determines the fixed-point
attractor. Applying the transformation matrix [A] in
Problems 4 and 6 iteratively to either pre-image matrix gives
the same fixed-point attractor. Applying the [A] from
Problem 1 iteratively to the dart [D] from Problems 1–5
gives a different fixed-point attractor from applying a
different [A] from Problem 4 iteratively to the same dart [D].

9. $[A] = \begin{bmatrix} 0.8\cos 60° & 0.8\cos 150° & 4 \\ 0.8\sin 60° & 0.8\sin 150° & 2 \\ 0 & 0 & 1 \end{bmatrix}$

$= \begin{bmatrix} 0.8\cos 60° & -0.8\sin 60° & 4 \\ 0.8\sin 60° & 0.8\cos 60° & 2 \\ 0 & 0 & 1 \end{bmatrix}$

$\begin{bmatrix} 0.8\cos 60° & -0.8\sin 60° & 4 \\ 0.8\sin 60° & 0.8\cos 60° & 2 \\ 0 & 0 & 1 \end{bmatrix}\begin{bmatrix} X \\ Y \\ 1 \end{bmatrix} = \begin{bmatrix} X \\ Y \\ 1 \end{bmatrix}$

$0.8X\cos 60° - 0.8Y\sin 60° + 4 = X$
$0.8X\sin 60° + 0.8Y\cos 60° + 2 = Y$
$(0.8\cos 60° - 1)X - (0.8\sin 60°)Y = -4$
$(0.8\sin 60°)X + (0.8\cos 60° - 1)Y = -2$

$\begin{bmatrix} 0.8\cos 60° - 1 & -0.8\sin 60° \\ 0.8\sin 60° & 0.8\cos 60° - 1 \end{bmatrix}\begin{bmatrix} X \\ Y \end{bmatrix} = \begin{bmatrix} -4 \\ -2 \end{bmatrix}$

$\begin{vmatrix} 0.8\cos 60° - 1 & -0.8\sin 60° \\ 0.8\sin 60° & 0.8\cos 60° - 1 \end{vmatrix}$
$= 0.64\cos^2 60° - 1.6\cos 60° + 1 + 0.64\sin^2 60°$
$= 1.64 - 1.6\cos 60°$

$\begin{bmatrix} X \\ Y \end{bmatrix} = \begin{bmatrix} 0.8\cos 60° - 1 & -0.8\sin 60° \\ 0.8\sin 60° & 0.8\cos 60° - 1 \end{bmatrix}^{-1}\begin{bmatrix} -4 \\ -2 \end{bmatrix}$

$= \dfrac{1}{1.64 - 1.6\cos 60°}\begin{bmatrix} 0.8\cos 60° - 1 & 0.8\sin 60° \\ -0.8\sin 60° & 0.8\cos 60° - 1 \end{bmatrix}\begin{bmatrix} -4 \\ -2 \end{bmatrix}$

$= \dfrac{1}{1.64 - 1.6\cos 60°}\begin{bmatrix} -3.2\cos 60° + 4 - 1.6\sin 60° \\ 3.2\sin 60° - 1.6\cos 60° + 2 \end{bmatrix}$

$X = \dfrac{-3.2\cos 60° + 4 - 1.6\sin 60°}{1.64 - 1.6\cos 60°} = \dfrac{2.4 - 0.8\sqrt{3}}{0.84} = 1.2075...$

$Y = \dfrac{3.2\sin 60° - 1.6\cos 60° + 2}{1.64 - 1.6\cos 60°} = \dfrac{1.6\sqrt{3} + 1.2}{0.84} = 4.7277...$

The fixed point $(X, Y) = (1.2075..., 4.7277...)$.

10. $[A] = \begin{bmatrix} 0.4 & -0.6928... & 4 \\ 0.6928... & 0.4 & 2 \\ 0 & 0 & 1 \end{bmatrix}$

$[D] = \begin{bmatrix} 6 & 8 & 10 & 8 \\ 1 & 2 & 1 & 6 \\ 1 & 1 & 1 & 1 \end{bmatrix}$

$[A][D] = \begin{bmatrix} 5.7071... & 5.8143... & 7.3071... & 3.0430... \\ 6.5569... & 8.3425... & 9.3282... & 9.9425... \\ 1 & 1 & 1 & 1 \end{bmatrix}$

$[A]([A][D]) = \begin{bmatrix} 1.7401... & 0.5458... & 0.4601... & -1.6711... \\ 8.5768... & 9.3653... & 10.7938... & 8.0853... \\ 1 & 1 & 1 & 1 \end{bmatrix}$

$[A]^2 = \begin{bmatrix} -0.32 & -0.5542... & 4.2143... \\ 0.5542... & -0.32 & 5.5712... \\ 0 & 0 & 1 \end{bmatrix}$

$[A]^2[D] = \begin{bmatrix} 1.7401... & 0.5458... & 0.4601... & -1.6711... \\ 8.5768... & 9.3653... & 10.7938... & 8.0853... \\ 1 & 1 & 1 & 1 \end{bmatrix}$

$[A]([A][D]) = [A]^2[D]$

11. $[A]^{50} = \begin{bmatrix} -0.0000... & -0.0000... & 1.2076... \\ 0.0000... & -0.0000... & 4.7277... \\ 0 & 0 & 1 \end{bmatrix}$

The numbers in the rotation and dilation part of the matrix
are approaching 0; the numbers in the translation part of the
matrix are approaching the coordinates of the fixed-point
attractor (1.2075..., 4.7277...). Raise the transformation
matrix to a higher power.

12. Answers will vary.

Exploration 110

1. Dilation by a factor of 0.7
   Rotation by 30° counterclockwise
   Translation by 5 units horizontally
   Translation by 4 units vertically

2. $[A] = \begin{bmatrix} 0.95\cos 20° & 0.95\cos 110° & 3 \\ 0.95\sin 20° & 0.95\sin 110° & 2 \\ 0 & 0 & 1 \end{bmatrix}$

$= \begin{bmatrix} 0.8927... & -0.3249... & 3 \\ 0.3249... & 0.8927... & 2 \\ 0 & 0 & 1 \end{bmatrix}$

3. $[D] = \begin{bmatrix} 6 & 8 & 10 & 8 \\ 1 & 2 & 1 & 6 \\ 1 & 1 & 1 & 1 \end{bmatrix}$

4. $[A][D] = \begin{bmatrix} 8.0313... & 9.4918... & 11.6021... & 8.1921... \\ 4.8422... & 6.3847... & 6.1418... & 9.9556... \\ 1 & 1 & 1 & 1 \end{bmatrix}$

$\approx \begin{bmatrix} 8.0 & 9.5 & 11.6 & 8.2 \\ 4.8 & 6.4 & 6.1 & 10.0 \\ 1 & 1 & 1 & 1 \end{bmatrix}$

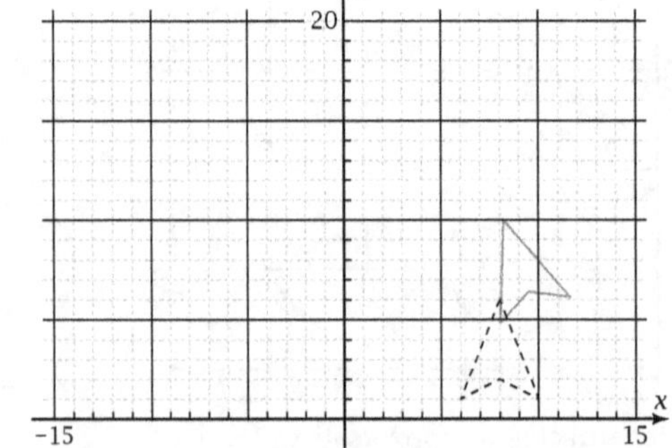

5.

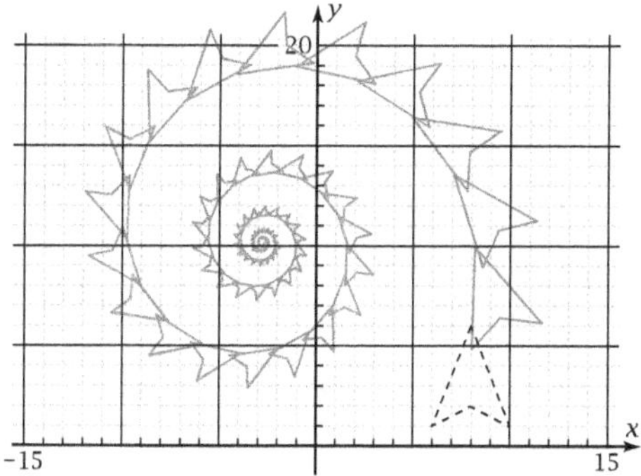

6. Images appear to be attracted to the point $(-3, 10)$.

7. $[A]^{100} = \begin{bmatrix} -0.0055... & 0.0020... & -2.8372... \\ -0.0020... & -0.0055... & 10.2088... \\ 0 & 0 & 1 \end{bmatrix}$

The fixed point is $(X, Y) \approx (-2.8372..., 10.2088...)$.

8. $[A] = \begin{bmatrix} 0.95\cos 20° & 0.95\cos 110° & 3 \\ 0.95\sin 20° & 0.95\sin 110° & 2 \\ 0 & 0 & 1 \end{bmatrix}$

$= \begin{bmatrix} 0.95\cos 20° & -0.95\sin 20° & 3 \\ 0.95\sin 20° & 0.95\cos 20° & 2 \\ 0 & 0 & 1 \end{bmatrix}$

$\begin{bmatrix} 0.95\cos 20° & -0.95\sin 20° & 3 \\ 0.95\sin 20° & 0.95\cos 20° & 2 \\ 0 & 0 & 1 \end{bmatrix}\begin{bmatrix} X \\ Y \\ 1 \end{bmatrix} = \begin{bmatrix} X \\ Y \\ 1 \end{bmatrix}$

$0.95X\cos 20° - 0.95Y\sin 20° + 3 = X$

$0.95X\sin 20° + 0.95Y\cos 20° + 2 = Y$

$(0.95\cos 20° - 1)X - (0.95\sin 20°)Y = -3$

$(0.95\sin 20°)X + (0.95\cos 20° - 1)Y = -2$

$\begin{bmatrix} 0.95\cos 20° - 1 & -0.95\sin 20° \\ 0.95\sin 20° & 0.95\cos 20° - 1 \end{bmatrix}\begin{bmatrix} X \\ Y \end{bmatrix} = \begin{bmatrix} -3 \\ -2 \end{bmatrix}$

$\begin{vmatrix} 0.95\cos 20° - 1 & -0.95\sin 20° \\ 0.95\sin 20° & 0.95\cos 20° - 1 \end{vmatrix}$

$= 0.9025\cos^2 20° - 1.9\cos 20° + 1 + 0.9025\sin^2 20°$

$= 1.9025 - 1.9\cos 20°$

$\begin{bmatrix} X \\ Y \end{bmatrix} = \begin{bmatrix} 0.95\cos 20° - 1 & -0.95\sin 20° \\ 0.95\sin 20° & 0.95\cos 20° - 1 \end{bmatrix}^{-1}\begin{bmatrix} -3 \\ -2 \end{bmatrix}$

$= \dfrac{1}{1.9025 - 1.9\cos 20°}\begin{bmatrix} 0.95\cos 20° - 1 & 0.95\sin 20° \\ -0.95\sin 20° & 0.95\cos 20° - 1 \end{bmatrix}\begin{bmatrix} -3 \\ -2 \end{bmatrix}$

$= \dfrac{1}{1.9025 - 1.9\cos 20°}\begin{bmatrix} -2.85\cos 20° + 3 - 1.9\sin 20° \\ 2.85\sin 20° - 1.9\cos 20° + 2 \end{bmatrix}$

$X = \dfrac{-2.85\cos 20° + 3 - 1.9\sin 20°}{1.9025 - 1.9\cos 20°} = -2.8010...$

$Y = \dfrac{2.85\sin 20° - 1.9\cos 20° + 2}{1.9025 - 1.9\cos 20°} = 10.1580...$

The fixed point is $(X, Y) = (-2.8010..., 10.1580...)$.

9. They add 3 to each element of the first row of the image and add 2 to each element of the second row of the image. The 0 0 1 in the bottom row of $[A]$ ensures that the third row of the image will be the same as the third row of the pre-image so that the transformation can be repeated many times.

10. So that $[E]$ can represent the current transformed image, which is always changing, while $[D]$ remains constant as the original pre-image, in case it is wanted again after the program is run.

11. Answers will vary.

Exploration 111

1. Each element in $[V_1] = [V_0][T]$ is given by $[V_0]$ multiplied by a column of $[T]$. But this turns out to be exactly the formula for calculating the number of viewers for that column's network next month.

2. $[V_1] = [V_0][T] = [34.07 \quad 22.08 \quad 41.85]$
   The A element of $[V_0][T]$ is $34.07 = 0.85 \cdot 35 + 0.13 \cdot 20 + 0.04 \cdot 43$, which is 85% of A's January viewers plus 13% of B's January viewers plus 4% of C's January viewers. But note that 85% is the percentage of A's January viewers that stayed with A; 13% is the percentage of B's January viewers that switched to A; and 4% is the percentage of C's January viewers that switched to A.

3. $[V_2] = [V_1][T] = [33.5039 \quad 23.582 \quad 40.9141]$
   represents the number of viewers at the beginning of March for the same reason that $[V_1] = [V_0][T]$ represents the number of viewers at the beginning of February. And
   $[V_2] = [V_1][T] = ([V_0][T] = [V_0]([T][T]) = [V_0][T]^2$.

4. $[V_{12}] = [V_0][T]^{12} = [33.3413... \quad 27.5120... \quad 37.1465...]$
   After one year, A will have about 33.3 million viewers, B will have about 27.5 million viewers, and C will have about 37.1 million viewers.

5. $[V_{100}] = [V_0][T]^{100} = [33.8078... \quad 27.8165... \quad 36.3755...]$
   After a large number of months, A will have about 33.8 million viewers, B will have about 27.8 million viewers, and C will have about 36.4 million viewers.

6. Answers will vary. Andrei A. Markov, 1856-1922, was a Russian mathematician. Markov chains are used to represent many physical and statistical processes.

7. Answers will vary.

Exploration 112

1. $[E] = \begin{bmatrix} 2 & 2 & -2 & -2 \\ 0 & 10 & 10 & 0 \\ 1 & 1 & 1 & 1 \end{bmatrix}$

2. $[A] = \begin{bmatrix} 0.7989... & 0.0418... & 0 \\ -0.0418... & 0.7989... & 3 \\ 0 & 0 & 1 \end{bmatrix}$

$[B] = \begin{bmatrix} 0.1846... & 0.2364... & 0 \\ -0.2364... & 0.1846... & 2 \\ 0 & 0 & 1 \end{bmatrix}$

$[C] = \begin{bmatrix} 0.2083... & -0.2158... & 0 \\ 0.2158... & 0.2083... & 3 \\ 0 & 0 & 1 \end{bmatrix}; [D] = \begin{bmatrix} 0 & 0 & 0 \\ 0 & 0.3 & 0 \\ 0 & 0 & 1 \end{bmatrix}$

3. $[A][E] \approx \begin{bmatrix} 1.6 & 2.0 & -1.2 & -1.6 \\ 2.9 & 10.9 & 11.1 & 3.1 \\ 1 & 1 & 1 & 1 \end{bmatrix}$

$[B][E] \approx \begin{bmatrix} 0.4 & 2.7 & 2.0 & -0.4 \\ 1.5 & 3.4 & 4.3 & 2.5 \\ 1 & 1 & 1 & 1 \end{bmatrix}$

$[C][E] \approx \begin{bmatrix} 0.4 & -1.7 & -2.6 & -0.4 \\ 3.4 & 5.5 & 4.7 & 2.6 \\ 1 & 1 & 1 & 1 \end{bmatrix}$

$[D][E] = \begin{bmatrix} 0 & 0 & 0 & 0 \\ 0 & 3 & 3 & 0 \\ 1 & 1 & 1 & 1 \end{bmatrix}$

4.

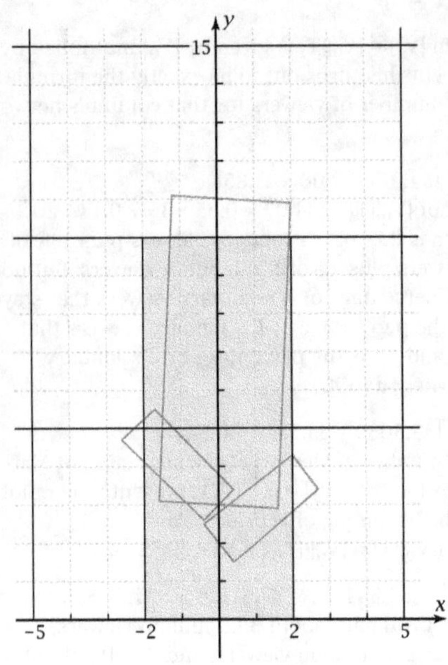

5. $[A]([A][E]) \approx \begin{bmatrix} 1.4 & 2.1 & -0.5 & -1.1 \\ 5.3 & 11.6 & 11.9 & 5.5 \\ 1 & 1 & 1 & 1 \end{bmatrix}$

6. $[A][A][E] = [A]^2[E] \approx \begin{bmatrix} 1.4 & 2.1 & -0.5 & -1.1 \\ 5.3 & 11.6 & 11.9 & 5.5 \\ 1 & 1 & 1 & 1 \end{bmatrix}$

7. Associative property

8. $[B][A][E] \approx \begin{bmatrix} 1.0 & 3.0 & 2.4 & 0.4 \\ 2.2 & 3.5 & 4.3 & 2.9 \\ 1 & 1 & 1 & 1 \end{bmatrix}$

$[A][B][E] \approx \begin{bmatrix} 0.4 & 2.3 & 1.8 & -0.2 \\ 4.2 & 5.6 & 6.4 & 5.0 \\ 1 & 1 & 1 & 1 \end{bmatrix}$

9. Commutative property

10. $[A][A][E] \approx \begin{bmatrix} 1.4 & 2.1 & -0.5 & -1.1 \\ 5.3 & 11.6 & 11.9 & 5.5 \\ 1 & 1 & 1 & 1 \end{bmatrix}$

$[A][B][E] \approx \begin{bmatrix} 0.4 & 2.3 & 1.8 & -0.2 \\ 4.2 & 5.6 & 6.4 & 5.0 \\ 1 & 1 & 1 & 1 \end{bmatrix}$

$[A][C][E] \approx \begin{bmatrix} 0.5 & -1.2 & -1.9 & -0.2 \\ 5.7 & 7.5 & 6.8 & 5.1 \\ 1 & 1 & 1 & 1 \end{bmatrix}$

$[A][D][E] \approx \begin{bmatrix} 0 & 0.1 & 0.1 & 0 \\ 3 & 5.4 & 5.4 & 3 \\ 1 & 1 & 1 & 1 \end{bmatrix}$

$[B][A][E] \approx \begin{bmatrix} 1.0 & 3.0 & 2.4 & 0.4 \\ 2.2 & 3.5 & 4.3 & 2.9 \\ 1 & 1 & 1 & 1 \end{bmatrix}$

$[B][B][E] \approx \begin{bmatrix} 0.4 & 1.3 & 1.4 & 0.5 \\ 2.2 & 2.0 & 2.3 & 2.5 \\ 1 & 1 & 1 & 1 \end{bmatrix}$

$[B][C][E] \approx \begin{bmatrix} 0.9 & 1.0 & 0.6 & 0.5 \\ 2.5 & 3.4 & 3.5 & 2.6 \\ 1 & 1 & 1 & 1 \end{bmatrix}$

$[B][D][E] \approx \begin{bmatrix} 0 & 0.7 & 0.7 & 0 \\ 2 & 2.6 & 2.6 & 2 \\ 1 & 1 & 1 & 1 \end{bmatrix}$

$[C][A][E] \approx \begin{bmatrix} -0.3 & -1.9 & -2.6 & -1.0 \\ 4.0 & 5.7 & 5.1 & 3.3 \\ 1 & 1 & 1 & 1 \end{bmatrix}$

$[C][B][E] \approx \begin{bmatrix} -0.3 & -0.2 & -0.5 & -0.6 \\ 3.4 & 4.3 & 4.3 & 3.4 \\ 1 & 1 & 1 & 1 \end{bmatrix}$

$[C][C][E] \approx \begin{bmatrix} -0.7 & -1.6 & -1.5 & -0.6 \\ 3.8 & 3.8 & 3.4 & 3.4 \\ 1 & 1 & 1 & 1 \end{bmatrix}$

$[C][D][E] \approx \begin{bmatrix} 0 & -0.6 & -0.6 & 0 \\ 3 & 3.6 & 3.6 & 3 \\ 1 & 1 & 1 & 1 \end{bmatrix}$

$[D][A][E] \approx \begin{bmatrix} 0 & 0 & 0 & 0 \\ 0.9 & 3.3 & 3.3 & 0.9 \\ 1 & 1 & 1 & 1 \end{bmatrix}$

$[D][B][E] \approx \begin{bmatrix} 0 & 0 & 0 & 0 \\ 0.5 & 1.0 & 1.3 & 0.7 \\ 1 & 1 & 1 & 1 \end{bmatrix}$

$[D][C][E] \approx \begin{bmatrix} 0 & 0 & 0 & 0 \\ 1.0 & 1.7 & 1.4 & 0.8 \\ 1 & 1 & 1 & 1 \end{bmatrix}$

$[D][D][E] = \begin{bmatrix} 0 & 0 & 0 & 0 \\ 0 & 0.9 & 0.9 & 0 \\ 1 & 1 & 1 & 1 \end{bmatrix}$

11.

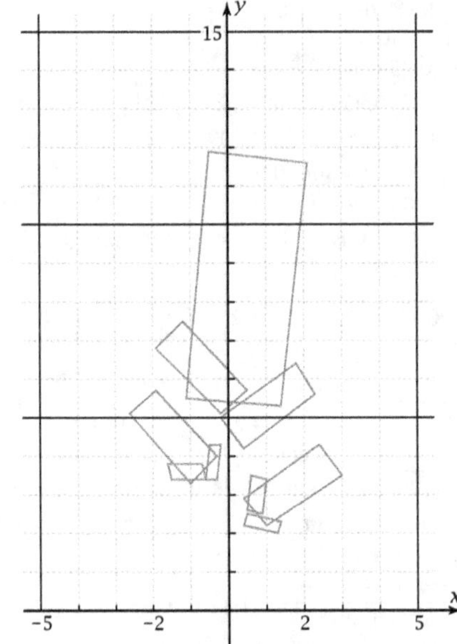

12. 3rd iteration: $4^3 = 64$ images
4th iteration: $4^4 = 256$ images
10th iteration: $4^{10} = 1{,}048{,}576$ images

13. Answers will vary.

Exploration 113

1. [A][E]: (1.8071…, 6.9107…)
    [C][Ans]: (−1.1147…, 4.8301…)
    [B][Ans]: (0.9359…, 3.1556…)
    [A][Ans]: (0.8798…, 5.4818…)
    [D][Ans]: (0, 1.6445…)
    [C][Ans]: (−0.3548…, 3.3427…)
    [C][Ans]: (−0.7953…, 3.6200…)
    [B][Ans]: (0.7088…, 2.8566…)
    [A][Ans]: (0.6859…, 5.2524…)
    [D][Ans]: (0, 1.5757…)

2. 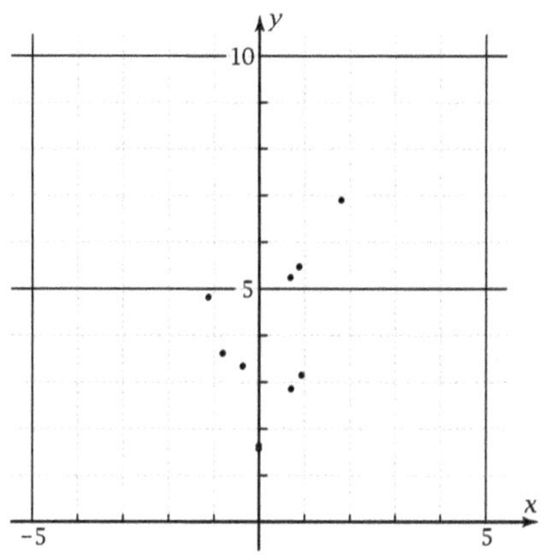

    The pattern is only vaguely hinted at and not yet clear.

3. Student activity

4. One run produces this graph:

    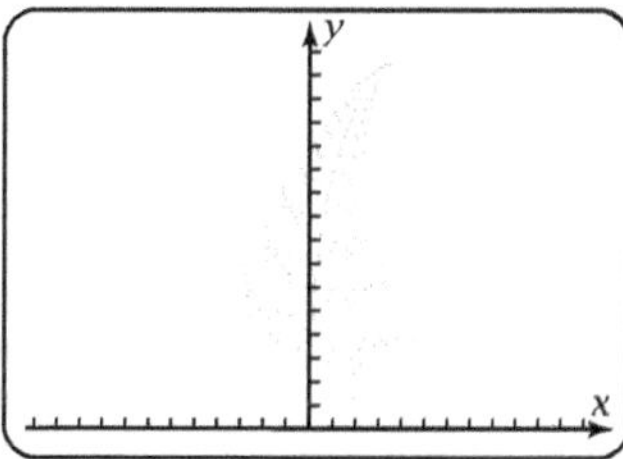

5. The pattern is starting to become clear. The points are definitely attracted to some regions rather than to others.

6. One run produces this graph:

    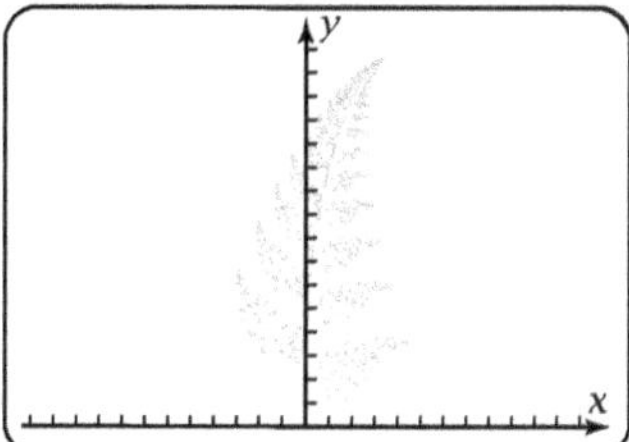

    The graph looks like the figure.

7. Student activity

8. The strange attractor does not depend on the pre-image.

9. Doing even a few iterations of rectangle transformations involves plotting thousands or millions of image rectangles.

10. One run produces this graph:

    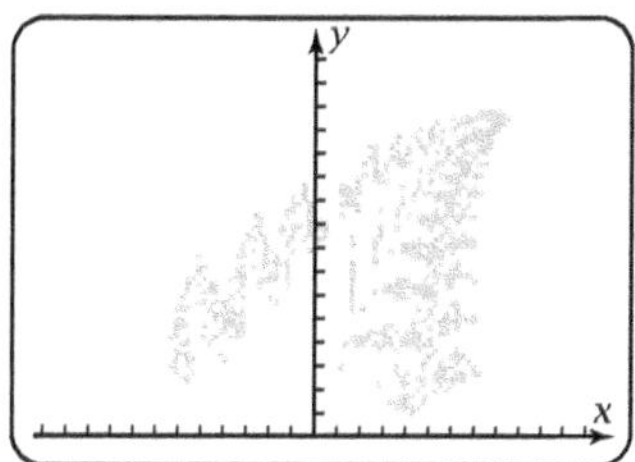

    The "leaf" has spread open.

11. "Strange attractor"

12. Answers will vary. The possible shapes of strange attractors are unlimited. If the design formed by the pre-image and its first iterations is fairly clear and simple, the resulting strange attractor will be very distinctive—but often quite surprising.

13. Answers will vary.

Exploration 114

1. 125 cubes

2. $125 = 5^3$ cubes

3. "3" is the dimension.

4. $N = \left(\dfrac{1}{r}\right)^3$

5. $\log N = \log\left(\dfrac{1}{r}\right)^3 = 3\log\dfrac{1}{r}$

    $3 = \dfrac{\log N}{\log\frac{1}{r}}$

6. $r = \dfrac{1}{5}, N = 25$

    $D = \dfrac{\log N}{\log\frac{1}{r}} = \dfrac{\log 25}{\log 5} = \dfrac{2\log 5}{\log 5} = 2$

7. 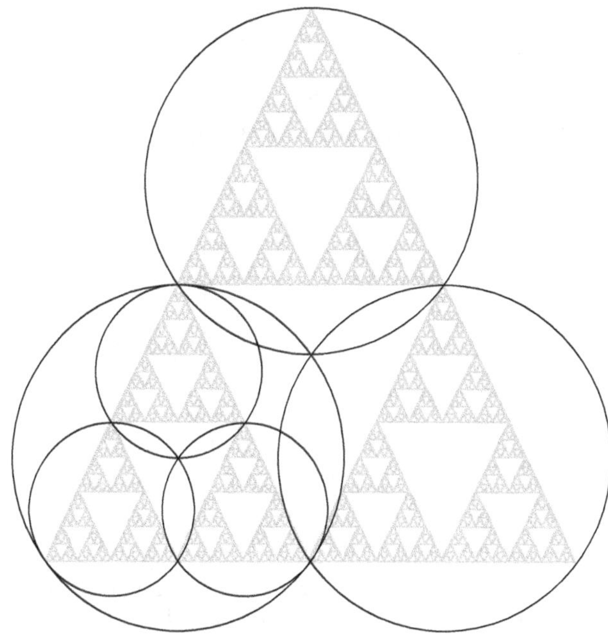

8. $r = \dfrac{1}{2}, N = 3$

$$D = \frac{\log N}{\log \frac{1}{r}} = \frac{\log 3}{\log 2} = 1.5849\ldots$$

9.

n	Total Area
0	300
1	225
2	168.75
10	16.8940...
100	$9.6216\ldots \times 10^{-11}$

As n grows infinite, the area $A = 300 \cdot 0.75^n$ approaches zero.

10. Answers will vary.

Exploration 115

1. Reducing by $\frac{1}{3}$ takes point $(12, 6)$ to point $(4, 2)$, so you must translate the upper point 8 units horizontally and 4 units vertically;

$$[A] = \begin{bmatrix} \frac{1}{3} & 0 & 8 \\ 0 & \frac{1}{3} & 4 \\ 0 & 0 & 1 \end{bmatrix}$$

2. Reducing by $\frac{1}{3}$ takes point $(12, -6)$ to point $(4, -2)$, so you must translate the upper point 8 units horizontally and -4 units vertically;

$$[B] = \begin{bmatrix} \frac{1}{3} & 0 & 8 \\ 0 & \frac{1}{3} & -4 \\ 0 & 0 & 1 \end{bmatrix}$$

3.
$$\begin{bmatrix} \frac{1}{3}\cos 60° & \frac{1}{3}\cos 150° & 0 \\ \frac{1}{3}\sin 60° & \frac{1}{3}\sin 150° & 0 \\ 0 & 0 & 1 \end{bmatrix}\begin{bmatrix} 12 \\ 6 \\ 1 \end{bmatrix}$$

$$= \begin{bmatrix} 4\cos 60° + 2\cos 150° \\ 4\sin 60° + 2\sin 150° \\ 1 \end{bmatrix} = \begin{bmatrix} 2 - \sqrt{3} \\ 2\sqrt{3} + 1 \\ 1 \end{bmatrix} = \begin{bmatrix} 0.2679\ldots \\ 4.4641\ldots \\ 1 \end{bmatrix}$$

Upper left end of the rotated and dilated segment is at the point $(2 - \sqrt{3}, 2\sqrt{3} + 1) = (0.2679\ldots, 4.4641\ldots)$.

$$[C] = \begin{bmatrix} \frac{1}{3}\cos 60° & \frac{1}{3}\cos 150° & 12 - (2 - \sqrt{3}) \\ \frac{1}{3}\sin 60° & \frac{1}{3}\sin 150° & 2 - (2\sqrt{3} + 1) \\ 0 & 0 & 1 \end{bmatrix}$$

$$= \begin{bmatrix} \frac{1}{6} & \frac{-\sqrt{3}}{6} & 10 + \sqrt{3} \\ \frac{\sqrt{3}}{6} & \frac{1}{6} & 1 - 2\sqrt{3} \\ 0 & 0 & 1 \end{bmatrix}$$

$$= \begin{bmatrix} 0.1666\ldots & -0.2886\ldots & 11.7320\ldots \\ 0.2886\ldots & 0.1666\ldots & -2.4641\ldots \\ 0 & 0 & 1 \end{bmatrix}$$

4.
$$\begin{bmatrix} \frac{1}{3}\cos(-60°) & \frac{1}{3}\cos 30° & 0 \\ \frac{1}{3}\sin(-60°) & \frac{1}{3}\sin 30° & 0 \\ 0 & 0 & 1 \end{bmatrix}\begin{bmatrix} 12 \\ -6 \\ 1 \end{bmatrix}$$

$$= \begin{bmatrix} 4\cos 60° - 2\cos 30° \\ -4\sin 60° - 2\sin 30° \\ 1 \end{bmatrix} = \begin{bmatrix} 2 - \sqrt{3} \\ -2\sqrt{3} - 1 \\ 1 \end{bmatrix} = \begin{bmatrix} 0.2679\ldots \\ -4.4641\ldots \\ 1 \end{bmatrix}$$

Lower left end of the rotated and dilated segment is at the point $(2 - \sqrt{3}, -2\sqrt{3} - 1) = (0.2679\ldots, -4.4641\ldots)$.

$$[D] = \begin{bmatrix} \frac{1}{3}\cos(-60°) & \frac{1}{3}\cos 30° & 12 - (2 - \sqrt{3}) \\ \frac{1}{3}\sin(-60°) & \frac{1}{3}\sin 30° & -2 - (-2\sqrt{3} - 1) \\ 0 & 0 & 1 \end{bmatrix}$$

$$= \begin{bmatrix} \frac{1}{6} & \frac{\sqrt{3}}{6} & 10 + \sqrt{3} \\ \frac{\sqrt{3}}{6} & \frac{1}{6} & -1 + 2\sqrt{3} \\ 0 & 0 & 1 \end{bmatrix}$$

$$= \begin{bmatrix} 0.1666\ldots & 0.2886\ldots & 11.7320\ldots \\ -0.2886\ldots & 0.1666\ldots & 2.4641\ldots \\ 0 & 0 & 1 \end{bmatrix}$$

5. $[E_1] = [A][E] = \begin{bmatrix} \frac{1}{3} & 0 & 8 \\ 0 & \frac{1}{3} & 4 \\ 0 & 0 & 1 \end{bmatrix}\begin{bmatrix} 12 & 12 \\ 6 & -6 \\ 1 & 1 \end{bmatrix} = \begin{bmatrix} 12 & 12 \\ 6 & 2 \\ 1 & 1 \end{bmatrix}$

$[E_2] = [B][E] = \begin{bmatrix} \frac{1}{3} & 0 & 8 \\ 0 & \frac{1}{3} & -4 \\ 0 & 0 & 1 \end{bmatrix}\begin{bmatrix} 12 & 12 \\ 6 & -6 \\ 1 & 1 \end{bmatrix} = \begin{bmatrix} 12 & 12 \\ -2 & -6 \\ 1 & 1 \end{bmatrix}$

$[E_3] = [C][E] = \begin{bmatrix} \frac{1}{6} & -\frac{\sqrt{3}}{6} & 10 + \sqrt{3} \\ \frac{\sqrt{3}}{6} & \frac{1}{6} & 1 - 2\sqrt{3} \\ 0 & 0 & 1 \end{bmatrix}\begin{bmatrix} 12 & 12 \\ 6 & -6 \\ 1 & 1 \end{bmatrix}$

$$= \begin{bmatrix} 12 & 12 + 2\sqrt{3} \\ 2 & 0 \\ 1 & 1 \end{bmatrix} = \begin{bmatrix} 12 & 15.4641\ldots \\ 2 & 0 \\ 1 & 1 \end{bmatrix}$$

$[E_4] = [D][E] = \begin{bmatrix} \frac{1}{6} & \frac{\sqrt{3}}{6} & 10 + \sqrt{3} \\ -\frac{\sqrt{3}}{6} & \frac{1}{6} & -1 + 2\sqrt{3} \\ 0 & 0 & 1 \end{bmatrix}\begin{bmatrix} 12 & 12 \\ 6 & -6 \\ 1 & 1 \end{bmatrix}$

$$= \begin{bmatrix} 12 + 2\sqrt{3} & 12 \\ 0 & -2 \\ 1 & 1 \end{bmatrix} = \begin{bmatrix} 15.4641\ldots & 12 \\ 0 & -2 \\ 1 & 1 \end{bmatrix}$$

6.

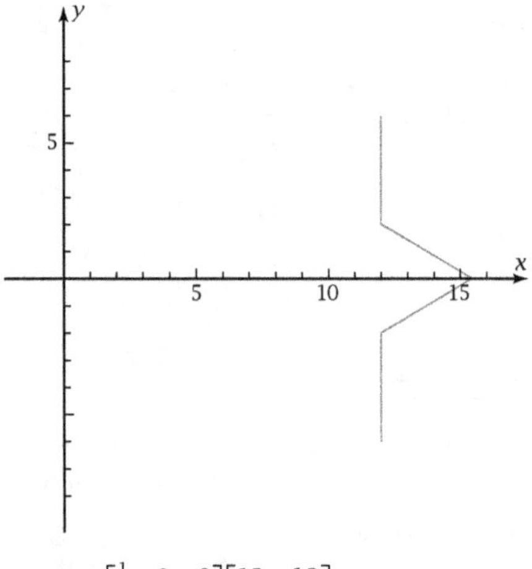

7. $[A][E_1] = \begin{bmatrix} \frac{1}{3} & 0 & 8 \\ 0 & \frac{1}{3} & 4 \\ 0 & 0 & 1 \end{bmatrix}\begin{bmatrix} 12 & 12 \\ 6 & 2 \\ 1 & 1 \end{bmatrix}$

$$= \begin{bmatrix} 12 & 12 \\ 6 & 4\frac{2}{3} \\ 1 & 1 \end{bmatrix} = \begin{bmatrix} 12 & 12 \\ 6 & 4.6666\ldots \\ 1 & 1 \end{bmatrix}$$

$[A][E_2] = \begin{bmatrix} \frac{1}{3} & 0 & 8 \\ 0 & \frac{1}{3} & 4 \\ 0 & 0 & 1 \end{bmatrix}\begin{bmatrix} 12 & 12 \\ -2 & -6 \\ 1 & 1 \end{bmatrix}$

$$= \begin{bmatrix} 12 & 12 \\ 3\frac{1}{3} & 2 \\ 1 & 1 \end{bmatrix} = \begin{bmatrix} 12 & 12 \\ 3.3333\ldots & 2 \\ 1 & 1 \end{bmatrix}$$

$$[A][E_3] = \begin{bmatrix} \frac{1}{3} & 0 & 8 \\ 0 & \frac{1}{3} & 4 \\ 0 & 0 & 1 \end{bmatrix}\begin{bmatrix} 12 & 12+2\sqrt{3} \\ 2 & 0 \\ 1 & 1 \end{bmatrix}$$

$$= \begin{bmatrix} 12 & 12+\frac{2\sqrt{3}}{3} \\ 4\frac{2}{3} & 4 \\ 1 & 1 \end{bmatrix} = \begin{bmatrix} 12 & 13.1547\ldots \\ 4.6666\ldots & 4 \\ 1 & 1 \end{bmatrix}$$

$$[A][E_4] = \begin{bmatrix} \frac{1}{3} & 0 & 8 \\ 0 & \frac{1}{3} & 4 \\ 0 & 0 & 1 \end{bmatrix}\begin{bmatrix} 12+2\sqrt{3} & 12 \\ 0 & -2 \\ 1 & 1 \end{bmatrix}$$

$$= \begin{bmatrix} 12+\frac{2\sqrt{3}}{3} & 12 \\ 4 & 3\frac{1}{3} \\ 1 & 1 \end{bmatrix} = \begin{bmatrix} 13.1547\ldots & 12 \\ 4 & 3.3333\ldots \\ 1 & 1 \end{bmatrix}$$

$$[B][E_1] = \begin{bmatrix} \frac{1}{3} & 0 & 8 \\ 0 & \frac{1}{3} & -4 \\ 0 & 0 & 1 \end{bmatrix}\begin{bmatrix} 12 & 12 \\ 6 & 2 \\ 1 & 1 \end{bmatrix}$$

$$= \begin{bmatrix} 12 & 12 \\ -2 & -3\frac{2}{3} \\ 1 & 1 \end{bmatrix} = \begin{bmatrix} 12 & 12 \\ -2 & -3.3333\ldots \\ 1 & 1 \end{bmatrix}$$

$$[B][E_2] = \begin{bmatrix} \frac{1}{3} & 0 & 8 \\ 0 & \frac{1}{3} & -4 \\ 0 & 0 & 1 \end{bmatrix}\begin{bmatrix} 12 & 12 \\ -2 & -6 \\ 1 & 1 \end{bmatrix}$$

$$= \begin{bmatrix} 12 & 12 \\ -4\frac{2}{3} & -6 \\ 1 & 1 \end{bmatrix} = \begin{bmatrix} 12 & 12 \\ -4.6666\ldots & -6 \\ 1 & 1 \end{bmatrix}$$

$$[B][E_3] = \begin{bmatrix} \frac{1}{3} & 0 & 8 \\ 0 & \frac{1}{3} & -4 \\ 0 & 0 & 1 \end{bmatrix}\begin{bmatrix} 12 & 12+2\sqrt{3} \\ 2 & 0 \\ 1 & 1 \end{bmatrix}$$

$$= \begin{bmatrix} 12 & 12+\frac{2\sqrt{3}}{3} \\ -3\frac{1}{3} & -4 \\ 1 & 1 \end{bmatrix} = \begin{bmatrix} 12 & 13.1547\ldots \\ -3.3333\ldots & -4 \\ 1 & 1 \end{bmatrix}$$

$$[B][E_4] = \begin{bmatrix} \frac{1}{3} & 0 & 8 \\ 0 & \frac{1}{3} & -4 \\ 0 & 0 & 1 \end{bmatrix}\begin{bmatrix} 12+2\sqrt{3} & 12 \\ 0 & 0 \\ 1 & 1 \end{bmatrix}$$

$$= \begin{bmatrix} 12+\frac{2\sqrt{3}}{3} & 12 \\ -4 & -4\frac{2}{3} \\ 1 & 1 \end{bmatrix} = \begin{bmatrix} 13.1547\ldots & 12 \\ -4 & -4.6666\ldots \\ 1 & 1 \end{bmatrix}$$

$$[C][E_1] = \begin{bmatrix} \frac{1}{6} & -\frac{\sqrt{3}}{6} & 10+\sqrt{3} \\ \frac{\sqrt{3}}{6} & \frac{1}{6} & 1-2\sqrt{3} \\ 0 & 0 & 1 \end{bmatrix}\begin{bmatrix} 12 & 12 \\ 6 & 2 \\ 1 & 1 \end{bmatrix}$$

$$= \begin{bmatrix} 12 & 12+\frac{2\sqrt{3}}{3} \\ 2 & \frac{4}{3} \\ 1 & 1 \end{bmatrix} = \begin{bmatrix} 12 & 13.1547\ldots \\ 0 & 1.3333\ldots \\ 1 & 1 \end{bmatrix}$$

$$[C][E_2] = \begin{bmatrix} \frac{1}{6} & -\frac{\sqrt{3}}{6} & 10+\sqrt{3} \\ \frac{\sqrt{3}}{6} & \frac{1}{6} & 1-2\sqrt{3} \\ 0 & 0 & 1 \end{bmatrix}\begin{bmatrix} 12 & 12 \\ -2 & -6 \\ 1 & 1 \end{bmatrix}$$

$$= \begin{bmatrix} 12+\frac{4\sqrt{3}}{3} & 12+2\sqrt{3} \\ \frac{2}{3} & 0 \\ 1 & 1 \end{bmatrix}$$

$$= \begin{bmatrix} 14.3094\ldots & 15.4641\ldots \\ 0.6666\ldots & 0 \\ 1 & 1 \end{bmatrix}$$

$$[C][E_3] = \begin{bmatrix} \frac{1}{6} & -\frac{\sqrt{3}}{6} & 10+\sqrt{3} \\ \frac{\sqrt{3}}{6} & \frac{1}{6} & 1-2\sqrt{3} \\ 0 & 0 & 1 \end{bmatrix}\begin{bmatrix} 12 & 12+2\sqrt{3} \\ 2 & 0 \\ 1 & 1 \end{bmatrix}$$

$$= \begin{bmatrix} 12+\frac{2\sqrt{3}}{3} & 12+\frac{4\sqrt{3}}{3} \\ \frac{4}{3} & 2 \\ 1 & 1 \end{bmatrix}$$

$$= \begin{bmatrix} 13.1547\ldots & 14.3094\ldots \\ 1.3333\ldots & 2 \\ 1 & 1 \end{bmatrix}$$

$$[C][E_4] = \begin{bmatrix} \frac{1}{6} & -\frac{\sqrt{3}}{6} & 10+\sqrt{3} \\ \frac{\sqrt{3}}{6} & \frac{1}{6} & 1-2\sqrt{3} \\ 0 & 0 & 1 \end{bmatrix}\begin{bmatrix} 12+2\sqrt{3} & 12 \\ 0 & -2 \\ 1 & 1 \end{bmatrix}$$

$$= \begin{bmatrix} 12+\frac{4\sqrt{3}}{3} & 12+\frac{4\sqrt{3}}{3} \\ 2 & \frac{2}{3} \\ 1 & 1 \end{bmatrix}$$

$$= \begin{bmatrix} 14.3094\ldots & 14.3094\ldots \\ 2 & 0.6666\ldots \\ 1 & 1 \end{bmatrix}$$

$$[D][E_1] = \begin{bmatrix} \frac{1}{6} & \frac{\sqrt{3}}{6} & 10+\sqrt{3} \\ -\frac{\sqrt{3}}{6} & \frac{1}{6} & -1+2\sqrt{3} \\ 0 & 0 & 1 \end{bmatrix}\begin{bmatrix} 12 & 12 \\ 6 & 2 \\ 1 & 1 \end{bmatrix}$$

$$= \begin{bmatrix} 12+2\sqrt{3} & 12+\frac{4\sqrt{3}}{3} \\ 0 & -\frac{2}{3} \\ 1 & 1 \end{bmatrix}$$

$$= \begin{bmatrix} 15.4641\ldots & 14.3094\ldots \\ 0 & -6.6666\ldots \\ 1 & 1 \end{bmatrix}$$

$$[D][E_2] = \begin{bmatrix} \frac{1}{6} & \frac{\sqrt{3}}{6} & 10+\sqrt{3} \\ -\frac{\sqrt{3}}{6} & \frac{1}{6} & -1+2\sqrt{3} \\ 0 & 0 & 1 \end{bmatrix}\begin{bmatrix} 12 & 12 \\ -2 & -6 \\ 1 & 1 \end{bmatrix}$$

$$= \begin{bmatrix} 12+\frac{2\sqrt{3}}{3} & 12 \\ -\frac{4}{3} & -2 \\ 1 & 1 \end{bmatrix} = \begin{bmatrix} 13.1547\ldots & 12 \\ -1.3333\ldots & -2 \\ 1 & 1 \end{bmatrix}$$

$$[D][E_3] = \begin{bmatrix} \frac{1}{6} & \frac{\sqrt{3}}{6} & 10+\sqrt{3} \\ -\frac{\sqrt{3}}{6} & \frac{1}{6} & -1+2\sqrt{3} \\ 0 & 0 & 1 \end{bmatrix}\begin{bmatrix} 12 & 12+2\sqrt{3} \\ 2 & 0 \\ 1 & 1 \end{bmatrix}$$

$$= \begin{bmatrix} 12+\frac{4\sqrt{3}}{3} & 12+\frac{4\sqrt{3}}{3} \\ -\frac{2}{3} & -2 \\ 1 & 1 \end{bmatrix}$$

$$= \begin{bmatrix} 14.3094\ldots & 14.3094\ldots \\ -0.6666\ldots & -2 \\ 1 & 1 \end{bmatrix}$$

$$[D][E_4] = \begin{bmatrix} \frac{1}{6} & \frac{\sqrt{3}}{6} & 10+\sqrt{3} \\ -\frac{\sqrt{3}}{6} & \frac{1}{6} & -1+2\sqrt{3} \\ 0 & 0 & 1 \end{bmatrix}\begin{bmatrix} 12+2\sqrt{3} & 12 \\ 0 & -2 \\ 1 & 1 \end{bmatrix}$$

$$= \begin{bmatrix} 12+\frac{4\sqrt{3}}{3} & 12+\frac{2\sqrt{3}}{3} \\ -2 & -\frac{4}{3} \\ 1 & 1 \end{bmatrix}$$

$$= \begin{bmatrix} 14.3094\ldots & 13.1547\ldots \\ -2 & -1.3333\ldots \\ 1 & 1 \end{bmatrix}$$

8.

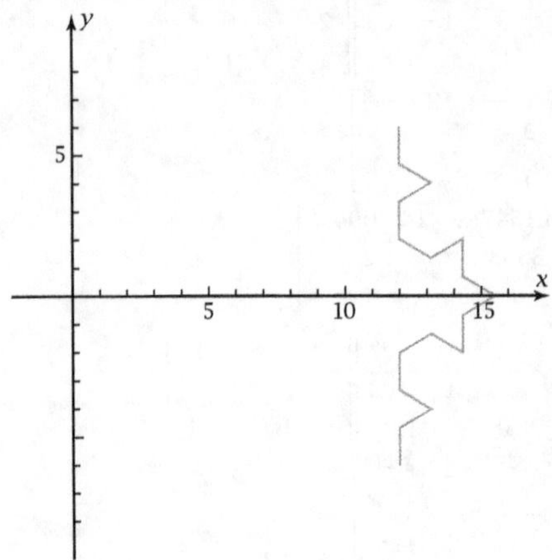

4.

Ruler	N	r	$\dfrac{1}{r}$
150	1	1	1
50	3.16	$\dfrac{1}{3}$	3
20	8.2	$\dfrac{1}{7.5}$	7.5
10	17.2	$\dfrac{1}{15}$	15
5	35.0	$\dfrac{1}{30}$	30
2	90.5	$\dfrac{1}{75}$	75

9. It can be divided into four segments, each of which is similar to the original. This subdividing can be continued indefinitely, always producing segments that are similar to the original.

5.

$\log \dfrac{1}{r}$	$\log N$
0	0
0.48	0.50
0.88	0.91
1.18	1.24
1.48	1.54
1.88	1.96

10. Second iteration: $12 \cdot \left(\frac{4}{3}\right)^2 = 21\frac{1}{3} = 21.3333\ldots$ units

    Third iteration: $12 \cdot \left(\frac{4}{3}\right)^3 = 28\frac{4}{9} = 28.4444\ldots$ units

    Fourth iteration: $12 \cdot \left(\frac{4}{3}\right)^4 = 37\frac{25}{27} = 37.9259\ldots$ units

    100th iteration: $12 \cdot \left(\frac{4}{3}\right)^{100} = 3.7415\ldots \times 10^{13}$ units

    As n grows infinite, the length $L = 12 \cdot \left(\frac{4}{3}\right)^n$ also grows infinite. The final length would be infinite.

11. $N = 1000$

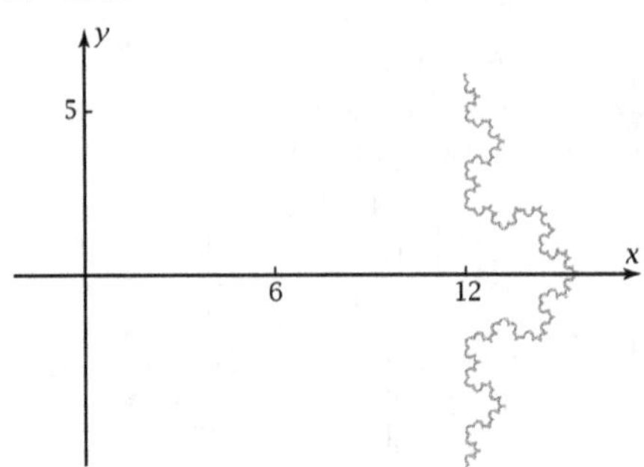

6.

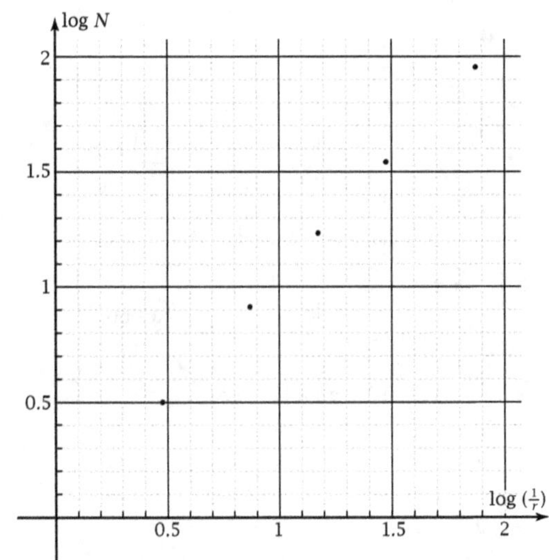

12. Answers will vary.

Exploration 116

1. The distance is 150 miles.

2. The river appears to be about 3.16 ruler lengths, or about 158 miles.

3. The distance appears to be about 164 miles, or 8.2 ruler lengths.

7. $\log N = 1.0443\ldots \log\frac{1}{r} + 0.0014\ldots,$
   correlation coefficient $= 0.9999\ldots$

8. $\log N = 1.0443\ldots \log 15 + 0.0014\ldots = 1.2296\ldots$
   $\Rightarrow N = 10^{1.2296\ldots} = 16.9694\ldots$. This is reasonably close to the 17.2 in the table.

9. $\log N = m \log\frac{1}{r} \Rightarrow m = \dfrac{\log N}{\log\frac{1}{r}}$. The dimension is $\approx 1.0443\ldots$, a fractional dimension slightly larger than 1. Hausdorff's definition says that if an object is cut into N identical pieces, each similar to the original object, and the ratio of the length of each piece to the length of the original object is r and the subdivisions can be carried on infinitely, then the dimension, D, of the object is $D = \dfrac{\log N}{\log\frac{1}{r}}$.

Note that in this case the river (and any smaller pieces of it) are not exactly similar; they are what is called **statistically self-similar,** which can be precisely defined, but which basically means that the pieces *resemble* the whole and *resemble* each other but are not necessarily *identical* to each other and not necessarily *exactly* similar to the whole.

10. $\dfrac{1}{r} = \dfrac{150 \text{ mi}}{1 \text{ in.}} = \dfrac{150 \text{ mi}}{1 \text{ in.}} \cdot \dfrac{12 \text{ in.}}{1 \text{ ft}} \cdot \dfrac{5280 \text{ ft}}{1 \text{ mi}} = 9{,}504{,}000$

$\log N = 1.0443\ldots \log 9{,}504{,}000 + 0.0014\ldots = 7.2888\ldots$

$\Rightarrow N = 10^{7.2888\ldots} = 19{,}448{,}102.3952\ldots$ pieces

$= 19{,}448{,}102.3952\ldots$ in. $= 306.9460\ldots$ mi

11. Answers will vary.

Exploration 117

1. $[D] = \begin{bmatrix} 0 & 0 \\ 0 & 10 \\ 1 & 1 \end{bmatrix}$

2. $[A] = \begin{bmatrix} 0.6\cos 30° & 0.6\cos 120° & 0 \\ 0.6\sin 30° & 0.6\sin 120° & 5 \\ 0 & 0 & 1 \end{bmatrix}$

$= \begin{bmatrix} 0.3\sqrt{3} & -0.3 & 0 \\ 0.3 & 0.3\sqrt{3} & 5 \\ 0 & 0 & 1 \end{bmatrix}$

$= \begin{bmatrix} 0.5196\ldots & -0.3 & 0 \\ 0.3 & 0.5196\ldots & 5 \\ 0 & 0 & 1 \end{bmatrix}$

$[B] = \begin{bmatrix} 0.6 & 0 & 0 \\ 0 & 0.6 & 0 \\ 0 & 0 & 1 \end{bmatrix}$

$[C] = \begin{bmatrix} 0.6\cos(-30°) & 0.6\cos 60° & 0 \\ 0.6\sin(-30°) & 0.6\sin 60° & 4 \\ 0 & 0 & 1 \end{bmatrix}$

$= \begin{bmatrix} 0.3\sqrt{3} & -0.3 & 0 \\ -0.3 & 0.3\sqrt{3} & 4 \\ 0 & 0 & 1 \end{bmatrix}$

$= \begin{bmatrix} 0.5196\ldots & -0.3 & 0 \\ -0.3 & 0.5196\ldots & 4 \\ 0 & 0 & 1 \end{bmatrix}$

3.

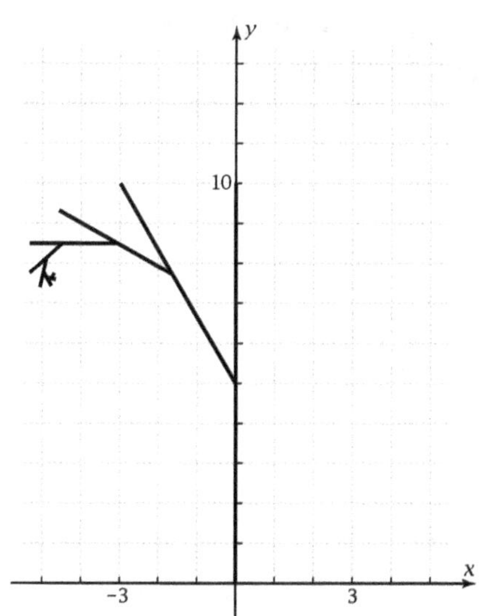

4. $[A]^{100} = \begin{bmatrix} -0.0000\ldots & -0.0000\ldots & -4.6762\ldots \\ 0.0000\ldots & -0.0000\ldots & 7.4880\ldots \\ 0 & 0 & 1 \end{bmatrix}$

The fixed point is $(-4.6762\ldots, 7.4880\ldots)$.

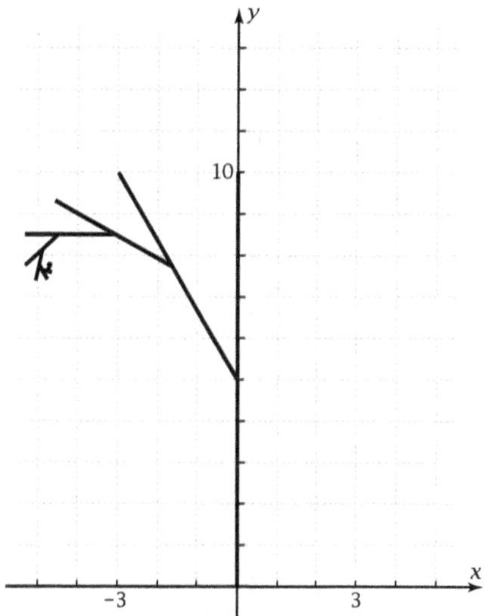

5. $[A][D] = \begin{bmatrix} 0.3\sqrt{3} & -0.3 & 0 \\ 0.3 & 0.3\sqrt{3} & 5 \\ 0 & 0 & 1 \end{bmatrix}\begin{bmatrix} 0 & 0 \\ 0 & 10 \\ 1 & 1 \end{bmatrix}$

$= \begin{bmatrix} 0 & -3 \\ 5 & 5 + 3\sqrt{3} \\ 1 & 1 \end{bmatrix}$

$[B][D] = \begin{bmatrix} 0.6 & 0 & 0 \\ 0 & 0.6 & 0 \\ 0 & 0 & 1 \end{bmatrix}\begin{bmatrix} 0 & 0 \\ 0 & 10 \\ 1 & 1 \end{bmatrix} = \begin{bmatrix} 0 & 0 \\ 0 & 6 \\ 1 & 1 \end{bmatrix}$

$[C][D] = \begin{bmatrix} 0.3\sqrt{3} & 0.3 & 0 \\ -0.3 & 0.3\sqrt{3} & 4 \\ 0 & 0 & 1 \end{bmatrix}\begin{bmatrix} 0 & 0 \\ 0 & 10 \\ 1 & 1 \end{bmatrix}$

$= \begin{bmatrix} 0 & 3 \\ 4 & 4 + 3\sqrt{3} \\ 1 & 1 \end{bmatrix}$

so

$[A][A][D] = \begin{bmatrix} 0.3\sqrt{3} & -0.3 & 0 \\ 0.3 & 0.3\sqrt{3} & 5 \\ 0 & 0 & 1 \end{bmatrix}\begin{bmatrix} 0 & -3 \\ 5 & 5 + 3\sqrt{3} \\ 1 & 1 \end{bmatrix}$

$= \begin{bmatrix} -1.5 & -1.5 - 1.8\sqrt{3} \\ 5 + 1.5\sqrt{3} & 6.8 + 1.5\sqrt{3} \\ 1 & 1 \end{bmatrix} \approx \begin{bmatrix} -1.5 & -4.6 \\ 7.6 & 9.4 \\ 1 & 1 \end{bmatrix}$

$[A][B][D] = \begin{bmatrix} 0.3\sqrt{3} & -0.3 & 0 \\ 0.3 & 0.3\sqrt{3} & 5 \\ 0 & 0 & 1 \end{bmatrix}\begin{bmatrix} 0 & 0 \\ 0 & 6 \\ 1 & 1 \end{bmatrix}$

$= \begin{bmatrix} 0 & -1.8 \\ 5 & 5 + 1.8\sqrt{3} \\ 1 & 1 \end{bmatrix} \approx \begin{bmatrix} 0 & -1.8 \\ 5 & 8.1 \\ 1 & 1 \end{bmatrix}$

$[A][C][D] = \begin{bmatrix} 0.3\sqrt{3} & -0.3 & 0 \\ 0.3 & 0.3\sqrt{3} & 5 \\ 0 & 0 & 1 \end{bmatrix}\begin{bmatrix} 0 & 3 \\ 4 & 4 + 3\sqrt{3} \\ 1 & 1 \end{bmatrix}$

$= \begin{bmatrix} -1.2 & -1.2 \\ 5 + 1.2\sqrt{3} & 8.6 + 1.2\sqrt{3} \\ 1 & 1 \end{bmatrix} \approx \begin{bmatrix} -1.2 & -1.2 \\ 7.1 & 10.7 \\ 1 & 1 \end{bmatrix}$

$$[B][A][D] = \begin{bmatrix} 0.6 & 0 & 0 \\ 0 & 0.6 & 0 \\ 0 & 0 & 1 \end{bmatrix}\begin{bmatrix} 0 & -3 \\ 5 & 5+3\sqrt{3} \\ 1 & 1 \end{bmatrix}$$

$$= \begin{bmatrix} 0 & -1.8 \\ 3 & 3+1.8\sqrt{3} \\ 1 & 1 \end{bmatrix} \approx \begin{bmatrix} 0 & -1.8 \\ 3 & 6.1 \\ 1 & 1 \end{bmatrix}$$

$$[B][B][D] = \begin{bmatrix} 0.6 & 0 & 0 \\ 0 & 0.6 & 0 \\ 0 & 0 & 1 \end{bmatrix}\begin{bmatrix} 0 & 0 \\ 0 & 6 \\ 1 & 1 \end{bmatrix} = \begin{bmatrix} 0 & 0 \\ 0 & 3.6 \\ 1 & 1 \end{bmatrix}$$

$$[B][C][D] = \begin{bmatrix} 0.6 & 0 & 0 \\ 0 & 0.6 & 0 \\ 0 & 0 & 1 \end{bmatrix}\begin{bmatrix} 0 & 3 \\ 4 & 4+3\sqrt{3} \\ 1 & 1 \end{bmatrix}$$

$$= \begin{bmatrix} 0 & 1.8 \\ 2.4 & 2.4+1.8\sqrt{3} \\ 1 & 1 \end{bmatrix} \approx \begin{bmatrix} 0 & 1.8 \\ 2.4 & 5.5 \\ 1 & 1 \end{bmatrix}$$

$$[C][A][D] = \begin{bmatrix} 0.3\sqrt{3} & 0.3 & 0 \\ -0.3 & 0.3\sqrt{3} & 4 \\ 0 & 0 & 1 \end{bmatrix}\begin{bmatrix} 0 & -3 \\ 5 & 5+3\sqrt{3} \\ 1 & 1 \end{bmatrix}$$

$$= \begin{bmatrix} 1.5 & 1.5 \\ 4+1.5\sqrt{3} & 7.6+1.5\sqrt{3} \\ 1 & 1 \end{bmatrix} \approx \begin{bmatrix} 1.5 & 1.5 \\ 6.6 & 10.2 \\ 1 & 1 \end{bmatrix}$$

$$[C][B][D] = \begin{bmatrix} 0.3\sqrt{3} & 0.3 & 0 \\ -0.3 & 0.3\sqrt{3} & 4 \\ 0 & 0 & 1 \end{bmatrix}\begin{bmatrix} 0 & 0 \\ 0 & 6 \\ 1 & 1 \end{bmatrix}$$

$$= \begin{bmatrix} 0 & 1.8 \\ 4 & 4+1.8\sqrt{3} \\ 1 & 1 \end{bmatrix} \approx \begin{bmatrix} 0 & 1.8 \\ 4 & 7.1 \\ 1 & 1 \end{bmatrix}$$

$$[C][C][D] = \begin{bmatrix} 0.3\sqrt{3} & 0.3 & 0 \\ -0.3 & 0.3\sqrt{3} & 4 \\ 0 & 0 & 1 \end{bmatrix}\begin{bmatrix} 0 & 3 \\ 4 & 4+3\sqrt{3} \\ 1 & 1 \end{bmatrix}$$

$$= \begin{bmatrix} 1.2 & 1.2+1.8\sqrt{3} \\ 4+1.2\sqrt{3} & 5.8+1.2\sqrt{3} \\ 1 & 1 \end{bmatrix} \approx \begin{bmatrix} 1.2 & 4.3 \\ 6.1 & 7.9 \\ 1 & 1 \end{bmatrix}$$

6.

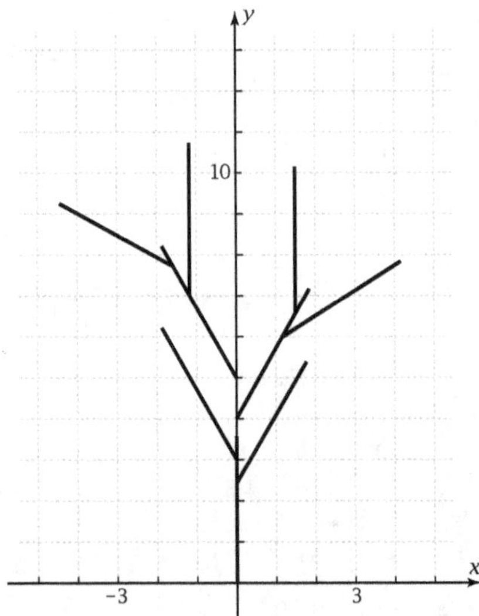

7. Graphs will vary but should resemble the figure in Problem 8, except with straight lines instead of dots.

8. The graph should look like the figure.

9. Answers will vary. A random point is chosen and plotted. One of the transformation matrices is randomly chosen, and its transformation is performed on the chosen point. Then one of the matrices is randomly chosen again and its transformation performed on the resulting point. The procedure is iterated as many times as desired. (The probability of choosing each transformation matrix is specified in advance).

10. "Strange attractor"

11. The fixed point is the tip of a "branch" of the tree.

12. 0th iteration: $3^0 \cdot 10 \cdot 0.6^0 = 10$ units
First iteration: $3^1 \cdot 10 \cdot 0.6^1 = 18$ units
Second iteration: $3^2 \cdot 10 \cdot 0.6^2 = 32.4$ units
Third iteration: $3^3 \cdot 10 \cdot 0.6^3 = 58.32$ units
100th iteration: $3^{100} \cdot 10 \cdot 0.6^{100} = 3.3670\ldots \times 10^{26}$ units
If the iterations were done forever, the length would become infinite.

13.

n	$r = 0.6^n$	$\dfrac{1}{r} = \left(\dfrac{5}{3}\right)^n$	$N = 3^n$
0	1	1	1
1	0.6	$1.\overline{6}$	3
2	0.36	$2.\overline{7}$	9
3	0.216	$4.\overline{629}$	27
4	0.1296	$7.7160\ldots$	81
5	0.07776	$12.8600\ldots$	243

14. $\log N = 2.1506\ldots \log \frac{1}{r}$, with intercept 0 and perfect correlation (of course).

15. If an object is cut into N identical pieces, each similar to the original, and the ratio of the length of each piece to the length of the original object is r and the subdivisions can be carried on infinitely, then the dimension, D, of the object is
$$D = \frac{\log N}{\log \frac{1}{r}}.$$
$$D = \frac{\log N}{\log \frac{1}{r}} = \frac{\log 3^n}{\log \left(\frac{5}{3}\right)^n} = \frac{n \log 3}{n \log \frac{5}{3}} = \frac{\log 3}{\log \frac{5}{3}} = 2.1506\ldots$$

16. "Fractal"

17. Answers will vary.

Analytic Geometry of Conic Sections and Quadric Surfaces

Exploration 118

1.

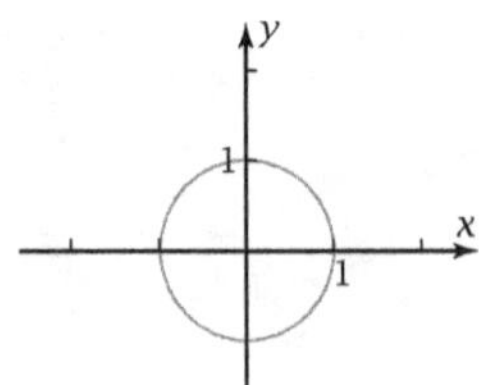

2. $x^2 = \cos^2 t,\ y^2 = \sin^2 t$
$x^2 + y^2 = \cos^2 t + \sin^2 t = 1$

3.

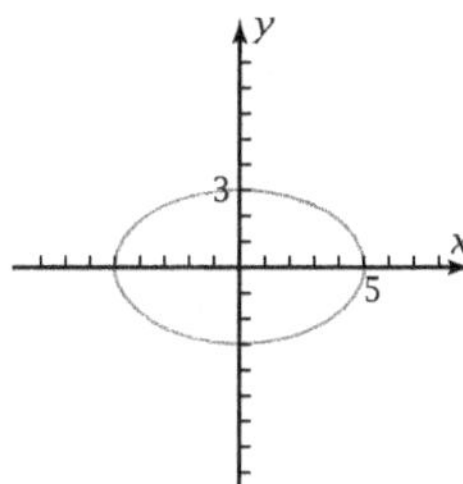

Horizontal dilation by 5, vertical dilation by 3

4.

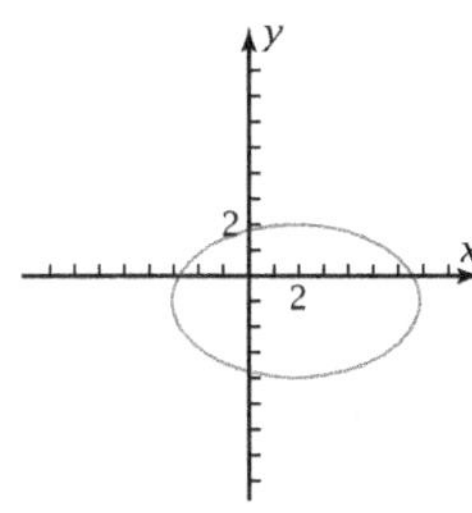

Horizontal translation by 2, vertical translation by -1

5.

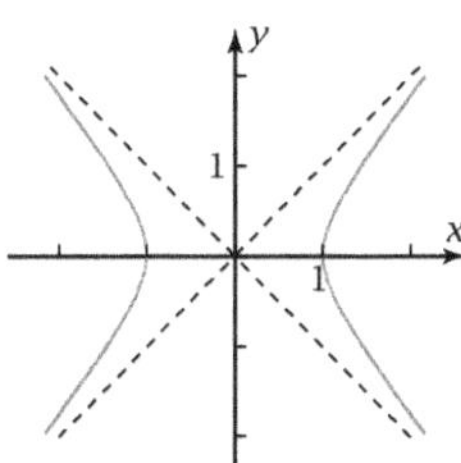

6. $x^2 = \sec^2 t,\ y^2 = \tan^2 t$
$x^2 - y^2 = \sec^2 t - \tan^2 t = 1$

7.

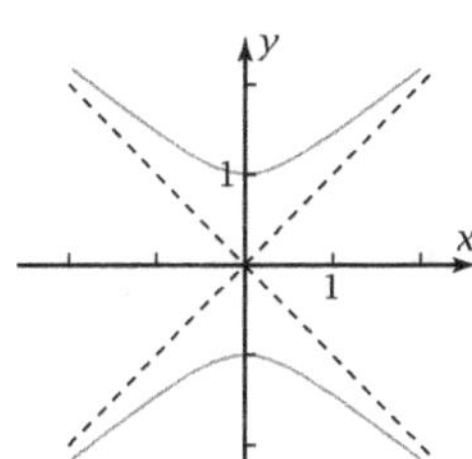

Reflection across the line $x = y$

8.

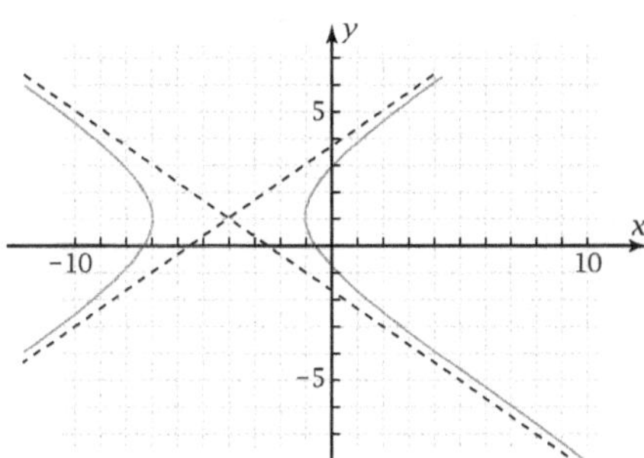

9. Answers will vary.

Exploration 119

1. Ellipse, center at the point (7, −4), x-radius $a = 2$, y-radius $b = 5$

2.

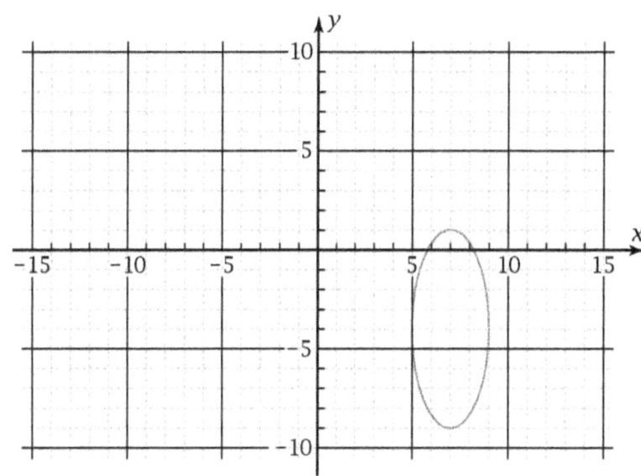

3. $x = 7 + 2 \cos t,\ y = -4 + 5 \sin t$

4. The graphs match.

5. $25(x - 7)^2 + 4(y + 4)^2 = 100$
$25x^2 - 350x + 1225 + 4y^2 + 32y + 64 = 100$
$25x^2 + 4y^2 - 350x + 32y + 1189 = 0$

6. The graphs match.

7. Hyperbola opening vertically, center at the point (−6, 1), transverse radius $a = 3$, conjugate radius $b = 4$, slope of asymptotes $m = \pm\frac{3}{4} = \pm 0.75$

8.

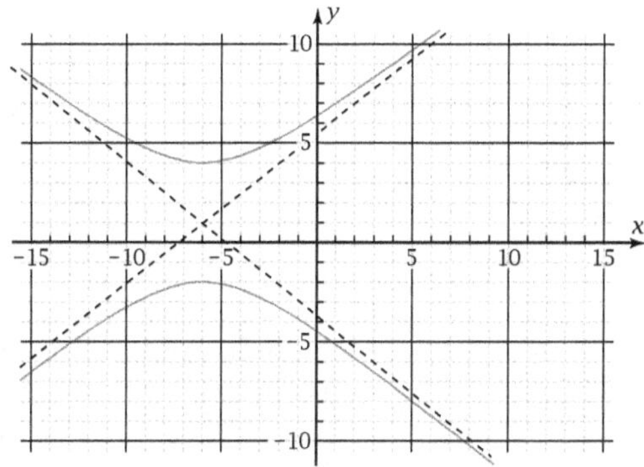

9. $x = -6 + 4 \tan t,\ y = 1 + 3 \sec t$
The graphs match.

10. $x = 1 + 6 \sec t,\ y = 2 + 4 \tan t$
The graphs match.

11. $\left(\dfrac{x-1}{6}\right)^2 - \left(\dfrac{y-2}{4a}\right)^2 = 1$
$4x^2 - 9y^2 - 8x + 36y - 176 = 0$

12. Answers will vary.

Exploration 120

1.

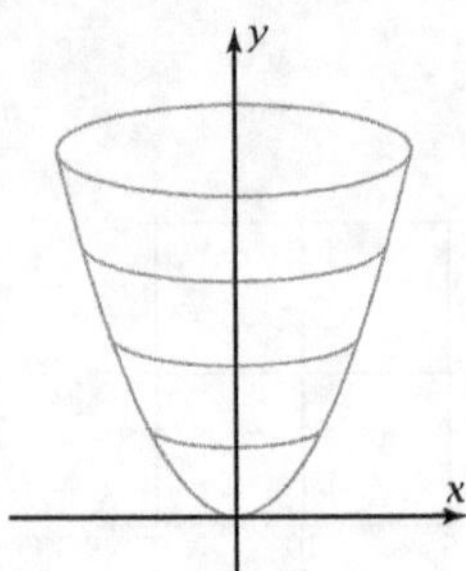

2.

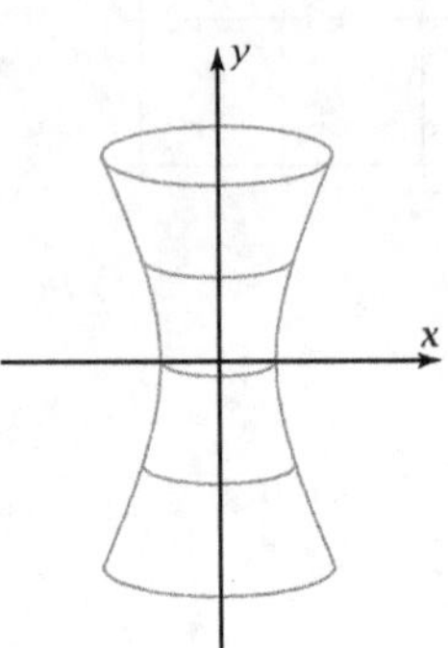

3.

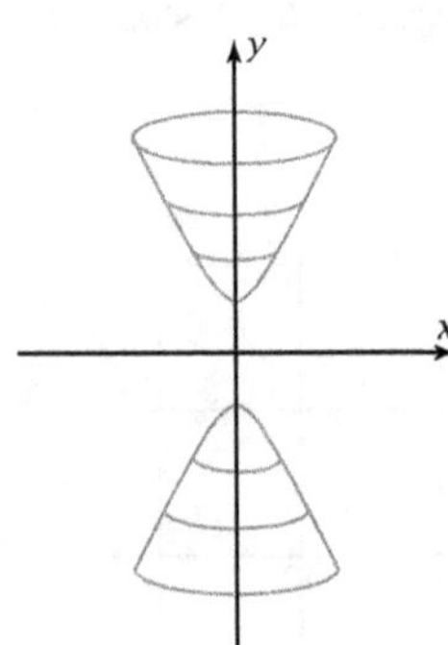

4. The hyperbola in Problem 2 has a single surface (or sheet), and that in Problem 3 has two disconnected surfaces.

5.

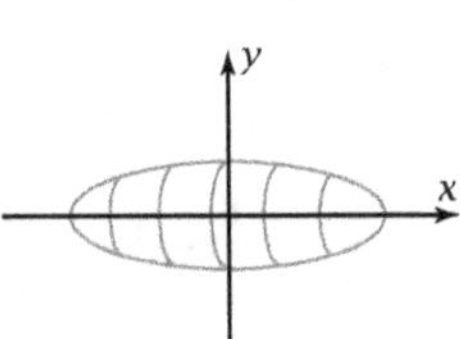

6.

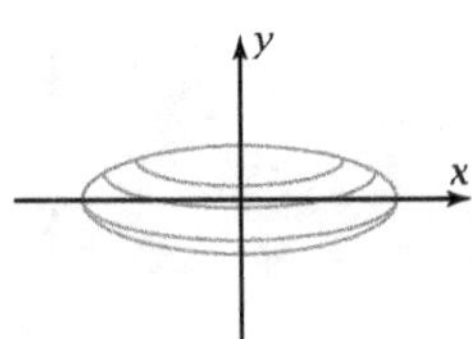

7. Answers will vary.

Exploration 121

1. $r = x = 0.8$ ft
   $h = y = -3(0.8) + 6 = 3.6$ ft
   $V = \pi r^2 h = \pi x^2 y = \pi \cdot 0.8^2 \cdot 3.6$
   $\quad = 2.304\pi = 7.2382...$ ft^3

2. $V = \pi r^2 h = \pi x^2 y = \pi \cdot 1^2 \cdot 3 = 3\pi = 9.4247...$ ft^3
   The resulting volume is $0.696\pi = 2.1865...$ ft^3 larger.

3. $V = \pi r^2 h = \pi x^2 y = \pi x^2(-3x + 6) = 6\pi x^2 - 3\pi x^3$

4.

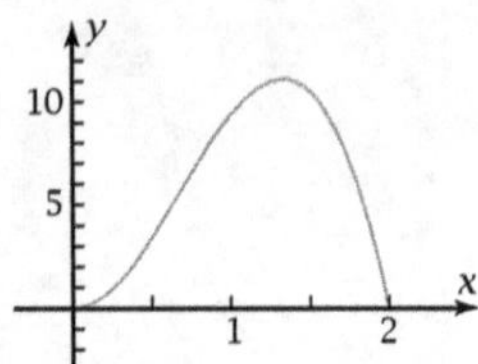

5. $r = \frac{4}{3},\ h = 2$
   Using the maximum feature on your grapher, maximum volume is $V = \frac{32}{9}\pi = 11.1701...$ ft^3 at $x = \frac{4}{3}$ ft.

6. The cylinder is neither tall and skinny nor short and fat. It's in-between.

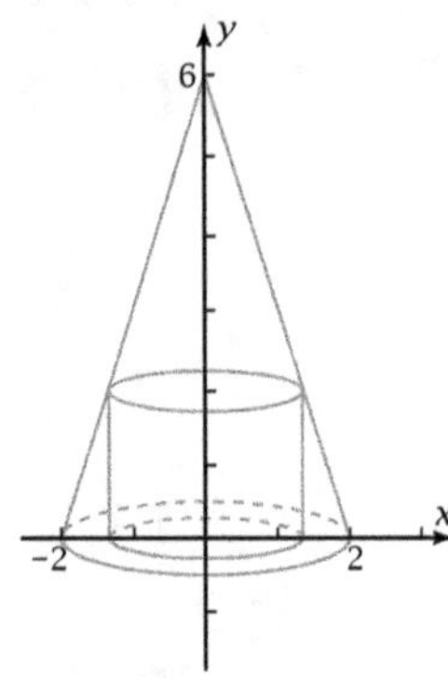

7. Answers will vary.

Exploration 122

1. $d_1 = 5,\ d_2 = 3,\ d_3 = 9$

2. $d_2 = 3 = 0.6 \cdot 5 = 0.6 d_1$

3. $d_2 + d_3 = 3 + 9 = 12$

4.

x	d_1	d_2	d_3	$d_2 = 0.6d_1$	$d_2 + d_3 = 12$
-3	13	7.8	4.2	7.8	12
0	10	6	6	6	12
6	4	2.4	9.6	2.4	12
-6	16	9.6	2.4	9.6	12

5. $d_2 = \sqrt{(0 - 3.6)^2 + (4.8 - 0)^2} = \sqrt{36} = 6$

   $d_3 = \sqrt{[0 - (-3.6)]^2 + (4.8 - 0)^2} = \sqrt{36} = 6$

6. $\left(\dfrac{5}{6}\right)^2 + \left(\dfrac{y}{4.8}\right)^2 = 1$

$$\Rightarrow y = \sqrt{4.8^2\left[1 - \left(\dfrac{5}{6}\right)^2\right]} = \sqrt{\dfrac{11 \cdot 4.8^2}{36}} = \sqrt{7.04}$$

$$d_2 = \sqrt{(5 - 3.6)^2 + (\sqrt{7.04} - 0)^2}$$
$$= \sqrt{1.4^2 + 7.04} = \sqrt{9} = 3$$
$$d_3 = \sqrt{[5 - (-3.6)]^2 + (\sqrt{7.04} - 0)^2}$$
$$= \sqrt{8.6^2 + 7.04} = \sqrt{81} = 9$$

7. $d_2 = 3 = 0.6 \cdot 5 = 0.6 \cdot d_1$

8. $d_2 = e \cdot d_1$

9. $d_2 + d_3 = 2a$

10. Answers will vary.

Exploration 123

1. $d_1 = d_2 = 5$

2.

x	d_1	d_2	Equal?
6	2	2	Yes
0	8	8	Yes
−4	12	12	Yes
7	1	1	Yes

3. $d_1 = d_2$
$(8 - x)^2 = (x - 6)^2 + y^2$
$y^2 = (8 - x)^2 - (x - 6)^2$
$y^2 = 28 - 4x$

4. Only one squared term

5. The graph agrees.

6. $e = 1$

7. $d_1 = 2$
$d_2 = 4 = 2 \cdot 2 = 2d_1$

8.

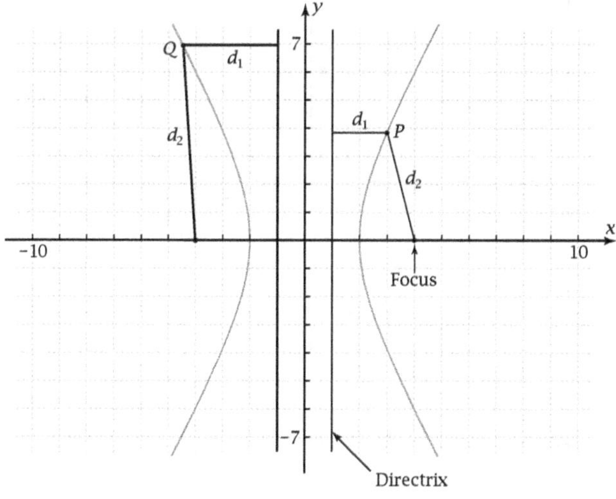

$d_1 = 5.5$
$d_2 = 11 = 2 \cdot 5.5 = 2d_1$

9. $e = 2$

10. Answers will vary.

Exploration 124

1. $9(4)^2 + 25y^2 = 225 \Rightarrow 25y^2 = 81 \Rightarrow y^2 = \dfrac{81}{25}$

$$\Rightarrow y = \pm\dfrac{9}{5} = \pm 1.8$$

The points $(4, \pm 1.8)$ are on the graph.

2. The path of the pencil follows the ellipse.

3. Hypotenuse $= \sqrt{4^2 + 3^2} = 5$. This is the same as the distance from the center to the vertex (the semi-major axis). So $3 = \sqrt{5^2 - 4^2}$.

4. $\sqrt{70^2 - 30^2} = 20\sqrt{10} = 63.2455...$

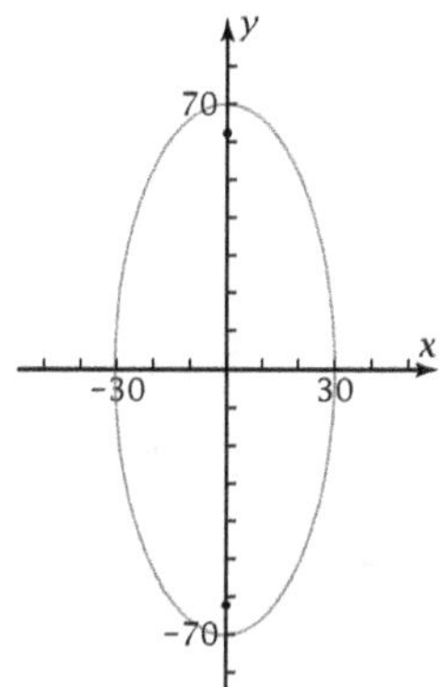

5. Answers will vary.

Exploration 125

1. The graph agrees.

2. $\left(\dfrac{x}{4}\right)^2 - \left(\dfrac{y}{3}\right)^2 = 1$
x-radius $= 4$, y-radius $= 3$

3. $m = \pm\dfrac{3}{4} = \pm\dfrac{x\text{-radius}}{y\text{-radius}}$

4. Transverse radius $a = x$-radius
Conjugate radius $b = y$-radius

5. The distance is 5, equal to the focal radius.

6. $\boxed{c^2 = a^2 + b^2}$

7.

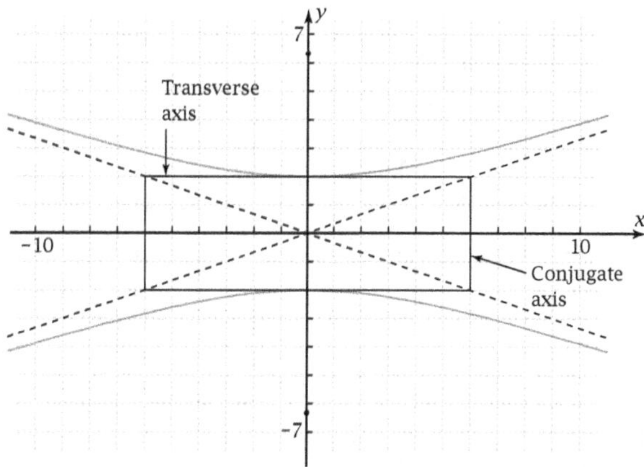

8. $c^2 = a^2 + b^2 = 22 + 62 = 40$
$c = \sqrt{40} = 6.3245...$
See the graph in Problem 7.

9. False

10. Answers will vary.

Exploration 126

1. $\left(\dfrac{x-0}{6}\right)^2 + \left(\dfrac{y-0}{10}\right)^2 = 1 \Rightarrow \dfrac{x^2}{36} + \dfrac{y^2}{100} = 1$
$\Rightarrow 25x^2 + 9y^2 = 900 \Rightarrow 25x^2 + 9y^2 - 900 = 0$

2. $c = \sqrt{a^2 - b^2} = \sqrt{10^2 - 6^2} = 8$; $e = \dfrac{c}{a} = \dfrac{4}{5}$;
$e = ad \Rightarrow d = \dfrac{a}{e} = \dfrac{25}{2} = 12.5$

3.

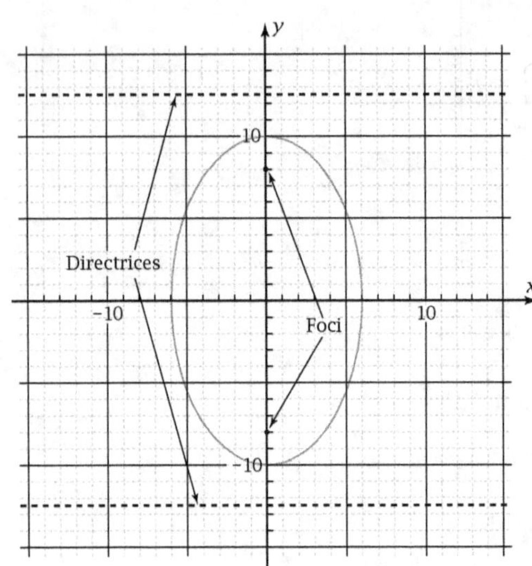

4.

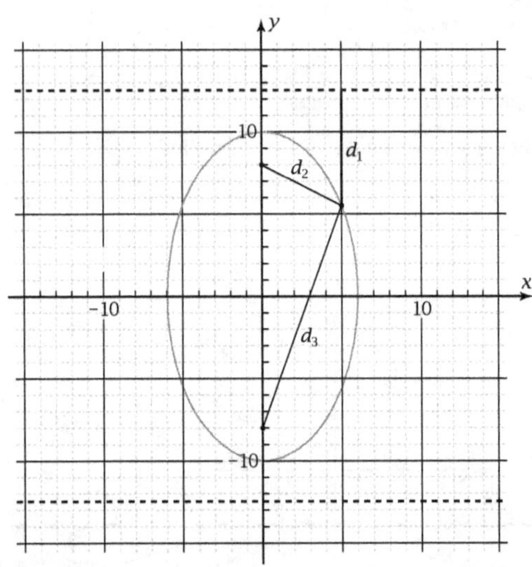

The points are approximately $(5, \pm 5.5)$. The positive y-value is shown.
$d_1 \approx 5.6$, $d_2 \approx 14.4$, and $d_3 = 7.0$.
$d_1 + d_2 = 5.6 + 14.4 = 20 = 2a$, supporting the two-focus property.
$\dfrac{d_1}{d_3} = \dfrac{5.6}{7.0} \approx 0.8 = e$, so $d_1 = ed_3$, supporting the focus-directrix property.

5. 1

6. Because $p = 0 - (-3) = 3$, the directrix, $y = -3 - p$, is $y = -6$.

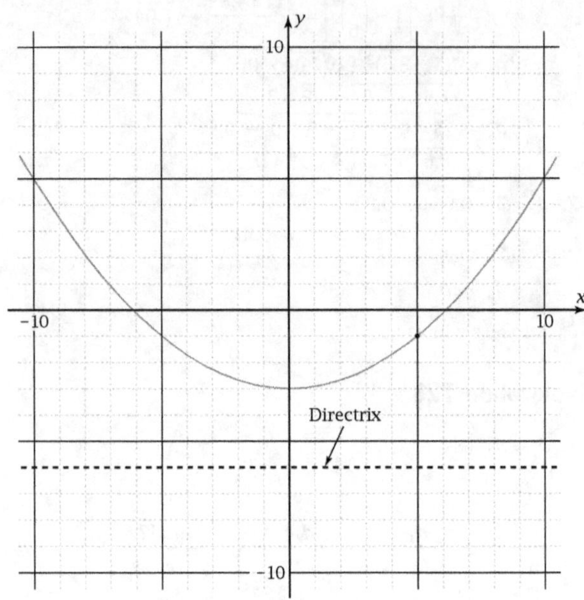

7. Let (x, y) be any point on the parabola, let d_1 be the distance from (x, y) to the focus, and let d_2 be the distance from (x, y) to the directrix. Then $d_1 = d_2$
$\Rightarrow \sqrt{(x-0)^2 + (y-0)^2} = |y - (-6)|$
$\Rightarrow x^2 + y^2 = (y+6)^2 = y^2 + 12y + 36$
$\Rightarrow y = \dfrac{1}{12}x^2 - 3$

8. $y = \frac{1}{12}(5)^2 - 3 = -\frac{11}{12} = -0.91\overline{6}$; the point $\left(5, -\frac{11}{12}\right)$ matches the graph, within the accuracy of the drawing.

9.

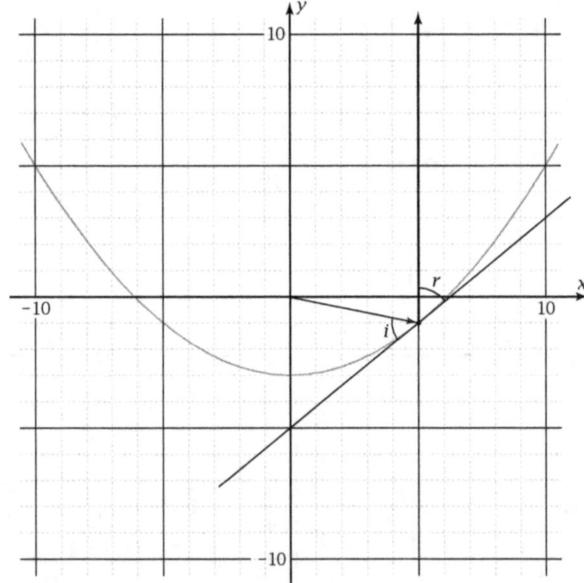

By measurement,
$m\angle r$ (reflected) $= m\angle i$ (incident) $\approx 50°$.

10. Answers will vary.

Exploration 127

1. $x^2 + 4y^2 - 4x - 16y - 16 = 0$
$\Rightarrow 4y^2 - 16y + (x^2 - 4x - 16) = 0$
$$\Rightarrow y = \frac{-(-16) \pm \sqrt{(-16)^2 - 4(4)(x^2 - 4x - 16)}}{2(4)}$$
$$= \frac{4 \pm \sqrt{-x^2 + 4x + 32}}{2}$$

The equation gives the same graph.

2. $x^2 + xy + 4y^2 - 4x - 16y - 16 = 0$
$\Rightarrow 4y^2 + (x - 16)y + (x^2 - 4x - 16) = 0$
$$y = \frac{-(x - 16) \pm \sqrt{(x - 16)^2 - 4(4)(x^2 - 4x - 16)}}{2(4)}$$
$$= \frac{-x + 16 \pm \sqrt{-15x^2 + 32x + 512}}{8}$$

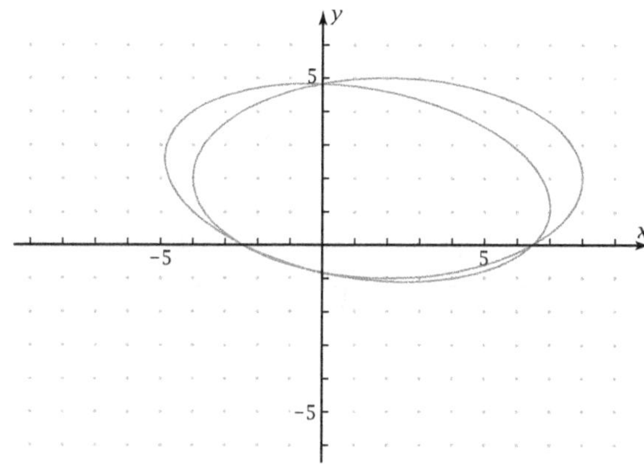

3. $x^2 - xy + 4y^2 - 4x - 16y - 16 = 0$
$\Rightarrow 4y^2 + (-x - 16)y + (x^2 - 4x - 16) = 0$
$$y = \frac{-(-x - 16) \pm \sqrt{(-x - 16)^2 - 4(4)(x^2 - 4x - 16)}}{2(4)}$$
$$= \frac{x + 16 \pm \sqrt{-15x^2 + 96x + 512}}{8}$$

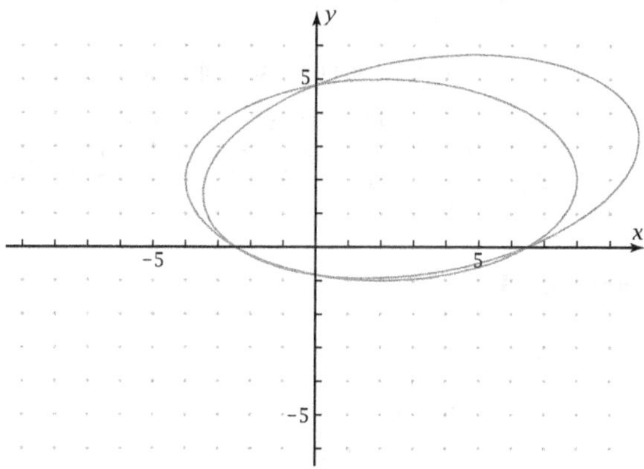

4. The slope of the axis is changed from negative to positive.

5. Problem 1: $(0)^2 - 4(1)(4) = -16 < 0$
Problem 2: $(1)^2 - 4(1)(4) = -15 < 0$
Problem 3: $(-1)^2 - 4(1)(4) = -15 < 0$

6. $B^2 - 4(1)(4) = 0 \Rightarrow B^2 = 16 \Rightarrow B = \pm 4$

7. $x^2 + 4xy + 4y^2 - 4x - 16y - 16 = 0$
$\Rightarrow 4y^2 + (4x - 16)y + (x^2 - 4x - 16) = 0$
$$y = \frac{-(4x - 16) \pm \sqrt{(4x - 16)^2 - 4(4)(x^2 - 4x - 16)}}{2(4)}$$
$$= \frac{-x + 4 \pm 2\sqrt{-x + 8}}{2}$$

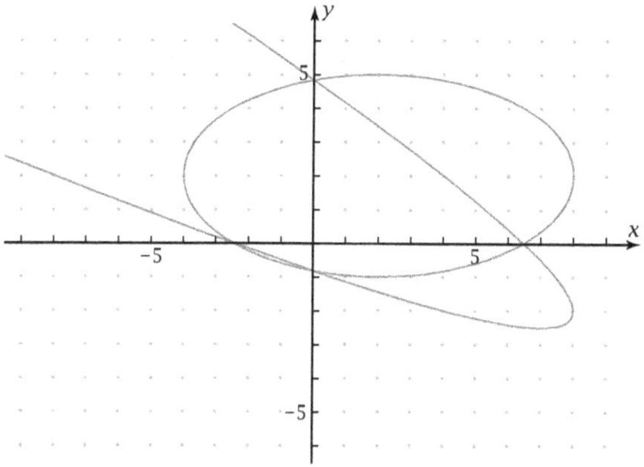

8. A parabola

9. $x^2 + 5xy + 4y^2 - 4x - 16y - 16 = 0$
$\Rightarrow 4y^2 + (5x - 16)y + (x^2 - 4x - 16) = 0$
$$y = \frac{-(5x - 16) \pm \sqrt{(5x - 16)^2 - 4(4)(x^2 - 4x - 16)}}{2(4)}$$
$$= \frac{-5x + 16 \pm \sqrt{9x^2 - 96x + 512}}{8}$$

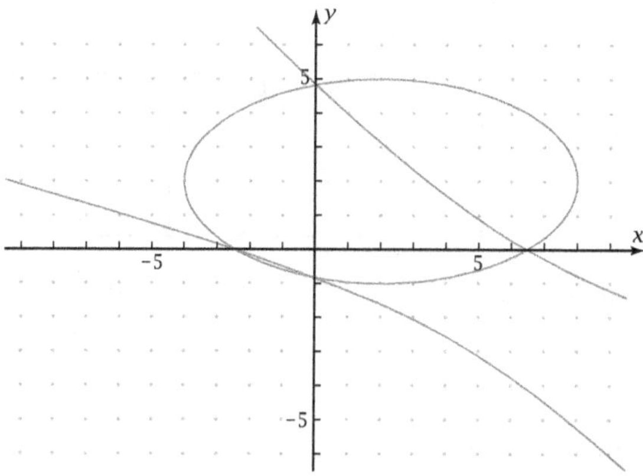

10. A hyperbola; $(5)^2 - 4(1)(4) = 9 > 0$

11. If the discriminant is negative, the graph is an ellipse (or a circle if $A = C$). If the discriminant is 0, the graph is a parabola. If the discriminant is positive, the graph is a hyperbola.

12. $(7)^2 - 4(3)(4) = 1$; hyperbola
$(-2)^2 - 4(5)(6) = -116$; ellipse
$(50)^2 - 4(10)(3) = 2380$; hyperbola
$(2)^2 - 4(1)(1) = 0$; parabola
$(3)^2 - 4(1)(1) = 5$; hyperbola

13. $B = 0$
Ellipse or circle: A and C have the same sign $\Leftrightarrow -4AC < 0$
Hyperbola: A and C have opposite signs $\Leftrightarrow -4AC > 0$
Parabola: either $A = 0$ or $C = 0 \Leftrightarrow -4AC = 0$

14. Answers will vary.

Exploration 128

1. $d_1 = \sqrt{(12-0)^2 + (5-0)^2} = 13$ mi
   $d_2 = \sqrt{(12-6)^2 + (5-0)^2} = \sqrt{61} \approx 7.8$ mi
   It is closer to Warehouse 2 by $13 - \sqrt{61} \approx 5.2$ mi.

2. $C_1 = (13 \text{ mi})(\$10/\text{mi}) = \$130.00$
   $C_2 = (\sqrt{61} \text{ mi})(\$20/\text{mi}) = \$156.20$
   The more remote warehouse is cheaper by
   $\$156.20 - \$130.00 = \$26.20$.

3. Any point (x, y) that is more than 4 miles from the point
   $(8, 0)$ works; that is, $(x-8)^2 + (y-0)^2 < 16$ (see Problem 6).

4. No question

5. $20d_2 \le 10d_1 \Rightarrow 2d_2 \le d_1$
   $\Rightarrow 2\sqrt{(x-6)^2 + (y-0)^2} \le \sqrt{(x-0)^2 + (y-0)^2}$
   $\Rightarrow 4[(x-6)^2 + y^2] \le x^2 + y^2$
   $\Rightarrow 4(x^2 - 12x + 36 + y^2) \le x^2 + y^2$
   $\Rightarrow 4x^2 - 48x + 144 + 4y^2 \le x^2 + y^2$
   $\Rightarrow 3x^2 + 3y^2 - 48x + 144 \le 0$

6. $3(x^2 - 16x + 64) + 3y^2 \le -144 + 3 \cdot 64$
   $\Rightarrow 3(x-8)^2 + 3y^2 \le 48$
   $\Rightarrow (x-8)^2 + (y-0)^2 \le 16 = 4^2$
   The region is inside (and on the boundary of) a circle with
   center at the point $(8, 0)$ and radius 4.

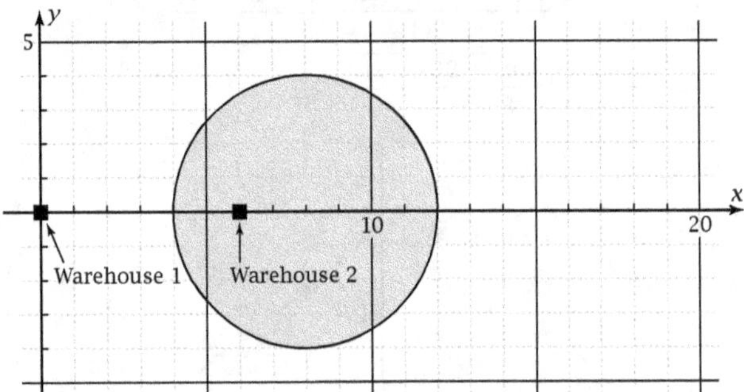

7. Answers will vary.

Exploration 129

1. $y = \pm\dfrac{24}{5} = \pm 4.8$ units

2. $\theta_2 = \tan^{-1}\dfrac{\frac{24}{8} - 0}{8 - 6} = \tan^{-1}\dfrac{12}{5} = 67.3801...°$

   $\theta_3 = \tan^{-1}\dfrac{\frac{24}{5} - 0}{6 - (-8)} = \tan^{-1}\dfrac{12}{35} = 18.9246...°$

3. $\theta_1 = 180° - \theta_2 - \theta_3 = 93.6952...°$

4. Supplement $= 180° - \theta_1 = 86.3047...°$
   $\frac{1}{2} \cdot$ Supplement $= 43.1523...°$

5. Constructing two angles of $43°$, you get:

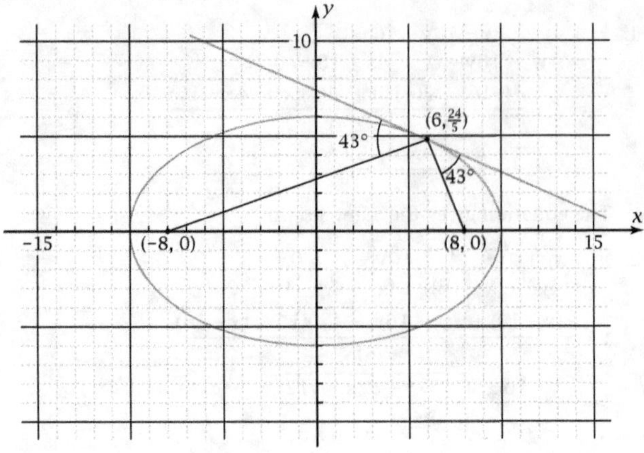

6. $\dfrac{180° - \theta_1}{2} + \theta_1 + \dfrac{180° - \theta_1}{2} = 180°$

7. It appears to be tangent. (In fact, you can show that it is
   tangent by using a method introduced in the next Exploration.)

8.

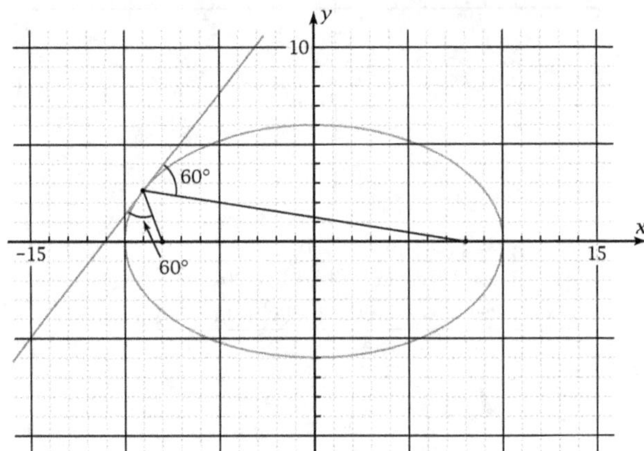

By direct measurement, each angle is approximately $60°$.

9. The lines from the foci to any point on the ellipse form equal
   angles with the tangent to the ellipse at that point. More
   physically, any line from one of the foci is reflected by the
   ellipse to the other focus.

Exploration 130

1. $\left(\dfrac{3}{5}\right)^2 + \left(\dfrac{y}{3}\right)^2 = 1 \Rightarrow \left(\dfrac{y}{3}\right)^2 = 1 - \left(\dfrac{3}{5}\right)^2 = \dfrac{16}{25}$
   $\Rightarrow \dfrac{y}{3} = \pm\dfrac{4}{5}$ (we use the positive value)
   $\Rightarrow y = \dfrac{12}{5} = 2.4$ units. The point is $\left(3, \dfrac{12}{5}\right)$.

2. As shown here,
$$m = \frac{\text{rise}}{\text{run}} = \frac{-4.5 \text{ units}}{10 \text{ units}} = -0.45$$

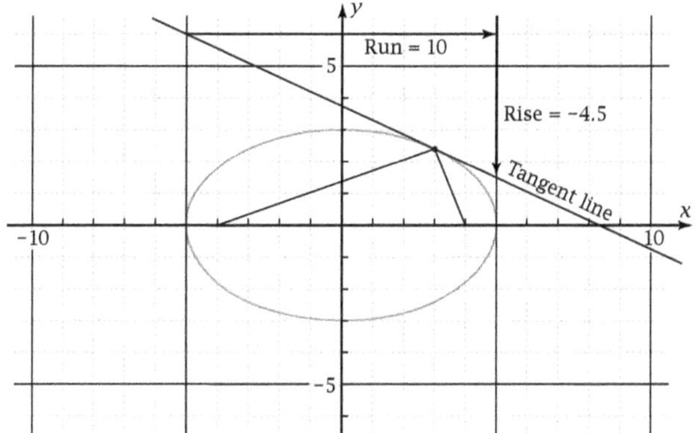

3. $y - \dfrac{12}{5} = -0.45(x - 3) \Rightarrow y = -0.45x + 3.75$

$\qquad = -\dfrac{9}{20}x + \dfrac{15}{4}$, or $9x + 20y = 75$

4. No question

5. Let l_1 be the line from point $(-4, 0)$, l_2 be the line from point $(4, 0)$, and l_T be the tangent line; let ϕ_1 be the positive acute angle between the x-axis and l_1, ϕ_2 be the positive obtuse angle between the x-axis and l_2, and ϕ_T be the positive angle from the x-axis to l_T; and let θ_1 be the acute angle between l_1 and l_T and θ_2 be the acute angle between l_2 and l_T. Then $\theta_1 = 180° - (\phi_T - \phi_1)$ and $\theta_2 = \phi_T - \phi_2$ (examine the sketch carefully to see why).

$$\tan\phi_1 = m_1 = \frac{\frac{12}{5} - 0}{3 - (-4)} = \frac{12}{35}$$

$$\tan\phi_2 = m_2 = \frac{\frac{12}{5} - 0}{3 - 4} = -\frac{12}{5}$$

$$\tan\phi_T = m_T = -0.45 = -\frac{9}{20}$$

So
$$\tan\theta_1 = \tan[180° - (\phi_T - \phi_1)]$$
$$= -\tan(\phi_T - \phi_1) = \tan(\phi_1 - \phi_T)$$
$$= \frac{\tan\phi_1 - \tan\phi_T}{1 + \tan\phi_T \tan\phi_1} = \frac{\frac{12}{35} - \left(-\frac{9}{20}\right)}{1 + \left(-\frac{9}{20}\right) \cdot \frac{12}{35}} = \frac{15}{16}$$

$$\tan\theta_2 = \tan(\phi_T - \phi_2) = \frac{\tan\phi_T - \tan\phi_2}{1 + \tan\phi_T \tan\phi_2}$$
$$= \frac{-\frac{9}{20} - \left(-\frac{12}{5}\right)}{1 + \left(-\frac{9}{20}\right)\left(-\frac{12}{5}\right)} = \frac{15}{16}$$

So $\theta_1 = \theta_2 = \tan^{-1}\frac{15}{16} = 43.1523...°$.

6. The rule for reflection off a curve is the same as for reflection off the tangent to the curve at that point: "angle of reflection = angle of incidence." Thus, the line from either focus to any point on the ellipse is reflected to the other focus.

7. $y - \dfrac{12}{5} = m(x - 3) \Rightarrow y = mx - 3m + \dfrac{12}{5}$

8. $9x^2 + 25y^2 = 225$

$$\Rightarrow 9x^2 + 25\left(mx - 3m + \frac{12}{5}\right)^2 = 225$$

$$\Rightarrow 9x^2 + 25\left(m^2x^2 - 6m^2x + \frac{24}{5}mx + 9m^2 - \frac{72}{5}m + \frac{144}{25}\right)$$
$$= 225$$

$$\Rightarrow 9x^2 + 25m^2x^2 - 150m^2x + 120mx$$
$$+ 225m^2 - 360m - 80 = 0$$

$$\Rightarrow (25m^2 + 9)x^2 + (-150m^2 + 120m)x$$
$$+ (225m^2 - 360m - 81) = 0$$

9. Try factoring using synthetic division. Recall that if $ax + b$ is a factor of $Ax^2 + Bx + C$, then b must be a factor of C. Factor $C = 225m^2 - 360m - 81 = 9(25m^2 - 40m - 9)$ $= 9(5m - 9)(5m + 1)$. So if $ax + b$ is a factor, then b must be ± 1, ± 3, $\pm(5m - 9)$, $\pm(5m + 9)$, or some product of those. Synthetic division (shown next for the case that *does* work) reveals that ± 1 and $+3$ do not work but that -3 does; i.e., $x - 3$ is a factor. (Recall that in synthetic division, to test $x - 3$, we use $+3$.)

$$
\begin{array}{r|ccc}
3 & 25m^2 + 9 & -150m^2 + 120m & 225m^2 - 360m - 81 \\
  &   & 75m^2 + 27   & -225m^2 + 360m + 81 \\
\hline
  & 25m^2 + 9 & -75m^2 + 120m + 27 & 0
\end{array}
$$

The factorization is
$(x - 3)[(25m^2 + 9)x + (-75m^2 + 120m + 27)]$, giving the solutions $x = 3$, or

$$x = \frac{75m^2 - 120m - 27}{25m^2 + 9} = 3 \cdot \frac{25m^2 - 40m - 9}{25m^2 + 9}.$$

The *other* solution should also yield $x = 3$.

$$3 \cdot \frac{25m^2 - 40m - 9}{25m^2 + 9} = 3 \Rightarrow 25m^2 - 40m - 9 = 25m^2 + 9$$

$$\Rightarrow 40m = -18 \Rightarrow m = -\frac{19}{20}$$

10. Answers will vary.

Exploration 131

1. Yes

2. Drawings will vary depending on the point chosen, but the answer should still be yes.

3. $d_1 = 2d_2$

$$\Rightarrow \sqrt{(x - 7)^2 + (y - 5)^2} = 2\sqrt{(x - 1)^2 + (y - 2)^2}$$
$$\Rightarrow x^2 - 14x + 49 + y^2 - 10y + 25$$
$$= 4(x^2 - 2x + 1 + y^2 - 4y + 4)$$
$$= 4x^2 - 8x + 4 + 4y^2 - 16y + 16$$
$$\Rightarrow 3x^2 + 3y^2 + 6x - 6y - 54 = 0$$
$$\Rightarrow x^2 + y^2 + 2x - 2y - 18 = 0$$

4. The x^2- and y^2-terms have the same coefficient.

5. $x^2 + y^2 + 2x - 2y - 18 = 0$

$$\Rightarrow y^2 - 2y + (x^2 + 2x - 18) = 0$$

$$\Rightarrow y = \frac{-(-2) \pm \sqrt{(-2)^2 - 4(1)(x^2 + 2x - 18)}}{2(1)}$$

$$= 1 \pm \sqrt{-x^2 - 2x + 19}$$

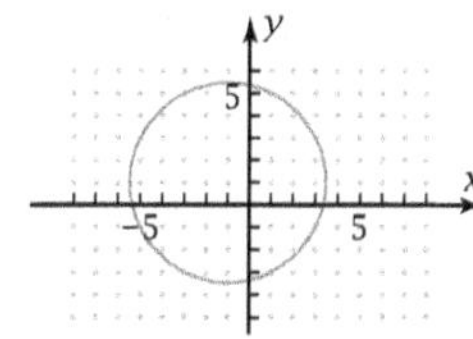

6. $x^2 + y^2 + 2x - 2y - 18 = 0$

$\Rightarrow \left(x^2 + 2x + 1\right) + \left(y^2 - 2y + 1\right) = 18 + 1 + 1 = 20$

$\Rightarrow (x + 1)^2 + (y - 1)^2 = \left(\sqrt{20}\right)^2$

Center $(-1, 1)$
radius $= \sqrt{20} = 4.4721\ldots$

7. Answers will vary.

Exploration 132

1. The graphs cross at about the point $(11, \pm5.8)$.

2. Major axis from $x = -3$ to $x = 15$,
   minor axis from $y = -7$ to $y = 7$;

   center $\left(\dfrac{-3 + 15}{2}, \dfrac{-7 + 7}{2}\right) = (6, 0)$,

   $a = \dfrac{15 - (-3)}{2} = 9$, $b = \dfrac{7 - (-7)}{2} = 7$;

   $\left(\dfrac{x - 6}{9}\right)^2 + \left(\dfrac{y - 0}{7}\right)^2 = 1$;

   $49(x - 6)^2 + 81y^2 = 3969 \Rightarrow 49x^2 + 81y^2 - 588x + 2205 = 0$

   Opening horizontally, center at point $(0, 0)$,
   $a = 5$, $m =$ slope of asymptotes $= \frac{3}{5}$,

   $\dfrac{b}{a} = m \Rightarrow b = ma = 3$;

   $\left(\dfrac{x - 0}{5}\right)^2 - \left(\dfrac{y - 0}{3}\right)^2 = 1$;

   $9x^2 - 25y^2 = 225 \Rightarrow 9x^2 - 25y^2 - 225 = 0$

3. $\left(\dfrac{x - 6}{9}\right)^2 + \left(\dfrac{y - 0}{7}\right)^2 = 1 \Rightarrow y = \pm\dfrac{7}{9}\sqrt{-x^2 + 12x + 45}$

   $\left(\dfrac{x - 0}{5}\right)^2 - \left(\dfrac{y - 0}{3}\right)^2 = 1 \Rightarrow y = \pm\dfrac{3}{5}\sqrt{x^2 - 25}$

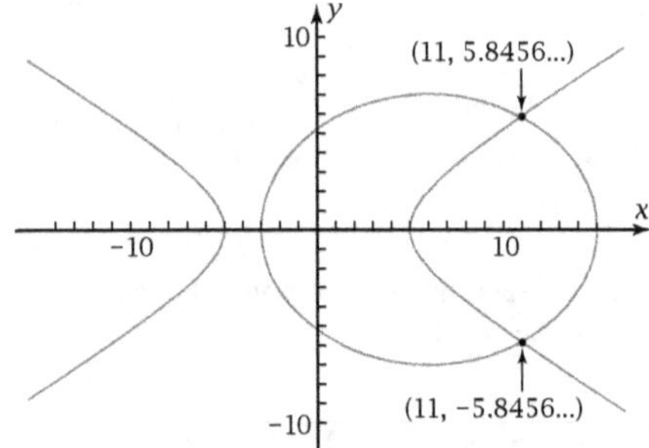

4. $y = \pm\dfrac{7}{9}\sqrt{-x^2 + 12x + 45}$ and $y = \pm\dfrac{3}{5}\sqrt{x^2 - 25}$

   $\Rightarrow \dfrac{7}{9}\sqrt{-x^2 + 12x + 45} = \pm\dfrac{3}{5}\sqrt{x^2 - 25}$

   $\Rightarrow 35\sqrt{-x^2 + 12x + 45} = \pm27\sqrt{x^2 - 25}$

   $\Rightarrow 1{,}225(-x^2 + 12x + 45) = 729(x^2 - 25)$

   $\Rightarrow -1{,}225 + 14{,}700x + 55{,}125 = 729x^2 - 18{,}225$

   $\Rightarrow 1{,}954x^2 - 14{,}700x - 73{,}350 = 0$

   $\Rightarrow x = \dfrac{-(-14{,}700) \pm \sqrt{(-14{,}700)^2 - 4(1{,}954)(-73{,}350)}}{2(1{,}954)}$

   $= \dfrac{3{,}675 \pm 90\sqrt{6{,}091}}{977} = 10.9509\ldots$ or $-3.4278\ldots$

   Substituting $-3.4278\ldots$ into the original equation gives two
   imaginary values for y:

   $y = \pm\dfrac{3}{5}\sqrt{(-3.4278)^2 - 25} = \pm2.1840i$

   Substituting $10.9509\ldots$ into the original equation gives

   $y = \pm\dfrac{3}{5}\sqrt{10.9509\ldots^2 - 25} = \pm5.8456\ldots$

5. Answers will vary.

Polar Coordinates, Complex Numbers, and Moving Objects

Exploration 133

1.

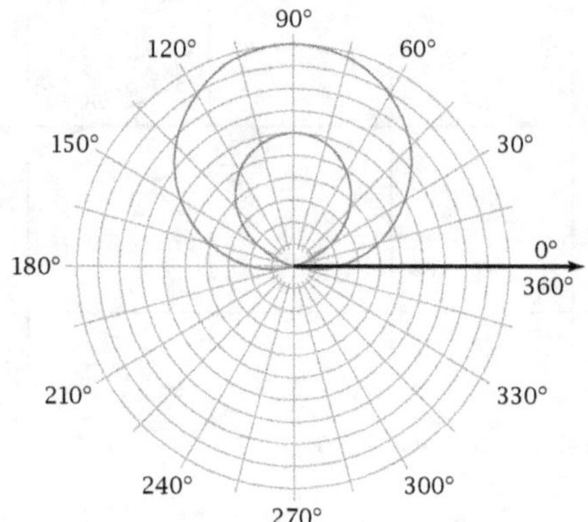

2. You go around to the correct angle θ and then plot the point
   back in the opposite direction.

3. The values agree when rounded to one decimal place.

4. The figures agree.

5. "Limaçon of Pascal"

6. $r = 3 + 7 \sin \theta = 0$
   $\sin \theta = -\dfrac{3}{7}$
   $\theta = -25.3769\ldots° + 360°n$ or $205.3769\ldots° + 360°n$

7. $\theta = 205.3769\ldots°$
   $\theta = 334.6230\ldots°$

8. The rays are tangent to the graph at the pole.

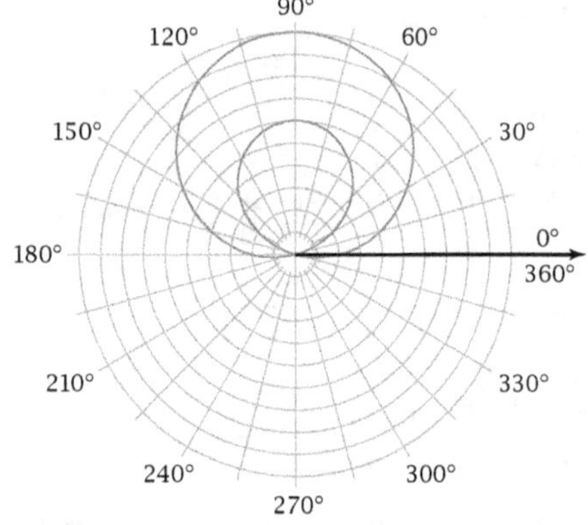

9. Answers will vary.

Exploration 134

1.

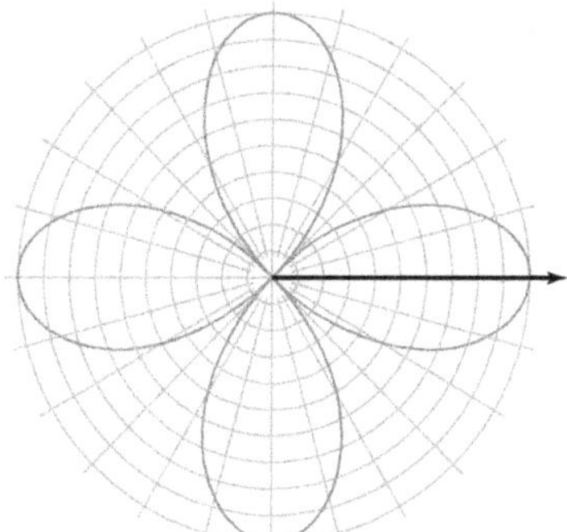

2.

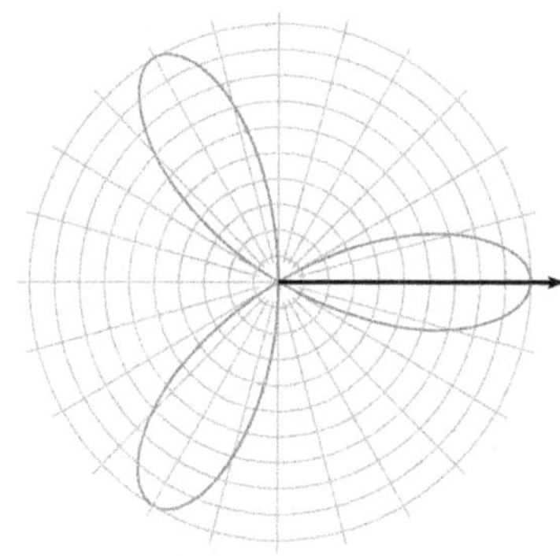

3. For $n = 4$, the rose has 8 leaves; for $n = 5$, the rose has 5 leaves. If n is even, the rose has $2n$ leaves. If n is odd, the number of leaves equals n. In this case, each leaf is traced twice as θ makes one complete revolution.

4.

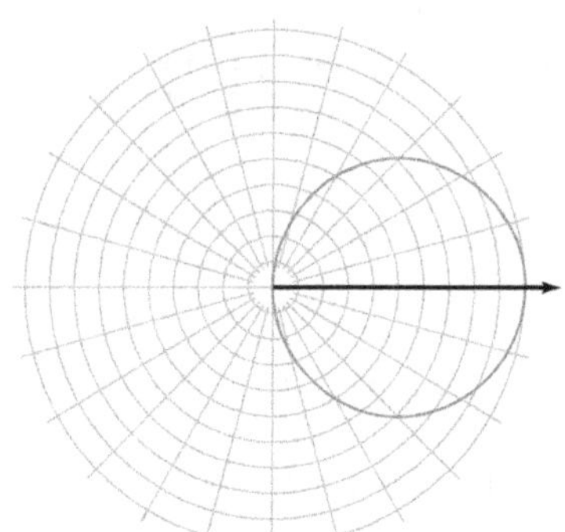

5. $r = 10 \cos \theta$
   $r^2 = 10r \cos \theta$
   $x^2 + y^2 = 10x$
   $x^2 + 10x + 25 + y^2 = 25$
   $(x - 5)^2 + y^2 = 5^2$

   Circle centered at point $(5, 0)$, with radius 5 and including the pole.

6. $r = a \cos n\theta$ has n leaves if n is odd, so $r = a \cos 1\theta$ has one leaf.

7. Answers will vary.

Exploration 135

1. Graph paused at $\theta \approx 65°$ (see Problem 5).

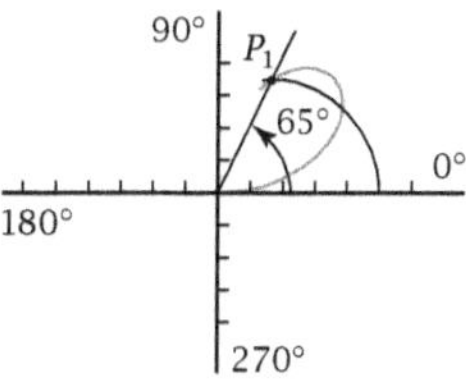

2. Graph paused at $\theta \approx 105°$ (see Problem 6).

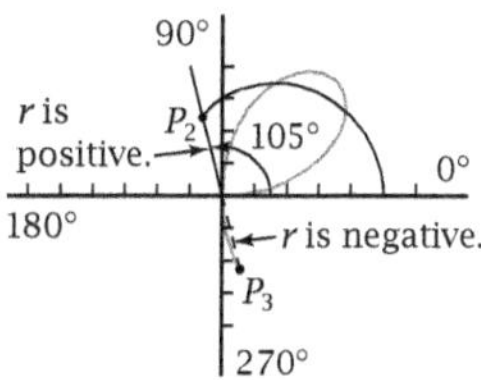

   At this angle, P_2 lies only on the limaçon, not on the rose.

3.

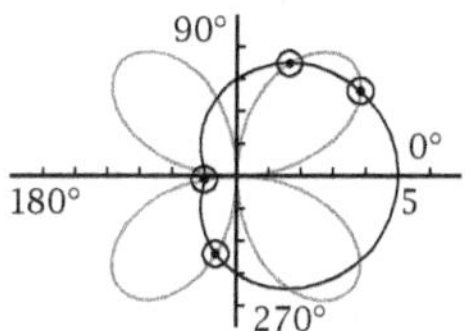

   The two points in the first quadrant and the two points in the third quadrant are true intersections. The two points in the second quadrant and the two points in the fourth quadrant are false intersections. At the false intersections, the r-values on the rose are negative.

4.

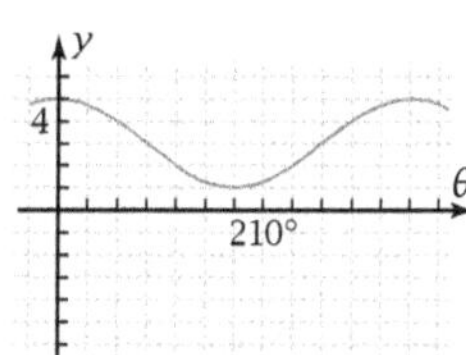

5.

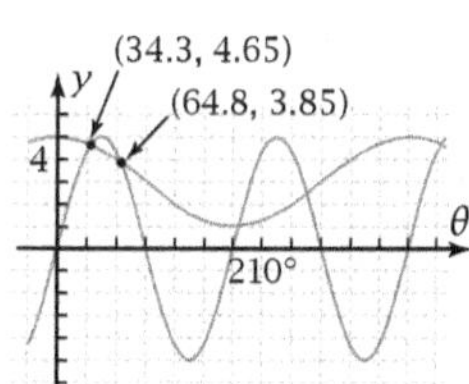

6. The polar graph shows that for P_2, $\theta \approx 105°$ and r has a positive value. So this point must be only on the limaçon, as the Cartesian graph shows.

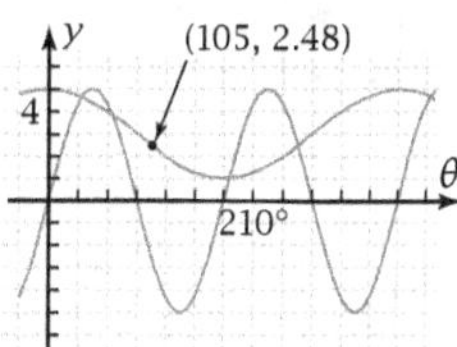

7. Answers will vary.

Exploration 136

1. $z_1 = 4 + 2i$, $z_2 = 1 + 3i$

2. $i = \sqrt{-1}$, so $i^2 = \left(\sqrt{-1}\right)^2 = -1$

3. $z_3 = z_1 z_2 = (4 + 2i)(1 + 3i)$
   $= 4 \cdot 1 + 4 \cdot 3i + 2i \cdot 1 + 2i \cdot 3i$
   $= 4 + 12i + 2i + 6i^2$
   $= 4 + 14i + 6(-1) = 4 + 14i - 6 = -2 + 14i$

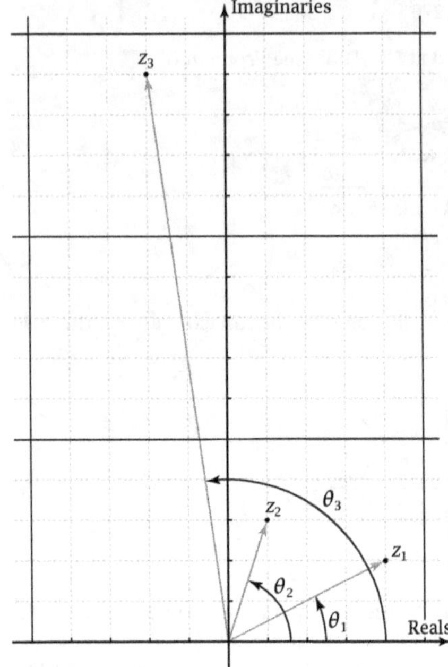

4. $\theta_1 \approx 26°$; $\theta_2 \approx 71°$; $\theta_3 = \arg z_1 z_2 \approx 98°$;
   $\theta_3 \approx \theta_1 + \theta_2$

5. $\theta_1 = \arg z_1 = \tan^{-1}\frac{2}{4} = 26.5650...°$;

   $\theta_2 = \arg z_2 = \tan^{-1}\frac{3}{1} = 71.5650...°$;

   $\theta_3 = \arg z_1 z_2 = \arctan\frac{14}{-2} = 180° + \tan^{-1}\frac{14}{-2}$
   $= 98.1301...° = \theta_1 + \theta_2$, so $\arg z_1 z_2 = \arg z_1 + \arg z_2$

6. $r_1 = |z_1| = \sqrt{4^2 + 2^2} = 2\sqrt{5}$;
   $r_2 = |z_2| = \sqrt{1^2 + 3^2} = \sqrt{10}$;
   $r_3 = |z_1 z_2| = \sqrt{(-2)^2 + 14^2} = 10\sqrt{2}$
   $= 2\sqrt{5} \cdot \sqrt{10} = r_1 r_2$, so $|z_1 z_2| = |z_1| \cdot |z_2|$

7. Answers will vary.

Exploration 137

1. $z_1 z_2 = (3 \text{ cis } 57°)(5 \text{ cis } 41°)$
   $= (3 \cos 57° + 3i \sin 57°)(5 \cos 41° + 5i \sin 41°)$
   $= 15 \cos 57° \cos 41° + 15i \cos 57° \sin 41°$
   $\quad + 15i \sin 57° \cos 41° + 15i^2 \sin 57° \sin 41°$
   $= 6.1656... + 5.3597...i + 9.4942...i - 8.532...$
   $= -2.0875... + 14.8540...i$
   $= \sqrt{(-2.0875...)^2 + 14.8540...^2} \text{ cis arctan}\frac{14.8540...}{-2.0875...}$
   $= \sqrt{225} \text{ cis}\left(180° + \tan^{-1}\frac{14.8540...}{-2.0875...}\right)$
   $= 15 \text{ cis } 98°$

2. $z_1 z_2 = (3 \text{ cis } 57°)(5 \text{ cis } 41°)$
   $= (3 \cdot 5)(\text{cis } 57° \cdot \text{cis } 41°)$
   $= 15(\cos 57° + i \sin 57°)(\cos 41° + i \sin 41°)$
   $= 15(\cos 57° \cos 41° + i \cos 57° \sin 41°$
   $\quad + i \sin 57° \cos 41° + i^2 \sin 57° \sin 41°)$
   $= 15[(\cos 57° \cos 41° - \sin 57° \sin 41°)$
   $\quad + i(\cos 57° \sin 41° + \sin 57° \cos 41°)]$
   $= 15[\cos(57° + 41°) + i \sin(57° + 41°)]$
   $= 15 \text{ cis}(57° + 41°)$

3. Multiply the moduli and add the arguments:
   $(r_1 \text{ cis } \theta_1)(r_2 \text{ cis } \theta_2) = r_1 r_2 \text{ cis}(\theta_1 + \theta_2)$

4. $\text{cis } 0° = \cos 0° + i \sin 0° = 1 + 0i = 1$
   A real number; complex conjugates

5. Answers will vary.

Exploration 138

1. $\vec{v} = 100 \cos 60° \cdot \vec{i} + 100 \sin 60° \cdot \vec{j} = 50 \cdot \vec{i} + 50\sqrt{3} \cdot \vec{j}$

2. $x = 50t$

3. $y = 50t\sqrt{3} - 16t^2$

4.

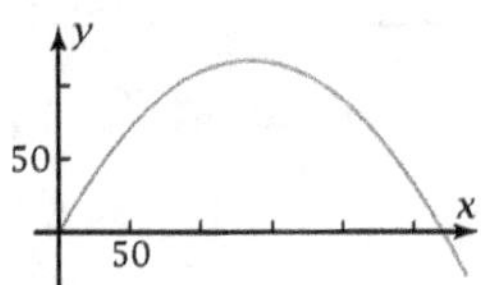

5. Graphically, $y \approx 0$ when $t \approx 5.4$ sec and $x \approx 270$ ft.
   (Algebraically, $y = 50t\sqrt{3} - 16t^2 = 2t(25\sqrt{3} - 8t) = 0$
   $\Rightarrow t = 0$ or $t = \frac{25\sqrt{3}}{8} = 5.4126$ sec,
   and $x = 50t = 50 \cdot \frac{25\sqrt{3}}{8} = \frac{125\sqrt{3}}{4} = 270.6329$ ft.)

6. Solve $y = 3 + 50\sqrt{3}t - 16t^2 = 0$ for t.
   $t = \frac{-50\sqrt{3} \pm \sqrt{(50\sqrt{3})^2 - 4(-16)(3)}}{2(-16)} = \frac{25\sqrt{3} \pm \sqrt{1923}}{16}$
   $t = \frac{25\sqrt{3} + \sqrt{1923}}{16} = 5.4470...$ sec

7. $x = 50\frac{25\sqrt{3} + \sqrt{1923}}{16} = 272.3540...$ ft

8. Solve $y = 100t \sin \alpha - 16t^2 = -3$ for t:
   $-16t^2 + 100t \sin \alpha - 3 = 0$
   $\Rightarrow t = \frac{-100 \sin \alpha \pm \sqrt{(100 \sin \alpha)^2 - 4(-16)(3)}}{2(-16)}$
   $= \frac{25 \sin \alpha + \sqrt{625 \sin^2 \alpha + 12}}{8}$
   $x = 100 \cos \alpha \cdot \frac{25 \sin \alpha + \sqrt{625 \sin^2 \alpha + 12}}{8}$
   $= 25 \cos \alpha \cdot \frac{25 \sin \alpha + \sqrt{625 \sin^2 \alpha + 12}}{2}$

α	t (sec)	x (ft)
60°	5.4470...	272.3540...
50°	4.8266...	310.2494...
40°	4.0635...	311.2870...
30°	3.1838...	275.7329...
20°	2.2220...	208.8005...

The time in the air decreases, but the distance first increases and then decreases. In fact, tracing the *x* function shows a maximum distance of 315.4857... ft at $\alpha = 44.7275...°$.

9. Answers will vary.

Exploration 139

1. $A = (3, 3 \tan t)$
   $B = (5, 5 \tan t)$
   $|B| = \sqrt{25 + 25\tan^2 t} = 5|\sec t|$
   The arc through B intersects the horizontal axis at two points, but define C as the intersection of the arc with the *positive* horizontal axis, so $C = (5|\sec t|, 0)$.
   $x = 5|\sec t|$
   $y = 3 \tan t$

2. 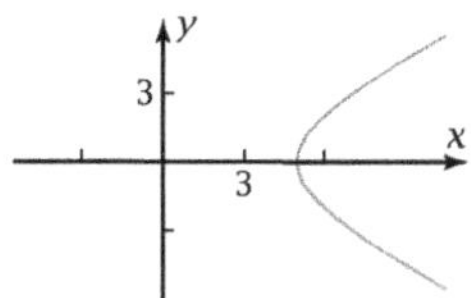

3. $x(0.85) = 5|\sec 0.85| = 7.5759...$
   $y(0.85) = 3 \tan 0.85 = 3.4149...$
   $(7.5759..., 3.4149...)$ appears to be the point P.

4. $\left(\dfrac{x}{5}\right)^2 - \left(\dfrac{y}{3}\right)^2 = \sec^2 t - \tan^2 t = 1$

5. $\left(\dfrac{7}{5}\right)^2 - \left(\dfrac{y}{3}\right)^2 = 1 \Rightarrow y = \pm 3\sqrt{\left(\dfrac{7}{5}\right)^2 - 1} = \pm\dfrac{6\sqrt{6}}{5} = 2.9393...$
   The points $(7, \pm 2.9393...)$ appear to be on the figure.
   $\left(\dfrac{2}{5}\right)^2 - \left(\dfrac{y}{3}\right)^2 = 1 \Rightarrow y = \pm 3\sqrt{\left(\dfrac{2}{5}\right)^2 - 1} = \pm\dfrac{3\sqrt{-21}}{5}$,
   each of which is imaginary. There is no point on the figure that has $x = 2$.

6. $\left(\dfrac{x}{5}\right)^2 - \left(\dfrac{y}{3}\right)^2 = 1$ is the equation of a hyperbola with center at point $(0, 0)$, horizontal semitransverse axis 5, and vertical semiconjugate axis 3.

7. For the parametric equation, $x \geq 5$; but for the Cartesian equation, $x \leq -5$ or $x \geq 5$.

8. Answers will vary.

Exploration 140

1. The triangle is inscribed in a circle, with one side along a diameter of the circle. Therefore, the triangle is a right triangle, with the right angle at A.

2. Let a be the length of the segment from the origin to point A.
   $|a| = 10 \sin t$
   To see this, note that the topmost vertex of the triangle in Problem 1 has measure t, that $|a|$ is the side opposite this angle, and that the hypotenuse has length 10; or just notice that the polar equation of the circle is $r = 10 \sin \theta$.

3. $\dfrac{10}{x} = \tan t \Rightarrow x = 10 \cot t$

4. $\dfrac{y}{a} = \sin t$
   $y = |a|\sin t = 10 \sin^2 t$

5. $x(0.6) = 10 \cot 0.6 = 14.6169...$
   $y(0.6) = 10 \sin^2 0.6 = 3.1882...$
   $(14.6169..., 3.1882...)$ appears to be point P.

6. 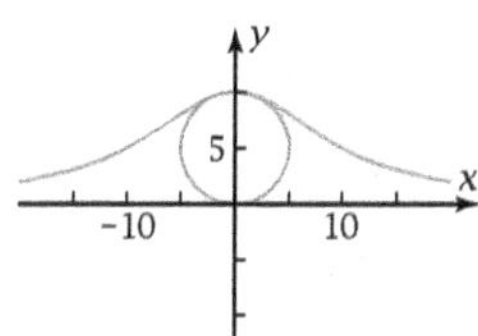

7. For $\pi < t < 2\pi$, the variable line extends in the opposite direction $(r < 0)$.

8. Answers will vary.

Exploration 141

1. a. The point is on the upper right branch. In fact, the point is on the upper right branch for $0 < t < \frac{\pi}{2}$. As long as $t > 0$, the ray eventually intersects the line $y = 2$, so the point lies above that line, so the curve asymptotically approaches $y = 2$ from above in the first quadrant as $t \to 0$ from above (approaches 0 but is greater than 0). For $t = 0$, there is no corresponding point (it's "at infinity").

   b. The point is $(0, 7)$.

   c. The point is on the upper left branch. As in part a, the curve asymptotically approaches $y = 2$ from above in the second quadrant as $t \to \pi$ from below. Again as in part a, for $t \to \pi$ the point is "at infinity"—there is no corresponding point.

2. 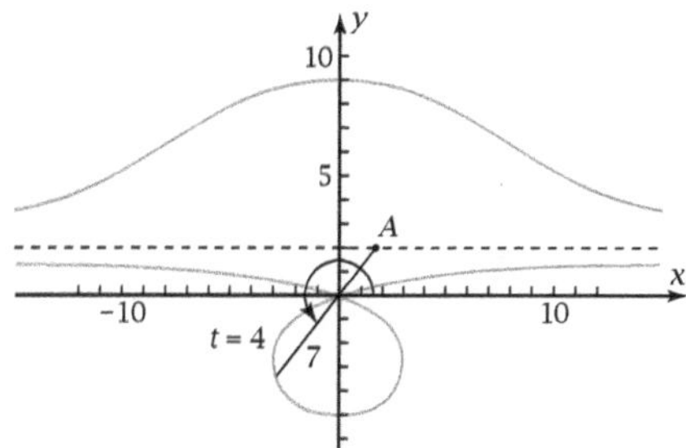

3. $x = 2 \cot t + 7 \cos t$

4. $y = 2 + 7 \sin t$

5. 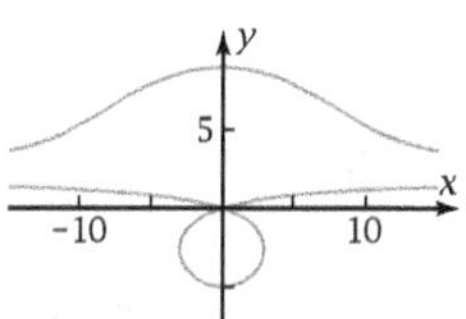

6. Draw a line at $t = 4$ radians $\approx 229°$. A line at this angle intersects the lower lobe of the conchoid at about the point $(-2.8, -3.3)$ and intersects the line $y = 2$ at about the point $(2, 1.7)$. By measuring, these points seem to be 7 units apart.

7. $(x^2 + 8^2)(8 - 2)^2 = 49 \cdot 8^2 \Rightarrow x = \pm\sqrt{\dfrac{3136}{36} - 64} = \pm\dfrac{4\sqrt{13}}{3}$
   $= \pm 4.8074....$ The points $(\pm 4.8074..., 8)$ appear to be on the curve.

---

8. $y = 2 + 7 \sin t \Rightarrow \sin t = \dfrac{y - 2}{7} \Rightarrow \sin^2 t = \dfrac{(y-2)^2}{49}$;

$x = 2 \cot t + 7 \cos t = \cos t\left(\dfrac{2}{\sin t} + 7\right)$

$\Rightarrow x^2 = \cos^2 t\left(\dfrac{2}{\sin t} + 7\right)^2 = (1 - \sin^2 t)\left(\dfrac{2}{\sin t} + 7\right)^2$

$= \left[1 - \dfrac{(y-2)^2}{49}\right]\left(\dfrac{14}{y-2} + 7\right)^2 = \left[\dfrac{49 - (y-2)^2}{49}\right]\left(\dfrac{7y}{y-2}\right)^2$

$= [49 - (y-2)^2]\left(\dfrac{y}{y-2}\right)^2 = 49\left(\dfrac{y}{y-2}\right)^2 - y^2$

$\Rightarrow x^2 + y^2 = 49\left(\dfrac{y}{y-2}\right)^2$

$\Rightarrow (x^2 + y^2)(y - 2)^2 = 49y^2$

Exploration 142

1. $\vec{v}_1 = 6 \cos t \cdot \vec{i} + 6 \sin t \cdot \vec{j}$

2. $\vec{v}_2 = 3 \cos \theta \cdot \vec{i} + 3 \sin \theta \cdot \vec{j}$

3. $A = t$ because of the "alternate interior angles" property of parallel lines.

4. $B = t$ because the two angles subtend equal arcs on circles with the same radius (the circles roll without slipping).

5. $\theta = -\pi$ when $t = 0$, so $\theta = 2t - \pi$.

6. $\vec{r}$ is an arrow from the origin to point P.

7. $\vec{r} = \vec{v}_1 + \vec{v}_2 = 6 \cos t \cdot \vec{i} + 6 \sin t \cdot \vec{j} + 3 \cos \theta \cdot \vec{i} + 3 \sin \theta \cdot \vec{j}$

8. $\vec{r} = 6 \cos t \cdot \vec{i} + 6 \sin t \cdot \vec{j}$
   $\quad + 3 \cos(2t - \pi) \cdot \vec{i} + 3 \sin(2t - \pi) \cdot \vec{j}$
   $= 6 \cos t \cdot \vec{i} + 6 \sin t \cdot \vec{j} - 3 \cos 2t \cdot \vec{i} - 3 \sin 2t \cdot \vec{j}$
   $= (6 \cos t - 3 \cos 2t) \cdot \vec{i} + (6 \sin t - 3 \sin 2t) \cdot \vec{j}$

9. $x = 6 \cos t - 3 \cos 2t$
   $y = 6 \sin t - 3 \sin 2t$

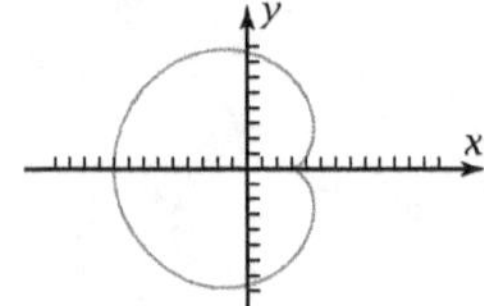

10. $x(1.22) = 6 \cos 1.22 - 3 \cos 2.44 = 4.3533...$
    $y(1.22) = 6 \sin 1.22 - 3 \sin 2.44 = 3.6982...$
    Point P does appear to be $(4.3533..., 3.6982...)$.

11. Answers will vary.

Exploration 143

1. $|\vec{v}_1| = 6 - 2 = 4$ cm, the difference of the radii.
   $\vec{v}_1 = 4 \cos t \cdot \vec{i} + 4 \sin t \cdot \vec{j}$

2. Note that θ is expressed as a positive angle going in the negative direction, so
   $\vec{v}_2 = 2 \cos(-\theta) \cdot \vec{i} + 2 \sin(-\theta) \cdot \vec{j}$
   $\quad = 2 \cos \theta \cdot \vec{i} - 2 \sin \theta \cdot \vec{j}$

3. Using the "corresponding angles" property of parallel lines, $A = t$. Because the circle's radius is three times that of the wheel, $B = 3t$ and $\theta = B - A = 2t$.

4. $\vec{v}_2 = 2 \cos \theta \cdot \vec{i} - 2 \sin \theta \cdot \vec{j}$
   $\quad = 2 \cos 2t \cdot \vec{i} - 2 \sin 2t \cdot \vec{j}$

5. $\vec{r} = \vec{v}_1 + \vec{v}_2$
   $= 4 \cos t \cdot \vec{i} + 4 \sin t \cdot \vec{j} + 2 \cos 2t \cdot \vec{i} - 2 \sin 2t \cdot \vec{j}$
   $= (4 \cos t + 2 \cos 2t) \cdot \vec{i} + (4 \sin t - 2 \sin 2t) \cdot \vec{j}$

6. $x = 4 \cos t + 2 \cos 2t$
   $y = 4 \sin t - 2 \sin 2t$
   The graphs match.

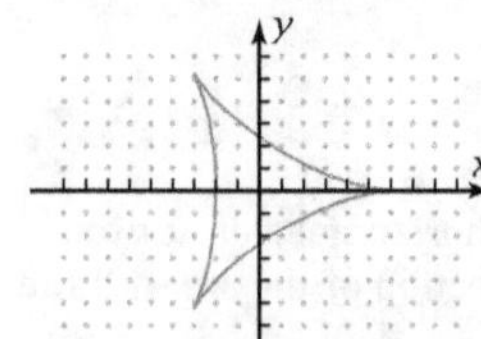

7. The graph resembles an uppercase delta—a triangular shape. *Hypo-* means "under," as opposed to *epi-*, which can mean "upon," "on," or "over." A cusp is a sharp "point" in a curve.

8. $x = 4 \cos t + \cos 2t$
   $y = 4 \sin t - \sin 2t$

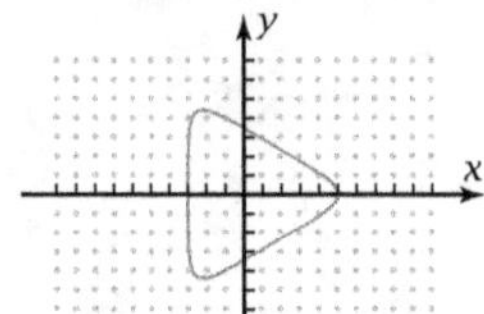

9. $x = 4 \cos t + 3 \cos 2t$
   $y = 4 \sin t - 3 \sin 2t$

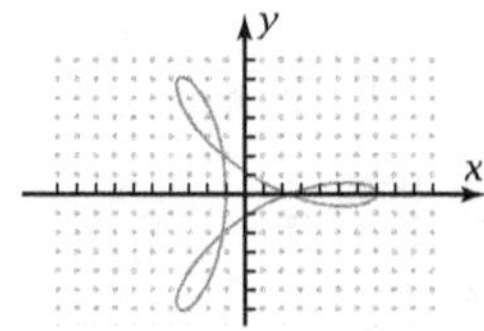

10. A hypocycloid with five cusps has the circle with radius five times that of the wheel.
    $x = 4.8 \cos t + 1.2 \cos 4t$
    $y = 4.8 \sin t - 1.2 \sin 4t$

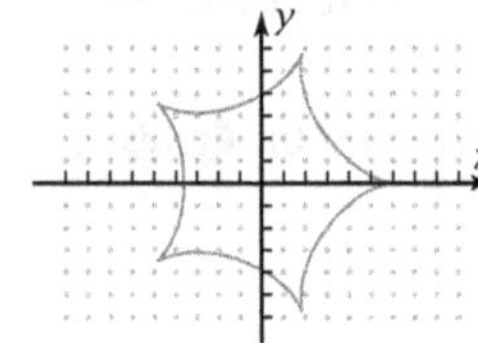

11. $x = 4.2 \cos t + 1.8 \cos \dfrac{4.2}{1.8} t = 4.2 \cos t + 1.8 \cos \dfrac{7}{3} t$

    $y = 4.2 \sin t - 1.8 \sin \dfrac{4.2}{1.8} t = 4.2 \sin t - 1.8 \sin \dfrac{7}{3} t$

    $0 \le t \le 6\pi$

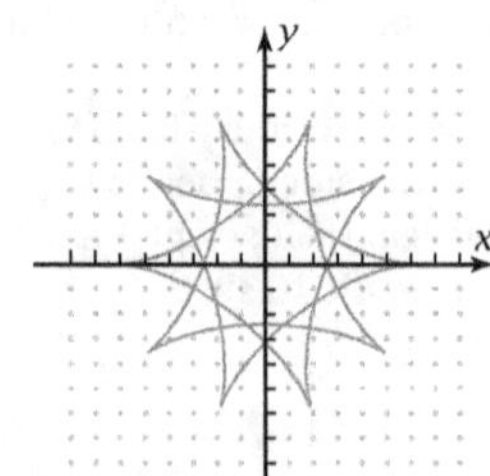

12. Answers will vary.

Exploration 144

1. $x = 2 \cos t$
   $y = 10 \sin t$

2. The graphs match.

3. $x = 2 \cos t + 15$
   $y = 10 \sin t$

4. $x = 2 \cos t + \dfrac{2}{\pi}t$
   $y = 10 \sin t$

5. $t = 0.5\pi \Rightarrow (x, y) = (1, 10)$
   $t = 2.5\pi \Rightarrow (x, y) = (5, 10)$

6. In one complete cycle (high point to high point), t increased by 2π and x increased by 4.

7. This is a vertical dilation by 5 of the path made by a point on the circumference of a circle of radius 2 rolling without slipping along the underside of the line $y = 2$ (or, after the dilation, $y = 10$). When the circle has rotated t radians, its center has traveled $2t$ units. Therefore, including the vertical dilation by 5, $x = 2 \cos t + 2t$, $y = 10 \sin t$.

8. Answers will vary but should start from the recognition that the path looks like a drawing of a spring.

9. Answers will vary.

Exploration 145

1. Student project

2. Student project

3. Student project

4. Answers will vary, but here is an example sketch of a catenary following $y = \cosh x - 1$, for $-2 \leq x \leq 2$, and the parabola $y = (\frac{\cosh 2 - 1}{4})x^2$:

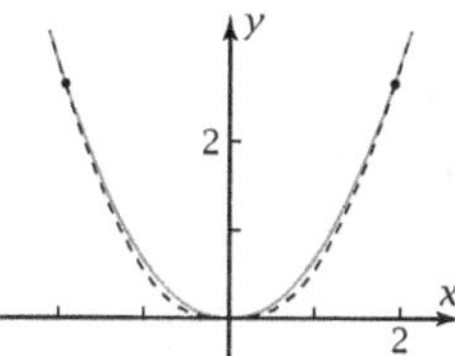

   The parabola is slightly "narrower" than the catenary and so is on the "inside."

5. Answers will vary.

Sequences and Series

Exploration 146

1. 56, 72. Two possibilities: The differences between the terms are 4, 6, 8, 10, and 12, so the next two differences are 14 and 16, and so the next two terms are $42 + 14 = 56$ and $56 + 16 = 72$. Or the terms are $1^2 + 1$, $2^2 + 2$, ..., $6^2 + 6$, so the next two terms are $7^2 + 7 = 56$ and $8^2 + 8 = 72$.

2. $t_5 = 30$; $t_{10} = 72 + 18 + 20 = 110$, or $t_{10} = 10^2 + 10 = 110$.

3.

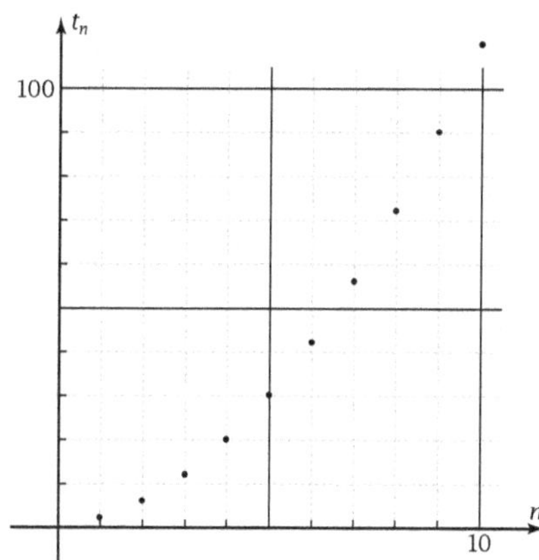

4. It consists of separate points rather than a continuous line.

5.
   t_n: 2, 6, 12, 20, 30, 42, ...
   n: 1, 2,   3,   4,   5,   6, ...

6. Square it and add it to the result:
   $6^2 + 6 = 42$
   $t_n = n^2 + n$

7.

n	t_n
1	2
2	6
3	12
4	20
5	30
6	42
7	56

   The values match.

8. $t_{1234} = \text{Y1}(1{,}234) = 1{,}234^2 + 1{,}234 = 1{,}523{,}990$

9. $n^2 + n = 11{,}340{,}056 \Rightarrow n^2 + n - 11{,}340{,}056 = 0$
   $$\Rightarrow n = \frac{-1 \pm \sqrt{1^2 - 4(1)(-11{,}340{,}056)}}{2(1)} = 3{,}367$$
   (because n must be positive); the 3367th term

10. Answers will vary.

Exploration 147

1. a. No; $10 = 5 + 5$ but $20 = 10 + 10$.

   b. Yes; $10 = 5 + 5$ and $15 = 10 + 5$.

   c. No; $10 = 5 + 5$ but $40 = 10 + 30$.

2. 5. It is the difference between successive terms:
   $5 = 10 - 5 = 15 - 10$.

3. a. Yes; $10 = 2 \cdot 5$ and $20 = 2 \cdot 10$.

   b. No; $10 = 2 \cdot 5$ but $15 = 1.5 \cdot 10$.

   c. No; $10 = 2 \cdot 5$ but $40 = 4 \cdot 10$.

4. 2. It is the ratio between successive terms:
   $10{:}5 = 20{:}10 = 2{:}1 = 2$.

5. 88, 108. Methods may vary. Two possibilities: The successive differences are 6, 8, 10, 12, 14, and 16, so the next two differences are 18 and 20, and so the next two terms are $70 + 18 = 88$ and $88 + 20 = 108$. Or the terms are $1 + 3, 4 + 6, \ldots, 49 + 21$, which is $1^2 + 3 \cdot 1, 2^2 + 3 \cdot 2, \ldots, 7^2 + 3 \cdot 7$, so the next two terms are $8^2 + 3 \cdot 8 = 88$ and $9^2 + 3 \cdot 9 = 108$ (see Problem 6).

6. $t_n = n^2 + 3n$

7.

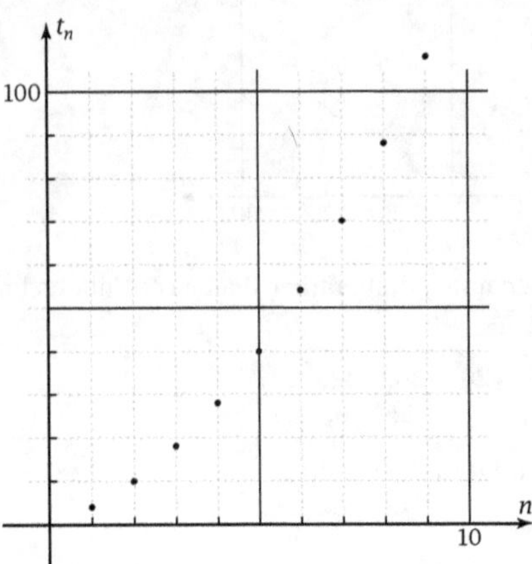

8. $t_{53} = 53^2 + 3 \cdot 53 = 2968$

9. $n^2 + 3n = 23{,}868 \Rightarrow n^2 + 3n - 23{,}868 = 0$
$$\Rightarrow n = \frac{-3 \pm \sqrt{3^2 - 4(1)(-23{,}868)}}{2(1)} = 153$$
(because n must be positive); the 153rd term

10. Answers will vary.

Exploration 148

1. Flame: 200 in., $200 + 3 = 203$ in., $203 + 3 = 206$ in., $206 + 3 = 209$ in.
Seeker: 195 in., $195(1.02) = 198.9$ in., $(198.9)(1.02) = 202.878$ in., $(202.878)(1.02) = 206.93556\ldots$ in.

2. Geometric

3. Arithmetic

4. Yes. In the sixth year, the Flame will be 215 in. long and the Seeker will be 215.2957... in. long. Thereafter, the Seeker will always be longer than the Flame.

5. $200 + (40 - 1) \cdot 3 = 317$ in.

6. $195(1.02)^{30-1} = 346.2897\ldots$ in.

7. $200 + (n - 1) \cdot 3 = 341 \Rightarrow n = \dfrac{341 - 200}{3} + 1 = 48$;
the 48th term

8. $195(1.02)^{n-1} = 1205.73\ldots \Rightarrow n = \dfrac{\log \frac{1205.73\ldots}{195}}{\log 1.02} + 1 = 93$;
the 93rd term

9. Answers will vary.

Exploration 149

1. 8, 11, 14, 17, 20, 23

2. $8 + 11 + \cdots + 23 = 93$

3. Student program.

4. 15,650

5. $t_n = n^2 + 1$
$S_5 = 60 = 2 + 5 + 10 + 17 + 26$

6. 42,975

7. 1000, $1000(1.06) = 1060$, $1060(1.06) = 1123.6$, $(1123.6)(1.06) = 1191.016$;
$1000 + 1060 + 1123.6 + 1191.016 = 4374.616$

8. $t_n = 1000(1.06)^{n-1}$
The program gives the same answer.

9. 79,058.1862...

10. $t_n = 800(0.9)^{n-1}$
$S_{10} = 5210.5724\ldots$, $S_{20} = 7027.3867\ldots$, $S_{50} = 7958.7697\ldots$, $S_{100} = 7999.7875\ldots$, and $S_{200} = 7999.9999\ldots$.

11. The partial sums get closer and closer to 8000.

12. Answers will vary.

Exploration 150

1. $20 - 13 = 27 - 20 = \cdots = 76 - 69 = 7$; $13 + 20 + \cdots + 76 = 445$

2. $13 + 76 = 20 + 69 = \cdots = 41 + 48 = 89$; 5 pairs;
$S_{10} = 5 \cdot 89 = 445$

3. $t_{11} = 76 + 7 = 83$;
$13 + 83 = 20 + 76 = \cdots = 41 + 55 = 96$
Now there are $5\frac{1}{2}$ pairs, and $5\frac{1}{2} \cdot 96 = 528 = 13 + 20 + \cdots + 83$.

4. $S_{300} = \dfrac{300}{2}(47 + 3{,}927) = 596{,}100$

5. $t_{47} = 54 + (47 - 1) \cdot 11 = 560$
$S_{47} = \dfrac{47}{2}(54 + 560) = 14{,}429$
The program gives the same answer.

6. $\dfrac{100{,}000}{2}(1 + 100{,}000) = 5{,}000{,}050{,}000$
At the current speed of calculators, the program would take too long.

7. $S_n = \dfrac{n}{2}(t_1 + t_n) = \dfrac{n}{2}[t_1 + [t_1 + d(n - 1)]]$
$$= \dfrac{n}{2}[2t_1 + d(n - 1)]$$

8. $d = 38.2 - 37 = 39.4 - 38.2 = 1.2$;
$$\dfrac{n}{2}[2 \cdot 37 + 1.2(n - 1)] = 86{,}697$$
$$\Rightarrow \dfrac{n}{2}(74 + 1.2n - 1.2) = 86{,}697$$
$$\Rightarrow 1.2n^2 + 72.8n - 173{,}394 = 0$$
$$\Rightarrow n = \frac{-72.8 \pm \sqrt{72.8^2 - 4(1.2)(-173{,}394)}}{2(1.2)} = 351$$
(because n must be positive); the 351st partial sum

9. Answers will vary.

Exploration 151

1. $\dfrac{21}{7} = \dfrac{63}{21} = \cdots = \dfrac{1701}{567} = 3$;
$7 + 21 + \cdots + 1701 = 2548$

2. $3S_6 = 21 + 63 + 189 + 567 + 1701 + 5103$. They are the same as the terms of the original series, except the first term is missing and there is an added term at the end.

3. They go to 0 by canceling each other out.

4. $S_6 - 3S_6 = t_1 - t_7 = 7 - 7 \cdot 3^{7-1} = 7(1 - 3^6)$

5. $-2S_6 = 7(1 - 3^6) \Rightarrow S_6 = \dfrac{7(1 - 3^6)}{-2} = 2548$

6. $S_n - rS_n = (t_1 + t_1 r + t_1 r^2 + \cdots + t_1 r^{n-1})$
$\qquad\qquad - (t_1 r + t_1 r^2 + \cdots + t_1 r^{n-1} + t_1 r^n)$
$\Rightarrow S_n(1 - r) = t_1 - t_1 r^n \Rightarrow S_n = \dfrac{t_1(1 - r^n)}{1 - r}$

7. $\dfrac{7(1 - 3^{15})}{1 - 3} = 50{,}221{,}171$. The program gives the same answer.

8. $S_{30} = \dfrac{1{,}000(1 - 1.06^{30})}{1 - 1.06} = 79{,}058.1862\ldots$
The program gives the same answer.

9. $S_{50} = \dfrac{800(1 - 0.9^{50})}{1 - 0.9} = 7958.7697\ldots;$
$S_{200} = \dfrac{800(1 - 0.9^{200})}{1 - 0.9} = 7999.9999\ldots;$
r^n approaches 0.

10. Because r^n approaches 0, S_n approaches
$\dfrac{800(1 - 0)}{1 - 0.9} = \dfrac{800}{1 - 0.9} = 8000.$
In general, S_n approaches $\dfrac{t_1}{1 - r}$.

11. Answers will vary.

Exploration 152

1. The coefficients match the values in the seventh row:

$$1$$
$$1 \quad 1$$
$$1 \quad 2 \quad 1$$
$$1 \quad 3 \quad 3 \quad 1$$
$$1 \quad 4 \quad 6 \quad 4 \quad 1$$
$$1 \quad 5 \quad 10 \quad 10 \quad 5 \quad 1$$
$$1 \quad 6 \quad 15 \quad 20 \quad 15 \quad 6 \quad 1$$
$$1 \quad 7 \quad 21 \quad 35 \quad 35 \quad 21 \quad 7 \quad 1$$

2. $35 = \dfrac{(\text{coefficient of } t_3)(\text{exponent of } a \text{ in } t_3)}{3} = \dfrac{21 \cdot 5}{3}$

3. The pattern works.

4. $t_3 = \dfrac{\overset{7}{\cancel{7}} \cdot 6}{2} a^5 b^2 = \dfrac{7}{1} \cdot \dfrac{6}{2} a^5 b^2$

5. $\dfrac{7 \cdot 6 \cdot 5 \cdot 4}{1 \cdot 2 \cdot 3 \cdot 4} = \dfrac{7 \cdot 6 \cdot 5 \cdot 4 \cdot (3 \cdot 2 \cdot 1)}{1 \cdot 2 \cdot 3 \cdot 4 \cdot (3 \cdot 2 \cdot 1)} = \dfrac{7!}{4!3!}$

6. $\dfrac{7!}{5!2!} = \dfrac{5040}{120 \cdot 2} = 21$

7. $t_{10} = \dfrac{13!}{9!4!} a^4 (-b)^9 = -715 a^4 b^9$

8. Answers will vary.

Exploration 153

1. $d = 8 - 5 = 3$ pushups
$5 + (10 - 1) \cdot 3 = 32$ pushups

2. $S_{10} = \dfrac{10}{2}(5 + 32) = 185$ pushups

3. $5 + 8 + 11 + 14 + 17 + 20 + 23 + 26 + 29 + 32 = 185$

4. $n = \dfrac{101 - 5}{3} + 1 = 33$rd workout

5. $S_n = 50(0.8^{1-1}) + 50(0.8^{2-1}) + 50(0.8^{3-1}) + \cdots$
$\qquad = 50 + 50(0.8) + 50(0.8^2) + \cdots$

6. $S_3 = 50 + 50(0.8) + 50(0.8^2) = 122$ mg

7. $249.9667\ldots$ mg

8. $S_n = 50\dfrac{1 - 0.8^n}{1 - 0.8} > 200$ mg $\Rightarrow 1 - 0.8^n > 0.8$
$\Rightarrow n > \dfrac{\log 0.2}{\log 0.8} = 7.2125\ldots$
At $n = 8$ days, $S_n = 208.0569\ldots$ mg.

9. As $n \to \infty$, $S_n = 50\dfrac{1 - 0.8^n}{1 - 0.8}$ is converging to
$50\dfrac{1 - 0}{1 - 0.8} = 50\dfrac{1}{1 - 0.8} = 250$ mg.

10. $P_1 = 0.10 \cdot \$200.00 = \$20.00$
$B_1 = \$200.00 - \$20.00 = \$180.00$

11. $P_2 = 0.10 \cdot \$180.00 = \$18.00$
$B_2 = \$180.00 - \$18.00 = \$162.00$
$P_3 = 0.10 \cdot \$162.00 = \$16.20$
$B_3 = \$162.00 - \$16.20 = \$145.80$

12. Geometric. The ratios between terms are constant:
$r = \dfrac{145.80}{162.00} = \dfrac{162.00}{180.00} = 0.9$

13. $B_{12} = B_1 r^{(12-1)} = 180 \cdot 0.9^{11} = 56.4859\ldots = \$56.49$

14. $B_n = B_1 r^{(n-1)} = 180 \cdot 0.9^{n-1} < 5 \Rightarrow n > \dfrac{\log\frac{5}{180}}{\log 0.9} + 1 = 35.0119\ldots$
$n = 36$ months
Check: $B_{35} = \$5.01$, $B_{36} = \$4.51$

15. $P_1 = 0.10 \cdot \$200.00 = \$20.00$
$P_2 = 0.10 \cdot B_1 = 0.10 \cdot \$180.00 = \$18.00$
$P_3 = 0.10 \cdot B_2 = 0.10 \cdot \$162.00 = \$16.20$
$S_3 = 20\dfrac{1 - 0.9^3}{1 - 0.9} = \$54.20$
$S_3 = \$20.00 + \$18.00 + \$16.20 = \$54.20$

16. As $n \to \infty$, $S_n = 20\dfrac{1 - 0.9^n}{1 - 0.9}$ converges to
$20\dfrac{1 - 0}{1 - 0.9} = 20\dfrac{1}{1 - 0.9} = \$200.00.$

17. Answers will vary.

Exploration 154

1. $S_2 = 1 + 0.6 + \dfrac{1}{2!}(0.6)^2 = 1.78;$
$S_3 = S_2 + \dfrac{1}{3!}(0.6)^3 = 1.816;$
$S_4 = S_3 + \dfrac{1}{4!}(0.6)^4 = 1.8214$

2. Y1=sum(seq(1/N!*0.6^N,N,0,X,1))

X	Y_1
4	1.8214
5	1.822
6	1.8221
7	1.8221
8	1.8221
9	1.8221
10	1.8221

They are approaching $e^{0.6}$.
Y1(10) = 1.8221188002949
and $e^{0.6}$ = 1.8221188003905 are the same until the tenth decimal place.

3. Y1=sum(seq(1/N!*X^N,N,0,10,1))

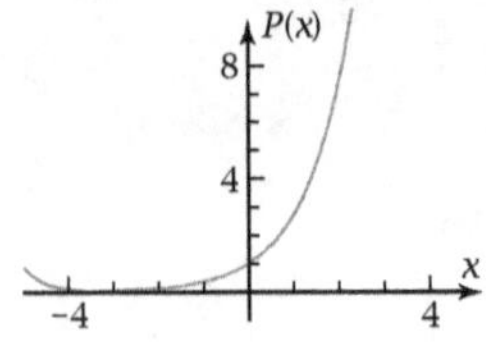

4.

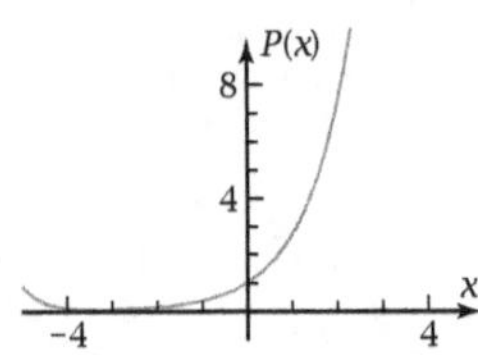

The graph supports the conjecture for $x \geq 3$ but not for $x < 3$.

5. Y1(1) = 2.7182818011464;
e^1 = 2.718 281 828 459 0;
error = -2.73126×10^{-8}

6. Y1(10) = 12,842.3051...; e^{10} = 22,026.4657...;
error = $-9{,}184.1606...$

7. Answers will vary.

Exploration 155

1. $\sin 0.6 \approx 0.6 - \dfrac{1}{3!}(0.6)^3 + \dfrac{1}{5!}(0.6)^5 - \dfrac{1}{7!}(0.6)^7 + \dfrac{1}{9!}(0.6)^9$
$= 0.5646424735...,\ \sin 0.6 = 0.564\,642\,473\,4...$
The error is approximately 1×10^{-10}.

$\cos 0.6 \approx 1 - \dfrac{1}{2!}(0.6)^2 + \dfrac{1}{4!}(0.6)^4 - \dfrac{1}{6!}(0.6)^6 + \dfrac{1}{8!}(0.6)^8$
$= 0.8253356166...,\ \cos 0.6 = 0.8253356149...$
The error is approximately 1.7×10^{-9}.

$e^{0.6} \approx 1 + 0.6 + \dfrac{1}{2!}(0.6)^2 + \dfrac{1}{3!}(0.6)^3 + \dfrac{1}{4!}(0.6)^4 = 1.8214$,

compared to the true value, $e^{0.6} = 1.8221188...$.
The error is approximately -7.188×10^{-4}.

2. $e^{ix} = 1 + ix + \dfrac{1}{2!}(ix)^2 + \dfrac{1}{3!}(ix)^3 + \dfrac{1}{4!}(ix)^4 + \cdots$

$= 1 + ix + \dfrac{1}{2!}(i^2x^2) + \dfrac{1}{3!}(i^2x^3) + \dfrac{1}{4!}(i^4x^4) + \cdots$

$= 1 + ix + \dfrac{1}{2!}(-x^2) + \dfrac{1}{3!}(-ix^3) + \dfrac{1}{4!}(x^4) + \cdots$

$= 1 + ix + \dfrac{1}{2!}x^2 - \dfrac{1}{3!}ix^3 + \dfrac{1}{4!}x^4 + \cdots$

3. $e^{ix} = 1 + ix - \dfrac{1}{2!}x^2 - \dfrac{1}{3!}ix^3 + \dfrac{1}{4!}x^4 + \cdots$

$= \left(1 - \dfrac{1}{2!}x^2 + \dfrac{1}{4!}x^4 - \cdots\right) + i\left(x - \dfrac{1}{3!}x^3 + \cdots\right)$

$= \cos x + i \sin x$

4. A complex number

5. $e^{-ix} = e^{i(-x)} = \cos(-x) + i \sin(-x) = \cos x + i \sin x$
Cosine is an even function, and sine is an odd function. They are complex conjugates.

6. a. $e^{3i}e^{3i}$

b. $3e^{2i}$

c. $1 + i = \sqrt{2}\left(\dfrac{1}{\sqrt{2}} + \dfrac{1}{\sqrt{2}} \cdot i\right)$

$= \sqrt{2}\left(\cos\dfrac{\pi}{4} + i \sin\dfrac{\pi}{4}\right) = \sqrt{2}e^{\frac{\pi}{4} \cdot i}$

d. $\sqrt{3} - i = 2\left(\dfrac{\sqrt{3}}{2} - \dfrac{1}{2} \cdot i\right)$

$= 2\left(\cos\dfrac{\pi}{6} + i \sin\dfrac{\pi}{6}\right) = 2e^{-\frac{\pi}{6} \cdot i}$

e. $i = 0 + i = \cos\dfrac{\pi}{2} + i \sin\dfrac{\pi}{2} = e^{\frac{\pi}{2} \cdot i}$

7. $e^{i\pi} = \cos\pi + i \sin\pi = -1 + 0 \cdot i = -1$

8. Using the result of Problem 6e,
$i^i = \left(e^{\frac{\pi}{2} \cdot i}\right)^i = e^{\frac{\pi}{2} \cdot i^2} = e^{-\frac{\pi}{2}} = 0.20787957635076....$

9. Answers will vary.

Polynomial and Rational Functions, Limits, and Derivatives

Exploration 156

1.

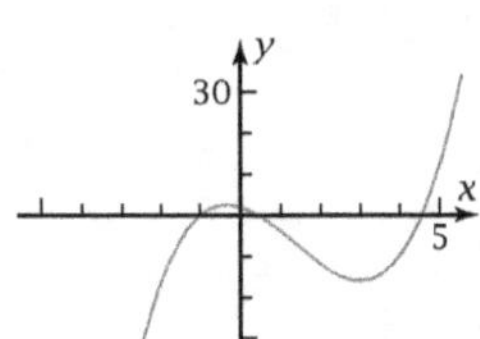

2. The graph crosses the axis at $x = -1$ and also at approximately $x \approx 0.4, 4.6$.

3. $f(x) = (x + 1)(x^2 - 5x + 2)$

4. $x = \dfrac{-(-5) \pm \sqrt{(-5)^2 - 4(1)(2)}}{2(1)} = \dfrac{5 \pm \sqrt{17}}{2} = 0.4384...,\ 4.5615...$
These values agree with the estimates in Problem 2.

5. $g(x)$ is just $f(x)$ translated upward by 16. However, note that f crosses the axis in three distinct points, whereas g crosses the axis only once and is tangent at one point.

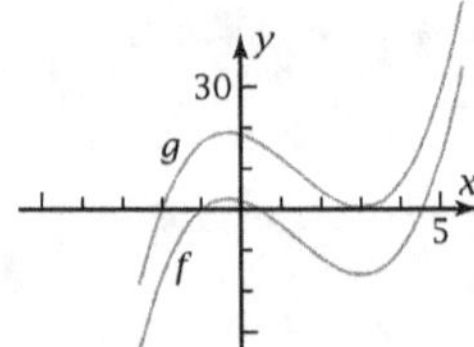

6. $g(x) = (x + 2)(x - 3)(x - 3)$
Roots are $x = -2, 3, 3$. Two of the roots are identical!

7.

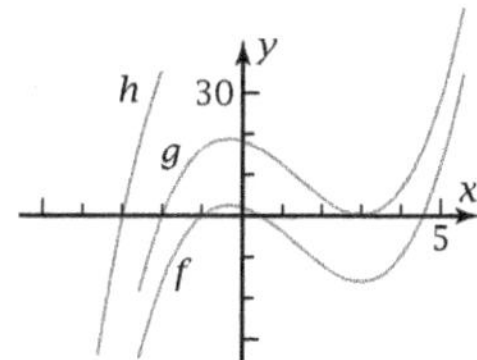

8. $h(x) = (x + 3)(x^2 - 7x + 18)$

$$x^2 - 7x + 18 = 0 \Rightarrow x = \frac{7 \pm \sqrt{-23}}{2} = 3.5 \pm 2.3979\ldots i$$

Two of the zeros are complex; the graph crosses the axis at only one real value of x. Numbers that make the polynomial equal zero may be complex values and therefore have no corresponding intercept.

9. Answers will vary.

Exploration 157

1.

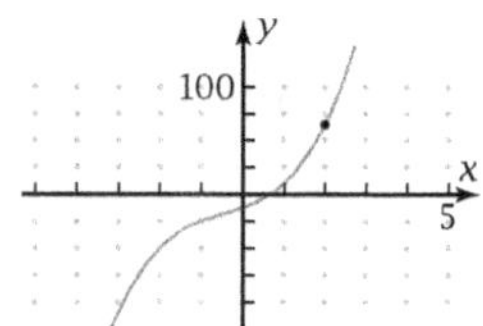

2. $f(2) = 52$ (See point on graph in Problem 1.)

3. Quotient $= 3x^2 + 10x + 31$; remainder $= 52$.

4. $f(2)$ equals the remainder.

5.
$$
\begin{array}{r|rrrr}
2 & 3 & 4 & 11 & -10 \\
  &   & 6 & 20 & 62 \\
\hline
  & 3 & 10 & 31 & 52
\end{array}
$$

6. The coefficients in Problem 3 are all the numbers in the bottom row in the synthetic substitution except the last one (which is the remainder).

7.
$$
\begin{array}{r|rrrr}
\frac{2}{3} & 3 & 4 & 11 & -10 \\
  &   & 2 & 4 & 10 \\
\hline
  & 3 & 6 & 16 & 0
\end{array}
$$

$f\left(\frac{2}{3}\right) = 0$, the graph crosses the axis at $x = \frac{2}{3}$.

8. $\dfrac{f(x)}{\left(x - \frac{2}{3}\right)} = 3x^2 + 6x + 15 = 3(x + 1 - 2i)(x + 1 + 2i)$

Roots are $-1 + 2i, -1 - 2i$.

9. Answers will vary.

Exploration 158

1. $f(x) = 5x^3 - 33x^2 + 58x - 24 = (5x - 3)(x - 2)(x - 4)$

Zeros are $\frac{3}{5}, 2, 4$.

2. $z_1 z_2 z_3 = \dfrac{24}{5} = 4.8$

3. The product of zeros, $z_1 z_2 z_3$, appears as the opposite of the constant term inside the parentheses.

4. The sum of zeros, $z_1 + z_2 + z_3 = \frac{33}{5} = 6.6$, appears as the opposite of the x^2-coefficient inside the parentheses.

5. The sum of the pairwise products of the zeros,

$$z_1 z_2 + z_1 z_3 + z_2 z_3 = \frac{58}{5} = 11.6$$

appears as the x-coefficient inside the parentheses.

6. $5 + (2 + \sqrt{3}) + (2 - \sqrt{3}) = 9$

$5(2 + \sqrt{3}) + 5(2 - \sqrt{3}) + (2 + \sqrt{3})(2 - \sqrt{3}) = 21$

$5(2 + \sqrt{3})(2 - \sqrt{3}) = 5$

$g(x) = (x - 5)(x - 2 - \sqrt{3})(x - 2 + \sqrt{3}) = x^3 - 9x^2 + 21x - 5$

7. $h(x) = 3g(x) = 3x^3 - 27x^2 + 63x - 15$

8. Functions have the same zeros.

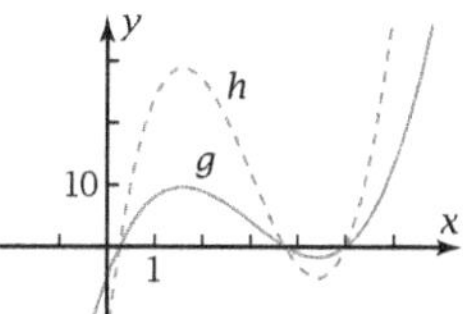

9. Answers will vary.

Exploration 159

1.
$$
\begin{array}{cccccc}
x & P(x) \\
1 & 3 \\
  &   & 9 \\
2 & 12 &   & -12 \\
  &   & -3 &   & 6 \\
3 & 9 &   & -6 &   & 6 \\
  &   & -9 &   & 0 &   & 6 \\
4 & 0 &   & 0 &   & 6 \\
  &   & -9 &   & 6 &   & 6 \\
5 & -9 &   & 6 &   & 6 \\
  &   & -3 &   & 12 &   & 6 \\
6 & -12 &   & 12 &   & 6 \\
  &   & 9 &   & 18 \\
7 & -3 &   & 18 \\
  &   & 27 \\
8 & 24 \\
\end{array}
$$

2. $P(x) = ax^3 + bx^2 + cx + d$

$P(1) = a + b + c + d = 3$

$P(2) = 8a + 4b + 2c + d = 12$

$P(3) = 27a + 9b + 3c + d = 9$

$P(4) = 64a + 16b + 4c + d = 0$

$$
\begin{bmatrix} 1 & 1 & 1 & 1 \\ 8 & 4 & 2 & 1 \\ 27 & 9 & 3 & 1 \\ 64 & 16 & 4 & 1 \end{bmatrix}
\begin{bmatrix} a \\ b \\ c \\ d \end{bmatrix}
=
\begin{bmatrix} 3 \\ 12 \\ 9 \\ 0 \end{bmatrix}
$$

$$
\begin{bmatrix} a \\ b \\ c \\ d \end{bmatrix}
=
\begin{bmatrix} 1 & 1 & 1 & 1 \\ 8 & 4 & 2 & 1 \\ 27 & 9 & 3 & 1 \\ 64 & 16 & 4 & 1 \end{bmatrix}^{-1}
\begin{bmatrix} 3 \\ 12 \\ 9 \\ 0 \end{bmatrix}
=
\begin{bmatrix} 1 \\ -12 \\ 38 \\ -24 \end{bmatrix}
$$

$P(x) = x^3 - 12x^2 + 38x - 24$

3. $P(5) = 125 - 12 \cdot 25 + 38 \cdot 5 - 24 = -9$

$P(6) = 216 - 12 \cdot 36 + 38 \cdot 6 - 24 = -12$

$P(7) = 343 - 12 \cdot 49 + 38 \cdot 7 - 24 = -3$

$P(8) = 512 - 12 \cdot 64 + 38 \cdot 8 - 24 = 24$

4. regEq $= x^3 - 12x^2 + 38x - 24$

5. $P(20) = 3936$

6. $x = 7.5770\ldots$

7. $P(x) = (x - 4)(x^2 - 8x + 6)$

8. $x^2 - 8x + 6 = 0$

$$\Rightarrow x = \frac{-(-8) \pm \sqrt{(-8)^2 - 4(1)(6)}}{2(1)} = 4 \pm \sqrt{10} = 0.8377\ldots, 7.1622\ldots$$

The answers agree.

9. Answers will vary.

Exploration 160

1.

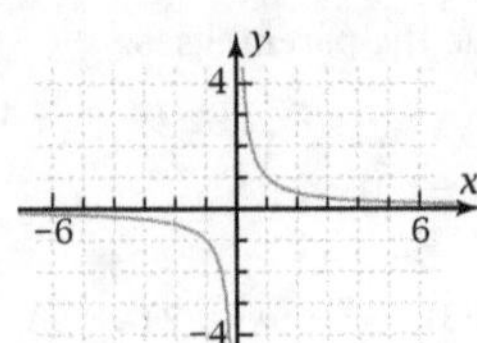

2. $f(x)$ goes to infinity as $x \to 0$.

3. Horizontal dilation by 2 and vertical dilation by 2 (or horizontal dilation by 4, or vertical dilation by 4).

4. Horizontal translation by 3

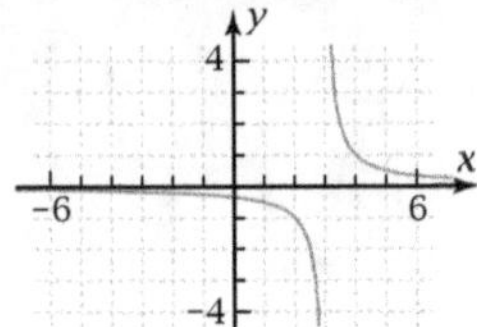

5.

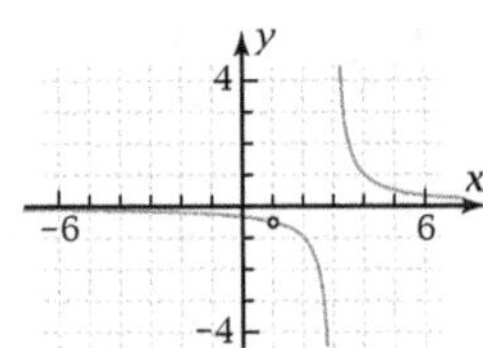

6. The graphs are identical, except that $r(x)$ is undefined at $x = 1$.

7. $r(x)$ is undefined at $x = 1$ because $x - 1$ appears in the denominator.

8. $r(x) = \dfrac{x - 1}{x^2 - 4x + 3} = \dfrac{x - 1}{(x - 1)(x - 3)}$

   $r(x) = \dfrac{1}{x - 3}$ if $x \neq 1$

9. If the numerator is nonzero at a point where the denominator is zero, then the graph will have a vertical asymptote. If the numerator is zero when the denominator is zero, then the graph may have a removable discontinuity. In this case, if zero appears "to a higher degree" in the denominator than in the numerator, then the graph will have a vertical asymptote; otherwise, the graph has a removable discontinuity.

10. Answers will vary.

Exploration 161

1.

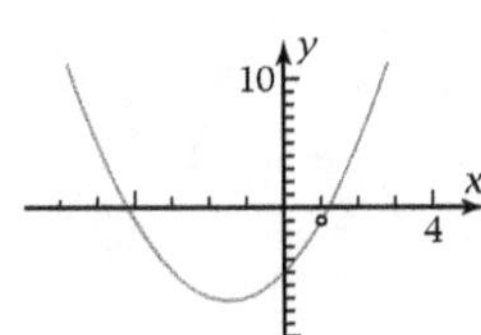

2. The graph has a removable discontinuity at $x = 1$.
   $f(x) \to -1$ as $x \to 1$ from both sides.

3. $f(x) = \dfrac{x^3 + 2x^2 - 8x + 5}{x - 1} = \dfrac{(x - 1)(x^2 + 3x - 5)}{x - 1}$
   $f(x) = x^2 + 3x - 5$ if $x \neq 1$.
   $f(1) = 1^2 + 3(1) - 5 = -1$, the limit.

4. "The limit of f of x as x approaches 1 is -1." As x gets closer and closer to 1, $f(x)$ gets closer and closer to -1.

5. $g(x)$ has a vertical asymptote at $x = 1$.

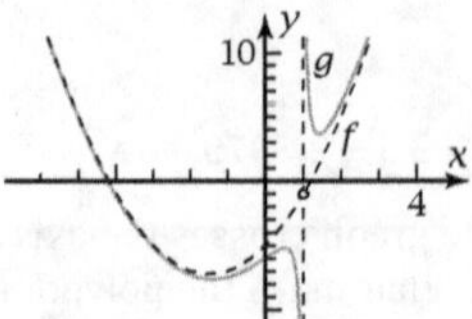

6. $g(x) \to -\infty$ as $x \to 1^-$ ("from below" or "from the left")
   $g(x) \to +\infty$ as $x \to 1^+$ ("from above" or "from the right")

7.
$$
\begin{array}{r|rrrr}
1 & 1 & 2 & -8 & 6 \\
  &   & 1 & 3 & -5 \\
\hline
  & 1 & 3 & -5 & 1
\end{array}
$$

   $g(x) = x^2 + 3x - 5 + \dfrac{1}{x - 1}$

   $f(x)$ and $g(x)$ have the same quotient polynomial.

8.

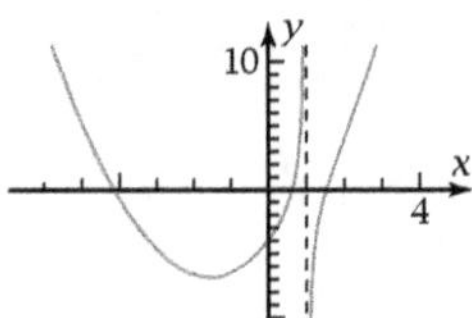

9.

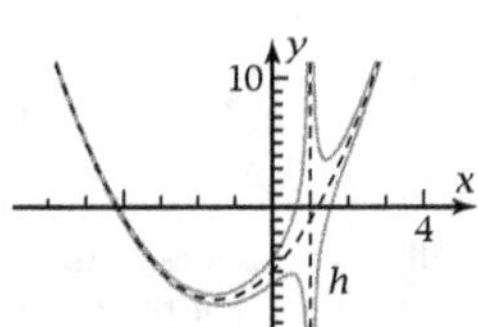

10.
$$
\begin{array}{r|rrrr}
1 & 1 & 2 & -8 & 4 \\
  &   & 1 & 3 & -5 \\
\hline
  & 1 & 3 & -5 & -1
\end{array}
$$

   $h(x) = x^2 + 3x - 5 - \dfrac{1}{x - 1}$

11. The polynomial is very close to the graph except near the asymptote.

12. An error in an earlier problem will propagate into later problems, possibly causing even more errors.

13. Answers will vary.

Exploration 162

1. $f(2) = 10$ mi, $f(2.1) = 10.361$ mi
$$
v_{av} = \frac{f(2.1) - f(2)}{2.1 - 2} = \frac{0.361 \text{ mi}}{0.1 \text{ min}} = 3.61 \text{ mi/min}
$$

2. $f(2) = 10$ mi, $f(2.001) = 10.003996001$ mi,
$$
v_{av} = \frac{f(2.001) - f(2)}{2.001 - 2} = \frac{0.003996001 \text{ mi}}{0.001 \text{ min}}
$$
$$
= 3.996001 \text{ mi/min}
$$

3. $v = 4$ mi/min

4. $r(x) = \dfrac{f(x) - 10}{x - 2} = \dfrac{x^3 - 10x^2 + 32x - 32}{x - 2}$

$\quad = \dfrac{(x - 2)(x^2 - 8x + 16)}{x - 2} = x^2 - 8x + 16 \text{ if } x \neq 2$

5. $\displaystyle\lim_{x \to 2} r(x) = \lim_{x \to 2} (x^2 - 8x + 16) = 4 \text{ mi/min}$

6. The line is tangent to the graph.

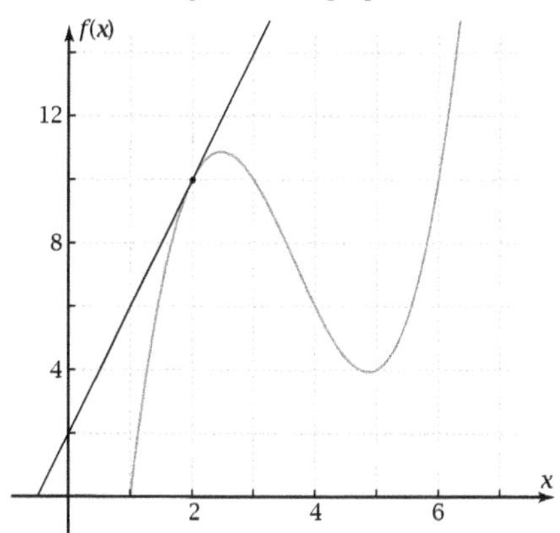

7. $f(3) = 11 \text{ mi}, \; r(x) = \dfrac{f(x) - 11}{x - 3} = \dfrac{x^3 - 10x^2 + 32x - 33}{x - 2}$

$\quad = \dfrac{(x - 3)(x^2 - 7x + 11)}{x - 3} = x^2 - 7x + 11 \text{ if } x \neq 2;$

$\displaystyle\lim_{x \to 3} r(x) = \lim_{x \to 3} (x^2 - 7x + 11) = -1 \text{ mi/min}$

Velocity is negative because Ella is falling back toward the planet.

8. $g(2) = 4, \; g(3) = -3$

This function seems to give the velocities.

9.

$$
\begin{array}{c|cccccc}
c & 1 & 0 & 0 & 0 & 0 & -c^5 \\
  &   & c & c^2 & c^3 & c^4 & c^5 \\
\hline
  & 1 & c & c^2 & c^3 & c^4 & 0
\end{array}
$$

$r(x) = \dfrac{(x - c)(x^4 + cx^3 + c^2x^2 + c^3x + c^4)}{x - c}$

$\quad = x^4 + cx^3 + c^2x^2 + c^3x + c^4 \text{ if } x \neq c$

10. $\displaystyle\lim_{x \to c} r(x) = \lim_{x \to c} (x^4 + cx^3 + c^2x^2 + c^3x + c^4)$

$\quad = (c)^4 + c(c)^3 + c^2(c)^2 + c^3(c) + c^4 = 5c^4$

11. Multiply the term by the exponent and then subtract 1 from the exponent.

12. $10 \cdot 17x^{10-1} = 170x^9$

13. $f(x) = x^3 - 10x^2 + 32x^1 - 22x^0;$

$\quad g(x) = 3 \cdot x^{3-1} - 2 \cdot 10x^{2-1} + 1 \cdot 32x^{1-1} - 0 \cdot 22x^{0-1}$

$\quad = 3x^2 - 20x + 32$

14. The derivative

15. Answers will vary.

Appendix: Programs for Graphing Calculators

Some Explorations require access to programs for graphing calculators. The programs listed and described in this appendix are specific to the Texas Instruments TI-83 graphing calculator but can be adapted for use with other graphers. The most time-efficient way to use a program is for you to program one grapher or to download the program from *www.keypress.com/pte* to one grapher, then allow members of the class to download the program. In this way, everyone starts the same program, minimizing the amount of debugging you might have to do otherwise. You are welcome to embellish the programs with more sophisticated features and outputs or even to write your own programs.

ITRANS, Explorations 107 and 117

This program graphically displays iterations of a single matrix transformation. Prior to executing the program, set the window and enter the transformation matrix as [A], a 3×3 matrix, and the pre-image matrix as [D], a 3×4 matrix. The pre-image matrix must have four columns, corresponding to four points in the pre-image. If there are fewer than four points, start repeating the points. When you run the program, the pre-image appears first. Each time you press the ENTER key, the next image appears. If the dilation is less than 1, the images converge to a fixed point. The images are stored in [E]. If you have done enough iterations, all the points in the final [E] should have approximately the same coordinates, the fixed point, a result of the fact that all four points in the pre-image converge to the same fixed point. You must press ON, then 1:QUIT to terminate the run of the program. To read the values in [E], set the mode so that two decimals are displayed.

ITRANS

```
ClrDraw
[D]→[E]
Lbl 1
Line([E](1,1),[E](2,1),[E](1,2),[E](2,2))
Line([E](1,2),[E](2,2),[E](1,3),[E](2,3))
Line([E](1,3),[E](2,3),[E](1,4),[E](2,4))
Line([E](1,4),[E](2,4),[E](1,1),[E](2,1)
[A][E]→[E]
Pause
Goto 1
```

BARNSLEY, Exploration 113

This program plots fractal images, a point at a time, on the calculator screen
using Barnsley's method. Before running the program, set the window. You will
normally want to use the same scales on both axes. Then enter up to four
transformation matrices as [A], [B], [C], and [D]. Enter into L_1 the probabilities of
selecting the transformations. The sum of the probabilities should equal 1. If, for
instance, there are three matrices and you want equal probabilities of each, enter
$\frac{1}{3}$, $\frac{1}{3}$, $\frac{1}{3}$, and 0. Then run the program. At the prompt for the initial pre-image point,
enter any x- and y-coordinates. At the next prompt, enter the number of points
that you want to plot. Usually, 1000 points gives a reasonable picture in two to three
minutes.

BARNSLEY

```
ClrDraw
{3,1}→dim([E]
Disp "PRE-IMAGE PT."
Prompt X,Y
X→[E](1,1)
Y→[E](2,1)
1→[E](3,1)
Disp "NUMBER OF PTS."
Input N
L₁(1)→L₂(1)
For(K,2,4,1):L₂(K-1)+L₁(K)→L₂(K):End
For(K,1,N,1)
rand→P
If P<L₂(1):Then:[A][E]→[E]
Goto 1:End
If P<L₂(2):Then:[B][E]→[E]:Goto 1:End
If P<L₂(3):Then:[C][E]→[E]:Goto 1:End
[D][E]→[E]
Lbl 1
Pt-On(Ans(1,1),Ans(2,1))
End
End
```

CONIC and CONIC2, Explorations 119 and 132

These programs plot the graphs of one or two (respectively) conic sections from the general equation $Ax^2 + Bxy + Cy^2 + Dx + Ey + F = 0$. Before running each program, set the window and clear out any equations in the y= menu. The coefficients of the equations are input as you run the program. The results are stored at the end of the y= menu in y_7, y_8, y_9, and y_0. When you are finished with the programs, you should go to the y= menu and clear them out. If you are using a low-resolution grapher, the graphs may not appear closed the way they are supposed to. Sometimes, changing the window slightly will fill in some of these gaps. For example, the first picture below represents the conic section viewed in a standard window, whereas the second image is displayed in a [-9.5,10,1,-10,10,1] window.

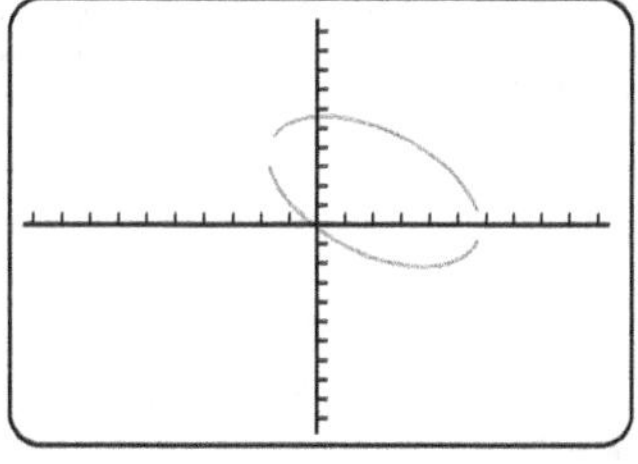

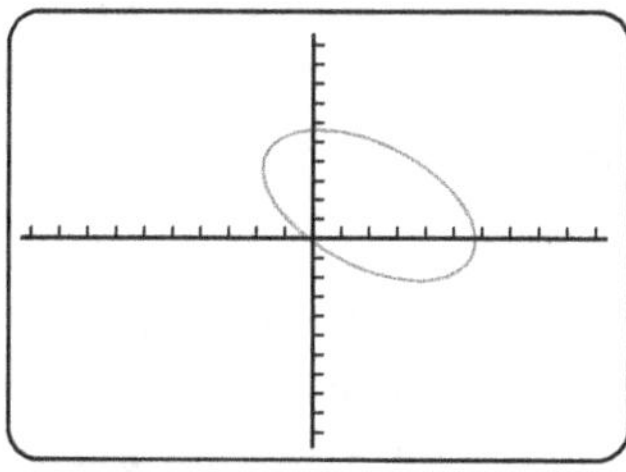

CONIC2 will enable you to graph a second equation on the same coordinate axes. After entering the coefficients of the first equation, follow the prompts to enter the coefficients of the second equation.

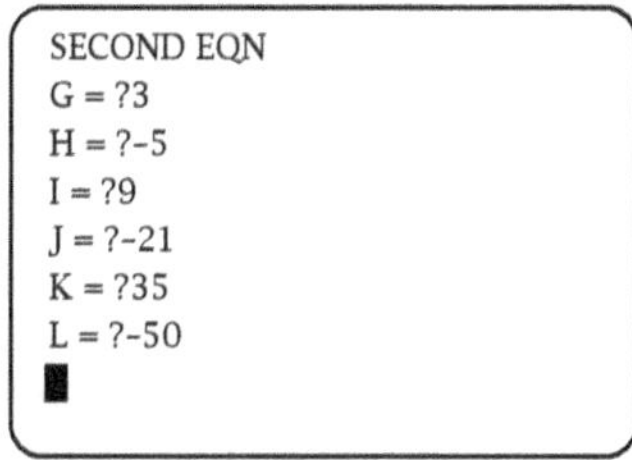

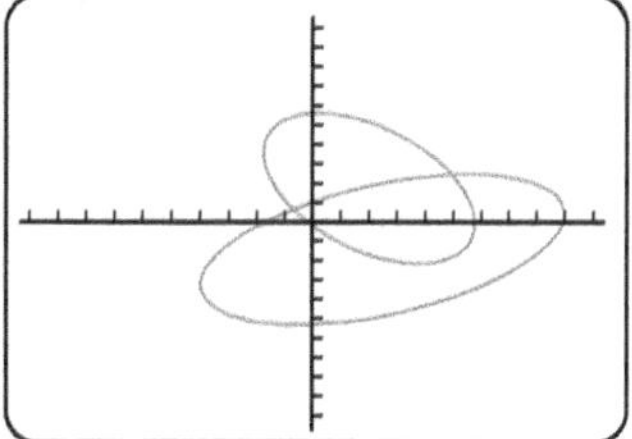

CONIC

```
ClrHome
Disp "AX²+BXY+CY²"
Disp "        +DX+EY+F=0"
Prompt A,B,C,D,E,F
FnOff
If C≠0:Goto 1
"-(AX²+DX+F)/(BX+E)"→Y7
""→Y8:Goto 2
Lbl 1
"(-BX-E+√((BX+E)²-4C(AX²+DX+F)))/(2C)"→Y7
"(-BX-E-√((BX+E)²-4C(AX²+DX+F)))/(2C)"→Y8
Lbl 2
DispGraph
```

CONIC2

```
ClrHome
Disp "AX²+BXY+CY²"
Disp "        +DX+EY+F=0"
Disp "FIRST EQN"
Prompt A,B,C,D,E,F
FnOff
If C≠0:Goto 1
"-(AX²+DX+F)/(BX+E)"→Y₇
""→Y₈:Goto 2
Lbl 1
"(-BX-E+√((BX+E)²-4C(AX²+DX+F)))/(2C)"→Y₇
"(-BX-E-√((BX+E)²-4C(AX²+DX+F)))/(2C)"→Y₈
Lbl 2
Disp "SECOND EQN"
Prompt G,H,I,J,K,L
If I≠0:Goto 3
"-(GX²+JX+L)/(HX+K)→Y₉
""→Y₀:Goto 4
Lbl 3
"(-HX-K+√((HX+K)²-4I(GX²+JX+L)))/(2I)"→Y₉
"(-HX-K-√((HX+K)²-4I(GX²+JX+L)))/(2I)"→Y₀
Lbl 4
DispGraph
```

Precalculus and Trigonometry Correlation Index

Main Topic	Exploration Number
Ambiguous case (SSA)	62, 63
Angle between vectors	88, 89, 90, 93
Angles, reference	12
Angles, standard position of	12
Arccosine, arcsine, arctangent	38
Area	
of a triangle	58, 59
of a triangle using cross product of vectors	93
of a triangle using Hero's formula	59
of a triangle using SAS	58, 59, 67
of a triangle using SSS	59
of an *n*-gon	67
Arithmetic sequences	148
Arithmetic series	150, 153
Barnsley's method	114, 117
Binomial formula	152
Binomial probability distribution	85
Binomial series	152
Boolean variables	3
Cardioids	142
Catenary	145
Circles in polar coordinates	134
Circular function parent graphs	29
Circular permutations	83
Combined translation, rotation, and dilation matrix	108
Complex number product proof	137
Complex numbers in exponential form	155
Component vectors	65
Composite argument property for cosine	46, 47
Composition of functions	8
Conchoid of Nicomedes	141
Conditional probability	84
Conic sections	
analytic properties of	122, 123, 124, 125, 126
applications of	128
Cartesian equations of	119
discriminant of	127
parametric equations of	118
Correlation coefficient	76, 77
Cosine	
and sine, linear combination of	45
composite argument property for	46, 47
direction	94, 95, 97
function definition by right triangle	13
function definition for any size angle	13
function graph	14
Counting principles for "and" and "or"	82
Cross product, determinant for finding	92
Cross product of two vectors	92, 93
Cubic function graphs	156
Determinant for finding cross product	92
Determinant of a matrix	104
Dilations of functions	4, 5, 6, 7, 10

Main Topic	Exploration Number
Dimension, fractal	114, 116
Direction angles	94
Direction cosines	94, 95, 97
Direction numbers	97
Discontinuities of rational functions	160
Discriminant of a conic section	127
Displacement vector	87
Distance between a line and a point in space	100
Distance between a plane and a point	99
Domain, restricted	3, 8
Dot product of two vectors	88, 90, 91, 93
Double argument properties	53
Eccentricity	126
Ellipse(s)	41, 118, 119, 122, 124, 126, 129, 130, 144
analytic properties of	122, 124, 126, 129
focus and directrix of	122
parametric equations for	41
reflecting property of	129, 130
tangent line to	130
two-foci property for	122, 124, 129
Equation of a circle from a geometric definition	131
Equation of a plane normal to a vector	91, 93, 98
Equations of functions from numerical patterns	71
Events, overlapping, and Venn diagrams	82
Exponential functions	2, 79, 80, 81
Exponential functions fit to data	79, 80, 81
Fitting a polynomial function to points	159
Fixed point attractor of iterated transformation matrix	108, 109, 110
Focus and directrix of an ellipse	122
Fractal dimension	114, 116
Fractal figures by random point plotting	113
Fractals, Hausdorff's dimension of	114
Function(s)	
composition of	8
cosine	14
cubic	156
exponential	2, 79, 80, 81
fitting to data	78, 79, 80, 81
graphical patterns in	68
hyperbolic cosine	145
inverse trigonometric	43
inverses of	2, 9
limits of	141
linear	2
logarithmic	72
logistic	73
numerical patterns in	69, 70, 71
numerically defined	1
piecewise	3

Main Topic	Exploration Number
polynomial	159
power	2, 78, 80
quadratic	2
quartic	2
rational	160
sine	13, 14, 15
sinusoidal	22, 23
sinusoidal, transformations of	23
trigonometric	16, 17, 34
velocity	162
Geometric sequences	148
Geometric series	151, 153
Graphical patterns in functions	68
Graphing vectors in three space	90, 91
Harmonic analysis	48, 49, 50, 51
Hausdorff's dimension of fractals	114
Hero's formula	59
Hyperbola	118, 119, 123, 125, 139
Hyperbola, analytic properties of	123, 125
Hyperbola construction	139
Hyperbolic cosine function	145
Hypocycloids	143
Identities, trigonometric	35, 36, 37
Inner product of two vectors	88, 90, 91, 93
Instantaneous velocity	162
Intercepts of a plane	91, 93
Intersection of a line and a plane in space	96
Intersection of events	84
Intersection of polar curves	135
Inverse(s)	
of functions	9
of matrices	105
trigonometric functions graphs	43
trigonometric relations graphs	42, 43, 44
trigonometric relations values	44
variation function	2
Iterated transformation matrix, fixed point attractor of	108, 109, 110
Iterated transformations	107, 108, 109, 117
Iterated transformations of a geometric figure	107, 108, 109
Koch's snowflake curve	115
Law of cosines	55, 56, 57
Law of cosines derivation	56
Law of sines	60, 61
Limaçon in polar coordinates	133
Limit(s)	
and curved asymptotes	161
of a function as x approaches a	161
Linear combination of cosine and sine	45
Linear function	2
Linear regression	74, 75, 78, 79, 80, 81
Logarithmic functions	72
Logistic function	73
Mathematical expectation	86
Matrix	
combined translation, rotation, and dilation	108, 109, 110
determinant of	104

Main Topic	Exploration Number
images and transformations	106
inverse	105
iterated transformation	108, 109, 110
Markov chain transition	111
multiplication	103
Moving objects	144, 162
Multiple transformation of a figure	112, 117
Navigation vectors	65
Normal vector to a plane	91, 93
Numerical patterns in functions	69, 70, 71
Outer product of two vectors	92, 93
Parabola	123, 126
Parabola, analytic properties of	123, 126
Parametric equations	
for ellipses	41, 118, 119
for moving objects	162
of a curve from a geometrical description	139, 141
of a curve from its geometrical properties	138
of a curve given by a geometrical construction	140, 142
of a curve using vectors	143
of a line in space	97
of conic sections	118
of hypocycloids	143
Parametric function trigonometric application	40
Partial sums of arithmetic series	150
Partial sums of geometric series	151
Period function transformations	11
Permutations	83
Piecewise function	3
Polar coordinates	133, 134
circles and roses in	134
limaçon in	133
Polar curves, intersection of	135
Polynomial function, rate of change of	162
Position vector	87
Positive and negative normal vectors to a plane	98
Power function	2, 79, 80
Power series for e^x	154, 155
Probability	
binomial distribution	85
conditional	84
counting principles	82
of various permutations	83
properties	84
union and intersection	84
Products of complex numbers	136, 137
Projectile motion	138
Projections of vectors	89, 90, 91, 93
Pythagorean properties for three dimensions	87
Pythagorean property for direction cosines	94
Quadratic function	2
Quadratic-quadratic system of equations	132
Quadric surfaces	120, 121
Quartic function	2
Radian measure of angles	27, 28
Rate of change of a polynomial function	162
Rational functions	160
Reference angles	12

Main Topic	**Exploration Number**	**Main Topic**	**Exploration Number**
Reflecting property of ellipses	129, 130	matrix, dilation and rotation	106
Reflections	10	matrix, translation	108, 109, 110
Regression, linear	74, 75, 78, 79, 80	of sinusoidal functions	23
		of tangent and secant graphs	26
Residual plots	79, 80, 81	of trigonometric expressions	54
Residuals	74, 75, 79, 80, 81	period function	11
		Translations	4, 5, 6, 7, 10, 108, 109, 110
Residuals, sums of squares of	74, 75		
Right-hand rule for cross product of two vectors	92	Transpose of a matrix	105
		Triangle, area of	58, 59
Right triangle measurement	18, 19, 20	Trigonometric equations	39
Rotated conic sections with an xy-term	127	Trigonometric expressions, transformations of	54
Rotating conic sections about one of its axes	120	Trigonometric function definitions by coordinate form	16
SAS, area of a triangle	58, 59, 67		
Scalar product of two vectors	88, 90, 93	Trigonometric function Pythagorean properties	34
Scalar projection of vectors	88, 90, 91, 93		
Sequences	146, 147, 148	Trigonometric function reciprocal properties	34
arithmetic	148	Trigonometric function values of angles using measurement	17
geometric	148		
patterns	147	Trigonometric identities	35, 36, 37
Series	149, 150, 151, 152, 153, 155	Trigonometric relations graphs, inverse	42, 43, 44
		Trigonometric relations values, inverse	44
arithmetic	150, 153	Two-foci property for an ellipse	122
binomial	152	Union of events	84
geometric	151, 153	Unit vector	88, 89, 90, 94, 95
power	155		
Sierpiński's triangle	114	Vector(s)	
Sine function definition by right triangle	13	addition	64
Sine function definition for any size angle	13	and parametric equations	143
Sine function graph	14	component	65
Sine function transformations	15	cross product of	92, 93
Sinusoidal applications	32, 33	displacement	87
Sinusoidal equations from graphs	24	dot product of	88, 90, 91, 93
Sinusoidal functions from angle measure	22	equation of a line in space	95, 96
Sinusoidal functions from data	21	equation of a plane normal to a	91, 93, 98
Sinusoids, given y, find x	30, 31	inner product of	88, 90, 91, 93
Sinusoids, sum or product of	48, 49, 50, 51, 52	navigation	65
		normal to a plane	91, 93
Snowflake curve	115	outer product of	92, 93
Springs and moving ellipses	144	position	87
SSA (ambiguous case)	62, 63	product	92, 93
SSS, area of a triangle	59	projection	89, 90
Standard equations of a line in space	97	properties of	90
Standard position of angles	12	right-hand rule for cross product of	92
Strange attractors	113, 117	scalar product of	88, 90, 91, 93
Substitution, synthetic	157	scalar projection of	88, 90, 91, 93
Sum and difference of two vectors	87, 90	sum and difference of	87, 90
Sum and product of the zeros of a polynomial	158	three-dimensional	87
Sum and product properties for sinusoids	52	unit	88, 89, 90, 94, 95
Sum or product of sinusoids with unequal periods	48, 49, 50, 51	using bearing	65
Sums of squares of residuals	74, 75	using components	65
Synthetic substitution	157	Velocity, average	162
Tangent and secant graphs	25, 26	Velocity function	162
transformations of	26	Velocity, instantaneous	162
Tangent line to an ellipse	130	Venn diagrams and overlapping events	82
Three-dimensional vectors	87, 101, 102	Von Koch's snowflake curve	115
Transformation(s)		Witch of Maria Agnesi	140
iterated	117	Zeros of a polynomial, sum and product of	158